FUNDAMENTALS OF DEFORMATION AND FRACTURE

International Union of Theoretical and Applied Mechanics

Fundamentals of Deformation and Fracture

Eshelby Memorial Symposium
Sheffield, 2–5 April 1984

Edited by
B. A. Bilby, K. J. Miller and J. R. Willis

The right of the University of Cambridge to print and sell all manner of books was granted by Henry VIII in 1534. The University has printed and published continuously since 1584

CAMBRIDGE UNIVERSITY PRESS
Cambridge
London New York New Rochelle
Melbourne Sydney

CAMBRIDGE UNIVERSITY PRESS
Cambridge, New York, Melbourne, Madrid, Cape Town, Singapore, São Paulo, Delhi

Cambridge University Press
The Edinburgh Building, Cambridge CB2 8RU, UK

Published in the United States of America by Cambridge University Press, New York

www.cambridge.org
Information on this title: www.cambridge.org/9780521105323

First published 1985
This digitally printed version 2009

A catalogue record for this publication is available from the British Library

Library of Congress Catalogue Card Number: 84-45713

ISBN 978-0-521-26735-9 hardback
ISBN 978-0-521-10532-3 paperback

CONTENTS

SCIENTIFIC COMMITTEE

Professor B. A. Bilby, F.R.S.	(Sheffield, U.K.)
Professor H. D. Bui	(Palaiseau, France)
Professor A. J. Carlsson	(Stockholm, Sweden)
Professor D. C. Drucker	(Illinois, U.S.A.) President of IUTAM
Professor M. Elices	(Madrid, Spain)
Professor K. J. Miller	(Sheffield, U.K.) Chairman
Professor H. Okamura	(Tokyo, Japan)
Professor D. M. Radenkovic	(Palaiseau, France)
Professor J. R. Rice	(Harvard, U.S.A.)

A broader Advisory Board to assist the Scientific Committee was composed of those organizing the sections which follow, together with

Professor A. S. Argon	(M.I.T., U.S.A.)
Sir Alan Cottrell, F.R.S.	(Cambridge, U.K.)
Professor D. S. Dugdale	(Sheffield, U.K.)
Professor Sir Charles Frank, O.B.E., F.R.S.	(Bristol, U.K.)
Professor F. R. N. Nabarro, M.B.E., F.R.S.	(Johannesburg, South Africa)
Professor Y. Yamamoto	(Tokyo, Japan)

The people responsible for the different sections were

	Introductory lecture	Professor B. A. Bilby, F.R.S. (Sheffield, U.K.)
I	Conservation laws and their applications	Professor J. R. Rice (Harvard, U.S.A.)
II	Inclusions, inhomogeneities and composites	Dr. A. Kelly, F.R.S. (Surrey, U.K.) Professor T. Mura (Northwestern, U.S.A.) Professor B. Budiansky (Sheffield, U.K.)

III Short cracks — Professor K. J. Miller (Sheffield, U.K.); Dr. L. M. Brown, F.R.S. (Cambridge, U.K.)

IV Dynamics of dislocations and cracks — Professor J. R. Willis (Bath, U.K.); Dr. C. Atkinson (London, U.K.)

IV Dynamics of dislocations and cracks — Professor J. R. Willis (Bath, U.K.); Dr. C. Atkinson (London, U.K.)

V Dislocation and point defects — Dr. R. Bullough (Harwell, U.K.); Professor Sir Peter Hirsch, F.R.S. (Oxford, U.K.)

VI Creep fracture — Professor M. F. Ashby, F.R.S. (Cambridge, U.K.); Professor G. W. Greenwood (Sheffield, U.K.)

LIST OF PARTICIPANTS

ACHENBACH, J. D.	Northwestern University, Evanston, Illinois, U.S.A.
AINSWORTH, R. A.	Central Electricity Generating Board, Berkeley, U.K.
ASHBY, M. F.	University of Cambridge, U.K.
ATKINSON, C.	Imperial College of Science and Technology, London, U.K.
BACON, D. J.	University of Liverpool, U.K.
BARNETT, D. M.	Stanford University, California, U.S.A.
BASINSKI, Z. S.	National Research Council of Canada, Ottawa, Ontario, Canada
BILBY, B. A.	University of Sheffield, U.K.
BLACKBURN, W. S.	Central Electricity Generating Board, (SER–SSD), Gravesend, Kent, U.K.
BROBERG, K. B.	Lund Institute of Technology, Sweden
BROOK, R.	University of Sheffield, U.K.
BROWN, L. M.	Cavendish Laboratory, University of Cambridge, U.K.
BROWN, M. W.	University of Sheffield, U.K.
BUI, H. D.	Electricité de France, Clamart, France
BULLOUGH, R.	Atomic Energy Research Establishment, Harwell, U.K.
CABLE, M.	University of Sheffield, U.K.
CARDEW, G. E.	University of Sheffield, U.K.
CARLSSON, A. J.	KTH, Stockholm, Sweden
COOK, D. B.	University of Sheffield, U.K.
DAVIDSON, D. L.	Southwest Research Institute, San Antonio, Texas, U.S.A.
DOMINGUEZ ABASCAL, J.	E.T.S.I.I., Sevilla, Spain
DUGDALE, D. S.	University of Sheffield, U.K.
EYRE, B. L.	University of Liverpool, U.K.
FOUND, M. S.	University of Sheffield, U.K.
FREUND, L. B.	Brown University, Providence, Rhode Island, U.S.A.

GANGLOFF, R. P.	Exxon Research and Engineering Company, Linden, N.J., U.S.A
GOLDTHORPE, M. R.	University of Sheffield, U.K.
GRANATO, A. V.	University of Illinois, Urbana, U.S.A.
GREENWOOD, G. W.	University of Sheffield, U.K.
GUIU, F.	Queen Mary College, London, U.K.
HASHIN, Z.	Tel-Aviv University, Israel
HAZZLEDINE, P. M.	University of Oxford, U.K.
HERRMANN, G.	Stanford University, California, U.S.A.
HIRSCH, P. B.	University of Oxford, U.K.
HOWARD, I. C.	University of Sheffield, U.K.
KELLY, A.	University of Surrey, Guildford, Surrey, U.K.
KENDALL, J.	University of Cambridge, U.K.
KFOURI, A. P.	University of Sheffield, U.K.
KRÖNER, E.	Institut für Theoret. und Agnew. Physik der Universität Stüttgart, Fed. Rep. of Germany
LAKES, R. S.	University of Iowa, U.S.A.
LAWS, N.	Cranfield Institute of Technology, Bedford, U.K.
LEE, J. K.	Michigan Technological University, U.S.A.
LILHOLT, H.	Risø National Laboratory, Roskilde, Denmark
LORETTO, M. H.	University of Birmingham, U.K.
MARSHALL, E. A.	University of Sheffield, U.K.
MIALON, P.	Electricité de France, Clamart, France
MILLER, K. J.	University of Sheffield, U.K.
MORI, T.	Tokyo Institute of Technology, Yokahama, Japan
MURA, T.	Northwestern University, Evanston, Illinois, U.S.A.
MURRELL, S. A. F.	University College London, U.K.
NABARRO, F. R. N.	University of the Witwatersrand, Johannesburg, South Africa
OGIN, S. L.	University of Cambridge, U.K.
OLVER, P. J.	University of Minnesota, U.S.A.
PAXTON, A. T.	University of Sheffield, U.K.
PEDERSEN, O. B.	Risø National Laboratory, Roskilde, Denmark
PRATT, P. L.	Imperial College of Science and Technology, London, U.K.
RICE, J. R.	Harvard University, Cambridge, Massachusetts, U.S.A.
RIEDEL, H.	Max-Planck-Institut für Eisenforschung, Düsseldorf, Fed. Rep. Germany

DE LOS RIOS, R. O.	University of Sheffield, U.K.
RITCHIE, R. O.	University of California, U.S.A.
ROSE, L. R. F.	Aeronautical Research Laboratories, Melbourne, Victoria, Australia
SINCLAIR, J. E.	Atomic Energy Research Establishment, Harwell, U.K.
SPENCER, A. J. M.	University of Nottingham, U.K.
STIGH, U.	Chalmers University, Gothenburg, Sweden
SWINGLER, J. N.	Central Electricity Generating Board, (SER–SSD), Gravesend, Kent, U.K.
TAYA, M.	University of Delaware, U.S.A.
WALPOLE, L. J.	University of East Anglia, Norwich, U.K.
WILLIS, J. R.	University of Bath, U.K.
WINTER, A. T.	Cavendish Laboratory, University of Cambridge, U.K.
YATES, J. R.	University of Sheffield, U.K.
YOFFE, E. H.	University of Cambridge, U.K.

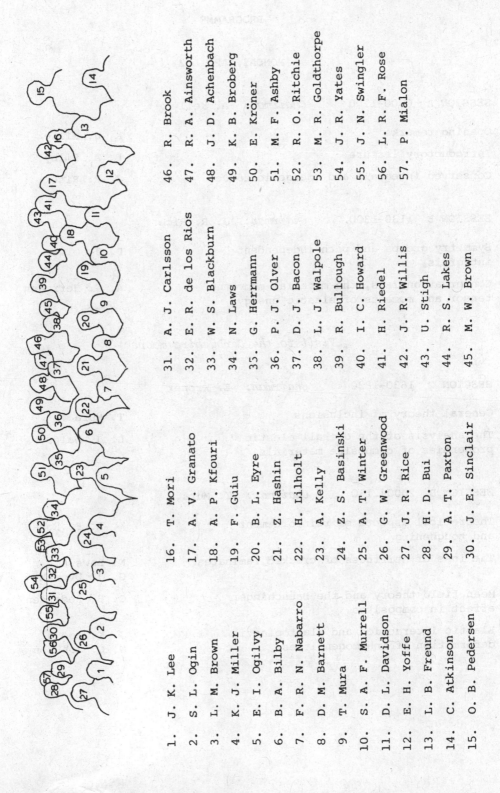

1. J. K. Lee
2. S. L. Ogin
3. L. M. Brown
4. K. J. Miller
5. E. I. Ogilvy
6. B. A. Bilby
7. F. R. N. Nabarro
8. D. M. Barnett
9. T. Mura
10. S. A. F. Murrell
11. D. L. Davidson
12. E. H. Yoffe
13. L. B. Freund
14. C. Atkinson
15. O. B. Pedersen
16. T. Mori
17. A. V. Granato
18. A. P. Kfouri
19. F. Guiu
20. B. L. Eyre
21. Z. Hashin
22. H. Lilholt
23. A. Kelly
24. Z. S. Basinski
25. A. T. Winter
26. G. W. Greenwood
27. J. R. Rice
28. H. D. Bui
29. A. T. Paxton
30. J. E. Sinclair
31. A. J. Carlsson
32. E. R. de los Rios
33. W. S. Blackburn
34. N. Laws
35. G. Herrmann
36. P. J. Olver
37. D. J. Bacon
38. L. J. Walpole
39. R. Bullough
40. I. C. Howard
41. H. Riedel
42. J. R. Willis
43. U. Stigh
44. R. S. Lakes
45. M. W. Brown
46. R. Brook
47. R. A. Ainsworth
48. J. D. Achenbach
49. K. B. Broberg
50. E. Kröner
51. M. F. Ashby
52. R. O. Ritchie
53. M. R. Goldthorpe
54. J. R. Yates
55. J. N. Swingler
56. L. R. F. Rose
57. P. Mialon

PROGRAMME

MONDAY, APRIL 2

SESSION A 0900-1100 *Chairman:* A. Kelly	
Opening remarks	K. J. Miller
Introductory lecture	B. A. Bilby
Conserved integrals and energetic forces	J. R. Rice
SESSION B 1130-1300 *Chairman:* J. R. Rice	
Symmetry groups and path-independent integrals	P. J. Olver
Conservation laws, the material momentum tensor and moments of elastic energy	A. G. Herrmann
Visit to the Derbyshire moors	
SESSION C 1630-1830 *Chairman:* E. Kröner	
General theory of inclusions	T. Mura
The analysis of the overall elastic properties of composite materials	L. J. Walpole
SESSION D 2000-2130 *Chairman:* T. Mura	
Theoretical aspects of reinforcement and toughening	L. R. F. Rose
The loss of stiffness of cracked laminates	N. Laws & G. J. Dvorak
Mean field theory and the Bauschinger effect in composites	O. B. Pedersen
Elastic interaction and elastoplastic deformation of inhomogeneities	J. K. Lee & W. C. Johnson

TUESDAY, APRIL 3

SESSION E 0900-1100 *Chairman:* C. Atkinson

The mechanics of dynamic crack growth in solids	L. B. Freund
Dislocation damping and phonon scattering	A. V. Granato

SESSION F 1130-1300 *Chairman:* J. R. Willis

Invariant integrals for some time dependent crack and heat conduction problems	C. Atkinson
Plastic deformations near a rapidly propagating crack tip	J. D. Achenbach & Z.-L. Li
What happens at fast crack growth?	K. B. Broberg

Visit to the Derbyshire moors

SESSION G 1630-1830 *Chairman:* M. F. Ashby

Creep crack growth in ductile alloys	A. S. Argon, C. W. Lau, B. Ozmat & D. M. Parks
Strength contributions during creep of composite materials	H. Lilholt
Suppression of grain boundary sliding by second phase particles	T. Mori

SESSION H 2000-2130 *Chairman:* M. F. Ashby

A continuum damage approach to creep crack growth	H. Riedel
Thermal expansion of polycrystalline aggregates	Z. Hashin
The roles of dislocation motion and vacancy fluxes in creep	G. W. Greenwood

WEDNESDAY, APRIL 4

SESSION I 0900-1100 *Chairman:* F. R. N. Nabarro

Dislocations in semiconductors	P. B. Hirsch
Dislocations in ionic crystals	P. L. Pratt
Image stress and the Bauschinger effect in dispersion-hardened alloys	L. M. Brown
Geometry and behaviour of prismatic dislocation loops and stacking fault tetrahedra	B. L. Eyre

SESSION J 1130-1300 *Chairman:* P. B. Hirsch

Direct observations of anisotropic and surface effects on dislocations	P. M. Hazzledine
Dislocations in anisotropic elastic media	D. J. Bacon
A dislocation method for solving 3-D crack and inclusion problems in linear elastic solids	G. K. Wong & D. M. Barnett

SESSION K 1400-1600 *Chairman:* R. Bullough

Interactions of point defects and dislocations	R. Bullough & J. R. Willis
Material forces and configurational forces in the interaction of elastic singularities	F. R. N. Nabarro
Material balance laws in one-dimensional strength-of-materials theories	G. Herrmann
On the elastic T-term	G. E. Cardew, M. R. Goldthorpe, I. C. Howard & A. P. Kfouri

SESSION L 1630-1730 *Chairman:* R. Bullough

Initiation and growth rates of short fatigue cracks	K. J. Miller

Symposium Dinner

THURSDAY, APRIL 5

SESSION M 0900-1100 *Chairman:* K. J. Miller

Role of internal stresses in the nucleation of fatigue cracks	L. M. Brown & S. L. Ogin
Environmental effects novel to the propagation of short fatigue cracks	R. P. Gangloff & R. O. Ritchie
Experimental mechanics of fatigue crack growth: the effect of crack size	D. L. Davidson & J. Lankford

SESSION N 1130-1300 *Chairman:* L. M. Brown

Cyclic deformation: the two phase model	A. T. Winter
Surface geometry in fatigued copper crystals	Z. S. Basinski & S. J. Basinski
Closing remarks	K. J. Miller

PAPERS NOT PRESENTED ORALLY

Similarity methods in fracture	G. I. Barenblatt & L. R. Botvina
A path-independent integral for mixed modes of fracture in linear thermoelasticity	H. D. Bui
Plasticity and creep mechanics from the viewpoint of the theory of elasticity	G. P. Cherepanov

POSTER SESSION

Cyclic deformation of molybdenum crystals at 400 K	J. A. Planell & F. Guiu
Experiments in Cosserat mechanics	R. S. Lakes
Short crack fatigue behaviour in a medium carbon steel	E. R. de los Rios, Z. Tang & K. J. Miller
Slip and cleavage from blunted cracks	J. E. Sinclair

PREFACE

This symposium had its origin in a proposal by Professor Miller to the International Union of Theoretical and Applied Mechanics, via the British National Committee for Theoretical and Applied Mechanics, for a meeting on short cracks. At that time, a few months after Jock Eshelby's untimely death, a number of colleagues were thinking about various ways in which he might be commemorated. After some discussions and suggestions, notably from Professors Argon, Mura, Rice and Willis, Professor Miller's original proposal was broadened to one for a meeting on the fundamentals of deformation and fracture. This was finally approved as an International Union of Theoretical and Applied Mechanics Symposium, J. D. Eshelby In Memoriam, by the General Assembly of IUTAM at its Cambridge meeting in August 1982. Later, the International Congress on Fracture and the International Congress on the Mechanical Behaviour of Materials kindly agreed to act as cosponsors of the symposium.

IUTAM symposia are usually designed to discuss a rather specific subject but on this occasion an up to date survey of a number of fields in which Eshelby had been interested was attempted, in order to show his influence on them. Eshelby's work affected so broad a span of scientific and technological endeavour that it was thought best to divide the meeting into a number of sections and seek the assistance of appropriate people to look after each topic. These section leaders were asked particularly to make recommendations about invitations and the acceptance of abstracts. Together with six other scientists, the section leaders formed a broader Advisory Board to assist the work of the Scientific Committee. Naturally, the latter alone must bear the responsibility for the final outcome. The names of the members of these two bodies are set out in the following pages. The editors wish to express their warmest thanks to the Scientific Committee and to the Advisory Board for their help. We also wish to thank those who kindly undertook to chair the various sessions, sometimes, unfortunately, at rather short notice.

It is required of meetings of this kind sponsored by IUTAM that they should not have parallel sessions and that the number of participants should be strictly limited. Accordingly, invitations could be issued initially to a few only of the many people who had found inspiration from working with Eshelby himself or from his published papers. Moreover, in view of his wide reputation, no active steps were taken to advertize the meeting. Despite this a number of requests to attend were received which had to be added to a reserve list of further people to be invited when a firm estimate of the final attendance could be made. As a result we know that a number of people who would have liked to attend were unfortunately disappointed, either because they received no invitation, or because their own came too late. We can only offer them our sincere apologies and note the comment that a quite different, much larger meeting, with no limitations on attendance, could have been arranged had resources been available. On the other hand we should thereby have certainly lost the unity, and possibly the informality and spontaneity of the discussion which, we believe, did characterize the meeting that took place. The number of papers which could be accepted was limited and some offered had to be rejected on that score. There also had to be some limitation on the range of subject matter allowed. A few authors accepted an invitation to present their work at a poster session; abstracts of some of these presentations appear in the symposium volume. The poster session was organized by Michael W. Brown and Eduardo de los Rios and we wish to express our thanks to them for carrying out this task. For various reasons, some other authors were able to make available here only abstracts or shortened versions of their work. A few papers which were not presented orally are also included in the symposium volume.

We are indebted to the International Union of Theoretical and Applied Mechanics for financial support and for the help and advice provided by the President, Professor Daniel Drucker, and the Secretary General, Professor Jan Hult. The meeting was organized from the Mechanical Engineering Department at the University of Sheffield and the main burden of administrative work fell on the shoulders of Bruce Bilby's secretary, Elaine Ogilvy. Other help was provided by Norma Parkes and Sheila Garvey. For the arrangements at the meeting itself we are indebted to our projectionist John Goodliffe, his assistants Alf Thompson and Richard Cummins, and to our photographer Derek Hallford. We have also to thank the manager of Earnshaw Hall and his staff for the excellent facilities provided. Visits to the Departments of Mechanical Engineering and of Metallurgy were organized through the courtesy of their respective Heads, Professors Miller and Argent; Ray Brook was active in arranging the latter. There are also many other colleagues and friends who in their several ways contributed over the two preparatory years to the work of organizing the meeting. To all these we express our warmest thanks. Finally we are indebted to Keith Miller for two characteristic and highly successful outings to the Derbyshire moors. On these the pace was too reminiscent of the Karakoram for some of the participants and one distinguished visitor was left alone for some time between the vanguard and the stragglers. Fortunately he was quite unaware of his predicament and modestly surprised by a rescue party sent out to meet him.

There must be an element of sadness about any memorial meeting, but we were able to recall repeatedly at the sessions and the dinner the dry humour of the man we knew; so much so indeed, that at times it seemed almost as if Jock were again with us. We feel sure that this is the atmosphere which he would have wished for the meeting and we hope that it will long continue so at other gatherings where he is remembered. To our great regret, however, the symposium had to contend with other sad events. Tragically, Alicia Golebiewska Herrmann died on December 7, 1983; her paper was read by George Herrmann. Several people missed the meeting through accident or illness. Kare Hellan had to withdraw because of a serious motoring accident and Ali Argon was prevented from attending by illness; Mike Ashby kindly agreed at short notice to present the paper. 'Bas' Basinski became unwell and had to leave the meeting early; our thanks are due to Mick Brown for making a presentation of this work. The symposium sent its best wishes for the speedy recovery of these casualties.

Despite the various problems the meeting was judged by the participants to be a most successful one and we hope that it will make its appropriate contribution to the commemoration of Jock Eshelby. We are sure that there will be many other occasions where he will be remembered, but we hope that this meeting, initiated by those who knew him closely for the latter half of his working life and held in the University where he then was, will have a special place among such memorials.

Bruce Bilby
Keith Miller
John Willis

John Douglas Eshelby

(Reproduced by kind permission of the American Society of Mechanical Engineers)

JOHN DOUGLAS ESHELBY, F.R.S.

The following Obituary appeared in the Times on January 2nd 1982 and is reproduced here by kind permission of Times Newspapers Limited.

WORK ON THEORY OF MATERIALS

Professor J. D. Eshelby, F.R.S., Professor of the Theory of Materials, Sheffield University from 1971, died on December 10. He was 64.

His death removes a great scholar known to a wide circle of scientists, engineers and mathematicians throughout the world. He was a shy and kindly man, concealing an inner reserve with occasional gruffness; a true gentleman, always ready to serve.

John Douglas Eshelby was born in Puddington, Cheshire, on December 21, 1916, the eldest son of Alan Douglas Eshelby. Because of ill health he missed his formal schooling from the age of 13 and lived at the family home in north Somerset, where he learned instead from tutors. So, as he used to say, he had to work many things out for himself, and perhaps this helped to make him such an original and creative thinker. Observant of people and things, he had a deep physical insight into the workings of nature around him. As a child, watching his father's diesel generator, he noticed how a moving belt retains its shape when struck; and recently he was to be seen studying the spider's web pattern of cracks in broken windows, while he pondered on the limitations of the present theory of elastic plates.

Through a contact with Professor Mott (now Sir Nevill) he went early to the University of Bristol and obtained a first in physics there in 1937. During the Second World War he served first at the Admiralty, degaussing ships, and then in the technical branch of the Royal Air Force, where he reached the rank of squadron leader. He flew sometimes in Sunderlands out of Pembroke Dock, and there is in the Science Museum some radar equipment that he helped to design.

He returned to Bristol in 1946, at an exciting time for solid state physics when rapid advances were being made in the theory of the deformation of crystals. The opportunity arose for him to take up theoretical research, and here he made his initial mark in dislocation theory, revealing quite suddenly to those around him a mastery of some of the most difficult problems of the time. He obtained his Ph.D. in 1950 and two years later spent a year at the University of Illinois.

There followed some ten years at the University of Birmingham, a period in 1963 as visiting professor at the Technische Hochschule, Stuttgart, and then two years at Cambridge, where he became a Fellow and College Lecturer at Churchill. In 1966 he went to the University of Sheffield, holding a readership and, from 1971, a personal chair in the theory of materials.

His work was a great part of his life. His general field was the theoretical physics of the deformation, strength and fracture of engineering materials, and his principal interests were lattice defects and continuum mechanics.

Though motivated by the desire to understand he kept a firm eye on application and had no time for useless erudition; like Willard Gibbs his object was to make things appear simple by "looking at them in the right way". With a keen discrimination he selected those worthwhile difficult problems which nevertheless had some chance of solution. Entirely unconcerned with personal advancement, he hoped only of his papers that each would be "a little gem".

And so it is. Many indeed are treasure houses, abounding in undeveloped asides on which others may later build, for often he did not elaborate. He regarded himself as a modest "supplier of tools for the trade", and he left to others their day to day use. His colleagues everywhere were always consulting him.

Eshelby was elected a Fellow of the Royal Society in 1974, being "distinguished for his theoretical studies of the micromechanics of crystalline imperfections and material inhomogeneities". He made major contributions to the theory of static and moving dislocations and of point defects. By an elegant use of the theory of the potential he obtained some remarkable results on the elastic fields of ellipsoidal inclusions and inhomogeneities.

In 1951 he introduced, in analogy with the Maxwell tensor, the elastic energy momentum tensor, which yields forces on elastic singularities. During his later years he was much concerned with this concept and its developments, which can provide parameters characterizing the singular fields.

In 1968 he published accounts of its application to the calculation of forces on static and moving cracks in elastic media. Related work, formulated for application also to plastic-elastic media, was published simultaneously and independently by J. R. Rice. Many others have made widespread use of these characterizing parameters in fracture mechanics, sometimes in a way to which Eshelby did not wholly subscribe.

Eshelby had a wide knowledge of theoretical physics and repeatedly applied ideas in one discipline to solve problems in another. He drew much inspiration from masters of the past and liked to regard some of his most important works as amusing applications of the theorem of Gauss.

But his scholarly interests went far beyond science. He read French, German and Russian and could find his way about a Chinese dictionary; indeed, he knew a great deal about languages and the ancient world and enjoyed holding his own in discussions with professionals in these fields. His dry jokes and sayings will long be remembered;

"It's obvious", he would say, "I forget exactly why". One of his great pleasures was to find good secondhand books.

Just before his death he was in correspondence with former colleagues about some implications of recent calculations he had made of forces on defects in liquid crystals; and also about cracks in metal fatigue. He was also preparing lectures to be given in California in the new year.

PUBLICATIONS OF J. D. ESHELBY

1. Dislocations as a cause of mechanical damping in metals 1949
Proc. Roy. Soc. Lond., A 197, 396-416.

2. Uniformly moving dislocations 1949
Proc. Phys. Soc. (London), A 62, 307-314.

3. Edge dislocations in anisotropic materials 1949
Phil. Mag., 40, 903-912.

4. The fundamental physics of heat conduction 1951
In: Proceedings of a General Discussion on Heat Transfer, Sept. 11-13, 1951, Institution of Mechanical Engineers - American Society of Mechanical Engineers, pp. 5-8.
London: Institution of Mechanical Engineers.

5.† The equilibrium of linear arrays of dislocations 1951
with F. C. Frank and F. R. N. Nabarro.
Phil. Mag., 42, 351-364.

6. The force on an elastic singularity 1951
Phil. Trans. R. Soc. Lond., A 244, 87-111.

7. Dislocations in thin plates 1951
with A. N. Stroh.
Phil. Mag., 42, 1401-1405.

8. Anisotropic elasticity with applications to dislocation theory 1953
with W. T. Read and W. Shockley.
Acta Met., 1, 251-259.

9. Screw dislocations in thin rods 1953
J. Appl. Phys., 24, 176-179.

10. The equation of motion of a dislocation 1953
Phys. Rev., 90, 248-255.

11. A tentative theory of metallic whisker growth 1953
Phys. Rev., 91, 755-756.

12. Geometrical and apparent X-ray expansions of a crystal containing lattice defects 1953
J. Appl. Phys., 24, 1249.

The four papers marked † (and two papers by other authors) have been published in a Russian translation by the Foreign Languages Publishing House, I.-L., Moscow, 1963, under the title J. Eshelby, "Kontinual'naya Teoriya Dislokatsii".

13. Distortion of a crystal by point imperfections 1954
J. Appl. Phys., 25, 255-261.

14. The elastic interaction of point defects 1955
Acta Met., 3, 487-490.

15. Note on the heating effect of moving dislocations 1956
with P. L. Pratt.
Acta Met., 4, 560-561.

16. Supersonic dislocations and dislocations in dispersive 1956
media.
Proc. Phys. Soc. (London), B 69, 1013-1019.

17.† The continuum theory of lattice defects. 1956
Solid State Physics, 3, 79-144.

18.† The determination of the elastic field of an ellipsoidal 1957
inclusion and related problems.
Proc. R. Soc. Lond., A 241, 376-396.

19. A note on the Gomer effect. Discussion to the paper on 1958
'Some observations on field emission from mercury whiskers'
by R. Gomer.
In: Growth and Perfection of Crystals,
ed. R. H. Doremus, B. W. Roberts and D. Turnbull, 130-132.
New York: Wiley.

20. Charged dislocations and the strength of ionic crystals 1958
with C. W. A. Newey, P. L. Pratt and A. B. Lidiard.
Phil. Mag., 3, 75-89.

21. The twist in a crystal whisker containing a dislocation 1958
Phil. Mag., 3, 440-447.

22. The elastic model of lattice defects 1958
Ann. der Phys., 1, 116-121.

23. Stress-induced ordering and strain ageing in low carbon 1959
steels

24. Scope and limitations of the continuum approach 1959
In: Internal Stresses and Fatigue in Metals,
ed. G. M. Rassweiler and W. L. Grube, pp. 41-58.
Amsterdam: Elsevier.

25.† The elastic field outside an ellipsoidal inclusion 1959
Proc. R. Soc. Lond., A 252, 561-569.

26. Elastic inclusions and inhomogeneities 1961
Prog. Solid Mech., 2, 89-140.

27. Dislocations in visco-elastic materials 1961
Phil. Mag., 6, 953-963.

28. The interaction of kinks and elastic waves 1962
Proc. R. Soc. Lond., A 266, 222-246.

29. The energy and line-tension of a dislocation in a hexagonal crystal 1962
with Y. T. Chou.
J. Mech. Phys. Solids, 10, 27-34.

30. The distortion and electrification of plates and rods by dislocations 1962
phys. stat. sol., 2, 1021-1028.

31. The distribution of dislocations in an elliptical glide zone 1963
phys. stat. sol., 3, 2057-2060.

32. On the elastic interaction between inclusions 1966
Acta Met., 14, 1306-1309.
(Appendix to a paper 'On the modulated structure of aged Ni-Al alloys' by A. J. Ardell and R. B. Nicholson.)

33. A simple derivation of the field of an edge dislocation 1966
Brit. J. Appl. Phys., 17, 1131-1135.

34. The velocity of a wave along a dislocation 1966
with T. Laub.
Phil. Mag., 14, 1285-1293.

35. The interpretation of terminating dislocations 1967
with T. Laub.
Canad. J. Phys., 45, 887-892.

36. Stress analysis: theory of elasticity 1968
In: Fracture Toughness,
ISI Publication 121, Ch. 2, pp. 13-29. London: The Iron and Steel Institute.

37. Stress analysis: fracture mechanics 1968
In: Fracture Toughness,
ISI Publication 121, Ch. 3, pp. 30-48. London: The Iron and Steel Institute.

38. The flow of energy into the tip of a moving crack 1968
with C. Atkinson.
Int. J. Fracture Mech., 4, 3-8.

39. Dislocations and the theory of fracture 1968
with B. A. Bilby.
In: Fracture, an advanced treatise,
ed. H. Liebowitz, 1, pp. 99-182. New York: Academic Press.

40. The elastic field of a crack extending non-uniformly under general anti-plane loading 1969
J. Mech. Phys. Solids, 17, 177-199.

41. The starting of a crack 1969
In: Physics of Strength and Plasticity, the Orowan 65th anniversary volume, ed. A. S. Argon, pp. 263-275. Cambridge, Massachusetts: M.I.T. Press.

42. Energy relations and the energy-momentum tensor in continuum mechanics 1970
In: Inelastic Behaviour of Solids, ed. M. F. Kanninen, W. F. Adler, A. R. Rosenfield and R. I. Jaffee, pp. 77-115. New York: McGraw-Hill.

43. The fracture mechanics of flint-knapping and allied processes 1971
with J. G. Fonseca and C. Atkinson.
Int. J. Fracture Mech., 7, 421-433.

44. Fracture Mechanics 1971
Science Progress, 59, 161-179.

45. Dislocation theory for geophysical applications 1973
Phil. Trans. R. Soc. Lond., A 274, 331-338.

46. The calculation of energy release rates 1975
In: Prospects of Fracture Mechanics, ed. G. C. Sih, H. C. van Elst and D. Broek, pp. 69-84. Leyden: Noordhoff.

47. Point defects 1975
In: The Physics of Metals 2. Defects (Vol. 2 of the Sir Nevill Mott 60th anniversary volumes), ed. P. B. Hirsch, pp. 1-42. Cambridge: Cambridge University Press.

48. The change of shape of a viscous ellipsoidal region embedded in a slowly deforming matrix having a different viscosity 1975
with B. A. Bilby and A. K. Kundu.
Tectonophysics, 28, 265-274.

49. The elastic energy-momentum tensor 1975
J. Elasticity, 5, 321-335.

50. The change of shape of a viscous ellipsoidal region embedded in a slowly deforming matrix having a different viscosity - some comments on a discussion by N. C. Gay 1976
with B. A. Bilby, M. L. Kolbuszewski and A. K. Kundu.
Tectonophysics, 35, 408-409.

51. Interaction and diffusion of point defects 1977
In: Vacancies '76,
ed. R. E. Smallman and E. Harris, pp. 3-10. London:
The Metals Society.

52. Boundary problems 1979
In: Dislocations in Solids, Vol. 1,
ed. F. R. N. Nabarro, pp. 167-221. Amsterdam:
North-Holland.

53. The force on a disclination in a liquid crystal 1980
Phil. Mag., A 42, 359-367.

54. The energy-momentum tensor of complex continua 1980
In: Continuum models of discrete systems,
ed. E. Kröner and K.-H. Anthony, pp. 651-665.
Waterloo: University of Waterloo Press.

55. Aspects of the theory of dislocations 1982
In: Mechanics of Solids, the Rodney Hill 60th
anniversary volume,
ed. H. G. Hopkins and M. J. Sewell, pp. 185-225.
Oxford: Pergamon Press.

56. Technical note: The stresses on and in a thin 1982
inextensible fibre in a stretched elastic medium
Engineering Fracture Mechanics, 16, 453-455.

INTRODUCTORY LECTURE

B. A. BILBY

Department of Mechanical Engineering, University of Sheffield, U.K.

ABSTRACT

The work of J. D. Eshelby is briefly reviewed and some indication given of its widespread influence. Several unpublished matters which had occupied him recently are also discussed.

1. INTRODUCTION

This is not the place to attempt a comprehensive account of Jock Eshelby's life and work, for even to say a few words about each of his papers (1-56) would use up my allotted time. Nor can I do more than refer most briefly to some only of the unpublished matters which were occupying his attention when he died. I shall therefore be highly selective, guiding my emphasis both by what is to come after here and by his central interests, particularly in later years.

Jock liked to think of himself as a "supplier of tools for the trade" and it is mainly about the tools that I shall speak. Perhaps I may thereby interest the craftsmen; of the toolmakers I crave forbearance. The main focus of Jock's work was on crystal defects and elastic singularities, and the forces that act upon them. One of his most important early contributions (6), to which his interest repeatedly returned (42), particularly in later years, was to show that these forces could be obtained by a powerful tool. The archetypal example of this tool is the Maxwell tensor in electrostatics. By this, the force is expressed as an integral over a surface S surrounding the defect. The tensor integrand has zero divergence except at singularities of the field and so the integral is independent of the particular surface S which is chosen. Jock applied this idea to the elastic field, referring to the "Maxwell tensor of elasticity" and the "elastic energy momentum tensor". Such tools are also commonly called conservation laws, or path independent integrals.

It is peculiarly apposite that we are here today in Earnshaw Hall, for Earnshaw's most well known theorem is that a point charge can have no stable equilibrium position under electrostatic forces alone, a result which comes about because where there is no charge the divergence of the electric field vanishes. (An alternative statement of the theorem is that a function harmonic within a surface S can have no maximum or minimum there.)

2. ENERGY MOMENTUM TENSORS

Noether's theorem (58) shows that if the Lagrangian density function of a field is invariant under the operations of a continuous group there exist expressions whose divergence vanishes. Of this and of other related matters we are to hear more later (e.g. (59-62,121,122)). Characteristically, Eshelby had his own way of looking at these questions ("Look at it my way; the right way!"). He believed in doing the simple problem first, but strove for generality and so gave his results wherever possible for finite deformation. Here he firmly advocated the use of the initial (cartesian) coordinates X_ℓ of a particle and the first Piola Kirchhoff or nominal stress tensor (63), $p_{\ell j} = \partial W/\partial u_{\ell,j} \neq p_{j\ell}$. W is the energy per unit initial volume and u_ℓ is the displacement, $u_\ell = x_\ell - X_\ell$, where x_ℓ are the final coordinates. The comma denotes differentiation with respect to X_j. When required, we write $\partial/\partial x_j$ explicitly, with the single exception of $X_{i,j} \equiv \partial X_i/\partial x_j$. If $J = \det(x_{m,n}) = j^{-1} = \rho_0/\rho$, ρ_0 and ρ being the initial and final densities, the initial and final elements of area ds_ℓ and dS_ℓ are related by $ds_\ell = JdS_k X_{k,\ell}$, $dS_\ell = jds_k x_{k,\ell}$; and the force df_ℓ on ds_k is given by $df_\ell = p_{\ell j} dS_j = \sigma_{\ell j} ds_j$ where $\sigma_{\ell j} = \sigma_{j\ell}$ is the Cauchy stress tensor. For a simple material $W = W(u_{i,j}, X_j)$, where the explicit appearance of X_j means that the material is inhomogeneous. Thus, in addition to the usual gradient $\partial W/\partial X_\ell$, there is the explicit derivative $(\partial W/\partial X_\ell)_{\exp}$, which is taken with all variables (e.g. $u_{i,j}$) other than X_j held constant, and so vanishes if there is no explicit dependence on X_ℓ.

To obtain the classical integrals F_ℓ, $L_{k\ell}$ or M, corresponding to invariance under translation, rotation or dilatation (64,65,66), Eshelby (42,46,49) applies the appropriate (Lie) operator $\partial/\partial X_\ell$, $X_1\partial/\partial X_2 - X_2\partial/\partial X_1$ (e.g. for L_{12}) or $X_\ell\partial/\partial X_\ell$ to W. Use of the Euler equations and some elementary algebra then gives the desired result. For a simple material the translation operator $\partial/\partial X_\ell$ yields, without further restriction

$$P_{\ell j,j} = (\partial W/\partial X_\ell)_{exp} \qquad (1)$$

where $$P_{\ell j} = W\delta_{\ell j} - (\partial W/\partial u_{i,j})u_{i,\ell} \neq P_{j\ell} . \qquad (2)$$

Thus $P_{\ell j,j} = 0$ for arbitrary non linear W provided only that there is homogeneity in the X_ℓ direction. For the X_3 rotation operator, $X_1\partial/\partial X_2 - X_2\partial/\partial X_1 = \partial/\partial\Phi$ (R,Φ,Z, cylindrical polars), we need in addition the relation $p_{a1}x_{a,2} = p_{a2}x_{a,1}$ for transverse isotropy about X_3 to get

$$(\partial W/\partial\Phi)_{exp} = (X_1P_{2j} - X_2P_{1j} + u_1p_{2j} - u_2p_{1j})_{,j} . \qquad (3)$$

We now have a vanishing divergence, giving the L_{12} integral (66), for non linear W, transverse isotropy and a "tree-trunk" inhomogeneity. The dilatation operator gives for $X_\ell(\partial W/\partial X_\ell)_{exp} = 0$, i.e. a sectored inhomogeneity, and in addition, for W homogeneous of degree N (66,67),

$$(X_\ell P_{\ell j} + [(N - \delta_{ss})/N]\, u_i\, \partial W/\partial u_{i,j})_{,j} = 0 . \qquad (4)$$

The linear case N = 2 gives the three dimensional (δ_{ss} = 3) and two dimensional (δ_{ss} = 2) M integrals (46,66).

This type of simple algebraic argument can readily be extended to a material of grade N. Most generally (42,54,57) the Lagrangian W is a function of any number of field variables A^α, α = 1, ... M, their derivatives $A^\alpha_{,i}$, $A^\alpha_{,ij}$, $A^\alpha_{,ijk}$ up to order N, and possibly also of X_m. (By making one of the variables the time t, the dynamic situation may be treated.) It is convenient to define the variational or Lagrange derivative $\delta W/\delta A^\alpha$,

$$\delta W/\delta A^\alpha = \partial W/\partial A^\alpha - (\partial W/\partial A^\alpha_{,j})_{,j} + (\partial W/\partial A^\alpha_{,aj})_{,aj} - (\partial W/\partial A^\alpha_{,abj})_{,abj} + \ldots , \tag{5}$$

the process being continued, with terms of alternating sign, as far as the W dependence allows. For $\delta W/\delta A^\alpha_{,pq\ldots r}$, replace A^α in Eqn (5) by $A^\alpha_{,pq\ldots r}$. The condition $\delta W/\delta A^\alpha = 0$ gives the Euler equations (equations of equilibrium, motion) for the field problem, and the variational derivative satisfies the useful relation

$$\partial W/\partial A^\alpha = \delta W/\delta A^\alpha + (\delta W/\delta A^\alpha_{,p})_{,p} . \tag{6}$$

After a little manipulation one can now show that

$$P_{\ell j,j} = (\delta W/\delta A^\alpha) A^\alpha_{,\ell} + (\partial W/\partial X_\ell)_{exp} \tag{7}$$

where the (canonical) energy momentum tensor $P_{\ell j}$ is

$$P_{\ell j} = W\delta_{\ell j} - Q_{\alpha j}[A^\alpha_{,\ell}] \tag{8}$$

with

$$Q_{\alpha j}[A^\alpha_{,\ell}] = (\delta W/\delta A^\alpha_{,j}) A^\alpha_{,\ell} + (\delta W/\delta A^\alpha_{,aj}) A^\alpha_{,a\ell} + \ldots + (\delta W/\delta A^\alpha_{,ab\ldots sj}) A^\alpha_{,ab\ldots s\ell} + \ldots \tag{9}$$

To avoid error in these manipulations the order of the suffixes needs to be preserved (54,68).

Eshelby gave several variants of his argument using cutting and rewelding operations on the system (and usually also on an exact replica of it) in order to show that (6,17,42,46,49,54)

$$F_\ell = \int_S P_{\ell j} dS_j \tag{10}$$

is the resultant configurational force on the defects within S: i.e., if all defects (singularities) within the surface S are displaced $\delta\xi_\ell$, the change in the total energy δE_{tot} of the system (including work done, if any, by the external forces or generalized forces) is $-F_\ell \delta\xi_\ell$. The argument has the merit that no discussion of what happens near the singularities is required and that a (first order) transfer of energy across S during the displacement process never has to be explicitly calculated. The external boundary conditions are thus irrelevant. On the whole, Eshelby regarded many aspects of the theory of generalized continua as but hopefully relevant to reality, although he identified a number of applications - to beam and plate theory (46,54), to plane strain (54) and to liquid crystals (53,54); a related extension to structures in Rayleigh-Bénard convection has been noted (69). The work on liquid crystals will be discussed later.

3. INITIAL AND FINAL COORDINATES

The proof that F_ℓ is the configurational force on a defect is straightforward in the Lagrangian description using the coordinates X_i that we have employed so far, but it is not easy to repeat it for an Eulerian description using x_i. The "trouser test", where a surface with initial normal along X_1 has final normal along X_2, illustrates the kind of difficulty which arises (46). Eshelby's procedure (49) was to ensure that $F_\ell = \int_S \Sigma_{\ell j} ds_j$ by setting $dF_\ell = P_{\ell j} dS_j = \Sigma_{\ell j} ds_j$ where ds_j is the final element of area. With w (= jW) the energy per unit final volume, Eqn (2) transforms to the expected analogue

$$\Sigma_{\ell j} = w\delta_{\ell j} - (\partial w/\partial(\partial u_i/\partial x_j))(\partial u_i/\partial x_\ell) \,. \tag{11}$$

On the other hand, if we note that $W(u_{i,j}, X_m)$ implies $w(u_i, \partial u_i/\partial x_j, x_m)$, the algebraic procedure leading to Eqn (1) or (8) yields

$$\partial\Sigma_{\ell j}/\partial x_j = (\partial w/\partial x_\ell)_{exp} = -\partial w/\partial u_\ell \tag{12}$$

and the Euler equations

$$(\partial/\partial x_j)(\partial w/\partial(\partial u_\ell/\partial x_j)) = \partial w/\partial u_\ell \quad (13)$$

Thus $$\partial\sigma_{\ell j}/\partial x_j = 0 \quad (14)$$

where $$\sigma_{\ell j} = \Sigma_{\ell j} + \partial w/\partial(\partial u_\ell/\partial x_j) \quad (15)$$

Transformation of the relation $p_{\ell j}dS_j = \sigma_{\ell j}ds_j$ confirms that, as the notation has implied, $\sigma_{\ell j}$ is the Cauchy stress tensor.

Much more symmetrical relations between the X_i and x_i descriptions are obtained if we use these quantities as the field variables rather than u_i, so that we have $W(x_{i,j}, X_m)$ or $w(X_{i,j}, X_m(x_i))$. The algebraic procedure then yields (49,70)

$$P^*_{\ell j} = W\delta_{\ell j} - (\partial W/\partial x_{i,j})x_{i,\ell} \ : \ \sigma_{\ell j} = w\delta_{\ell j} - (\partial w/\partial X_{i,j})X_{i,\ell} \quad (16)$$

It follows at once from Eqn (16) that

$$P^*_{\ell j} = P_{\ell j} - p_{\ell j} \neq P^*_{j\ell} \quad (17)$$

Thus, when integrated over S, $P^*_{\ell j}$ gives the configurational force when, occasionally, any real body force is a necessary component of the model of the defect, e.g., for a charged dislocation in an ionic crystal in an electrostatic field (42): of course, if, as is usual, there is no body force, either $P_{\ell j}$ or $P^*_{\ell j}$ may be used.

The duality evident in Eqn (16) extends, of course, to other relations (42,57,70). Thus we have, with Hamel's and Cosserat's forms for $\sigma_{\ell j}$,

$$\sigma_{\ell j} = -jX_{i,\ell}\partial W/\partial X_{i,j} = -2jc_{i\ell}\partial W/\partial c_{ij} \quad (18)$$

$$P^*_{\ell j} = -Jx_{i,\ell}\partial w/\partial x_{i,j} = +2JC_{i\ell}\partial w/\partial C_{ij} \quad (19)$$

where $c_{\ell a} = X_{s,\ell}X_{s,a} = \delta_{\ell a} - 2e_{\ell a}$ and $C_{\ell a} = x_{s,\ell}x_{s,a} = \delta_{\ell a} + 2E_{\ell a}$ are the Almansi-Hamel and Green - St. Venant strain measures. Again, with $S_{mn} = \partial W/\partial E_{mn} = X_{m,s}p_{sn} = S_{nm}$, the second Piola Kirchhoff tensor, and $s_{mn} = \partial w/\partial e_{mn}$,

$$P^*_{\ell j} = W\delta_{\ell j} - C_{\ell a}S_{aj} \quad : \quad \sigma_{\ell j} = w\delta_{\ell j} - c_{\ell a}s_{aj} \tag{20}$$

For isotropy the relation $p_{aj}x_{a,\ell} = p_{a\ell}x_{a,j}$ ensures that $C_{\ell a}S_{aj} = S_{\ell a}C_{aj}$ so that $P^*_{\ell j} = P^*_{j\ell}$. Thus P^* commutes with C and S, i.e. P^*, C and S are coaxial (70). Standard methods of field theory can be applied (49) to symmetrize $P_{\ell j}$ but the symmetric tensor $P^*_{\ell j}$ for isotropic elasticity appears to have no counterpart for other fields (42). Note that $P^*_{\ell j}$ is not symmetric for anisotropy or for linear isotropy (49).

4. OTHER INTEGRALS

Eshelby found, in addition to the standard integrals, many other forms, e.g., a "more elaborate Günther type" (57): the expression

$$(X_k X_\ell - (\tfrac{1}{2})X_i X_i \delta_{k\ell})T_{\ell j} \tag{21}$$

has zero divergence if $T_{\ell j,j} = 0$, $T_{\ell j} = T_{j\ell}$ and $T_{\ell\ell} = 0$. For isotropy $P^*_{\ell j}$ satisfies the first two conditions and also the third if $W\delta_{\ell\ell} = (\partial W/\partial x_{i,j})x_{i,j}$, i.e. if W is homogeneous of degree 3 or 2 in three or two dimensions, respectively. He was also aware (57) that the (unphysical) relation $3\lambda + 7\mu = 0$ could sometimes ensure a zero divergence in linear elasticity. In linear isotropic antiplane and plane strain he found (46) infinite classes of integrals depending on any two functions ϕ_1 and ϕ_2 such that $\phi_1 + i\phi_2 = f(x_1 + ix_2)$, associating this (57) with the conformal property of the transformation thus defined. The integrands are $\phi_\ell P_{\ell j}$ for anti-plane strain and

$$\sigma_\ell P_{\ell j} - (\tfrac{1}{2})[(x_1\phi_2)_{,j}(P_{21}-P_{12}) - (P_{21}-P_{12})_{,j}x_1\phi_2]$$

for plane strain. For the latter state he also noted that $p'_{ij} = \varepsilon_{3ik}P_{kj}$

was an ordinary elastic field. Thus, if u_i' is its displacement and u_i^0, p_{ij}^0 any other elastic field, $u_i^0 p_{ij}' - u_i' p_{ij}^0$ has zero divergence by the Rayleigh-Betti reciprocal theorem (46). He was at one time most interested in the sequence of transformations $u_i, p_{ij} \to P_{ij} \to u_i', p_{ij}'$, noting that if the u_i are given by the standard Muskhelishvili (71) functions $\phi(z), \psi(z)$ then the $\Phi(z)$ and $\Psi(z)$ for u_i' are obtained by the changes $\phi \to \Phi$, $\psi \to \Psi$, such that $\Phi_z = (1-\nu)\phi_z^2/i\mu$, $\Psi_z = 2(1-\nu)\phi_z\psi_z/i\mu$ (the expression in (46) is in error by a factor of 2). Here Φ_z, Ψ_z denote the z derivatives. The configurational force $\underline{F}$ from the $P_{\ell j}$ of p_{ij} corresponds to a body force $\underline{f}$ from p_{ij}' where $\underline{f}$ is $\underline{F}$ rotated $\pi/2$ counterclockwise: e.g. (46), a crack tip in the X_1 direction at 0 with $F_1 = G$ becomes a positive edge dislocation of Burgers vector $b = (1-\nu)G/2\mu$ plus a point force $f_2 = G$.

5. TWO UNPUBLISHED APPLICATIONS

Two applications of path independent integrals on which Eshelby was working at the time of his death are dealt with in detail in papers here. The first (72) is interesting in showing how internal stresses can produce a crack extension force G due to an image effect. In fatigue a dislocation structure may form in a narrow band running obliquely from the surface. Regarded on a suitable scale this (persistent slip) band is a self-stressed sliver of thickness h, with a longitudinal residual strain ε_t, on whose borders fatigue cracks initiate, running parallel to the band. The inclusion method (18) gives the following model: mark out where the band is to be, cut it out, give it a stress free strain $-\varepsilon_t$, apply $+\varepsilon_t$ elastically, and replace it, holding it with a total surface force $f = E'h\varepsilon_t$ (and a balancing remote one). Here $E' = E/(1-\nu^2)$. Reweld, except where the crack is to be. When f is annulled by -f at the surface, a G appears, which for a slim band may be calculated using the M integral by regarding -f as a point force (46,73,74). The G is independent of the sign of ε_t. Eshelby concluded (75) that there was little bias favouring nucleation on a particular side of the band but made a numerical error. In fact (76), the theory shows that crack initiation on the side of the band making an obtuse angle with the surface is strongly favoured. However, preliminary numerical calculations which also take account of forces normal to the band in the inclusion model indicate that these may redress this inbalance (76), or even reverse it.

A second application (77) enables the constant term in the Williams expansion (78), widely known as the T term (79), to be obtained by evaluating path independent integrals. Application of a point force f at the tip (at 0) of an unloaded crack, together with suitable tractions elsewhere, provides a stress field with $p^f \sim f/r$, while the loaded crack field alone has $p^c \sim K/(2\pi r)^{\frac{1}{2}} + T + \ldots$. Evaluation of J (i.e. F_1) integrals gives $J^c \sim K^2$, $J^{f+c} \sim K^2 + fT$. Thus T can be found. Of course, if f is applied without suitable tractions elsewhere there may be an image K^f and T^f and a slightly more elaborate procedure is necessary (77).

6. CRACKS AND DYNAMICS

The general procedure leading to Eqn (8) can be made to look after the dynamic situation (6,17,42). With $\lambda,\mu = 0,1,2,3$; $i,j = 1,2,3$; $X_0 = t$, $u_0 = 0$ and the Lagrangian $L(u_\lambda, u_{\lambda,\mu}, X_\lambda) = T - W(u_{i,j}, X_m)$, $T = (\frac{1}{2})\rho_0 \dot{u}_i \dot{u}_i$, we get the 4 x 4 dynamic energy momentum tensor $D_{\lambda\mu}$,

$$D_{\lambda\mu} = -L\delta_{\lambda\mu} + (\partial L/\partial u_{\sigma,\mu})u_{\sigma,\lambda}$$

$$\text{with} \quad D_{\lambda\mu,\mu} = -(\partial L/\partial X_\lambda)_{\exp} \tag{22}$$

and

$$D_{\ell j} = (W-T)\delta_{\ell j} - p_{ij}u_{i,\ell} \tag{23}$$

as the spatial part. $D_{oo} = W+T = H$, the Hamiltonian, $D_{oj} = -p_{ij}\dot{u}_i = s_j$, the energy flux vector and $-P_{\ell o} = -\rho\dot{u}_i u_{i,\ell} = g_\ell$ the quasi (field, pseudo) momentum. One interpretation of g_ℓ is obtained (17,42) by considering the motion of a mesh (of mass $\varepsilon^3\rho_0$ and initially of side ε) of the moving curvilinear embedded coordinate system $x^{m'} = \delta^{m'm}X_m$. Then $g_\ell = G_\ell - \pi_\ell$, where $G_\ell = \varepsilon^3\rho_0\dot{u}_\ell$ is the ordinary momentum of the mesh and π_ℓ the generalized (Lagrange) momentum $\pi_\ell = \delta_{\ell s'}\partial T_m/\partial\dot{x}^{s'}$ for the coordinates $x^{s'}$: here $T_m = (\frac{1}{2})\varepsilon^3\rho_0\dot{\underline{x}}.\dot{\underline{x}}$ is the kinetic energy of the mesh. For a homogeneous material with time independent properties Eqn (22) gives $D_{\ell j,j} = \dot{g}_\ell$ and $-s_{j,j} = \dot{H}$, expressing the conservation of quasi momentum and of energy. More general forms are obtained if the elastic medium interacts with another system through a potential $U(u_{i,j}, x_m)$, e.g. because there are electrons interacting via a deformation potential, but even so, under conditions frequently met, the ordinary and quasi momentum within a bounding surface S both increase at the same rate (17,42). When defects are present, results can often be obtained correctly (42) from the assumption (80) that the force on a defect and quasi momentum are related in the same way as ordinary force and momentum. An example is given below.

Applications of the F_1 (or J) integral to crack problems were made independently by several workers (38,39,81,82,83). Eshelby considered both the static (38,39) and dynamic (38) situations, for the latter emphasizing that there was no generally valid path independent integral and for the former treating the crack both as a singularity in its own right, and also as a distribution of dislocations.

The relation between $D_{\ell j}$ and the integral

$$F_1 = \lim_{S\to 0}\int_S H_{1j}dS_j \quad \text{with} \quad H_{\ell j} = (W+T)\delta_{\ell j} - p_{ij}u_{i,\ell} \tag{24}$$

giving the force on a crack moving with velocity $v_\ell = v\delta_{\ell 1}$ can be made to follow (42) from the above assumption (80) about quasi momentum. For if S is a fixed surface surrounding the crack tip (pointing along X_1)

$$\int_S D_{1j}dS_j = F_1 + \frac{d}{dt}\int g_1 dV , \tag{25}$$

since $\int_S D_{ij}dS_j$ is the flux of quasi momentum into S. If we can assume that $u_i = u_i^0(X_1-vt,X_2) + u_i'(X_1,X_2,t)$ where $u_i' \ll u_i$ near $X_1 = vt$, $X_2 = 0$, we have, for sufficiently small S, $\dot{u}_i = -vu_{i,1}$ and can calculate the increase in the quasi momentum from the convected flow,

$$(d/dt) \int_V g_1 dV = - v \int_S g_1 dS_1 \qquad (26)$$

Here $vg_1 = -\rho\dot{u}_i u_{i,1} = +\rho\dot{u}_i\dot{u}_i = 2T$, so that $F_1 = \int_S (D_{1j} + 2T\delta_{1j})dS_j$, which gives Eqn (24).

Eshelby wrote several articles about static cracks (36,37, 39,44) and considered a number of dynamic crack problems (38,40-44). The work on crack dynamics has been briefly reviewed elsewhere (84) and will receive extensive discussion at this meeting. He had much earlier (2,6,10) considered moving dislocations and treated their equation of motion, a rather more difficult problem. As he commented (42), the expression $F_\ell = \lim(S\to 0) \int_S H_{\ell j} dS_j$, diverges for a kink, crowdion or dislocation (as r^{-1}, r^{-3} and $\ln r$) because these defects have effective masses proportional to r_0^{-1}, r_0^{-3} and $\ln[R(t)/r_0]$, where r_0 is a cut-off radius and R(t) depends on the history of the dislocation's motion - it is (6), peculiarly "haunted by its past". In contrast, the crack tip has no effective mass and the force can be worked out on the basis of continuum mechanics, without recourse to an atomic treatment. He emphasized by using Peierls-Nabarro models of the crack, dislocation, crowdion or kink that the quasi-momentum balance equation $\int_S D_{\ell j} dS_j = (d/dt) \int_S g_\ell dV$ is not upset by the presence of a defect, so there is no total net force on it. What we usually do (42) is to calculate "one of the partial forces on it which add up to zero".

Eshelby was the first to obtain the energy flow into an arbitrarily moving crack tip (40,41) (the dynamic energy release rate) and so to establish the energy balance equation from which the motion of the tip can be determined. He also used it to discuss crack branching (42). Kostrov (85) had earlier obtained the dynamic stress intensity factor and formulated an equation of motion in terms of it; of course, by an appropriate choice of 'modulus of cohesion' $K_c(v)$, his criterion can be made to correspond to an energy balance criterion.

7. CONTINUOUS DISTRIBUTIONS OF DISLOCATIONS

One must distinguish the F_1 and J integrals (86-89). In J, W is the density of stress working and u_i is the shape displacement of field plasticity or of the theory of continuous distributions of dislocations (90-93). Thus J may be evaluated along a path through a plastic field. The plastic deformation may be of a special kind so that it appears that a strain energy function exists: then J is path independent and equivalent to F_1 (42,86). However, Bilby and Eshelby (94) defined another integral giving the resultant force on all dislocations within S, including those representing a crack tip, if any,

$$Q_\ell = \int_S (W\delta_{\ell j} - p_{ij}\beta^E_{\ell i})\, dS_j \tag{27}$$

This can be evaluated over a path or surface in a plastic region. Here $\beta^E_{\ell i}$ is the elastic distortion tensor (90,91). Q_ℓ has also been formulated independently (89,95) and received some application (96,97) under the name of J_{ext}. Eshelby made a number of contributions to the general continuum theory of dislocations (6,17,24,26) and used distributions of dislocations throughout his work. In (24) incidentally, he wrote the configurational force F_x on a lattice defect in the form most widely used by engineers for the J integral. An interesting paper concerns terminating dislocations (35). In the theory of continuous distributions of dislocations, this is a problem related to Cartan's theorem on the conservation of affine displacement, or to the generalized Bianchi identities, which translate into the node theorem (90-93). Characteristically, Eshelby noted the analogy between the angular dislocation (98) and the Winkelstrom of Grassmann (99). At the end of a terminating dislocation (100,101) there arises (35) a "naturally occurring case of a continuous distribution of dislocations". To discuss this, and as an additional source function for internal stress, the rational dislocation element was introduced (35), in analogy with the rational current element of Heaviside (102).

8. INCLUSIONS, INHOMOGENEITIES AND POINT DEFECTS

It is natural now to turn to another of Eshelby's major contributions, the work on the theory of inclusions, inhomogeneities and point defects. His study of the elastic energy momentum tensor arose from his concern with the energies and interactions of lattice defects. He wrote extensively on point defects (6,12,13,14,17,22,23, 24,47,51) drawing attention inter alia to the effects of their image fields in finite bodies and to the interpretation of X-ray and density measurements. As early as 1951 (6) Eshelby noted the implications of different boundary conditions (103,104) on the use of $P_{\ell j}$ to calculate interaction forces, remarking (42) that $P_{\ell j}$ had guided him (6) in his selection of rival expressions for dislocation point defect interactions. He also discussed in some detail the slipping boundary condition (22) for interacting spheres. He commented (42) on the classical point defect formula (for a defect idealized as a centre of dilatation), $\underline{F} = -\Delta V \text{ grad } P$. Here ΔV is the volume change produced by the introduction of the defect, as measured by the expansion of the outer surface of the body; and P the hydrostatic pressure there would be at the centre of the defect if it were not there. This formula, he noted, is valid for non-linear theory (as may be proved by integrating $P_{\ell j}$ over a large surface S surrounding the defect) provided that P is taken to be not the (for a non-linear theory, meaningless) actual hydrostatic pressure P due to the surface loads and other defects, but that "which would be calculated from the applied conditions prevailing over the large surface S on the assumption that the linear theory is valid everywhere". A later didactic interest in point defect theory was the improvement of the common somewhat dubious treatments of the jump frequency formula. This he effected (47,51) by devising a simple correct derivation of Vineyard's expression (105).

The ellipsoid and inclusion papers (18,25,26,31,32,56) will be thoroughly discussed elsewhere here. The general theorems depend on potential theory and a geometrical property of the ellipsoid most widely known through its application to magnetisation by Poisson (106). For linear isotropic elasticity (18), if an ellipsoidal region in an infinite medium undergoes a constant stress free strain e_{ij}^{T}, the constrained final strain e_{ij}^{C} within the ellipsoid is also constant. This result may be extended to the anisotropic case (18,107-110). It may be exploited (18) to study the perturbation of a uniform field by an ellipsoidal inhomogeneity having elastic properties different from the matrix and the field of an inhomogeneous inclusion. Results about the field outside the ellipsoid can also be obtained (25). More generally (26,110-112), if the stress free transformation displacement is a polynomial in x_i of degree N, the displacement is also a polynomial of degree N inside the constrained ellipsoid. Thus it is possible to find the perturbation of a non uniform stress field of quite general type caused by an ellipsoidal cavity or inhomogeneity. As is evident from many other papers here, the application of this work on ellipsoidal inclusions and inhomogeneities has become very extensive.

Because it provided a useful but tractable solution to a finite deformation problem, Eshelby was particularly interested in one application which was anticipated in his earlier papers but not undertaken until required by a colleague interested in the homogenization of glass (113). It also has application to the determination of strain in rocks and to the deformation of suspensions and of materials containing voids. Use of the linear viscous analogy and the fact that an ellipsoid remains an ellipsoid after homogeneous deformation enables the finite deformation of an ellipsoidal inhomogeneity in a uniform incompressible viscous flow to be studied without obtaining the complete solution to the flow field (48,50,114,115,133,134). Intriguing results involving stationary, tumbling and oscillating inclusions arise when the applied flow has vorticity (116,117). These stem from the non-linear differential equations which arise and illustrate nicely simple aspects of the Lie theory (117-119). In connection with problems of inclusions moving in liquids Eshelby was also much exercised (49) over the stationary dissipation criteria sometimes used, e.g. (120), to select the observed motion from a class of possible motions. He noted that these criteria were equivalent to requiring the vanishing of the (viscous) F_ℓ and L_{kj} integrals for a surface embracing the inclusion, since they give minus half the total rate of change of the total dissipation rate with respect to the change of position or orientation of the inclusion. The viscous J (F_1) integral, generalized further to the (non-linear) creep case, is also used in fracture mechanics as the parameter C* (123-127).

9. DISLOCATIONS AND KINKS

Throughout all Eshelby's work there runs the theme of dislocations and their stress fields. Only a passing mention can be made here of most of this work. His early study of dislocation damping (1) led to interests in moving dislocations (2,6,10,16,17,27) and heat flow (4,15). He wrote a general paper on anisotropic elasticity (8) and considered many problems involving dislocation statics (3,7,9,11,19, 21,29,31,33,39). He has left us two long surveys (52,55), and a paper (45) devoted to geophysical applications. (A survey by Eshelby inevitably contains much intriguing and suggestive thought, and these are no exception.) Two papers (20,30) concern charged dislocations in ionic crystals, a situation where he noted that $P^*_{\ell j}$ was appropriate for the calculation of the force on a defect (42). The classical paper with Frank and Nabarro (5) on dislocation arrays is one of the most elegant. In this the positions of the dislocations are obtained as the roots of polynomials satisfying appropriate differential equations; the paper has had wide application.

The treatments of the velocity of a wave along a dislocation (34) and of the interaction of kinks and elastic waves (28), like some of his work on moving dislocations and cracks (1,6,10,17,40-43), show his real power as a theoretical physicist; sophisticated techniques are concisely applied with ruthless economy to obtain the desired results. Eshelby (28) found that in estimating classically the retarding force on a moving kink interacting with the thermal vibrations it was necessary to include higher order terms in the radiation reaction. He calculated a retarding force in an isotropic flux of transverse waves which was proportional to the kink velocity v, the energy density of the waves and the square of the kink width. Then, by quantizing the acoustic field and studying the effect on the kink momentum of the creation and annihilation of phonons, he confirmed this result. He also showed that Leibfried's suggestion (128), that the zero point energy should contribute to the retarding force on the same basis as the thermal component, was incorrect.

10. LIQUID CRYSTALS

Finally we discuss a topic which interested Eshelby greatly in his last years. As indicated above he hoped that the calculation of forces on disclinations in liquid crystals would provide one of the more realistic examples of the use of the $P_{\ell j}$ for generalized continua. In the Oseen-Frank continuum theory of nematic crystals the energy density $w(\nu_i, \partial\nu_i/\partial x_j)$ is a function only of the (unit) director ν_i and the director gradient $\partial\nu_i/\partial x_j$. Here the x_i are final coordinates. The widely accepted developments of the static theory (129-131) treat the material as incompressible and introduce the Ericksen stress tensor $t_{\ell j} \neq t_{j\ell}$,

$$t_{\ell j} = (w-P_0)\delta_{\ell j} - (\partial w/\partial(\partial\nu_i/\partial x_j))(\partial\nu_i/\partial x_\ell) \qquad (28)$$

where P_0 is a constant hydrostatic pressure. This gives the (real) force df_i on a final element ds_j in the material through the relation $df_i = t_{ij}ds_j$. With $w(\nu_i, \partial\nu_i/\partial x_j)$ we have, in the X_i, $W(\nu_i, \nu_{i,j}, u_{i,j})$ and $W = Jw$, where ultimately $J = 1$. By applying the algebraic procedure leading to Eqn (8), either by working with $w(\nu_i, \partial\nu_i/\partial x_j)$ directly or by use of $W(\nu_i, \nu_{i,j}, u_{i,j})$ and transformation to x_i, Eshelby (53,54) "was rather disconcerted to find that" the energy momentum tensor differed from the $t_{\ell j}$ of Eqn (28) only in that it lacked the term P_0. Since P_0 makes no contribution when integrating over s, the implication is that any calculation giving a configurational force on a singularity in a nematic crystal implies the existence on it also of a real (Ericksen) force. Eshelby drew attention (53,54) to errors which were thereby implicit in a number of calculations based on a widely used special case where the governing equation $\nabla^2\phi = 0$ for the director orientation ($\nu_1 = \cos\phi$, $\nu_2 = \sin\phi$, $\nu_3 = 0$) is linear but the Ericksen stress tensor is not. This work began a controversy about the relation between real and configurational forces. Eshelby consulted a number of colleagues about it, but the matter was still unresolved at the time of his death (132). His last published comment (54) was that "the basic reason for the different status of the force on a dislocation in a solid and a disclineation in a nematic liquid crystal is that the former moves through the medium, the latter with it".

11. CONCLUDING REMARKS

We are all aware of the stature of this man we have lost. He was a great scholar and original thinker, drawing continually on a wide knowledge of other branches of theoretical physics in all aspects of his work. Perhaps his unusual education, which necessarily forced him from an early age often to work things out for himself, and so to probe deeply into them, accounts for some of his originality and thoroughness. It is sad that, in the present situation, there is now ever less time for the fundamental studies so typical of his own, which are the only sure basis for true scientific advance. He will long be remembered for his dry wit and kindly human qualities. His published work is his true monument, but we hope that the present symposium will help to add in some small way to it, by indicating something of the great influence he has had over a wide field of scientific and technological endeavour.

REFERENCES

The references (1-56) of this paper are the publications of J. D. Eshelby listed on pages 5-9 of the present volume.

(57) Eshelby, J. D. Unpublished work.
(58) Noether, E. (1918). Invariante Variations-probleme. (F. Klein zum fünfzigjährigen Doktorjubiläum). Göttinger Nachrichten, (Math-Phys. Klasse), 2, 235-257.

(59) Rice, J. R. (1984). Conserved integrals and energetic forces. In: Fundamentals of deformation and fracture, IUTAM symposium Sheffield 1984, J. D. Eshelby In Memoriam, ed. B. A. Bilby, K. J. Miller & J. R. Willis, pp. 33-56. Cambridge: Cambridge University Press.

(60) Olver, P. J. (1984). Symmetry groups and path-independent integrals. Ibid., pp. 57-71.

(61) Herrmann, A. G. (1984). Conservation laws, the material momentum tensor and moments of elastic energy. Ibid., p. 73.

(62) Herrmann, G. (1984). Material balance laws in one-dimensional strength-of-materials theories. Ibid., pp. 461-463.

(63) Hill, R. (1957). On uniqueness and stability in the theory of finite elastic strain. J. Mech. Phys. Solids, 5, 221-241.

(64) Günther, H. (1962). Über einige Randintegrale der Elastomechanik. Abh. Braunschw. wiss. Ges., 14, 53-72.

(65) Knowles, J. K. & Sternberg, E. (1972). On a class of conservation laws in linearized and finite elastostatics. Arch. Rat. Mech. Ann., 44, 187-211.

(66) Budiansky, B. & Rice, J. R. (1973). Conservation laws and energy-release rates. J. Appl. Mech., 40, 201-203.

(67) Chen, F. H. K. & Shield, R. T. (1977). Conservation laws in elasticity of the J-integral type. Z. angew. Math. Phys., 28, 1-22.

(68) Atkinson, C. & Leppington, F. G. (1974). Some calculations of the energy-release rate G for cracks in micropolar and couple-stress elastic media. Int. J. Fract., 10, 599-602.

(69) Bilby, B. A. (1983). Note on the paper: A grain boundary in cellular structures near the onset of convection by P. Manneville & Y. Pomeau. Phil. Mag., A 48, 620-621.

(70) Chadwick, P. (1975). Applications of an energy momentum tensor in non linear elastostatics. Journal of Elasticity, 5, 249-258.

(71) Muskhelishvili, N. I. (1963). Some basic problems of the mathematical theory of elasticity, trans. J. R. M. Radok from 4th Russian edition, Moscow 1954. Groningen: Noordhoff.

(72) Brown, L. M. & Ogin, S. L., with an appendix by the late J. D. Eshelby (1984). Role of internal stresses in the nucleation of fatigue cracks. In: Fundamentals of deformation and fracture, IUTAM symposium Sheffield 1984, J. D. Eshelby In Memoriam, ed. B. A. Bilby, K. J. Miller & J. R. Willis, pp. 357-367. Cambridge: Cambridge University Press.

(73) Freund, L. B. (1978). Stress intensity factor calculations based on a conservation integral. Int. J. Solids Structures, 14, 241-250.

(74) Bilby, B. A. (1979). The Tewksbury lecture: Putting fracture to work. In: Fourth Tewksbury symposium, Fracture at work, Melbourne 1979, ed. D. S. Mansell & G. H. Vasey, pp. 1.1-1.44. Melbourne: University of Melbourne. (Reprinted in J. Mater. Sci. (1980), 15, 535-556).

(75) Eshelby, J. D. (1981). Letter to Dr. L. M. Brown, 19.11.1981.

(76) Bilby, B. A. & Cardew, G. E. (1984). Unpublished work.

(77) Cardew, G. E., Goldthorpe, M. R., Howard, I. C. & Kfouri, A. P. (1984). On the elastic T term. In: Fundamentals of deformation and fracture, IUTAM symposium Sheffield 1984, J. D. Eshelby In Memoriam, pp. 465-476. Cambridge: Cambridge University Press.
(78) Williams, M. L. (1957). On the stress distribution at the base of a stationary crack. J. Appl. Mech., 24, 109-114.
(79) Rice, J. R. (1974). Limitations to the small scale yielding approximation for crack tip plasticity. J. Mech. Phys. Solids, 22, 17-26.
(80) Nabarro, F. R. N. (1951). The interaction of screw dislocations and sound waves. Proc. R. Soc. Lond., A 209, 278-290.
(81) Cherepanov, G. P. (1967). Crack propagation in continuous media. Appl. Math. Mech., 31, 503-512. (Prikl. Mat. Mekh., 31, 476-488).
(82) Rice, J. R. (1968). A path independent integral and the approximate analysis of strain concentration by notches and cracks. J. Appl. Mech., 35, 379-386.
(83) Rice, J. R. (1968). Mathematical analysis in the mechanics of fracture. In: Fracture: An advanced treatise, ed. H. Liebowitz, 2, pp. 191-311. N.Y.: Academic Press.
(84) Bilby, B. A. (1972). Moving cracks. In: Amorphous materials. Papers presented to the third international conference on the physics of non-crystalline solids, Sheffield 1970, ed. R. W. Douglas & B. Ellis, pp. 489-506. New York: Interscience.
(85) Kostrov, B. V. (1966). Unsteady propagation of longitudinal shear cracks. Appl. Math. Mech., 30, 1241-1248. (Prikl. Mat. Mekh., 1966, 30, 1042-1049).
(86) Bilby, B. A. (1977). Fracture. In: Fracture 1977 (ICF4), Volume 4, Waterloo, Canada, June 19-24, pp. 31-48. Waterloo: University of Waterloo Press.
(87) Chell, G. G. & Heald, P. T. (1975). The path dependence of the J-contour integral. Int. J. Fract., 11, 349-351.
(88) Rice, J. R. (1975). Discussion: "The path dependence of the J-contour integral", by G. G. Chell & P. T. Heald. Int. J. Fracture, 11, 352-353.
(89) Chell, G. G. & Heald, P. T. (1976). Response: Discussion of "The path dependence of the J-contour integral", by J. R. Rice. Int. J. Fracture, 12, 172-174.
(90) Bilby, B. A. (1960). Continuous distributions of dislocations. Progr. Solid Mech., 1, 331-398.
(91) Kröner, E. (1958). Kontinuumstheorie der Versetzungen und Eigenspannungen. Berlin: Springer-Verlag.
(92) Bilby, B. A. (1968). Geometry and continuum mechanics. In: Mechanics of generalized continua, IUTAM Symposium Freudenstadt 1967, ed. E. Kröner, pp. 180-199. Berlin: Springer-Verlag.
(93) Kondo, K. (1968). On the two main currents of the geometrical theory of imperfect continua. Ibid, pp. 200-213.
(94) Bilby, B. A. (1973). Dislocations and cracks. In: Papers presented at the third international conference on fracture, Munich April 1973, Part XI, pp. 1-20. Düsseldorf: Verein Deutscher Eisenhüttenleute.

(95) Miyamoto, H. & Kageyama, K. (1978). Extension of the J-integral to the general elastic-plastic problem and suggestion of a new method for its evaluation. In: Proceedings of the first international conference on numerical methods in fracture mechanics Swansea 1978, ed. A. R. Luxmoore & D. R. J. Owen, pp. 479-486. Swansea: University College Swansea.
(96) Miyamoto, H. & Kageyama, K. (1979). Fundamental study on the J_{ext} integral applied to elasto-plastic fracture mechanics. In: Recent research on mechanical behavior of solids (Hiroshi Miyamoto volume), K. Saito, Secretary of editorial committee, pp. 229-236. Tokyo: University of Tokyo Press.
(97) T. Mura (1979). On the extended J-integral. Ibid., pp. 237-243.
(98) E. H. Yoffe (1960). The angular dislocation. Phil. Mag., 5. 161-175.
(99) Grassmann, H. G. (1845). Neue Theorie der Elektrodynamik. Ann. der Physik, 64, 1-18.
(100) Li, J. C. M. (1964). Stress field of a dislocation segment. Phil. Mag., 10, 1097-1098.
(101) Brown, L. M. (1964). The self stress of dislocations and the shape of extended nodes. Phil. Mag., 10, 441-466.
(102) Heaviside, O. (1888). The mutual action of a pair of rational current-elements. The Electrician, Vol. 18, p.229 (Reprinted in: Electrical papers, 2, 500-502. London: Macmillan 1892).
(103) Bilby, B. A. (1950). On the interactions of dislocations and solute atoms. Proc. Phys. Soc. A 63, 191-200.
(104) Leibfried, G. (1949). Über die auf eine Versetzung wirkenden Kräfte. Z. Phys., 126, 781-789.
(105) Vineyard, G. H. (1957). Frequency factors and isotope effects in solid state rate processes. J. Phys. Chem. Solids, 3, 121-127.
(106) Korringa, J., Lin, I-H. & Mills, R. L. (1978). General theorems about homogeneous ellipsoidal inclusions. Am. J. Phys., 46, 517-521.
(107) Willis, J. R. (1970). The solution of asymmetric problems of elasticity by Fourier transforms. Adams Prize Essay, University of Cambridge.
(108) Kinoshita, N. & Mura, T. (1971). Elastic fields of inclusions in anisotropic media. phys. stat. sol. (a), 5, 759-768.
(109) Walpole, L. J. (1977). The determination of the elastic field of an ellipsoidal inclusion in an anisotropic medium. Math. Proc. Camb. Phil. Soc., 81, 283-289.
(110) Kunin, I. A. & Sosnina, E. G. (1972). Ellipsoidal inhomogeneity in an elastic medium. Sov. Phys. Dokl., 16, 534-540. (Dokl. Akad. Nauk. SSSR, (1971), 199, 571-574).
(111) Asaro, R. J. & Barnett, D. M. (1975). The non-uniform transformation strain problem for an anisotropic ellipsoidal inclusion. J. Mech. Phys. Solids, 23, 77-83.
(112) Mura, T. & Kinoshita, N. (1978). The polynomial eigenstrain problem for an anisotropic ellipsoidal inclusion. phys. stat. sol. (a), 48, 447-450.
(113) Cable, M. (1968). The physical chemistry of glassmaking. In: Proc. Int. Congr. on Glass, pp. 163-178. London: Society of Glass Technology.

(114) Howard, I. C. & Brierley, P. (1976). On the finite deformation of an inhomogeneity in a viscous liquid. Int. J. Engng Sci., 14, 1151-1159.
(115) Budiansky, B., Hutchinson, J. W. & Slutsky, S. (1982). Void growth and collapse in viscous solids. In: Mechanics of solids, the Rodney Hill 60th anniversary volume, ed. H. G. Hopkins & M. J. Sewell, pp. 13-45. Oxford: Pergamon Press.
(116) Kolbuszewski, M. L. (1977). Inclusions, inhomogeneities and holes in slow flow. Ph.D. Thesis, University of Sheffield.
(117) Bilby, B. A. & Kolbuszewski, M. L. (1977). The finite deformation of an inhomogeneity in two-dimensional slow viscous incompressible flow. Proc. R. Soc. Lond., A 335, 335-353.
(118) Lie, S. (1888). Theorie der Transformations gruppen, vol. I, p.59. Leipzig: Teubner.
(119) Lie, S. (1891). Differential gleichungen. Reprinted 1967. New York: Chelsea.
(120) Jeffery, G. B. (1922). The motion of ellipsoidal particles immersed in a viscous fluid. Proc. R. Soc. Lond., A 102, 161-179.
(121) Atkinson, C. (1984). Invariant integrals for some time dependent crack and heat conduction problems. In: Fundamentals of deformation and fracture, IUTAM symposium Sheffield 1984, J. D. Eshelby In Memoriam, ed. B. A. Bilby, K. J. Miller & J. R. Willis, pp. Cambridge: Cambridge University Press.
(122) Freund, L. B. (1984). The mechanics of dynamic crack growth in solids. Ibid. pp.
(123) Landes, J. D. & Begley, J. A. (1976). A Fracture Mechanics Approach to Creep Crack Growth. In: Mechanics of Crack Growth. ASTM STP 590, pp. 128-148. Philadelphia: American Society for Testing and Materials.
(124) Webster, G. A. (1975). The application of fracture mechanics to creep cracking. In: Conference on mechanics and physics of fracture, Cambridge, January 1975, pp. 18/1-18/9. London: Institute of Physics and the Metals Society.
(125) Riedel, H. (1978). Cracks loaded in Anti-Plane Shear under Creep Conditions. Z. Metallkde, 69, 755-760.
(126) Riedel, H. & Rice, J. R. (1980). Tensile cracks in creeping solids. In: Fracture Mechanics. ASTM STP 700, pp. 112-130. Philadelphia: American Society for Testing and Materials.
(127) Harper, M. P. & Ellison, E. G. (1977). The use of the C* parameter in predicting creep crack propagation rates. J. Strain Anal., 12, 167-179.
(128) Leibfried, G. (1950). Über den Einfluss thermisch angeregter Schallwellen auf die plastische Deformation. Z. Phys., 127, 344-356.
(129) Ericksen, J. L. (1976). Equilibrium theory of liquid crystals. Adv. liq. Cryst., 2, pp. 233-298.
(130) Chandrasekhar, S. (1977). Liquid Crystals. Cambridge: Cambridge University Press.
(131) Leslie, F. M. (1983). Some topics in continuum theory of nematics. Phil. Trans. R. Soc. Lond., A 309, 155-165.

(132) Nabarro, F. R. N. (1984). Material forces and configurational forces in the interaction of elastic singularities. In: Fundamentals of deformation and fracture, IUTAM symposium Sheffield 1984, J. D. Eshelby In Memoriam, ed. B. A. Bilby, K. J. Miller & J. R. Willis, pp. 455-459. Cambridge: Cambridge University Press.

(133) Bilby, B. A. & Kolbuszewski, M. L. (1976). Discussion of paper: N. C. Gay & R. E. P. Fripp. Phil. Trans. R. Soc. Lond., A 283, 125-126.

(134) Lisle, R. J., Rondeel, H. E., Doorn, D., Brugge, J. & van de Gaag, P. (1983). Estimation of viscosity contrast and finite strain from deformed elliptical inclusions. J. Struct. Geol., 5, 603-609.

CONSERVED INTEGRALS AND ENERGETIC FORCES

JAMES R. RICE

Harvard University, Cambridge, MA 02138, USA

ABSTRACT

Conserved integrals of the Eshelby type representing energetic forces on dislocations, inclusions, voids, cracks and the like are reviewed and related to invariant transformations. Applications are discussed based on path independence for 2D integrals of J and M type and on the Maxwell reciprocity satisfied by energetic forces. Such concepts have had wide use in crack mechanics to aid analysis of near tip fields and provide elegant short-cut solutions of boundary value problems. Here new applications of path independence to dislocations show that the M integral when centered on a dislocation line is equal to the prelogarithmic energy factor and, also, that simple expressions involving the factor result for the image force drawing a dislocation to a V-notch tip and to a bicrystal interface. A review of reciprocity for energetic forces reveals a wide range of applications to such topics as weight functions for elastic crack analysis, the structure of inelastic constitutive relations, and compliance methods in nonlinear fracture mechanics, and in a new application there is developed the full 2D interaction effects between a crack tip and a nearby dislocation in a crystal of general anisotropy.

1. INTRODUCTION

Eshelby (1,2) was the first to associate an energetic force on an elastic defect with a conserved integral of elastostatic field quantities over a surrounding surface (or contour for 2D fields). His concern was with "any source of internal stress" such as an inclusion or dislocation whose motion translated an incompatible and possibly singular field along with its core. The emergence of the same integral for the energy release rate in crack mechanics (3,4,5) together with the discovery of further non-translational conserved integrals (6,7) with energetic force interpretations (8,9) has now led to an enormous literature. I do not attempt to completely summarize it here but, rather, focus on two primary issues. The first relates to the fact that the integrals involved are of conservation type, i.e., independent of the surface chosen. It developed that this feature, in the form of 2D path independence, could be exploited to great advantage in the development of nonlinear crack mechanics (5,10,11,12) and allowed also elegant short-cut solutions to some elastic crack boundary value problems (10,13, 14). I present new applications of path independence here to an interpretation of the "M" integral and the determination of forces on dislocations near notch tips and grain interfaces. The second issue is that the energetic forces discussed relate to strain energy changes and must satisfy reciprocity relations of the Maxwell type in thermodynamics. This has proven to be the source of some remarkable and unexpected inter-relations which I summarize in the final section. The topics discussed there range from elastic weight functions to compliance testing and the structure of inealstic constitutive relations; they also include some new results on the energetic forces on dislocations near a crack tip, complementing those derived from path independent integrals.

The translational integrals are

$$J_\alpha = \int_\Gamma (W n_\alpha - n_\beta \sigma_{\beta\gamma} u_{\gamma,\alpha})\, ds \tag{1}$$

where Γ is some closed surface (or contour in 2D) and ds is an element of area (or arc length). Here $\underline{n}$ is the outer normal to Γ, $W = W(\underline{\nabla}\underline{u})$ is the elastic strain energy density per unit volume of an unstressed reference configuration, $\underline{u}(\underline{x})$ is displacement of the particle which was at $\underline{x}$ in the reference configuration, and $\sigma_{\alpha\beta} = \partial W/\partial u_{\beta,\alpha}$. When the surface Γ surrounds a solute $\underline{J}$ is the drift force, and when Γ surrounds a straight dislocation line as a contour in a 2D field $\underline{J}$ is the combined glide/climb force on the dislocation (1,2). Similarly for planar crack growth in a 2D field J (= J_1, the component in the direction of crack growth) on any contour Γ surrounding the tip, starting on one traction free crack surface and ending on the other, is the Irwin energy release rate G (4,5), i.e., the energetic force conjugate to crack area.

Related to the force interpretations, $\underline{J} = \underline{0}$ when Γ is any closed surface (contour in 2D) surrounding homogeneous defect-free material. This property $\underline{J} = \underline{0}$ may be understood alternatively as a consequence of Noether's theorem (15) and follows from the fact that for translationally homogeneous materials the volume integral in the elastic variational principle,

$$\delta\left[\int_V W\, dV\right] + [\text{boundary terms}] = 0 \quad , \tag{2}$$

is invariant when we change positional coordinates and displacement from the unprimed to a primed set by

$$x'_\alpha = x_\alpha + \varepsilon_\alpha \qquad u'_\alpha(\underline{x}') = u_\alpha(\underline{x}) \tag{3}$$

where $\underline{\varepsilon}$ is arbitrary. The invariance discussed means that

$$\int_{V'} W[\underline{\underline{\nabla}}\underline{u}'(\underline{x}')]\, dV' - \int_V W[\underline{\underline{\nabla}}\underline{u}(\underline{x})]\, dV = 0 \tag{4}$$

Apparently Günther (6) first applied the Noether procedure to elastostatics and, in results discovered independently by Knowles and Sternberg (7), established the conservation integrals which result if invariance in the above sense applies for all or some of the transformations defined by

$$\begin{aligned} x'_\alpha &= x_\alpha + \varepsilon_\alpha + (\underline{\omega} \times \underline{x})_\alpha + \gamma x_\alpha \\ u'_\alpha(\underline{x}') &= u_\alpha(\underline{x}) + [\underline{\omega} \times \underline{u}(\underline{x})]_\alpha + [(m-n)/m]\gamma u_\alpha(\underline{x}) \end{aligned} \tag{5}$$

where $\underline{\varepsilon}$, $\underline{\omega}$ and γ are infinitesimal. Here $\underline{\omega}$ refers to rotation and if there is invariance under the action of $\underline{\omega}$ there results the three conservation integrals customarily denoted by $\underline{L} = \underline{0}$. Of more interest to applications in the next section is that generated by self similar scale change by the factor γ. Then we will have invariance if the medium is homogeneous along rays from the coordinate origin and if W is homogeneous of degree m in $\underline{\underline{\nabla}}\underline{u}$ (e.g., $m = 2$ for linear elasticity); n is the number of spatial dimensions of the "volume" denoted by V. For example the consequence for 2D linear elasticity is then that

$$M_o = \int_\Gamma x_\alpha (W n_\alpha - n_\beta \sigma_{\beta\gamma} u_{\gamma,\alpha})\, ds \tag{6}$$

satisfies $M_o = 0$ when Γ is a closed contour surrounding no singularity or defect. The subscript o (for coordinate origin) on M_o is to remind that the integral depends on the origin for $\underline{x}$. For example, if the integral is evaluated on a given path Γ in one case with origin at $\underline{x}_P$ and in another with origin at $\underline{x}_Q$, then

$$M_P = M_Q + (\underline{x}_Q - \underline{x}_P)_\alpha J_\alpha \quad . \tag{7}$$

Budiansky and Rice (8) showed that the integrals $\underline{L}$ and M, when taken on surfaces enclosing voids, could be interpreted as energetic forces: $\underline{L}$ is associated with erosion/addition of material so as to rotate the void boundary about the coordinate origin and M with self-similar erosion relative to the origin. Eshelby (13) noted the remarkable ability of path-independence of M to resolve some elastic crack boundary value problems, and such was pursued further by Freund (14), Ouchterlony (16,17) and Kubo (18) in application to various 2D point force (3D line force) loadings. The next section similarly makes use of M, but for dislocations.

Before going on to specifics let us study the formal working of Noether's (15) procedure for elasticity, writing the transformations of Eqn (5) as

$$x'_\alpha = x_\alpha + y_\alpha(\underline{x}) \ , \qquad u'_\alpha(\underline{x}') = u_\alpha(\underline{x}) + v_\alpha(\underline{x}) \ , \tag{8}$$

where $\underline{y}$ and $\underline{v}$ are infinitesimal, and assuming that Eqn (4) holds to first order in them. Then note that

$$\int_{V'} W[\underline{\underline{\nabla}}\underline{u}'(\underline{x}')]\, dV' = \int_V W[\underline{\underline{\nabla}}\underline{u}'(\underline{x})]\, dV + \int_\Gamma n_\alpha y_\alpha W\, ds \tag{9}$$

and that for any variation $\delta\underline{u}$ from an elastic field

$$\int_V \delta W\, dV = \int_V \sigma_{\alpha\beta}\delta u_{\beta,\alpha}\, dV = \int_\Gamma n_\alpha \sigma_{\alpha\beta}\delta u_\beta\, ds \ . \tag{10}$$

Thus, if we insert Eqn (9) into Eqn (4) and use the last expression with

$$\delta u_\beta = u'_\beta(\underline{x}) - u_\beta(\underline{x}) = v_\beta - y_\gamma u_{\beta,\gamma} \tag{11}$$

to deal with the difference between the remaining integrals over V, there results

$$\int_\Gamma [n_\alpha y_\alpha W + n_\alpha \sigma_{\alpha\beta}(v_\beta - y_\gamma u_{\beta,\gamma})]\, ds = 0 \ . \tag{12}$$

This expresses the conserved integrals discussed; identifying $\underline{y}$ and $\underline{v}$ with the terms in Eqn (5) the integral can be rewritten

$$J_\beta \varepsilon_\beta + L_\beta \omega_\beta + M\gamma = 0 \ , \tag{13}$$

implying that the integrals J_β, L_β and M thereby defined vanish when there is invariance in the sense of Eqn (4) relative to the respective coefficient ε_β, ω_β or γ. Perhaps the simplest context for realizing the energetic force interpretations of the integrals is to consider a traction-free void (8). When Γ surrounds the void the integral of Eqn (12) is conserved but not generally zero. If now we shrink Γ to the void surface and realize that the resulting integral is to first order the negative of the energy change when we erode a layer $n_\alpha y_\alpha$ from the void surface (see (19,20) for discussion of energy changes in such processes), then rewriting the integral as on the left side of Eqn (13) shows that $\underline{J}$, $\underline{L}$ and M are the energetic forces conjugate to translation $\underline{\varepsilon}$, rotation $\underline{\omega}$ and self-similar scaling γ.

2. DISLOCATIONS NEAR NOTCH TIPS AND INTERFACES

2.1 M_d IS THE DISLOCATION ENERGY FACTOR

Suppose that an indefinitely long straight dislocation line of Burgers vector $\underline{b}$ lies along the x_3 axis of a rectilinearly anisotropic solid (e.g., single crystal) which sustains a 2D deformation field of combined plane and anti-plane strain in the x_1x_2 plane. The dislocation line pierces the x_1x_2 plane at the point with coordinates c_1, c_2 (Fig. 1a). It is well known, and follows from dimensional considerations and linearity, that the stress field is of the form

$$\sigma_{\alpha\beta} = D_{\alpha\beta}(\theta)/r + \bar{\sigma}_{\alpha\beta}(x_1, x_2) \tag{14}$$

where r and θ are polar coordinates at the core site and $\bar{\sigma}_{\alpha\beta}$ is the combined non-singular result of applied loadings and image effects. With $\underline{h} = (-\sin\theta, \cos\theta, 0)$ denoting the unit vector in the direction of increasing θ, it is elementary to show from stress equilibrium equations that the functions $\underline{D}$ of θ are such that $h_\alpha(\theta) D_{\alpha\beta}(\theta)$ is is independent of θ. Further, if one introduces the positive definite symmetric prelogarithmic "energy factor" tensor $K_{\alpha\beta}$ of anisotropic elastic dislocation theory (21,22,23) it is easy to show that

$$h_\alpha(\theta) D_{\alpha\beta}(\theta) = 2K_{\beta\alpha} b_\alpha \; . \tag{15}$$

This expression will be motivated by what follows; in it $\underline{b}$ is the Burgers vector of the dislocation. The energy factor arises further in, and derives its name from, the expression for the strain energy (per unit length in the x_3 direction) of the unloaded but dislocated body with core cut-off at r_o:

$$U = K_{\alpha\beta} b_\alpha b_\beta \ln(L/r_o) + \text{terms which} \to 0 \text{ as } r_o \to 0 \; . \tag{16}$$

In this expression L depends on the outer dimensions of the dislocated body. Barnett and Swanger (23) explain how to extract the tensor $\underline{K}$ numerically in terms of the elastic moduli of an anisotropic crystal. For the isotropic case

$$K_{11} = K_{22} = \mu/4\pi(1-\nu), \quad K_{33} = \mu/4\pi, \quad \text{other } K_{\alpha\beta} = 0 \; . \tag{17}$$

The logarithmic dependence on r_o in Eqn (16) may be understood via the Clapeyron expression for U: introduce a cut along the ray $\theta = \text{constant}$ from $r = 0$ to $r = R(\theta)$, at the outer boundary, and displace its surfaces by $\underline{b}$ by introduction of tractions $h_\alpha \sigma_{\alpha\beta}$ on the cut; the work done is

$$U = \frac{1}{2} \int_{r_o}^{R(\theta)} (h_\alpha \sigma_{\alpha\beta}) b_\beta \, dr = \frac{1}{2} \int_{r_o}^{R} [(h_\alpha D_{\alpha\beta}/r) b_\beta + \ldots] \, dr$$

$$= K_{\alpha\beta} b_\alpha b_\beta \, \ell n \frac{R}{r_o} + \ldots \qquad (18)$$

where the dots denote bounded terms. Note that the independence of $h_\alpha D_{\alpha\beta}$ on θ is essential to the form of the final result.

Now, from Eshelby's work (1,2) we know that if we evaluate $\underline{J}$ on a circuit Γ (Fig. 1a) surrounding the dislocation we get the energetic force $\underline{f}$ on the dislocation, defined such that $\delta U = -f_\alpha \delta c_\alpha$ at fixed outer-boundary displacements and related to $\bar{\sigma}_{\alpha\beta}$ by the Peach-Koehler force expression. Thus, with such choice of Γ we have

$$J_\alpha = f_\alpha = e_{\alpha\lambda 3} \, \bar{\sigma}_{\lambda\beta}(c_1, c_2) b_\beta \qquad (\alpha = 1,2) \qquad (19)$$

where $\underline{e}$ is the alternating tensor.

What is M for a dislocation? Plainly the result depends on the origin chosen for $\underline{x}$ in Eqn (6). Let us choose this origin at the dislocation line itself, denoting the corresponding M by M_d, and choose for path Γ a circle of radius r centered on the dislocation. Then

$$M_d = r^2 \int_0^{2\pi} [W(r,\theta) - n_\alpha(\theta) \sigma_{\alpha\beta}(r,\theta) \partial u_\beta(r,\theta)/\partial r] \, d\theta \qquad (20)$$

where here $\underline{n}$ denotes a radially directed unit vector. We note that M_d must be the same for any path Γ surrounding the dislocation and hence, in particular, this expression for M_d is independent of r. We can therefore evaluate it by letting $r \to 0$ and in that limit we see that only the $1/r$ singular stress terms in Eqn (14) can contribute to

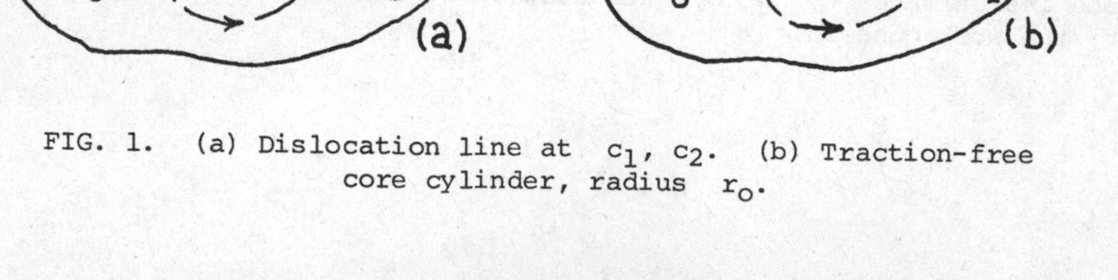

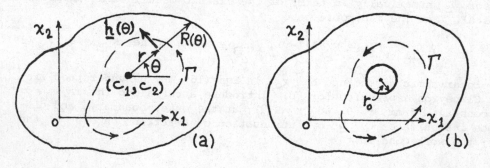

FIG. 1. (a) Dislocation line at c_1, c_2. (b) Traction-free core cylinder, radius r_o.

it; those terms and the corresponding $1/r$ singular terms in $u_{\alpha,\beta}$ cause each term within [...] above to behave as $1/r^2$ and hence to survive in the expression for M_d as $r \to 0$. The bounded terms $\bar{\sigma}_{\alpha\beta}$, $\bar{u}_{\alpha,\beta}$ do not so contribute, and hence we have reached the conclusion that M_d is a path invariant quantity associated with a dislocation and which can be evaluated solely in terms of the stress/deformation field which that dislocation would induce in an unbounded and otherwise unloaded body. Thus M_d can depend only on $\underline{b}$.

I will now show that M_d is, in fact, nothing other than the dislocation energy factor:

$$M_d = K_{\alpha\beta} b_\alpha b_\beta \quad . \tag{21}$$

We begin with a core model slightly different from that implied by Eqn (18) and, instead, regard the core radius r_o as the radius of a traction free circular cylindrical hole centered on the dislocation line (Fig. 1b). We shall always be interested in the case $r_o \lll L$, by which it will be understood that an intermediate range of r exists satisfying

$$r_o \ll r \ll L$$

within which the stress field is given by the first term of Eqn (14), $\sigma_{\alpha\beta} = D_{\alpha\beta}(\theta)/r$, with numerically negligible additional contributions from the core perturbations (whose effective size scale will be of order r_o; see below) and from $\bar{\sigma}_{\alpha\beta}$. This assures us that M_d has effectively the same numerical value (e.g., take the path Γ in the intermediate region) as for the coreless dislocation, and exactly so as $r_o \to 0$. However, M_d remains independent of path Γ and if we shrink Γ onto the core cylinder itself, on which tractions $n_\alpha \sigma_{\alpha\beta} = 0$, we have

$$M_d = r_o^2 \int_0^{2\pi} W(r_o, \theta)\, d\theta \ . \tag{22}$$

Now, let us enlarge the core by eroding a layer of material of thickness δr_o from the core cylinder while the outer boundary of the dislocated body is held fixed. The strain energy change is minus the energy of the layer removed (19,20), and thus

$$\delta U = -\int_0^{2\pi} W(r_o, \theta)\, \delta r_o r_o \, d\theta = - M_d \, \delta r_o / r_o \ . \tag{23}$$

Thus $r_o \, \partial U/\partial r_o = -M_d$, which is independent of r_o in the r_o range considered. If, however, we provisionally accept that for the present core model U will be given by an expression like that in Eqn (16), then $r_o \, \partial U/\partial r_o$ is seen to be nothing other than the energy factor, and Eqn (21) is proven. We have now only to understand that the same form for U as in Eqn (16), specifically the same energy factor coefficient, does indeed result for the present core model of a traction free cylinder as for model implied by Eqn (18). Let us note that the effect of introducing the cylindrical hole is that the solution for reverse tractions $n_\alpha D_{\alpha\beta}(\theta)/r_o$ on a cylinder of radius r_o must be added to the coreless solution of Eqn (14). Since these reverse tractions amount to zero net force on the cylinder, the stress field which they produce contains terms which, relative to boundary traction values at $r = r_o$, decay with r at least as fast as $(r_o/r)^2$; see (21). Thus the total stress field will have the form

$$\sigma_{\alpha\beta} = D_{\alpha\beta}(\theta)/r + \bar{\sigma}_{\alpha\beta}(x_1,x_2) + D^{(2)}_{\alpha\beta}(\theta)r_o/r^2 + D^{(3)}_{\alpha\beta}(\theta)r_o^2/r^3 + \ldots \qquad (24)$$

and the calculation of U given by the first equality in Eqn (18) is now exact for the cylindrical hole core model. The added terms in $\sigma_{\alpha\beta}$ above are readily seen to contribute a bounded amount to U, independently of r_o, and hence the structure of U will be precisely as in the last line of Eqn (18). Hence the provisional assumption above that the same energy factor applies for the cylindrical hole core model is seen to be correct, and we see therefore that M_d equals the energy factor.

2.2 ATTRACTION OF A DISLOCATION TO A NOTCH TIP

Fig. 2 shows a dislocation line at distance ρ from the tip of a V-notch in a bicrystal. We consider the case in which there is no externally applied loading, so that if we can evaluate the image forces on the dislocation we know how much it is attracted to the notch tip. When the notch has the form of a flat crack on a bicrystal boundary, this problem corresponds to a fundamental element of the Rice and Thomson (24) analysis of brittle versus ductile response, as that analysis would be extended to failure on grain interfaces. In keeping with this sort of application, we consider ρ to be small enough compared to overall notch and body dimensions that the bicrystal may be considered of infinite extent, with the V-notch extending to infinity.

Two contours Γ and Γ' are shown and it is evident the integral M_d, corresponding to the coordinate origin at the notch tip, must be the same for each contour. We can see, however, that $M_o = 0$ on contour Γ': The integrand of M_o vanishes identically on the traction free notch surfaces ($\underline{n}$ is perpendicular to $\underline{x}$ there). The stress singularity at a V-notch tip is less strong than $1/r$ (and never stronger than $1/\sqrt{r}$), where here r means distance from the tip, and hence by shrinking the radius of the small circular arc at the tip to zero, there is no contribution to M_o from that part of Γ' either. Finally, by letting the radius of the large arc expand to infinity and

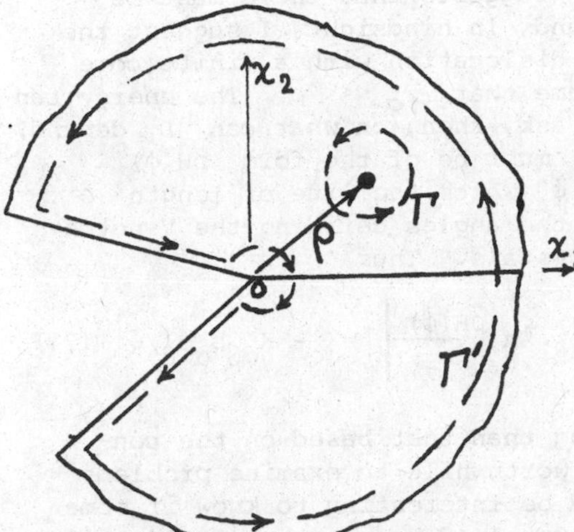

FIG. 2. Dislocation near notch tip.

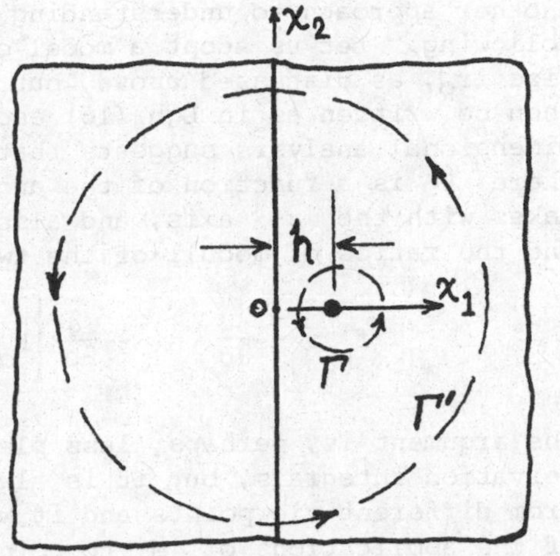

FIG. 3. Dislocation near bicrystal interface.

recognizing that, owing to the traction free notch, the stress field falls off at large r more rapidly than $1/r$, there is again seen to be no contribution to M_o.

The conclusion is that $M_o = 0$ on Γ', and thus on Γ also. However, by application of Eqn (7), this means that

$$(M_o)_\Gamma = c_1(J_1)_\Gamma + c_2(J_2)_\Gamma + (M_d)_\Gamma = 0 \quad . \tag{25}$$

Thus, using Eqns (21) and (19), we see that the radial component of "image" force attracting the dislocation into the tip is

$$f_\rho \equiv (c_1 f_1 + c_2 f_2)/\rho = -K_{\alpha\beta} b_\alpha b_\beta/\rho \ , \tag{26}$$

where $\underline{K}$ is evaluated for the crystal in which the dislocation resides. This is a remarkable result; the force component f_ρ depends on neither the angular range of the V-notch nor on the properties of the second material constituting the bicrystal.

Special versions of Eqn (26) have appeared previously. Rice and Thomson (24) derived by a novel method an expression which can be put in the same form for the force on a dislocation in a radially gliding orientation near the tip of a crack in an isotropic and homogeneous solid. Also, Asaro (25) subsequently showed by conventional 2D anisotropic elasticity methods that Eqn (26) applies for a dislocation near the tip of a crack in a homogeneous anisotropic solid, a problem addressed earlier by Atkinson (26).

The generality of Eqn (26) suggests that there must be another approach to understanding it and, in hindsight, I suggest the following. Let us adopt a model of a dislocation with a finite core size r_o, as discussed above, but assume that $r_o <<< \rho$. The energy can then be written as in Eqn (16) and we ask, then, on what can L depend? Dimensional analysis suggests that L must be of the form $\rho H(\phi)$ where H is a function of the angle ϕ which the line of length ρ makes with the x_1 axis, and also of the angles defining the V-notch and the ratios of moduli of the two crystals. Thus

$$f_\rho = -\frac{\partial U}{\partial \rho} = -\frac{\partial}{\partial \rho}\left[K_{\alpha\beta}b_\alpha b_\beta \,\ell n\,\frac{\rho H(\phi)}{r_o}\right] = -K_{\alpha\beta}b_\alpha b_\beta/\rho. \quad (27)$$

The argument is, perhaps, less pleasing than that based on the conservation integrals, but it is always worthwhile to examine problems from different viewpoints and it would be interesting to know if some of the applications of M to point force loads on cracks (13,14,16,18) could be approached in a similar fashion.

2.3 DISLOCATION NEAR AN INTERFACE

Fig. 3 shows a bicrystal with a line dislocation at distance h from the boundary. We shall use conservation integrals to derive the same expression for the image force on such a dislocation as presented by Barnett and Lothe (27). We assume that h is small compared to crystal dimensions so that the bicrystal can be assumed to be infinite.

Again the value of M_o, based on a coordinate origin on the interface, is the same for path Γ' as for Γ. We know from Eqn (7) that

$$(M_o)_\Gamma = (M_d)_\Gamma + (J_1)_\Gamma\, h \quad . \quad (28)$$

The value of M_o on path Γ' is the same for any such contour encircling the dislocation and hence, for a large circle whose radius increases to infinity. In that case the leading far field stress term, which decays as $1/r$ and is the only term contributing to M_o as $r\rightarrow\infty$, must coincide with the solution for a dislocation on the interface. Hence

$$(M_o)_{\Gamma'} = M_d^{int} \quad (29)$$

where the superscript "int" refers to the interface dislocation of the same Burgers vector. Thus, writing $(J_1)_\Gamma = f$ for the horizontal (and only non-zero) component of force on the dislocation, we have

$$f = -(K_{\alpha\beta} - K^{int}_{\alpha\beta})b_\alpha b_\beta/h , \quad (30)$$

which is the Barnett and Lothe result.

3. FORCES AND RECIPROCITY

It is, I think, not well recognized how powerful are the notions that energetic forces can be associated with defect motion or other forms of structural rearrangement within a solid and that these forces can be related to local fields at the defect site. For example, within linear elasticity the Peach-Koehler force f on a dislocation is given in terms of stress at the dislocation site by Eqn (19) and the Irwin crack extension force, for growth involving a single mode of crack tip deformation, by

$$G = k^2/M \quad (31)$$

where k is the crack tip stress intensity factor for that mode and M is an appropriate modulus or combination of moduli. To see the strength of these ideas, a few short developments are given here. All but the last two subsections presume 2D deformation fields.

3.1 WEIGHT FUNCTIONS

Here is presented a brief synopsis of my interpretation (28) of the Bueckner (29) weight function theory. Let a 2D elastic body contain a crack of length a (Fig. 4, ignoring the dislocation) and be loaded by two systems of applied force, one with intensity measured by Q_1 and the other by Q_2 (only one load system is illustrated in the figure). Both systems induce the same loading mode

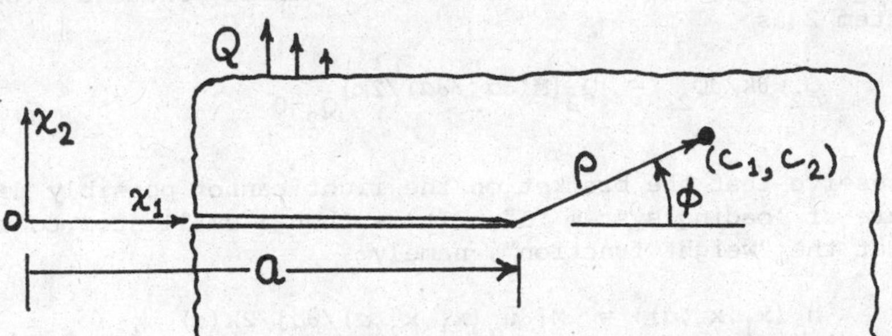

FIG. 4. Solid with loadings proportional to Q (generalized force) containing crack of length a and dislocation at c_1, c_2.

at the crack tip, say, mode I (tension). The Q's may be regarded as generalized forces; with them we may associate generalized displacements q_1 and q_2 which will be defined by differently weighted averages of the displacement field in the body such that $Q_1\delta q_1 + Q_2\delta q_2$ denotes the work of applied loading per unit thickness of the 2D body. Then the energy U per unit thickness satisfies

$$\delta U = Q_1\delta q_1 + Q_2\delta q_2 - (k^2/M)\delta a \quad , \tag{32}$$

where the meaning of G as an energetic force and its expression by Eqn (31) have been used. It is understood that k is homogeneously linear in Q_1 and Q_2.

We suppose that we know the complete solution for the displacement throughout the body and the stress intensity factor when load system 1 is applied, but nothing about the solution for system 2. In particular this means that we know q_2 and k as functions of a and (linearly) of Q_1 when $Q_2 = 0$. The power of such information is shown by use of the reciprocal relations that follow from δU being an exact differential. First introducing the Legendre transform of Eqn (32) to

$$\delta(U - Q_1 q_1 - Q_2 q_2) = -q_1\delta Q_1 - q_2\delta Q_2 - (k^2/M)\delta a \quad , \tag{33}$$

we find the particularly useful reciprocal relation

$$\partial q_2(Q_1,Q_2,a)/\partial a = 2[k(Q_1,Q_2,a)/M]\,\partial k(Q_1,Q_2,a)/\partial Q_2 \quad . \tag{34}$$

Setting $Q_2 = 0$, it is seen that all terms in this equation except $\partial k/\partial Q_2$ are known and thus we find that we can solve for $\partial k/\partial Q_2$ (which is independent of the Q's since k is linear in them).

Thus we can write the stress intensity induced by non-zero load system 2 as

$$Q_2\,\partial k/\partial Q_2 = Q_2[M(\partial q_2/\partial a)/2k]_{Q_2=0} \quad . \tag{35}$$

But we observe that the bracket on the right cannot possibly depend on the nature of loading system "1". It is then a small step to understand that the "weight function", namely

$$h_\alpha(x_1,x_2,a) = M[\partial u_\alpha(x_1,x_2,a)/\partial a]/2k(a) \quad , \tag{36}$$

computed from the displacement field u_α and stress intensity k due to any particular loading, is universal. It is the same (within unessential rigid motions) for all possible loadings. Once determined from the solution for one particular loading, we find k for any second loading by summing the scalar product of $\underline{h}$ with the applied forces of that loading. Once k is thereby determined for the second loading, we may find $\underline{u}$ for that loading by multiplying through by $k(a)$ in Eqn (36) and then integrating on a. The procedure gives the complete displacement field $\underline{u}$ for the second loading if it is known before crack introduction. This is a powerful and widely used method.

3.2 FORCES ON DISLOCATIONS

Let Q and q be a conjugate generalized force and displacement set acting on a 2D elastic body containing a line dislocation at c_1, c_2 (Fig. 4, now ignoring the crack). Then

$$\delta U = Q\,\delta q - f_1 \delta c_1 - f_2 \delta c_2 \tag{37}$$

with $f_1 = \bar{\sigma}_{2\beta} b_\beta$, $f_2 = -\bar{\sigma}_{1\beta} b_\beta$. After a Legendre transformation on q, similar to Eqn (33), we can read off the reciprocal relations

$$\partial q(Q,c_1,c_2)/\partial c_\alpha = \partial f_\alpha (Q,c_1,c_2)/\partial Q \ ,$$
$$\partial f_1 (Q,c_1,c_2)/\partial c_2 = \partial f_2 (Q,c_1,c_2)/\partial c_1 \ . \tag{38}$$

Using the expressions for the f's just given, the last relation requires that

$$(\partial \bar{\sigma}_{1\beta}/\partial c_1 + \partial \bar{\sigma}_{2\beta}/\partial c_2) b_\beta = 0 \ . \tag{39}$$

But $\bar{\sigma}_{\alpha\beta}$ consists additively of the stress field induced by load Q, say $\hat{\sigma}_{\alpha\beta}$, plus the image field which is independent of Q. Thus if Eqn (39) is to be satisfied for arbitrary dislocation vectors it is seen to be necessary that

$$\partial \hat{\sigma}_{1\beta}/\partial c_1 + \partial \hat{\sigma}_{2\beta}/\partial c_2 = 0 \ . \tag{40}$$

That is, the existence of δU as a perfect differential in conjunction with the formulae for force on a dislocation requires that the stress field due to loading Q satisfy the equations of equilibrium. This is an unexpected connection.

We see also from the first of Eqns (38) that the effect of change of dislocation position on the displacement is given by

$$\partial q(Q,c_1,c_2)/\partial c_\alpha = e_{\alpha\lambda 3} \hat{\sigma}_{\lambda\beta}(c_1,c_2) b_\beta / Q \ , \tag{41}$$

which is closely related to Eqn (60) to follow.

3.3 DISLOCATION NEAR A CRACK TIP

Consider a geometry like that in Fig. 2 with the V-notch collapsed to a flat crack and, for simplicity, assume that the material is homogeneous. Fig. 4 shows the configuration. Crack length is denoted by a and position of the near tip dislocation by (c_1, c_2), all measured from a fixed origin not at the crack tip. The cracked body with near tip dislocation is loaded by three independent loading systems of strength Q_1, Q_2, Q_3 which are now understood to be chosen so that any combination of opening, in-plane sliding or anti-plane sliding can be induced near the tip.

We have

$$\delta U = Q_\alpha \delta q_\alpha - G\delta a - f_\alpha \delta c_\alpha \tag{42}$$

(summation convention, $\alpha = 1,2,3$ in the first product and $\alpha = 1,2$ in the last) where f_α is given in terms of $\bar{\sigma}_{\alpha\beta}$ as before, Eqn (19), and where it is known from anisotropic crack mechanics as presented by Stroh (22) and Barnett and Asaro (30) that

$$G = (8\pi)^{-1} K^{-1}_{\alpha\beta} k_\alpha k_\beta \tag{43}$$

with

$$k_\alpha = \lim_{x_1 \to a+} [(2\pi)^{1/2}(x_1 - a)^{1/2} \sigma_{2\alpha}(x_1, x_2 = 0)] \tag{44}$$

defining the stress intensity factors; $\underline{K}^{-1}$ is the inverse of the energy factor tensor. Let us suppose, reasonably, that we know the stress intensity factors and stress fields $\sigma_{\alpha\beta}$ induced in the cracked body by each of the loadings Q_1, Q_2, Q_3 acting singly, in the absence of any dislocation. We wish to find the effect of the dislocation on the stress intensity factors (hence on G) and also find the "image" contribution to the forces $\underline{f}$ themselves; the radial component f_ρ is given already by Eqn (26) but that will be rederived here. This is a generalization to anisotropic media and to fuller determination of dislocation forces of a problem posed and solved on the basis of reciprocity by Rice and Thomson (24). I shall outline its solution here and then proceed to give full details when the dislocation is very close to the crack tip.

Regard Q_1, Q_2, Q_3, a, c_1, c_2 as the independent variables in all differentiations. Then reciprocity requires that

$$\begin{aligned} \partial G/\partial c_\mu &= (4\pi)^{-1} K^{-1}_{\alpha\beta} k_\alpha \partial k_\beta/\partial c_\mu \\ &= \partial f_\mu/\partial a = e_{\mu\lambda 3} b_\beta \partial \bar{\sigma}_{\beta\lambda}/\partial a \,. \end{aligned} \tag{45}$$

Recognizing that $\partial k_\beta/\partial c_\mu$ is independent of the Q's and that $\partial k_\alpha/\partial Q_\gamma$ (denoted $R_{\gamma\alpha}$ below) are known functions of a only, whereas $\partial\bar{\sigma}_{\beta\lambda}/\partial Q_\gamma$ are known functions of a, c_1 and c_2 (the same as if the dislocation were absent), we have after differentiation with respect to Q_γ and multiplication by KR^{-1} that

$$\partial k_\beta/\partial c_\mu \quad = \quad 4\pi K_{\beta\alpha} R^{-1}_{\alpha\gamma} e_{\mu\lambda 3} b_\delta [\partial(\partial\bar{\sigma}_{\delta\lambda}/\partial Q_\gamma)/\partial a] \quad . \tag{46}$$

The right hand side is a known function of a, c_1 and c_2 and is independent of the Q's. Hence, by integration beginning at values of c_1, c_2 large enough that the k's vanish, we can get the dependence of k_β on a, c_1, c_2 when all the Q's vanish. If this solution for the k's is now substituted into the previous equation, we have an expression for $\partial f_\mu/\partial a$ when all the Q's vanish, from which we can solve for f_μ by an integration beginning at such large negative a that the f's effectively vanish.

In this way all the desired information can in principle be extracted and, curiously, such is done without posing any standard elastic boundary value problem for a dislocation in a cracked elastic solid.

The details are now presented for the case of a dislocation which is much closer to the crack tip than distances like overall crack length and body dimensions. In that case it suffices to regard the crack as semi-infinite and to treat all parameters denoting interaction between the crack and dislocation (e.g., like the k's and f's when $Q=0$) as functions of $c_1 - a$ and c_2 only. We can then further regard $R_{\alpha\gamma}$ as changing sufficiently little with a over distance scales of interest to be taken as effectively constant. For the stress field induced in the region of interest near the crack when the body is loaded but there is no dislocation, it suffices to represent the stress state only by the standard leading crack tip singular term

$$\bar{\sigma}_{\delta\lambda}(c_1, c_2; a) \quad = \quad k_\alpha \, C^\alpha_{\delta\lambda}(\phi)/\sqrt{2\pi\rho} \tag{47}$$

where $\rho^2 = (c_1 - a)^2 + c_2^2$, $\tan\phi = c_2/(c_1 - a)$. This represents the well understood inverse square root singular crack tip field for the anisotropic solid considered (22,30). Consistently with the definition of stress intensity factors, the functions $\underset{\sim}{C}$ giving the angular dependence of the crack tip field are normalized such that $C^\alpha_{\lambda 2}(0) = \delta_{\alpha\lambda}$ (Kronecker delta). Now, since $k_\alpha = Q_\gamma R_{\gamma\alpha}$ when there is no dislocation, we see that then

$$R^{-1}_{\alpha\gamma}(\partial\bar{\sigma}_{\delta\lambda}/\partial Q_\gamma) \quad = C^\alpha_{\delta\lambda}(\phi)/\sqrt{2\pi\rho} \tag{48}$$

and, of course, the equality remains valid even when the dislocation is present for reasons already discussed. Now set $\mu = 1$ in Eqn (46) and observe that $\partial/\partial a = -\partial/\partial c_1$. Then upon integration in c_1 we obtain

$$k_\beta = -(2\pi)^{3/2} K_{\beta\alpha} b_\delta C^\alpha_{\delta 2}(\phi)/\sqrt{\rho} \tag{49}$$

for the stress intensity factors induced by a dislocation $\underline{b}$ at position ρ, ϕ near the crack tip in an unloaded body. We can now insert this expression into Eqn (45), still presuming $\underline{Q} = 0$, set $\mu = 1$, use $\partial/\partial a = -\partial/\partial c_1$, and integrate to get

$$\begin{aligned} f_1 &= -G = -(8\pi)^{-1} K^{-1}_{\alpha\beta} k_\alpha k_\beta \\ &= -K_{\alpha\beta} b_\lambda b_\mu C^\alpha_{\lambda 2}(\phi) C^\beta_{\mu 2}(\phi)/\rho \end{aligned} \tag{50}$$

for the crack energy release rate and parallel component of image force in the unloaded solid. The corresponding results can then be written at once for the loaded solid since we know the results for forces and stress intensity factors due to the loadings $\underline{Q}$.

The expression for f_1 enables calculation of both components f_1 and f_2 of image force, since we know f_ρ from Eqn (26). However, for completeness, f_ρ is derived here in the following steps. Observe that ρf_1 is independent of ρ, so that

$$0 = f_1 + \rho \partial f_1/\partial\rho = f_1 + (c_1 - a)\partial f_1/\partial c_1 + c_2 \partial f_1/\partial c_2 , \tag{51}$$

and rewrite the last term using the second reciprocal relation of Eqns (38). Hence

$$\begin{aligned} 0 &= f_1 + (c_1 - a)\partial f_1/\partial c_1 + c_2 \partial f_2/\partial c_1 \\ &= \partial[(c_1 - a) f_1 + c_2 f_2]/\partial c_1 \end{aligned} \tag{52}$$

so that the bracketed term, which equals ρf_ρ, is independent of c_1. Since by dimensional considerations that term is at most a function of ϕ, this means that it is constant. We can evaluate the constant by setting $\phi = 0$, along which line f_ρ must coincide with f_1 as given in Eqn (50). Thus there results

$$f_\rho = f_1(\rho, \phi=0) = -K_{\alpha\beta} b_\alpha b_\beta/\rho \tag{53}$$

since $C^\alpha_{\lambda 2}(0) = \delta_{\alpha\lambda}$ in Eqn (50) with $\phi = 0$.

The fact derived above that $G + f_1 = 0$ for the unloaded but dislocated solid follows also from the J_1 integral taken on an outer contour surrounding the dislocation and crack tip. $J_1 = 0$ on that contour (evident by expanding it to large radius), but J_1 gets contributions $G + f_1$ when the contour is wrapped, respectively, around the crack tip and dislocation. The relation follows also from invariance of the total elastic energy when δa and δc_1 are changed equally in Eqn (42) as written for the unloaded solid.

The angular dependence $C^{\lambda}_{\alpha\beta}(\phi)$ of the crack tip stress $\sigma_{\alpha\beta}$ for mode λ was introduced in the previous discussion, and the dependence $D_{\alpha\beta}(\theta)$ for the dislocation stress field was introduced in Section 2.1, Eqn (14). Since, as Bilby and Eshelby (31) emphasized, cracks can be represented as continuous arrays of dislocations, these fields are not independent. I leave it for the interested reader to verify that if we write the singular dislocation stress field as

$$\sigma_{\alpha\beta} = b_{\mu} D^{\mu}_{\alpha\beta}(\theta)/r \,, \tag{54}$$

then

$$C^{\lambda}_{\alpha\beta}(\phi) = \frac{1}{2\pi} K^{-1}_{\lambda\mu} \int_{0}^{\phi} [\sin\theta \sin(\phi-\theta)]^{-1/2} D^{\mu}_{\alpha\beta}(\theta)\, d\theta \tag{55}$$

for $\phi > 0$. A minus sign should precede the integral for $\phi < 0$. It helps in getting started to note that associated with the singular crack tip stress field, the displacement discontinuity $\Delta\underline{u}$ between crack surfaces is

$$\Delta u_{\alpha} = (1/\pi)\, K^{-1}_{\alpha\beta} k_{\beta} \sqrt{\rho/2\pi} \tag{56}$$

at distance ρ from the tip, which is necessary for consistency between Eqns (43) and (44).

3.4 STRUCTURE OF INELASTIC CONSTITUTIVE RELATIONS

Following earlier work (32,33) we may represent an increment of inelastic deformation within a 3D macroscopic sample of material by a set of local incremental variables $d\xi$, marking the advance of dislocation loops or more macroscopic measures of crystalline shear, microcracks, phase boundaries, and the like. Eshelby (1,2,20), Rice (33) and others have shown how to identify energetic forces with these and we may write

$$\delta\Phi = S_{\alpha\beta}\delta E_{\alpha\beta} - \langle F\delta\xi\rangle \tag{57}$$

where $\underline{S}$ and $\underline{E}$ are work conjugate macroscopic stress and strain tensors, Φ the strain energy per unit reference volume, and $Fd\xi$ denotes the inner product of all structural rearrangements $d\xi$ with their conjugate energetic forces F, and $\langle \ldots \rangle$ denotes the average of these products over a unit reference volume. Specific forms of $\langle Fd\xi\rangle$ for crystalline slip within grains of a polycrystalline aggregate are given in (32,33).

For example, if inelasticity results in a particular case from the glide motion of dislocation lines within a sample of material, then

$$\langle F d\xi \rangle = \frac{1}{V} \int_L f \delta c \, ds \tag{58}$$

where L denotes all such dislocation lines within a representative sample of volume V, δc is the incremental glide of the dislocation line normal to itself and f is the force resolved in the direction of glide. It is well known that within the conventional linear elastic treatment of dislocations, f may be written as the sum of two terms. The first represents the force associated with the same dislocation array in an unloaded solid ($\underline{S} = \underline{0}$); it depends on Peach-Koehler like contributions from the internal stresses due to other segments of dislocation line within the sample and also, at points where a dislocation line has curvature, on the core size. The second term is linear in the applied stress $\underline{S}$ and given by a Peach-Koehler term based on that part $\hat{\underline{\sigma}}$ of the local stress field due to application of $\underline{S}$ at a fixed dislocation configuration.

In plastic constitutive studies one often averages out discrete dislocations, representing their effects as amounts of shear γ^λ on each of the possible slip systems $\lambda = 1,2,3,4,\ldots$ within the crystalline element at a local point of, say, a polycrystalline aggregate. In that case

$$\langle F \delta\xi \rangle = \frac{1}{V} \int_V \left(\sum_\lambda \tau^\lambda \delta\gamma^\lambda \right) dV \quad , \tag{59}$$

with the summation extending over all slip systems at each point. In that case the thermodynamic "forces" τ^λ are, when the elasticity is treated as linear, the sum of those "forces" in the plastically sheared but unloaded solid plus the shear stress $\hat{\tau}^\lambda$ resolved onto slip system λ due to the local stress $\hat{\underline{\sigma}}$ created by application of $\underline{S}$ at fixed plastic shears (33,34). The τ's have also been specified precisely for finite elastic distortions (32,35). Other representations of structural rearrangement can be considered (33). For example, in deformation due to advance of a phase interface, Eshelby's (20) representation of force in terms of a jump in the integrand for $\underline{J}$ at the interface effectively defines F.

Now, any infinitesimal change $d\underline{E}$ in $\underline{E}$ may be regarded as the sum of that due to a change $d\underline{S}$ in $\underline{S}$ at fixed structural arrangement plus that due to a change $d\xi$ in structural arrangement at fixed $\underline{S}$. We may define the latter increment as the plastic part $d^p\underline{E}$ of a general deformation increment. It represents the residual strain in an infinitesimal load-unload cycle, providing that the unloading is elastic (here taken to mean at fixed structural arrangement). There follows then the reciprocal relation

$$d^p E_{\alpha\beta} = \left\langle \frac{\partial F}{\partial S_{\alpha\beta}} \, d\xi \right\rangle \tag{60}$$

where the derivatives of energetic forces with respect to macroscopic stress are taken at fixed structural arrangement. This represents an extension of Maxwell reciprocity to cases where some of the variables $d\xi$ have only a significance as incremental quantities, there being no definable variable ξ of which $d\xi$ may be said to represent an infinitesimal increase; see (32,33,36) for derivation.

Eqn (60) provides a formalism for relating structural rearrangement to macroscopic strain. In general the summing over sites and volume averaging implied cannot be carried out exactly but there is a literature on approximate and numerical ways of achieving this in the case of crystalline slip; see (33) for a summary.

In the specific models for plastic flow noted earlier, we evidently get $d^pE_{\alpha\beta}$ by replacing f by $\partial f/\partial S_{\alpha\beta}$ and τ^λ by $\partial\tau^\lambda/\partial S_{\alpha\beta}$ on the right sides of Eqns (58) and (59), the derivatives on $\underset{\sim}{S}$ being taken at fixed dislocation or plastic shear configuration and hence involving only that part of f or τ^λ traceable to the Peach-Koehler force or resolved shear stress, respectively, due to the local stress field $\hat{\underset{\sim}{\sigma}}$ induced by $\underset{\sim}{S}$. When the material sample is elastically homogeneous, e.g., a single crystal, $\hat{\underset{\sim}{\sigma}} = \underset{\sim}{S}$ and the expressions for plastic strain increment reduce to those widely quoted in the literature but, as is less widely realized, are not precisely true for an elastically heterogeneous solid such as a polycrystal (34).

Eqn (60) also leads to an interesting attribute of macroscopic constitutive relations in cases for which structural rearrangements can be characterized locally by scalar variables (like increments of shear on active crystallographic slip systems) whose rate of advance $d\xi/dt \equiv r$ is stress state dependent only via a dependence of each $d\xi/dt$ on the conjugate F (e.g., shear rate related to the resolved shear stress on a slip system). Such encompasses the accepted Schmid resolved shear stress phenomenology for crystal plasticity (32,33,34,35,36). In all such cases a macroscopic flow potential is readily shown to exist. It is defined by

$$\Omega = \left\langle \int^F r(F)\, dF \right\rangle , \tag{61}$$

and the plastic part of the instantaneous macroscopic strain rate is related to it by

$$d^pE_{\alpha\beta}/dt = \partial\Omega(\underset{\sim}{S}, \text{structural arrangement})/\partial S_{\alpha\beta} . \tag{62}$$

For a more extended discussion the reader is referred to (32,34,37).

3.5 CRACKS AND COMPLIANCE

Consider now a tensile loaded non-linear elastic solid containing a 3D planar crack so that only mode I is involved. Let $\underset{\sim}{Q}$ and

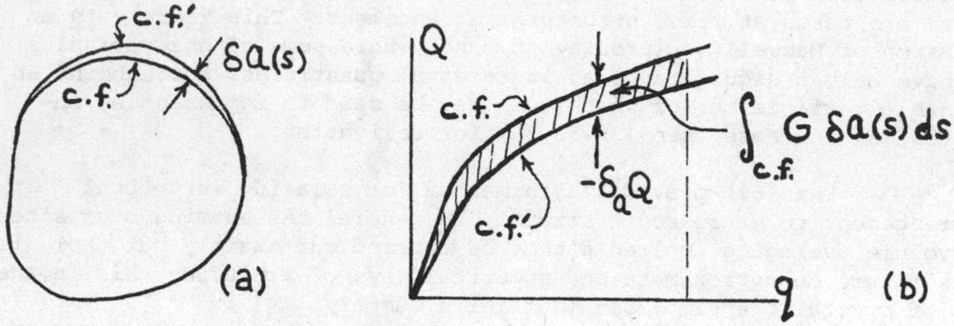

FIG. 5. (a) Planar tensile crack; c.f. = initial position of crack front, c.f.' = position after crack advance by $\delta a(s)$, s is arc length along c.f. (b) Alteration of force (Q) versus displacement (q) diagram.

q be conjugate force and displacement. Then, if "c.f." denotes crack front, Fig. 5a, ds arc length along it, and $\delta a(s)$ the advance of the crack normal to itself (all measured relative to the undeformed state of the cracked body), then G may be defined locally along the crack front at any location s by requiring that

$$\delta U = Q\delta q - \int_{c.f.} G\,\delta a\,ds \tag{63}$$

be valid to first order in $\delta a(s)$ for arbitrary distributions of crack advance along c.f. It then follows by similar reciprocity arguments as those leading to Eqn (60) that the change $\delta_a Q$ in Q when crack advance δa occurs at fixed q is

$$\delta_a Q = -\int_{c.f.} \frac{\partial G(q,s;c.f.)}{\partial q}\,\delta a(s)\,ds \quad . \tag{64}$$

Here, to confirm the notation, if c.f.' denotes the alteration of c.f. after crack advance by $\delta a(s)$ along c.f., then

$$\delta_a Q = Q(q,c.f.') - Q(q,c.f.) \quad . \tag{65}$$

Thus if we determine complete load versus displacement relations for two crack front positions c.f.' and c.f., $\delta_a Q$ is determined as a function of Q and

$$\int_{c.f.} G\,\delta a\,ds = -\int_0^q \delta_a Q\,dq = \int_0^Q \delta_a q\,dQ \tag{66}$$

where

$$\delta_a q = q(Q,c.f.') - q(Q,c.f.) \quad . \tag{67}$$

The integration region is shown in Fig. 5b.

These expressions summarize the Irwin relations between compliance changes and G. They are widely used in elastic (possibly non-linear) fracture mechanics where they are perhaps more familiar in 2D form (with $Q\delta q$ redefined as work per unit thickness) as

$$G = -\int_0^q \frac{\partial Q(q,a)}{\partial a}\,dq = \int_0^Q \frac{\partial q(Q,a)}{\partial a}\,dQ \quad . \tag{68}$$

The same relations between compliance changes with crack size and G are also widely used in a version of elastic-plastic fracture mechanics based on plastic "deformation" theory (38,39,40,41). There G is understood not to be interpretable in terms of energy changes in crack growth under applied load, but rather only as the crack tip J integral for a loaded but nongrowing crack, and it is understood that all force versus displacement relations used in Eqns (66,68) are to be generated by loading from an unstressed state with the considered crack configuration. Eshelby (20) put it nicely when he suggested that all works well in this procedure "... if we do not call the material's bluff by unloading...", although perhaps it is more a matter of the material calling our bluff!

In any event the approach to elastic-plastic fracture as outlined enables J to be inferred from load versus displacement data as obtained in the laboratory or estimated from simple calculations (40,41). The integral itself, no longer valid as an energy release rate, is instead interpreted as a parameter characterizing the intensity of the near tip deformation field (10); this is sensible since the path Γ for evaluation of J can be taken arbitrarily close to the tip. Thus, as long as there is a one-parameter form for the very near tip deformation field at a tensile crack tip (e.g., the "HRR" field (11,12)), we can use J as the parameter, and as long as that field dominates over size scales large enough to envelop that of fracture micro-mechanisms at the tip, we can use J in the "resistance curve" sense to correlate the early increments of ductile crack advance and sometimes to predict crack instability (41,42,43). There is now a substantial literature on this topic and also its extension to viscoplastic crack analysis, contained in large part in the Special Technical Publication series of ASTM over the last decade, and it is not further summarized here.

ACKNOWLEDGEMENT

The preparation of this paper was supported by the NSF Materials Research Laboratory at Harvard University.

REFERENCES

(1) Eshelby, J.D. (1951). The force on an elastic singularity. Phil. Trans. Roy. Soc., A244, 87-112.

(2) Eshelby, J.D. (1956). The continuum theory of lattice defects. In: Prog. Solid State Physics, ed. F. Seitz and D. Turnbull, 3, pp. 79-114. N.Y.: Academic Press.

(3) Sanders, J.L. (1960). On the Griffith-Irwin fracture theory. J. Appl. Mech., 27, 352-352.

(4) Cherepanov, G.P. (1967). Crack propagation in continuous media. Appl. Math. Mech. (translation of PMM), 31, 476-488.

(5) Rice, J.R. (1968). Mathematical analysis in the mechanics of fracture. In: Fracture: An Advanced Treatise, ed. H. Liebowitz, 2, pp. 191-311. N.Y.: Academic Press.

(6) Günther, W. (1962). Über einige Randintegrale der Elastomechanik. Abh. Braunschw. Wiss. Ges., 14, 53-72.

(7) Knowles, J.K. and Sternberg, E. (1972). On a class of conservation laws in linearized and finite elastostatics. Arch. Rat. Mech. Anal., 44, 187-211.

(8) Budiansky, B. and Rice, J.R. (1973). Conservation laws and energy release rates. J. Appl. Mech., 40, 201-203.

(9) Eshelby, J.D. (1975). The elastic energy-momentum tensor. J. Elasticity, 5, 321-335.

(10) Rice, J.R. (1968). A path independent integral and the approximate analysis of strain concentration by notches and cracks. J. Appl. Mech., 35, 379-386.

(11) Hutchinson, J.W. (1968). Singular behavior at the end of a tensile crack in a hardening material. J. Mech. Phys. Solids, 16, 13-31.

(12) Rice, J.R. and Rosengren, G.F. (1968). Plane strain deformation near a crack tip in a power law hardening material. J. Mech. Phys. Solids, 16, 1-12.

(13) Eshelby, J.D. (1975). The calculation of energy release rates. In: Prospects of Fracture Mechanics, ed. G.C. Sih, pp. 69-84. Leyden: Noordhoff.

(14) Freund, L.B. (1978). Stress intensity factor calculations based on a conservation integral. Int. J. Solids Structures, 14, 241-250.

(15) Noether, E. (1918). Invariante Variations-Probleme. Nachr. Ges. Göttingen, Math.-Phys. Klasse, 235. [Translation by Tavel, M. (1971). Milestones in mathematical physics: Noether's Theorem. Transport Theory and Stat. Phys., 1, 183-207.]

(16) Ouchterlony, F. (1978). Some stress intensity factors for self-similar cracks derived from path-independent integrals. J. Elasticity, 8, 259-271.

(17) Ouchterlony, F. (1980). Symmetric cracking of a wedge by transverse displacements. J. Elasticity, 10, 215-223.

(18) Kubo, S. (1982). An application of M-integral to cracks in dissimilar elastic materials. Int. J. Fracture, 20, R27-R30.

(19) Rice, J.R. and Drucker, D.C. (1967). Energy changes in stressed bodies due to void and crack growth. Int. J. Fracture Mech., 3, 19-27.

(20) Eshelby, J.D. (1970). Energy relations and the energy-momentum tensor in continuum mechanics. In: Inelastic Behavior of Solids, ed. M.F. Kanninen et al., pp.77-115. N.Y.: McGraw-Hill.
(21) Eshelby, J.D., Read, W.T. and Shockley, W. (1953). Anisotropic elasticity with applications to dislocation theory. Acta Met., 1, 251-259.
(22) Stroh, A.N. (1958). Dislocations and cracks in anisotropic elasticity. Phil. Mag., 3, 625-646.
(23) Barnett, D.M. & Swanger, L.A. (1971). The elastic energy of a straight dislocation in an infinite anisotropic elastic medium. Phys. Stat. Sol. (b), 48, 419-428.
(24) Rice, J.R. & Thomson, R. (1974). Ductile versus brittle behavior of crystals. Phil. Mag., 29, 73-97.
(25) Asaro, R.J. (1975). An image force theorem for a dislocation near a crack in an anisotropic elastic medium. J. Phys. F: Metal Phys., 5, 2249-2255.
(26) Atkinson, C. (1966). The interaction between a dislocation and a crack. Int. J. Fracture Mech., 2, 567-575.
(27) Barnett, D.M. & Lothe, J. (1974). An image force theorem for dislocations in anisotropic bicrystals. J. Phys. F: Metal Phys., 4, 1618-1635.
(28) Rice, J.R. (1972). Some remarks on elastic crack tip stress fields. Int. J. Solids Structures, 8, 751-758.
(29) Bueckner, H.F. (1970). A novel principle for the computation of stress intensity factors. Z. Angew. Math. Mech., 50, 529-546.
(30) Barnett, D.M. & Asaro, R.J. (1972). The fracture mechanics of slit-like cracks in anisotropic elastic media. J. Mech. Phys. Solids, 20, 353-366.
(31) Bilby, B.A. & Eshelby, J.D. (1968). Dislocations and the theory of fracture. In: Fracture: An Advanced Treatise, ed. H. Liebowitz, 1, pp. 99-182. N.Y.: Academic Press.
(32) Rice, J.R. (1971). Inelastic constitutive relations for solids: an internal variable theory and its application to metal plasticity. J. Mech. Phys. Solids, 19, 433-455.
(33) Rice, J.R. (1975). Continuum mechanics and thermodynamics of plasticity in relation to microscale deformation mechanism. In: Constitutive Equations in Plasticity, ed. A.S. Argon, pp. 23-79. Cambridge (USA): M.I.T. Press.
(34) Rice, J.R. (1970). On the structure of stress-strain relations for time dependent plastic deformation in metals. J. Appl. Mech., 37, 728-737.
(35) Hill, R. & Rice, J.R. (1972). Constitutive analysis of elastic/plastic crystals at arbitrary strain. J. Mech. Phys. Solids, 20, 401-413.
(36) Hill, R. & Rice, J.R. (1973). Elastic potentials and the structure of inelastic constitutive laws. SIAM J. Appl. Math., 25, 448-461.
(37) Clifton, R.J. (1983). Dynamic plasticity. J. Appl. Mech., 50, 941-952.
(38) Broberg, K.B. (1971). Crack growth criteria and non-linear fracture mechanics. J. Mech. Phys. Solids, 19, 407-418.

(39) Begley, J.A. & Landes, J.D. (1972). The J integral as a fracture criterion. In: Fracture Toughness, Spec. Tech. Publ. 514, pp. 1-23. Philadelphia: ASTM.
(40) Rice, J.R. (1976). Elastic-plastic fracture mechanics. In: The Mechanics of Fracture, ed. F. Erdogan, Appl. Mech. Div.-Vol. 19, pp. 23-53. N.Y.: ASME.
(41) Hutchinson, J.W. (1983). Fundamentals of the phenomenological theory of nonlinear fracture mechanics. J. Appl. Mech., 50, 1042-1051.
(42) Paris, P.C., Tada, H., Zahoor, A. and Ernst, H. (1979). Instability of the tearing mode of elastic-plastic crack growth. In: Spec. Tech. Publ. 668, pp. 5-36. Philadelphia: ASTM.
(43) Hutchinson, J.W. & Paris, P.C. (1979). Stability analysis of J controlled crack growth. In: Spec. Tech. Publ. 668, pp. 37-64. Philadelphia: ASTM.

SYMMETRY GROUPS AND PATH-INDEPENDENT INTEGRALS

P.J. OLVER

School of Mathematics, University of Minnesota, Minneapolis, MN, USA 55455

ABSTRACT

Noether's general theorem gives a one-to-one correspondence between nontrivial conservation laws or path independent integrals for the Euler-Lagrange equations of some variational problem and the generalized variational symmetries of the variational problem itself, provided that it satisfies certain nondegeneracy assumptions. Here we give a brief introduction to the theory of generalized symmetries and their connections with conservation laws. Applications are given to the classification of conservation laws for the equations of two dimensional elasticity, especially the linear isotropic and anisotropic cases.

1. INTRODUCTION

One of Professor Eshelby's lasting contributions to the study of dislocations and fracture mechanics was his discovery in 1951 (4-7) of the celebrated energy-momentum tensor. It was the first example in a collection of four (seven in three dimensions) well-known and important path independent integrals that arise in both finite and linear elasticity, the applications of which are well documented in the other contributions to this memorial volume.

Subsequently, Günther(10) and Knowles and Sternberg (13) firmly established the group theoretic origins of these integrals by showing how they arise from the invariance of the underlying variational problem, under groups of translations, rotations and scaling symmetries, through a straight-forward application of Noether's Theorem relating symmetry groups to conservation laws. Although Knowles and Sternberg made claims that these are the only path independent integrals arising in this fashion, a closer analysis of their work shows that they employed only a limited version of the full power of Noether's Theorem. Indeed, in her widely quoted, but less widely appreciated paper (16), Emmy Noether not only gave the means to construct conservation laws from ordinary geometrical symmetry groups of the type studied by Günther and Knowles and Sternberg, she introduced the important concept of generalized symmetry groups, whose transformations depend on the deformation gradients and possibly higher order derivatives of the relevant dependent variables, AND showed that

ALL path independent integrals could be constructed from a knowledge of the generalized symmetries of the variational problem. Moreover, the fundamental infinitesimal methods introduced by Sophus Lie in his study of symmetry groups of differential equations, (14) - see also (17,21) - provide a systematic computational method of finding all such symmetry groups, and hence all path independent integrals. (A number of symbol-manipulating computer programs are now being developed to compute these symmetries - see (22,23) for instance - although as yet I am unaware of their extension to computing the corresponding integrals.)

To the best of my knowledge, despite the fact that Noether's Theorem has been available for well over 60 years, there was no attempt to apply this powerful result to any of the equations of elasticity until my own complete classification of the first order symmetries and conservation laws for the equations of two and three dimensional elasticity (18,20). The results are surprising. In three dimensions there are, in addition to the seven well-known conservation laws, six additional laws arising from generalized symmetries, except in a special case (when the Lamé moduli satisfy $7\mu + 3\lambda = 0$) where 13 additional laws result from generalized symmetries and ordinary conformal symmetries. (For $7\mu + 3\lambda \neq 0$, these further laws still give rise to interesting divergence identities.) In two dimensions, as Eshelby recognized (6,7), and as indicated in recent work of Tsamasphyros and Theocauis, (24), there are whole families of conservation laws depending on arbitrary analytic functions.

The present paper consists of two parts. First we will review the general theory of generalized symmetries of differential equations, and Noether's general theorem relating these to conservation laws. One qustion that is of importance in the classification of conservation laws or symmetries is the question of triviality. Usually one is only interested in nontrivial conservation laws, but the issue of their precise relationship to nontrivial symmetries has not been dealt with adequately in the literature to date. Here we announce the result that for systems satisfying certain nondegeneracy assumptions there is a one-to-one correspondence between nontrivial generalized symmetries of a variational problem, and nontrivial conservation laws of the corresponding Euler-Lagrange equations. (Interestingly, according to some very recent results (17,25) this theorem is intimately related to the question of when a system of differential equations can be put into Kovalevskaya form.)

The second part of this paper deals with the applications of the general form of Noether's Theorem to the classification of path independent integrals for the equations of two dimensional elasticity. A complete analysis has only been completed for the linear case, but it is shown that for both isotropic and anisotropic linear elasticity, there are families of path independent integrals depending on pairs of arbitrary analytic functions. The precise structure of these integrals, though, does depend on whether or not the material is "equivalent" to an isotropic material or not. This latter analysis is based on a partial solution to the equivalence problem for linear, two dimensional elasticity: when are two problems equivalent under a linear change of variables in both the independent and dependent variables? The solution, perhaps surprising, indicates that for strongly elliptic problems there are, under this general notion of equivalence, only two independent invariants among the 16 elasticities. The case of nonlinear, two-dimensional elasticity has yet to be fully analyzed. However, striking similarities between the symmetry equations and the equations for conformal symmetries for Riemannian metrics leads to the conjecture that even in this case there will again be whole families of path independent integrals depending on two arbitrary analytic functions.

Of course, while this problem of classification of path independent integrals has some intrinsic interest, the real question is whether these new integrals have genuine applications to problems in fracture mechanics, dislocation theory, scattering of waves in elastic media and so on. Unfortunately, lack of time has precluded my addressing this important question in these papers, but it is an area that well deserves a concerted investigation, the results of which I hope to report on at a later date.

My thanks go to John Ball, who originally sparked my interest in applying Noether's Theorem to the problems of elasticity; and Professors Bilby, Miller and Rice for inviting me to participate in this conference.

2. SYMMERY GROUPS OF DIFFERENTIAL EQUATIONS

Let $x=(x^1,\ldots,x^p)$, $u=(u^1,\ldots,u^q)$ be the independent and dependent variables in a system of differential equations

$$\Delta_i(x,u^{(n)})=0 \quad , \quad i=1,\ldots,\ell \; , \tag{1}$$

where $u^{(n)}$ denotes the partial derivatives $u^i_J=\partial^k u^i/\partial x^{j_1}\ldots\partial x^{j_k}$ of orders $0\le k\le n$. A GEOMETRICAL SYMMETRY GROUP of the system is a connected (local) group of transformations $g:(x,u)\mapsto(\tilde{x},\tilde{u})$ with the property that if $u=f(x)$ is a solution, and $\tilde{u}=\tilde{f}(\tilde{x})$ is the function obtained from f by transforming its graph by the group element g, then $\tilde{f}=g\cdot f$ is also a solution. For instance, if G is the group of rotations $g_\epsilon:(x,u)\mapsto(x\cos\epsilon+u\sin\epsilon\;,\;-x\sin\epsilon+u\cos\epsilon)$, then

$\tilde{f} = g_\epsilon f$ is obtained from f by rotating the graph of f through the angle ϵ.

Each one-parameter subgroup of G is characterized by its infinitesimal generator $\underline{v} = \Sigma\, \xi^i(x,u)\partial/\partial x^i + \Sigma\, \varphi_j(x,u)\partial/\partial u^j$, the group elements being recovered by solving the system of ordinary differential equations $dx^i/d\epsilon = \xi^i$, $du^j/d\epsilon = \varphi_j$, ϵ being the group parameter. (For the rotation group $\underline{v} = u\,\partial/\partial x - x\,\partial/\partial u$.) Since G acts on functions, it also acts on their derivatives - this defines the prolonged group action $pr^{(n)} g: (x,u^{(n)}) \mapsto (\tilde{x},\tilde{u}^{(n)})$, where $\tilde{u}^{(n)}$ are the derivatives of $\tilde{f}(\tilde{x})$. Similarly, an infinitesimal generator $\underline{v}$ prolongs to the space of derivatives:

$$pr\,\underline{v} = \underline{v} + \Sigma\, \varphi_j^J\, \partial/\partial u_J^j$$

where

$$\varphi_j^J = D_J(\varphi_j - \sum_i \xi^i u_i^j) + \sum_i \xi^i u_{J,i}^j$$

in which $u_i^j = \partial u^j/\partial x^i$, $u_{J,i}^j = \partial u_J^i/\partial x^i$, and $D_J = D_{j_1} D_{j_2} \cdots D_{j_k}$ is the total derivative (treating u as a function of x).

THEOREM. If the system (1) is nondegenerate (see below), then G is a symmetry group if and only if

$$pr\,\underline{v}(\Delta_i) = 0\ ,\quad i = 1,\ldots,\ell\ , \tag{2}$$

whenever $\Delta = 0$.

The infinitesimal condition (2) for invariance yields a large number of elementary differential equations for the coefficient functions ξ^i, φ_j of $\underline{v}$. In practice, these can always be solved, and hence the most general symmetry group of the system can be systematically computed - see (14,17,21) for examples.

DEFINITION. A system of differential equations is NONDEGENERATE if it satisfies

a) MAXIMAL RANK. The Jacobian matrix of Δ with respect to all variables $x,u^{(n)}$ is of rank ℓ everywhere.

b) LOCAL SOLVABILITY. If $x_o,u_o^{(n)}$ are any fixed values satisfying $\Delta(x_o,u_o^{(n)})=0$, then there exists a solution $u=f(x)$ of the system with $u_o^{(n)}=f^{(n)}(x_o)$.

A result of Nirenberg (15) shows that quasi-linear elliptic systems are nondegenerate on a dense subset of $\{(x,u^{(n)}): \Delta(x,u^{(n)})=0\}$ which is enough for the preceding theorem to be valid. A second class of nondegenerate systems are the so-called NORMAL ANALYTIC systems. A system $\Delta(x,u^{(n)})=0$ is normal if at each point $(x_o,u_o^{(n)})$ it has at least one non-characteristic direction. (In particular, elliptic systems are certainly normal as are almost all systems arising in physical problems.) The local solvability comes from the Cauchy-Kovalevskaya theorem, since normality is equivalent to the system being in Kovalevskaya form

$$\partial^n u^i/\partial t^n = K_i(y,t,u^{(n)}) \quad , \quad i=1,\ldots,q$$

under a change of independent variables

If we allow the coefficients ξ^i,φ_j of the infinitesimal generator $\underline{v}$ to depend on derivatives of u, we have a GENERALIZED SYMMETRY. It is not difficult to see that we can assume, without loss of generality, that the symmetry is in EVOLUTIONARY FORM $\underline{v}_K=\Sigma K_j(x,u^{(m)})\partial/\partial u^j$, in which $K_j=\varphi_j-\Sigma\xi^i u_i^j$. The corresponding group transformations are obtained by solving the system of evolution equations

$$\partial u^j/\partial\epsilon = K_j(x,u^{(m)}) \text{ , } u(x,0)=f(x) \text{ ,} \tag{3}$$

with $f_\epsilon(x) = g_\epsilon . f(x) = u(x,\epsilon)$. Again, $\underline{v}_K$ generates a symmetry group of (1), meaning that whenever $f(x)$ is a solution, so is $f_\epsilon(x)$ for all ϵ , if and only if (2) is satisfied for all solutions of $\Delta = 0$. (This requires that the prolonged systems $\Delta^{(n)} = 0$ obtained by differentiating: $D_J\Delta_i = 0$, are all nondegenerate - this is essentially equivalent, (17), to the system being normal). A symmetry $\underline{v}_K$ is TRIVIAL if $K = 0$ on solutions of $\Delta = 0$; two symmetries are EQUIVALENT if their difference is trivial, and we are really interested in equivalence classes of nontrivial symmetries.

3. CONSERVATION LAWS AND PATH-INDEPENDENT INTEGRALS

Given a system of differential equations (1), a CONSERVATION LAW is a divergence expression

$$\text{Div } P = \sum_{i=1}^{p} D_i P_i = 0 \ , \ P = P(x,u^{(k)}) \ , \tag{4}$$

which vanishes on all solutions $u = f(x)$ of the system. Each conservation law in two dimensions $(p = 2 \ , \ (x^1,x^2) = (x,y)$, provides a path independent integral, namely by Green's theorem

$$\oint_C P(x,u^{(k)})dy - Q(x,u^{(k)})dx = 0 \tag{5}$$

for all closed curves C provided $u = f(x)$ is a solution to the system. (In three dimensions, we end up with a "surface independent" integral $\int P \cdot dS$.)

These are two types of TRIVIAL CONSERVATION LAWS: I) If $P = 0$ itself for all solutions of the system, then $\text{Div } P = 0$ for all solutions too; II) If $\text{Div } P = 0$ for ALL functions $u = f(x)$, then the law is also trivial. An example of a conservation law of this latter type is $D_x(u_y) + D_y(-u_x) = 0$ - see (16) for a complete classification.

Two laws P and $\tilde{P}$ are EQUIVALENT if their difference $P - \tilde{P}$ is a sum of trivial laws of the two types. As with symmetries, we are interested in classifying equivalence classes of nontrivial conservation laws.

Under the assumption of nondegeneracy, (4) is equivalent to the existence of functions $K_i^J(x,u^{(m)})$ such that

$$\mathrm{Div}\, P = \Sigma\, K_i^J\, D_J \Delta_i \,.$$

A simple integration by parts shows that there is an equivalent conservation law $\tilde{P}$ in CHARACTERISTIC FORM.

$$\mathrm{Div}\, \tilde{P} = \Sigma\, K_i \cdot \Delta_i \,, \tag{6}$$

where $K_i = \Sigma(-D)_J K_i^J$ is the CHARACTERISTIC of $\tilde{P}$ (and hence P). A characteristic is TRIVIAL if $K_i = 0$ on all solutions of the system (1); it can be seen that in (6) the characteristic $K = (K_1, \ldots, K_\ell)$ is uniquely defined up to addition of a trivial characteristic, so we should really talk about equivalence classes of characteristics as well.

THEOREM. For a normal system of differential equations, there is a one-to-one correspondence between (equivalence classes of) nontrivial conservation laws and (equivalence classes of) nontrivial characteristics.

In other words, a nontrivial characteristic uniquely determines a nontrivial conservation law and vice versa. See (17,25) for proofs. The requirement that the system be normal turns out to be essential; if this is not the case then either the local solvability condition for some prolongation $\Delta^{(m)} = 0$ is NOT satisfied, or there is a nontrivial relation of the form

$$\Sigma\, \mathcal{D}_\nu \Delta_\nu = 0 \tag{7}$$

among the equations, the $\mathcal{D}_\nu$ being certain differential operators, (8). If the Δ_ν actually arise as the Euler-Lagrange equations of some variational problem, then (7) means that Noether's second theorem, (16), is applicable, and there are nontrivial symmetry groups depending on arbitrary functions which give rise to only trivial conservation laws.

4. NOETHER'S THEOREM

We now suppose that our system of differential equations are the Euler-Lagrange equations

$$E_i(W) = \delta\mathcal{W}/\delta u^i = 0 \quad , \quad i = 1,\dots,q \tag{8}$$

for some variational problem $\mathcal{W}_{\Omega_o}[u] = \int_{\Omega_o} W(x,u^{(n)})dx$. If $\mathcal{W}$ satisfies the Legendre-Hadamard condition, the system is nondegenerate.

DEFINITION. A generalized vector field $\underline{v}_K$ is a VARIATIONAL SYMMETRY of $\mathcal{W}$ if for every $\Omega \subset \Omega_o \subset \mathbb{R}^p$ and every solution $u_\epsilon(x)$ of (3),

$$\mathcal{W}_\Omega[u_\epsilon] = \mathcal{W}_\Omega[u_o] + \int_0^\epsilon \mathcal{B}_{\partial\Omega}[u_{\epsilon'}]d\epsilon' \tag{9}$$

where

$$\mathcal{B}_{\partial\Omega}[u] = \int_{\partial\Omega} B(x,u^{(k)})\cdot dS$$

depends only on the boundary behavior of u on $\partial\Omega$.

Thus, up to the addition of boundary terms, $\mathcal{W}$ is invariant under the group action of $\underline{v}_K$. (Another way of stating definition 4 is that $\mathcal{W}[u]$ is a conservation law for the evolution equations (3), i.e. $D_\epsilon\mathcal{W} + \text{Div } X = 0$ for some flux $X(x,u^{(k)})$.)

LEMMA. A vector field $\underline{v}_K$ is a variational symmetry of $\mathcal{W}$ if and only if

$$\text{pr } \underline{v}_K(W) = \text{Div } B \tag{10}$$

for some p-tuple $B = (B_1,\dots,B_p)$.

This is the form of variational symmetry proposed by Bessel-Hagen (1); Noether (16) omitted B (and hence the corresponding boundary term in (9)), but could no longer just consider evolutionary vector fields. (In this case (10) has the extra term $-L \text{ Div } \xi$.) A simple computation shows that

$$\text{pr } \underline{v}_K(L) = \Sigma\, K_i E_i(W) + \text{Div } A\ ,$$

where $A = (A_1,\dots,A_p)$ depends on K and W; the explicit form of A is not required. As a result we immediately have Noether's Theorem.

THEOREM. There is a one-to-one correspondence between (equivalence classes of) nontrivial conservation laws and (equivalence classes of) variational symmetries; namely K is the characteristic of a conservation law if and only if $\underline{v}_K$ is a variational symmetry.

Every variational symmetry of $\mathcal{W}$ is a symmetry of the Euler-Lagrange equations $E(W) = 0$ in the sense of section 2, the most common counter-examples being groups of scaling transformations $(x,u) \longmapsto (\lambda x, \lambda^\alpha u)$. One practical method of finding variational symmetries, thus, is to first compute all symmetries of the Euler-Lagrange equations using Theorem 1, and then check which of these satisfy the additional variational condition (10). (There are, however, more direct ways of doing this - see (13).)

5. TWO DIMENSIONAL ELASTICITY

For simplicity, we treat the path-independent integrals for two-dimensional homogeneous hyperelastic materials in the abscence of body forces. Thus the variational problem is

$$\mathcal{W} = \int_\Omega W(\nabla\underline{u})\,dx\,dy$$

in which W is the stored energy function, $\underline{x} = (x,y)$ the material coordinates, $\underline{u} = (u,v)$ the deformation, so $\nabla\underline{u} = (u_x,u_y;v_x,v_y) \equiv (p,q;r,s)$ is the deformation gradient. The Euler-Lagrange equations are the second order system

$$E_u(W) = D_x W_p + D_y W_q = 0 ,$$

$$E_v(W) = D_x W_r + D_y W_s = 0 .$$

The goal is to analyze all first order path independent integrals, i.e. those of the form

$$\oint P\,dy - Q\,dx$$

in which P and Q are functions of $\underline{x},\underline{u},\nabla\underline{u}$.

For computational purposes, it is advantageous to begin by looking at those integrals in which P and Q depend solely on the deformation gradient $\nabla\underline{u}$ (see (20).) In this case, Noether's theorem says that $P(\nabla\underline{u})$, $Q(\nabla\underline{u})$ are the components of a conservation law if and only if there exists functions $K(\nabla\underline{u})$, $L(\nabla\underline{u})$ such that

$$\begin{aligned} P_p &= \alpha K + \beta L , \quad P_q + Q_p = \alpha' K + \beta' L , \quad Q_q = \alpha'' K + \beta'' L , \\ P_r &= \beta K + \gamma L , \quad P_s + Q_r = \beta' K + \gamma' L, \quad Q_s = \beta'' K + \gamma'' L , \end{aligned} \tag{11}$$

where

$$\alpha = W_{pp}\ ,\ \beta = W_{pr}\ ,\ \gamma = W_{rr}\ ,\ \alpha' = 2W_{pq}\ ,\ \beta' = W_{ps} + W_{qr}\ ,$$
$$\gamma' = 2W_{rs}\ ,\ \alpha'' = W_{qq}\ ,\ \beta'' = W_{qs}\ ,\ \gamma'' = W_{ss}\ .$$

These equations bear a remarkable similarity to a "vector version" of the equations for a conformal symmetry of a Riemannian metric, (3), and are thus called the two-dimensional VECTOR CONFORMAL EQUATIONS. To date no progress has been made with their solution for general nonlinear W, although the following CONJECTURE seems plausible: For the two-dimensional vector conformal equations there are an infinite family of solutions, hence an infinite family of path independent integrals with P,Q depending only on $\nabla\underline{u}$. (In the analogous situation for two dimensional Riemannian metrics, this result is true, since every such metric is conformal to a flat metric, in which case, each complex analytic function provides a conformal symmetry.) Finally, if P,Q are a solution of (1), then $\underline{v} = K\partial_u + L\partial_v$ is a (generalized) variational symmetry of $\mathcal{W}$.

For quadratic $W(\nabla\underline{u})$, leading to the equations of linear two dimensional elasticity, the situation is much better understood. Here the coefficients $\alpha,\ldots,\gamma''$ in the vector conformal equations are related to the more usual elasticity constants $c_{ijk\ell}$ according to the rule

$$\alpha = c_{1111},\ \beta = c_{1121}\ ,\ \gamma = c_{2121},\ \alpha' = 2c_{1112},\beta' = c_{1122} + c_{1221}\ ,$$
$$\gamma' = 2c_{2122}\ ,\ \alpha'' = c_{1212}\ ,\ \beta'' = c_{1222}\ ,\ \gamma'' = c_{2222}\ .$$

Of course, the $c_{ijk\ell}$'s obey additional symmetry properties when arising from a theory of linear elasticity (7) leading to the following relations among $\alpha,\ldots,\gamma$:

$$2\beta = \alpha'\ ,\ 2\beta'' = \gamma'\ ,\ \gamma = \alpha''\ ,$$

but these do not appear to be especially relevant to the subsequent analysis.

Let $\nabla P = (P_p,P_q,P_r,P_s)^T$, and similarly for ∇Q. Eliminating K and L from (11), we obtain a system of the form

$$M\nabla P = N\nabla Q \qquad (12)$$

in which M and N are 4×4 matrices whose entries depend on the constants $\alpha,\ldots,\gamma$. (The precise expressions are easy to write down, but rather messy.) The nature of the solutions to (12) depends on the structure of the eigenvalues of the matrix $M^{-1}N$; in general it can be shown that $M^{-1}N$ has two pairs of complex conjugate eigenvalues $k_1 \pm i\ell_1$ and $k_2 \pm i\ell_2$. If these pairs are distinct, then $M^{-1}N$ is diagonalizable, otherwise we will (usually) have a nondiagonal Jordan canonical form. The structure of the space of path-independent integrals is different in each case.

THEOREM. Given the matrix $M^{-1}N$ there exist two independent complex-linear combinations ξ and η of p,q,r,s such that the complex function $F = F(\nabla \underline{u}) = P + iQ$ is a conservation law if and only if

a) In case $M^{-1}N$ is diagonalizable, with distinct eigenvalues

$$F = F_1(\xi) + F_2(\eta)$$

b) In case $M^{-1}N$ is not diagonalizable

$$F = \xi F_1(\eta) + \overline{F_1(\eta)} + F_2(\eta) \,,$$

where F_1, F_2 are analytic in η.

(The combinations ξ, η can be constructed from the eigenvectors to the matrix $M^{-1}N$.)

It is thus important to know whether a given elastic material is in the diagonalizable or nondiagonalizable case. The only case analyzed to date, the case of linear isotropic materials is nondiagonalizable (20). To attempt any analysis of nonisotropic materials it is necessary to simplify the constants $c_{ijk\ell}$ as much as possible.

6. THE EQUIVALENCE PROBLEM

In its general form, the EQUIVALENCE PROBLEM is connected with the question of when two variational problems $\mathcal{V}$ and $\mathcal{V}'$ are the same under a change of both independent and dependent variable. (See section 6 of (5) for an "elementary" case.). Here we are first interested in the special two-dimensional quadratic equivalence problem: When are

$$\int W(\nabla \underline{u})\, d\underline{x} \quad \text{and} \quad \int \tilde{W}(\nabla \underline{\tilde{u}})\, d\underline{\tilde{x}} \,,$$

for $W, \tilde{W}$ quadratic in the deformation gradient, equivalent under a linear change of variables:

$$\underline{\tilde{x}} = A\underline{x} \,,\; \underline{\tilde{u}} = B\underline{u} \;? \tag{13}$$

Equivalently, when are two sets of elasticities $c_{ijk\ell}$ and $\tilde{c}_{ijk\ell}$

related under such a change of variables? Included in the equivalence problem is the problem of finding simple (or canonical) forms of variational problems: Given $\int W(\nabla \underline{u})d\underline{x}$, find A,B such that (13) makes $\int \tilde{W}(\nabla \tilde{\underline{u}})d\tilde{\underline{x}}$ as simple as possible. Even the relatively easy case of two independent and two dependent variables presents a number of difficulties, and so far I have only incomplete, but nevertheless intriguing results.

DEFINITION. A two-dimensional problem is called QUASI-ISOTROPIC if its elasticities have the form

$$c_{1111} = c_{2222} = 2\mu + \lambda \ , \ c_{1212} = c_{2121} = c_{1221} = c_{2112} = \mu \ ,$$
$$c_{1122} = c_{2211} = \nu \ ,$$

for some constants μ, λ, ν , with all unspecified elasticities vanishing.

In particular, if $\nu = \lambda$, then the material is isotropic. Note that actually only two of the constants μ, λ, ν are arbitrary, since by rescaling $\underline{u}$ we can always arrange that $\mu = 1$, say. It is easy to check that a quasi-isotropic material is strongly elliptic if and only if

$$\mu > 0 \ , \ -4\mu - \lambda < \nu < 2\mu + \lambda \ .$$

THEOREM. Every quadratic $W(\nabla \underline{u})$ satisfying the Legendre-Hadamard condition and sufficiently close to the linear isotropic case is equivalent to a strongly elliptic quasi-isotropic problem. See * below.

The phrase "close to isotropic" means that the elasticities $c_{ijk\ell}$ do not differ too much from the isotropic elasticities. I conjecture that the theorem remains true if this condition is dropped i.e. any strongly elliptic quadratic $W(\nabla \underline{u})$ is equivalent to a quasi-isotropic one, but to prove this looks rather complicated. The present proof relies on some Lie-algebraic tools and a generalization of Frobenius' theorem due to Hermann (12). Unfortunately, the proof is nonconstructive; it gives no clues as to how to find the requisite matrices A,B such that (12) transforms $W(\nabla \underline{u})$ into a quasi-isotropic problem. The values of λ, ν , however, can be found using invariant theory.

Note that in principle the theorem says that there are only two independent elasticities in two-dimensional elasticity, up to the above generalized notion of equivalence. (Contrast this with the more standard six independent elasticities (11) when one just uses the basic symmetry relations on the subscripts $c_{ijk\ell}$.)

Returning to our classification of path independent integrals it is not to difficult to prove that for a strongly elliptic quasi-isotropic problem, the eigenvalues of the matrix $M^{-1}N$ of section 5 are distinct unless $\nu = \lambda$ or $\nu = -2\mu - \lambda$. The first is the isotropic case, the second equivalent under a reflection $(u,v) \longmapsto (u,-v)$.

PROPOSITION. If $W(\nabla \underline{u})$ is strongly elliptic, then $M^{-1}N$ is diagonalizable unless $W(\nabla u)$ is equivalent to an isotropic material.

(Of course, this is subject to the establishment of our earlier conjecture.) Thus the isotropic materials are distinguished by the structure of their space of path independent integrals. (This is probably only true in two dimensional elasticity!)

7. FURTHER INTEGRALS.

So far we have concentrated on path-independent integrals in which $\oint P dy - Q dx$ is such that $P + iQ = F$ depends only on $\nabla \underline{u}$. If we now relax the requirements so that F depends on $\underline{x}, \underline{u}$ and $\nabla \underline{u}$, then in (20) it was shown that for linear isotropic elasticity all such integrals were given as follows.

THEOREM For linear, isotropic two-dimensional elasticity, every path independent integral is given by a linear combination on the following

$$2\mu(2\mu+\lambda)\xi\partial F_1/\partial\eta + i(\mu+\lambda)\overline{F}_1 + F_2 ,$$

$$i(\tilde{w}\eta - w\tilde{\eta}) , \quad (4\mu(2\mu+\lambda)w - (\mu+\lambda)iz\eta)\eta ,$$

where $z = x + iy$, $w = u + iv$, $\xi = (u_x - v_y) + i(u_y + v_x)$,

$\eta = \mu(v_y - u_x) + i(2\mu+\lambda)(u_x + v_y)$; and $F_1(\bar{z},\eta)$, $F_2(\bar{z},\eta)$ are analytic in their arguments, and $\tilde{w}(x,y)$ is an arbitrary solution of the system with corresponding $\tilde{\eta}(x,y)$.

A similar result holds for anisotropic materials, but I have not completed the details in it.

*Note added in proof: More recent investigations seem to indicate that the geometry of this "quadratic equivalence problem" is more complicated than indicated here, owing to the degeneracy of the group action on the subspace of isotropic elasticities. It now appears that there is an additional three-dimensional subspace of "degenerate" elasticities, which includes the isotropic cases, which cannot be included in the statement of the theorem. These and further results on the equivalence problem and applications to elasticity will be reported on in a subsequent publication.

REFERENCES

(1) Bessel-Hagen, E.,(1921). Uber die Erhaltingssatze der Elektrodynamik. Math. Ann. 84 258-276.

(2) Courant, R. & Hilbert, D., (1953). Methods of Mathematical Physics, vol II. London: Wiley-Interscience.

(3) Eisenhart, L.P.,(1926). Riemannian Geometry. Princeton, N.J.: Princeton University Press.

(4) Eshelby, J.D.,(1951). The force on an elastic singularity. Phil. Trans. Roy. Soc., A244, 87-112.

(5) Eshelby, J.D., (1956). The continuum theory of lattice defects. In Solid State Physics, vol 3. New York: Academic Press.

(6) Eshelby, J.D.,(1975). The calculation of energy release rates. In: Prospects of Fractures Mechanics, ed. G. C. Sih, pp. 69-84. Leyden: Noordhoff.

(7) Eshelby, J.D.,(1975) The elastic energy-momentum tensor. J. of Elasticity, 5, 321-225.

(8) Finzi, A.,(1947). Sur les systemes d'equations aux derivees partielles, qui, comme les systemes normaux, comportent autant d'equations que de fonctions inconnues. Proc. Kon. Ned. Akad. van Wetenschappen, Amsterdam, 50, 136-150, 288-297.

(9) Gardner, R.B., (1983). Differential geometric methods interfacing control theory. In: Differential Geometric Control Theory, e.d. R.W. Brockett et al. Bassel: Birkhauser.

(10) Günther, W., (1962). Uber einige Randintegrale der Elastomechanik. Abh. Braunschw. Wiss. Ges. 14, 53-72.

(11) Gurtin, M.E., (1977). The Linear Theory of Elasticity. In: Handbuch der Physik VIa/2, pp. 1-295. Berlin: Springer-Verlag.

(12) Hermann, R., (1964). Cartan connections and the equivalence problem for geometric structures. Contributions to Diff. eqs. 3, 199-248.

(13) Knowles, J.K. & Sternberg, E., (1972). On a class of conservation laws in linearized and finite elastostatics. Arch. Rat. Mech. Anal. 44, 187-211.

(14) Lie, S., (1891). Vorlesungen uber Differentialgleichungen mit bekannten infinitesimalen Transformationen. Leipzig: B.G. Teubner.

(15) Nirenberg, L., (1973). Lectures on Partial Differential Equations. Providence, R.I.: American Mathematical Society.

(16) Noether, E., (1918). Invariante Variations probleme. Kgl. Ger. Wiss. Nachr. Gottingen, Math-Phys. Kl. 2 235-257.

(17) Olver, P.J., (1980). Applications of Lie Groups to Differential Equations, Oxford University Lecture Notes, 1980. (to appear in Springer-Verlag Graduate Texts in Mathematics Series).

(18) Olver, P.J., (1983). Group theoretic classification of conservation laws in elasticity. In Systems of Nonlinear Partial Differential Equations, ed. J.M. Ball. Lancaster: D. Reidel.

(19) Olver, P.J., (1983). Conservation laws and null divergences. Math. Proc. Camb. Phil. Soc., 94, 529-540.

(20) Olver, P.J., (1984). Conservation laws in elasticity I: General results, II: Linear homogeneous isotropic elastostatics. Arch. Rat. Mech. Anal., to appear.

(21) Ovsiannikov, L.V., (1982). Group Analysis of Differential Equations. London: Academic Press.

(22) Schwarz, F., (1982). A REDUCE package for determining Lie symmetries of ordinary and partial differential equations, Computer Physics Comm. 27, 179-186.

(23) Steinberg, S., (1983). Symmetries of differential equations. University of New Mexico, Preprint.

(24) Tsamasphyros, G.T. & Theocaris, P.S., (1982). A new concept of path independent integrals for plane elasticity. J. Elasticity 12, 265-280.

(25) Vinogradov, A.M., (1984). Local symmetries and conservation laws. Acta Math. Appl. (to appear).

CONSERVATION LAWS, THE MATERIAL MOMENTUM TENSOR AND MOMENTS OF ELASTIC ENERGY

ALICIA GOLEBIEWSKA HERRMANN*

in life of the

Division of Applied Mechanics
Stanford University
Stanford, CA 94305

The central role in conservation laws of elasticity is played by Eshelby's energy-momentum tensor. It is asymmetric and considerable attention has been devoted in the past to the problem of its symmetrization. It appeared that it is relatively straightforward to define a symmetric tensor for nonlinear isotropic bodies, while in the linear case, the desired property had to be introduced by additional manipulations.

In the present paper the derivation of the material momentum tensor for the general nonlinear continuum (the static counterpart of Eshelby's energy-momentum tensor) is proposed in a manner completely analogous to the Eshelby approach for linear elastostatics. The simplified form for an isotropic (but still nonlinear) body is then obtained. This form allows a general discussion of the relation between eigenvalues and eigenvectors of the Cauchy stress and the material momentum tensors.

From the nonlinear expression for the material momentum tensor the appropriate form for linear statics is derived next. This tensor is new and inherently symmetric for an isotropic body. The relation between the two eigenvalue problems for stress and material momentum tensor is discussed both in general and for particular examples.

Based on the above discussion, the physical interpretation of the material momentum tensor for the deformable continuum (with and without defects) is proposed.

Finally, the derivation of all known conservation laws in local form based on moments of the strain energy is presented.

* Readers will be very sad to learn of the untimely death on December 7, 1983, of Alicia Golebiewska Herrmann. Her paper was presented by George Herrmann.

GENERAL THEORY OF INCLUSIONS

T. MURA

Dept. of Civil Engineering and Materials Research Center
Northwestern University, Evanston, Ill. 60201 USA

ABSTRACT

A general theory of inclusions is presented to summarize Eshelby's work (1,2,3) on inclusions and inhomogeneities. Further development along these lines is reviewed, including sliding inclusions, elastic anisotropy, non-uniform eigenstrains, interaction among inhomogeneities, half-space problems, and dynamic problems.

1. INTRODUCTION

"Eigenstrain" is a generic name given to such nonelastic strains, such as those due to thermal expansion, phase transformations, initial strains, plastic, and misfit strains. Eshelby referred to eigenstrains as stress-free transformation strains. "Eigenstress" is a generic name given to self-equilibriated internal stresses caused by one or several of these eigenstrains in bodies which are free of external forces and surface constraints.

The case of an infinitely extended material is considered, since it is of a particular interest for the mathematical simplicity of the solution as well as for its practical importance. For infinitesimal deformations considered in this paper, the total strain ε_{ij} is regarded as the sum of the elastic strain e_{ij} and the eigenstrain ε^*_{ij},

$$\varepsilon_{ij} = e_{ij} + \varepsilon^*_{ij}. \tag{1}$$

The total strain must be compatible, or

$$\varepsilon_{ij} = (\tfrac{1}{2})(u_{i,j} + u_{j,i}), \tag{2}$$

where u_i is the displacement and $u_{i,j} = \partial u_i/\partial x_j$. The elastic strain is related to stress σ_{ij} by Hooke's law,

$$\sigma_{ij} = C_{ijk\ell}e_{k\ell}, \tag{3}$$

where $C_{ijk\ell}$ are the elastic moduli. The summation rule for repeated indices is employed. The equations for equilibrium are

$$\sigma_{ij,j} = 0. \tag{4}$$

Suppose $\varepsilon^*_{ij}(\underset{\sim}{x})$ is a periodic function with amplitude $\bar{\varepsilon}^*_{ij}(\xi)$, where $\underset{\sim}{\xi}$ is the wave vector corresponding to the given period of the wave,

$$\varepsilon^*_{ij}(\underset{\sim}{x}) = \bar{\varepsilon}^*_{ij}(\underset{\sim}{\xi}) \exp(i\underset{\sim}{\xi}\cdot\underset{\sim}{x}), \tag{5}$$

with $\underset{\sim}{\xi}\cdot\underset{\sim}{x} = \xi_i x_i$. The solution of (1) ~ (4) corresponding to this periodic function may also be expressed in the form

$$u_i(\underset{\sim}{x}) = \bar{u}_i(\underset{\sim}{\xi}) \exp(i\underset{\sim}{\xi}\cdot\underset{\sim}{x}). \tag{6}$$

Then, $\bar{u}_i(\underset{\sim}{\xi})$ is obtained from (1) ~ (4) as

$$\bar{u}_i(\underset{\sim}{\xi}) = X_j N_{ij}(\underset{\sim}{\xi})/D(\underset{\sim}{\xi}), \tag{7}$$

where

$$X_i = -iC_{ijk\ell}\bar{\varepsilon}^*_{k\ell}\xi_j, \tag{8}$$

N_{ij} are cofactors of the matrix

$$\underset{\sim}{K}(\xi) = \begin{pmatrix} K_{11} & K_{12} & K_{13} \\ K_{21} & K_{22} & K_{23} \\ K_{31} & K_{32} & K_{33} \end{pmatrix}, \tag{9}$$

$$K_{ik} = C_{ijk\ell}\xi_j\xi_\ell \tag{10}$$

and $D(\underset{\sim}{\xi})$ is the determinant of $\underset{\sim}{K}(\underset{\sim}{\xi})$.

If ε^*_{ij} is given by the Fourier integral form,

$$\varepsilon^*_{ij}(\underset{\sim}{x}) = \int_{-\infty}^{\infty} \bar{\varepsilon}^*_{ij}(\underset{\sim}{\xi}) \exp(i\underset{\sim}{\xi}\cdot\underset{\sim}{x})d\underset{\sim}{\xi} \tag{11}$$

with

$$\bar{\varepsilon}^*_{ij}(\underset{\sim}{\xi}) = (2\pi)^{-3}\int_{-\infty}^{\infty} \varepsilon^*_{ij}(\underset{\sim}{x}) \exp(-i\underset{\sim}{\xi}\cdot\underset{\sim}{x})d\underset{\sim}{x}, \tag{12}$$

we have from (7)

$$u_i(\underset{\sim}{x}) = -i \int_{-\infty}^{\infty} C_{j\ell mn}\bar{\varepsilon}^*_{mn}(\underset{\sim}{\xi})\xi_\ell N_{ij}(\underset{\sim}{\xi})D^{-1}(\underset{\sim}{\xi}) \exp(i\underset{\sim}{\xi}\cdot\underset{\sim}{x})d\underset{\sim}{\xi} \tag{13}$$

or from (12)

$$u_i(\underset{\sim}{x}) = -(2\pi)^{-3}\frac{\partial}{\partial x_\ell}\int_{-\infty}^{\infty}\int_{-\infty}^{\infty} C_{j\ell mn}\,\varepsilon^*_{mn}(\underset{\sim}{x}')N_{ij}(\underset{\sim}{\xi})D^{-1}(\underset{\sim}{\xi})$$

$$\times \exp\{i\underset{\sim}{\xi}\cdot(\underset{\sim}{x}-\underset{\sim}{x}')\}d\underset{\sim}{\xi}d\underset{\sim}{x}', \tag{14}$$

where $d\underset{\sim}{\xi} = d\xi_1 d\xi_2 d\xi_3$ and $d\underset{\sim}{x}' = dx_1'dx_2'dx_3'$.

When Green's functions $G_{ij}(\underset{\sim}{x}-\underset{\sim}{x}')$ are defined as

$$G_{ij}(\underset{\sim}{x}-\underset{\sim}{x}') = (2\pi)^{-3}\int_{-\infty}^{\infty} N_{ij}(\underset{\sim}{\xi})D^{-1}(\underset{\sim}{\xi})\exp\{i\underset{\sim}{\xi}\cdot(\underset{\sim}{x}-\underset{\sim}{x}')\}d\underset{\sim}{\xi}, \tag{15}$$

(14) can be written as

$$u_i(\underset{\sim}{x}) = -\frac{\partial}{\partial x_\ell}\int_{-\infty}^{\infty} C_{j\ell mn}\,\varepsilon^*_{mn}(\underset{\sim}{x}')G_{ij}(\underset{\sim}{x}-\underset{\sim}{x}')d\underset{\sim}{x}' \tag{16}$$

or

$$u_i(\underset{\sim}{x}) = \int_{-\infty}^{\infty} C_{j\ell mn}\,\varepsilon^*_{mn}(\underset{\sim}{x}')\frac{\partial}{\partial x'_\ell}G_{ij}(\underset{\sim}{x}-\underset{\sim}{x}')d\underset{\sim}{x}'. \tag{16'}$$

For the isotropic medium, expression (15) is evaluated as

$$G_{ij}(\underset{\sim}{x}-\underset{\sim}{x}') = \frac{1}{16\pi\mu(1-\nu)|\underset{\sim}{x}-\underset{\sim}{x}'|}\left[(3-4\nu)\delta_{ij} + \frac{(x_i-x_i')(x_j-x_j')}{|\underset{\sim}{x}-\underset{\sim}{x}'|^2}\right], \tag{17}$$

where μ and ν are shear modulus and Poisson's ratio, respectively.

2. ESHELBY'S SOLUTION

Eshelby evaluated (16) when ε^*_{ij} is distributed uniformly in an ellipsoidal domain Ω in an isotropic medium. Ω is called an ellipsoidal inclusion (Fig. 1). Eshelby's important conclusion was that the strain ε_{ij} defined by (1) (and therefore the stress) is uniform inside the inclusion. Thus, we can write in Ω

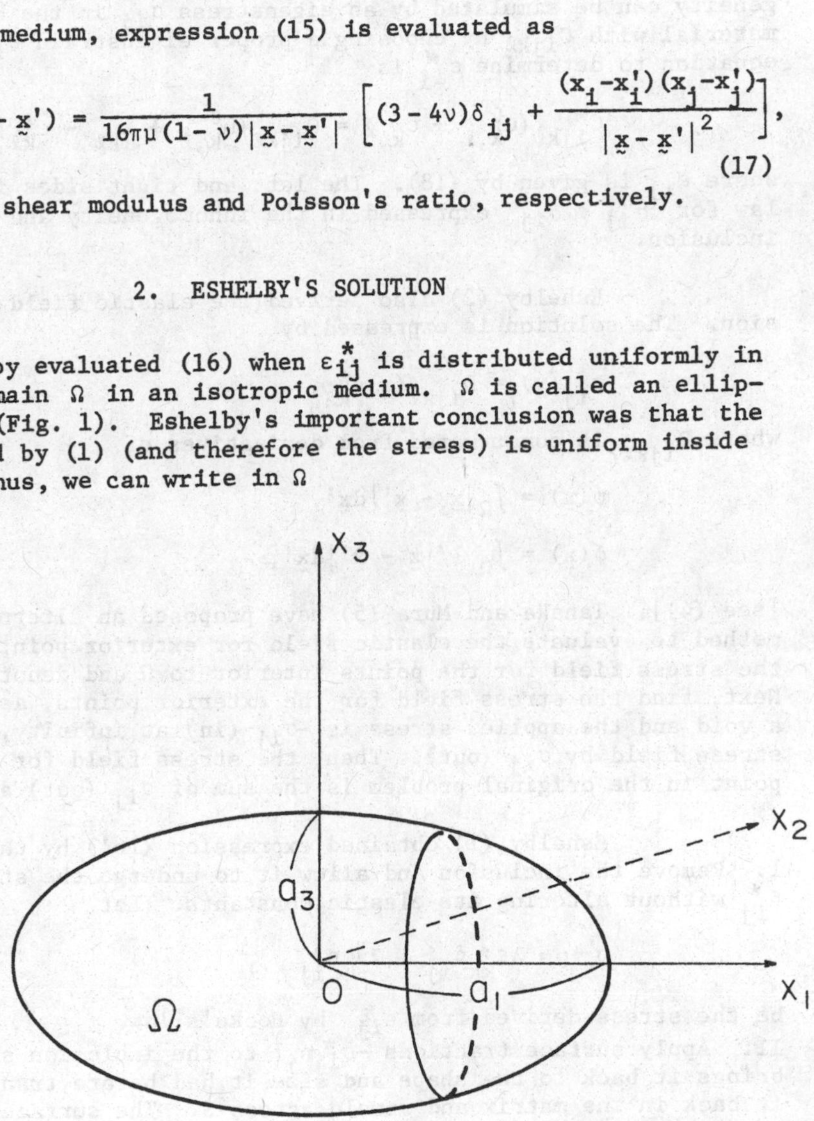

FIG. 1. Ellipsoidal inclusion

$$\varepsilon_{ij} = S_{ijk\ell}\,\varepsilon^*_{k\ell}, \tag{18}$$

where $S_{ijk\ell}$ (the components of Eshelby's tensor) are constant and depend only on the ratio of a_1, a_2, a_3 and ν.

This uniformity of stress field in Ω leads to the so-called equivalent inclusion method for the inhomogeneity problem. If an infinite body with elastic moduli $C_{ijk\ell}$ contains an ellipsoidal inhomogeneity with elastic moduli $C^*_{ijk\ell}$ and is subjected to a uniform applied stress $\sigma^0_{ij} = C_{ijk\ell}u^0_{k,\ell}$ at infinity, the stress disturbance due to the inhomogeneity can be simulated by an eigenstress σ_{ij} in the homogeneous material with $C_{ijk\ell}$ by choosing a proper eigenstrain ε^*_{ij} in Ω. The equation to determine ε^*_{ij} is

$$C^*_{ijk\ell}(u^0_{k,\ell} + \varepsilon_{k\ell}) = C_{ijk\ell}(u^0_{k,\ell} + \varepsilon_{k\ell} - \varepsilon^*_{k\ell}) \quad \text{in } \Omega \tag{19}$$

where ε_{ij} is given by (18). The left and right sides in (19) are Hooke's law for $\sigma^0_{ij} + \sigma_{ij}$ expressed in the inhomogeneity and in the equivalent inclusion.

Eshelby (2) also derived the elastic field outside the inclusion. The solution is expressed by

$$\varepsilon_{ij}(\underset{\sim}{x}) = D_{ijk\ell}(\underset{\sim}{x})\varepsilon^*_{k\ell}, \tag{20}$$

where $D_{ijk\ell}$ is constructed from derivatives of

$$\psi(\underset{\sim}{x}) = \int_\Omega |\underset{\sim}{x} - \underset{\sim}{x}'|\,d\underset{\sim}{x}',$$
$$\phi(\underset{\sim}{x}) = \int_\Omega 1/|\underset{\sim}{x} - \underset{\sim}{x}'|\,d\underset{\sim}{x}', \tag{21}$$

[see (4)]. Tanaka and Mura (5) have proposed an alternative simpler method to evaluate the elastic field for exterior points. First, obtain the stress field for the points interior to Ω and denote it by σ_{ij} (in). Next, find the stress field for the exterior points, assuming that Ω is a void and the applied stress is $-\sigma_{ij}$ (in) at infinity. Denote the stress field by σ_{ij} (out). Then, the stress field for the exterior point in the original problem is the sum of σ_{ij} (out) and σ_{ij} (in).

Eshelby (1) obtained expression (16') by the following steps:
I. Remove the inclusion and allow it to undergo the stress-free strain ε^*_{ij} without altering its elastic constants. Let

$$\sigma^*_{ij} = \lambda\varepsilon^*_{kk}\delta_{ij} + 2\mu\,\varepsilon^*_{ij} \tag{22}$$

be the stress derived from ε^*_{ij} by Hooke's law.
II. Apply surface tractions $-\sigma^*_{ij}n_j$ to the inclusion surface S. This brings it back to the shape and size it had before transformation. Put it back in the matrix and reweld across S. The surface forces have now become a layer of body force spread over S.

III. Let these body forces relax, or, what comes to the same thing, apply a further distribution $\sigma^*_{ij}n_j$ over S. The body is now free of external force but in a state of self-stress (eigenstress), because of the transformation of the inclusion.

The result that a uniform eigenstrain induces a uniform final strain in an ellipsoidal inclusion constrained by infinite surroundings can be extended to the anisotropic case (1,6). Eshelby (3) also showed (explicitly for isotropy) that if the stress-free transformation strain in an ellipsoid is a polynomial in x_1, x_2, x_3 of degree N then so also is the constrained strain inside it. Asaro and Barnett (7) have proved this result for a similar transformation strain in an anisotropic linear elastic medium. Mura and Kinoshita (8) have further shown that the final stress and strain become inhomogeneous polynomials of degree N, (N - 2),... if the eigenstrain is a homogeneous polynomial of degree N.

Eshelby's result (1) was immediately applied to the calculation of elastic constants of a polycrystal from a single crystal by Kröner (9) and also to plastic deformation of polycrystals by Kröner (10), and Budiansky and Wu (11). Tanaka and Mori (12) explained the dispersion hardening by the average back stress $-f\sigma^{\infty}_{ij}$, where f is the volume fraction of particles and σ^{∞}_{ij} is the Eshelby solution for an isolated inclusion. Recently Mori and Osawa (13) and Okabe, Mochizuki and Mori (14) have used the polynomial solution of degree 4 of Asaro and Barnett (7) for explanation of softening of work-hardened Al-Si crystals and low-temperature creep and low-temperature recovery creep of a Cu-SiO_2 alloy. Okabe et al. (15) have measured internal friction of Al-Si alloys and found a third power dependence of particle size of Si predicted by use of Eshelby's result and diffusion of the misfit atoms.

3. SLIDING INCLUSIONS

Eshelby's solution corresponds to a perfectly bonded inclusion which results in continuity of the displacement and interfacial traction across the interface S,

$$[u_i] = 0, \tag{23}$$

$$[\sigma_{ij}]n_j = 0, \tag{24}$$

where $[\cdot]$ = [out] - [in], and n_i is the normal vector on S. When condition (23) is relaxed, we have a sliding inclusion.

Consider an anisotropic infinite body, undergoing a uniform (constant) eigenstrain ε^*_{ij} in an ellipsoidal domain Ω bounded by

$$x_1^2/a_1^2 + x_2^2/a_2^2 + x_3^2/a_3^2 = 1. \tag{25}$$

We investigate a solution of the elastic field when sliding takes place along the inclusion interface. It is assumed that the inclusion interface cannot sustain any shear tractions. This free sliding inclusion is subjected to the following conditions:

$$\sigma_{ij,j} = 0, \tag{26}$$

$$\sigma_{ij} = C_{ijk\ell}(u_{k,\ell} - \varepsilon^*_{k\ell}) \quad \text{in } \Omega, \tag{27}$$

$$[\sigma_{ij}]n_j = 0 \quad \text{on } S, \tag{28}$$

$$\sigma_{ij}n_j = \alpha n_i \quad \text{on } S, \tag{29}$$

$$[u_i]n_i = 0 \quad \text{on } S, \tag{30}$$

$$[u_i] = -b_i \quad \text{on } S. \tag{31}$$

Condition (29), with α an unknown constant, means that the interface is free of shear traction. Conditions (30) and (31) correspond to the continuity condition for the normal displacement on S and the discontinuity condition for the tangential displacement on S, respectively. Vector b_i is an unknown slip vector.

Recently, Mura and Furuhashi (16) have found an unexpected result for the sliding inclusion. It is found that when an ellipsoidal inclusion undergoes a shear eigenstrain coinciding with the axes of the inclusion and the inclusion is free to slip along the interface, the stress field vanishes everywhere in the inclusion and the matrix. Their result is obtained from the sum of Volterra's solution for the Somigliana dislocation b_i on S and Eshelby's solution. The same result is obtained by the following simple derivation. As Eshelby did, the inclusion is removed from the matrix and allowed to undergo the stress-free strain ε^*_{ij}. Then, the displacement in this inclusion is

$$u_i = \varepsilon^*_{ij}x_j - \omega_{ij}x_j, \tag{32}$$

where ω_{ij} is a rotation, $\omega_{ij} = -\omega_{ji}$, and $\varepsilon^*_{ij} = 0$ for $i = j$. An arbitrary point with coordinates x_i is transformed to x'_i point by (32),

$$x'_i = x_i + \varepsilon^*_{ij}x_j - \omega_{ij}x_j. \tag{33}$$

The inverse transformation is obtained as

$$x_i = x'_i - (\varepsilon^*_{ij} - \omega_{ij})x'_j, \tag{33'}$$

when terms of order higher than or equal to 2 are neglected, assuming $\varepsilon^*_{ij} - \omega_{ij}$ to be small quantities.

An ellipsoid expressed by (25) is transformed by (33') into

$$(x'_1)^2/a_1^2 + (x'_2)^2/a_2^2 + (x'_3)^2/a_3^2 - 2x'_1x'_2\{(\varepsilon^*_{12} - \omega_{12})/a_1^2$$

$$+ (\varepsilon^*_{12} + \omega_{12})/a_2^2\} - 2x_2'x_3'\{(\varepsilon^*_{23} - \omega_{23})/a_2^2 + (\varepsilon^*_{23} + \omega_{23})/a_3^2\}$$
$$- 2x_3'x_1'\{(\varepsilon^*_{31} - \omega_{31})/a_3^2 + (\varepsilon^*_{31} + \omega_{31})/a_1^2\} = 1 \tag{34}$$

when the higher order terms are neglected. It is obvious that Eqn (25) is transformed to the identical ellipsoid

$$(x_1')^2/a_1^2 + (x_2')^2/a_2^2 + (x_3')^2/a_3^2 = 1 \tag{35}$$

if we can choose ω_{ij} as

$$\omega_{12} = \varepsilon^*_{12}(1/a_1^2 + 1/a_2^2)/(1/a_1^2 - 1/a_2^2),$$
$$\omega_{23} = \varepsilon^*_{23}(1/a_2^2 + 1/a_3^2)/(1/a_2^2 - 1/a_3^2), \tag{36}$$
$$\omega_{31} = \varepsilon^*_{31}(1/a_3^2 + 1/a_1^2)/(1/a_3^2 - 1/a_1^2).$$

Namely, the ellipsoid, Fig. 2(a), is transformed into the identical ellipsoid as shown in Fig. 2(b) by the displacement (32) with (36).

This means that there exists no misfit between the matrix and the inclusion except for remaining slip b_i on S. Therefore, no

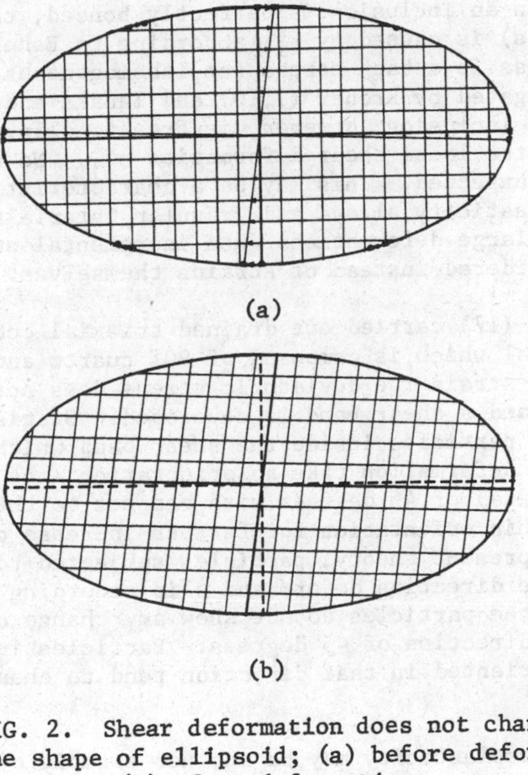

(a)

(b)

FIG. 2. Shear deformation does not change the shape of ellipsoid; (a) before deformation, (b) after deformation.

stress is necessary to put the inclusion back in the matrix since the interface slip is allowable. The displacement in the matrix is zero and that of inclusion is tangential to the interface, namely

$$u_i n_i = 0 \quad \text{on } S \tag{37}$$

when (36) is satisfied. The slip vector b_i in (31) is nothing more than u_i in (32). All conditions (26) ~ (31) are satisfied.

The choice of (36) is only possible when $a_1 \neq a_2 \neq a_3$. If $a_2 = a_3$, for instance, ε^*_{23} must vanish. Since we have assumed that $\varepsilon^*_{ij} \pm \omega_{ij}$ are small quantities, we are constrained by the condition

$$\begin{aligned} &|2\varepsilon^*_{12}a_1a_2/(a_1^2 - a_2^2)| \ll 1, \\ &|2\varepsilon^*_{23}a_2a_3/(a_2^2 - a_3^2)| \ll 1, \\ &|2\varepsilon^*_{31}a_3a_1/(a_3^2 - a_1^2)| \ll 1. \end{aligned} \tag{38}$$

If the ellipsoid degenerates into a spheroid, the corresponding strain component must vanish. The linear approximation employed here is valid for small strain ε^*_{ij} and larger differences of the semi-axes of the ellipsoid.

When an inclusion is perfectly bonded, the eigenstress (residual stress) is caused by ε^*_{ij} according to Eshelby's calculation (1). This stress is a back stress for inhomogeneous plastic deformation ε^*_{ij}, as investigated by Kröner (9,10) and Tanaka and Mori (12), among others. If the inclusion, however, is free to slip, no back stress is accumulated by the local shear deformation ε^*_{ij}. No resistance for shear deformation is expected. This may be a characteristic of deformation seen in superplasticity alloys and granular materials. The theory is valid even for large deformations when incremental strains (or strain rates) are considered instead of strains themselves.

Oda (17) carried out drained triaxial compression tests on a granular material which is composed of 90% quartz and 10% feldspar. After 25% axial strain the deviatoric stress does not increase for further straining and a shear band is developed. Statistically, most of the ellipsoidal particles inside the shear band which have been randomly oriented before deformation take an orientation (the direction of major axes of ellipsoids) of 45 degrees with respect to the axis of compression and keep this orientation for further increase of the axial strain. In view of the present theory, particles subjected to the maximum shear in the 45 degree direction deform and slip according to (32) without any constraint and the particles do not show any change of shape and orientation in that direction of 45 degrees. Particles in the shear band which are not oriented in that direction tend to change to the 45 degree direction.

Superplasticity may be another example of the present result.

Some eutectoid alloys with fine grains elongate up to a few thousands of pct before fracture at high temperature above half the melting point (18). In view of the present theory, these fine grains deform as a group which has an ellipsoidal shape regarded as an ellipsoidal inclusion. These inclusions deform and slip according to (32).

As the deformation proceeds considerably further, the orientation and shape of the inclusion do not change although individual grains change their orientations and shapes. Since no back stress is expected from the deformation, very large elongation may be possible. The free slip on the interface means no resistance on the interface. However, in practice there might be some resistance for the slip. Reality may be between Eshelby's perfect bonding and the free slip treated here. When a strain increment (or strain rate) is small, Eshelby's stress is not large enough to cause the stress relaxation by slip. On the other hand, if the strain rate is large, the relaxation by slip also meets a difficulty to follow since the relaxation by slip accompanies a diffusion process of matter. This is why an intermediate strain rate gives a maximum elongation to failure as reported by Langdon (19), among others.

When ε^*_{11}, ε^*_{22}, ε^*_{33} are given in the free sliding inclusion, the situation is not simple as in the case of shear eigenstrains. The stress field in Ω takes the form of an infinite series of $\underset{\sim}{x}$. When the ellipsoid is spherical, the stress field becomes equal to the sum of quadratic polynomials and a constant term as shown by Mura and Furuhashi (16). This special case can be expected from Ghahremani's result (20) for a sliding spherical inhomogeneity under a uniform tension at infinity.

Figure 3, which has been calculated by Jasiuk et al. (21), shows a stress distribution inside a sliding spheroidal inclusion ($a_1 = a_2 = a_3/2$) when $\varepsilon^*_{11} = \varepsilon^*_{22} = 0$, ε^*_{33} = const. For comparison, Eshelby's result for a perfect bonding is also shown in Fig. 3. It is seen that the stress field for the sliding inclusion significantly deviates from the uniform stress field of Eshelby's inclusion.

4. ANISOTROPIC INCLUSIONS

Most inclusion and inhomogeneity problems (6,7,8,22,23,24,41) solved after Eshelby are devoted to elastic anisotropy and polynomial eigenstrains under the perfect bonding condition on the interface. It is found that Eshelby's prediction about the correlation between the degree of polynomials of the interior stress field and that of eigenstrains is preserved even for elastically anisotropic media.

The integral (14) for the displacement field has been further integrated by Mura and Cheng (23) when an ellipsoidal inclusion Ω has a general eigenstrain $\varepsilon^*_{ij}(\underset{\sim}{x})$ in an anisotropic medium. The result is

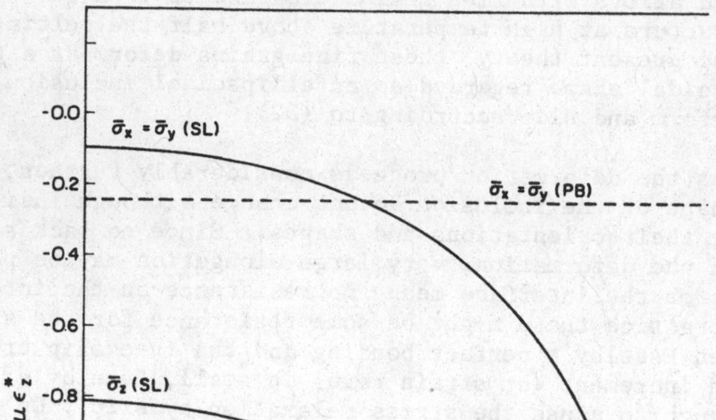

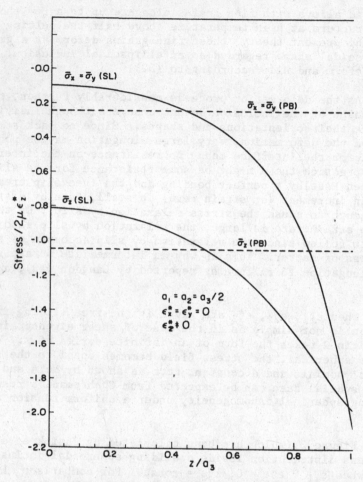

FIG. 3. Stress field in an ellipsoid. The dotted lines are for the perfect bonding and the solid lines are for the sliding case.

$$u_i(\underset{\sim}{x}) = -(8\pi^2)^{-1} \int_0^{2\pi} d\phi \Big[\int_0^R r\,dr \frac{\partial}{\partial z} \varepsilon^*_{nm}(\underset{\sim}{x}') - (\bar{\zeta}\cdot y)\{\varepsilon^*_{nm}(\underset{\sim}{x}')\}_{r=R}\Big]_{z=\bar{\zeta}\cdot \underset{\sim}{y}} \times \int_{S^*} C_{k\ell mn} N_{ik}(\bar{\xi}) D^{-1}(\bar{\xi}) \bar{\xi}_\ell\, \zeta\, dS(\bar{\xi}), \tag{39}$$

where changing of variables (see Figs. 4 and 5),

$$x_1/a_1 = y_1, \quad x_2/a_2 = y_2, \quad x_3/a_3 = y_3$$
$$x_1'/a_1 = y_1', \quad x_2'/a_2 = y_2', \quad x_3'/a_3 = y_3'$$

$$a_1\bar{\xi}_1 = \zeta_1, \quad a_2\bar{\xi}_2 = \zeta_2, \quad a_3\bar{\xi}_3 = \zeta_3 \tag{40}$$

$$\zeta_1/\zeta = \bar{\zeta}_1, \quad \zeta_2/\zeta = \bar{\zeta}_2, \quad \zeta_3/\zeta = \bar{\zeta}_3$$

$$\zeta = (\zeta_1^2 + \zeta_2^2 + \zeta_3^2) = \{(a_1\bar{\xi}_1)^2 + (a_2\bar{\xi}_2)^2 + (a_3\bar{\xi}_3)^2\}^{\frac{1}{2}},$$

$$\bar{\xi}_1^2 + \bar{\xi}_2^2 + \bar{\xi}_3^2 = 1$$

transforms Ω into a unit sphere S^2, and

$$d\underset{\sim}{x}' = a_1a_2a_3dy_1'dy_2'dy_3' = a_1a_2a_3\, r\, dr d\phi dz,$$
$$R = (1 - z^2)^{\frac{1}{2}}, \qquad z = \bar{\underset{\sim}{\zeta}}\cdot\underset{\sim}{y}'. \tag{41}$$

Furthermore $dS(\bar{\underset{\sim}{\zeta}})$ is a surface element on S^2 for a given $\bar{\underset{\sim}{\zeta}}$, S^* is a subdomain of S^2 satisfying condition

$$\bar{\underset{\sim}{\zeta}}\cdot\underset{\sim}{y} \leq 1 \tag{42}$$

and is shown in Fig. 5 by the unshaded surface on S^2. When a point $\underset{\sim}{x}$ is inside Ω ($\underset{\sim}{y}$ is inside S^2), the condition (42) is identically satisfied and, therefore, S^* becomes identical to S^2.

When ε_{ij}^* is a polynomial, integral (39) leads to the results obtained by Asaro and Barnett (7) and Mura and Kinoshita (8).

5. INTERACTIONS AND DYNAMICS

The interaction between two inhomogeneities under a uniform applied stress at infinity can be reduced to a problem of two inclusions in the homogeneous medium. The eigenstrains in the inclusions necessary

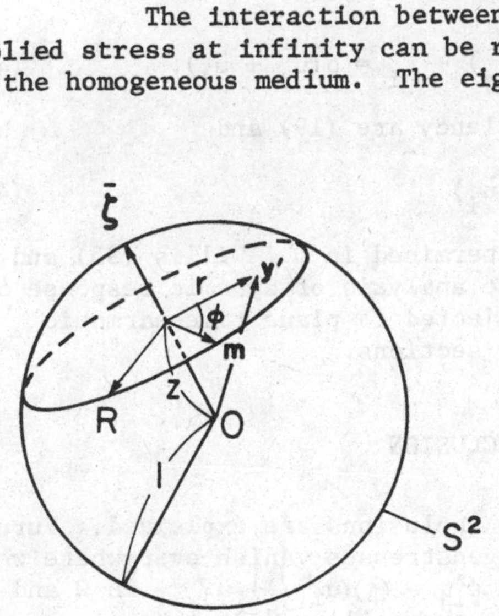

FIG. 4. Volume element $d\underset{\sim}{x}'$ in Ω is written as $a_1a_2a_3\, r\, dr d\phi dz$.

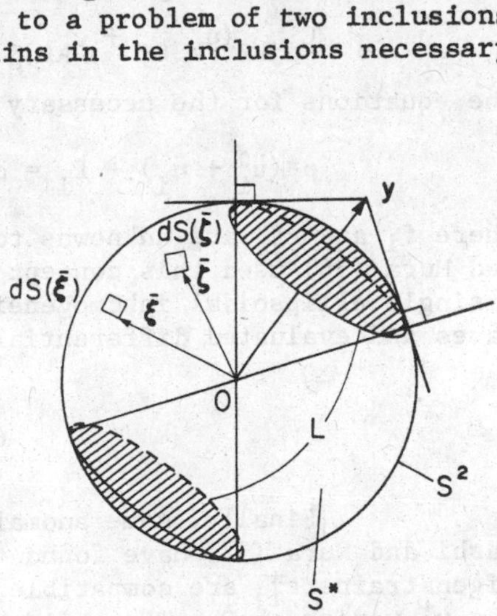

FIG. 5. Subspace S^* on S^2 is denoted by the unshaded part.

for this equivalent inclusion method are calculated by Moschovidis and Mura (25) in series forms.

When an ellipsoidal inclusion is located near the surface of a half-space, expression (16') must be used and $G_{ij}(\underset{\sim}{x} - \underset{\sim}{x}')$ is replaced by $G_{ij}(\underset{\sim}{x}, \underset{\sim}{x}')$, which is Green's function in the half-space found by Mindlin (26). Seo and Mura (27) gave numerical examples for a case of uniform dilatational eigenstrains.

For the inhomogeneity problem in a half-space the Boussinesq method using displacement functions seems to be more effective than the equivalent inclusion method. Examples are seen in the papers by Tsutsui and Saito (28), Tsuchida et al. (29) and Tsuchida and Mura (30). This method is also applied to a two-confocal spheroidal inhomogeneity by Mikata and Taya (40).

When an inclusion is not ellipsoidal, the stress field in the inclusion is not uniform as seen in (31 ~ 33) for a cuboidal inclusion with uniform eigenstrains.

The equivalent inclusion method can also be applied to the dynamic response of an ellipsoidal inhomogeneity by a slight modification as suggested by Mura (34). The equation of motion in the inhomogeneity is

$$C^*_{ijk\ell}(u^0_{k,\ell j} + \varepsilon_{k\ell,j}) = \rho^*(\ddot{u}^0_i + \ddot{u}_i), \tag{43}$$

where ρ^* is the density of Ω. This equation is replaced by the equations of motion in the homogeneous material with a fictitious body force f_i and eigenstrain ε^*_{ij} in Ω,

$$C_{ijk\ell}(u^0_{k,\ell j} + \varepsilon_{k\ell,j} - \varepsilon^*_{k\ell,j}) + f_i = \rho(\ddot{u}^0_i + \ddot{u}_i). \tag{44}$$

The equations for the necessary equivalency are (19) and

$$\rho^*(\ddot{u}^0_i + \ddot{u}_i) + f_i = \rho(\ddot{u}^0_i + \ddot{u}_i) \tag{45}$$

where f_i and ε^*_{ij} are unknowns to be determined in Ω. Willis (35) and Fu and Mura (36) used this concept for the analysis of dynamic response of a single ellipsoidal inhomogeneity subjected to plane time-harmonic waves and evaluated differential cross sections.

6. CONCLUSION

Finally, some anomalies of inclusions are explained. Furuhashi and Mura (37) have found that eigenstresses vanish everywhere when eigenstrains ε^*_{ij} are compatible, i.e., $\varepsilon^*_{ij} = (\frac{1}{2})(u^*_{i,j} + u^*_{j,i})$ in Ω and the u^*_i vanish on S. These eigenstrains are called "impotent eigenstrains." This interesting result leads to the conclusion that there is no unique eigenstrain in the equivalent inclusion method when the inhomogeneity is a void.

If an infinite number of inclusions Ω_p (p = 1,2,...) are three-dimensionally distributed in an infinite body, the stress σ_{ij} is not the sum of stresses caused by individual inclusions. Furuhashi et al. (38) have proved that this summation does not converge as if the summation of lights coming from an infinite number of stars at night does not converge, unless some additional considerations are taken into account (39).

ACKNOWLEDGMENT

This research was supported by U.S. Army Research Grant DAAG29-81-K-0090.

REFERENCES

(1) Eshelby, J. D. (1957). The determination of the elastic field of an ellipsoidal inclusion, and related problems, Proc. Roy. Soc. A241, 376-396.
(2) Eshelby, J. D. (1959). The elastic field outside an ellipsoidal inclusion, Proc. Roy. Soc. A252, 561-569.
(3) Eshelby, J. D. (1961). Elastic inclusions and inhomogeneities, in Progress in Solid Mechanics 2, eds. I. N. Sneddon and R. Hill, North-Holland, Amsterdam, 89-140.
(4) Mura, T. (1982). Micromechanics of defects in solids. Martinus Nijhoff Pub., The Hague.
(5) Tanaka, K. & Mura, T. (1982). A theory of fatigue crack initiation at inclusions, Metall. Trans. 13A, 117-123.
(6) Kinoshita, N. & Mura, T. (1971). Elastic fields of inclusions in anisotropic media, phys. stat. sol. (a) 5, 759-768.
(7) Asaro, R. J. & Barnett, D. M. (1975). The non-uniform transformation strain problem for an anisotropic ellipsoidal inclusion, J. Mech. Phys. Solids, 23, 77-83.
(8) Mura, T. & Kinoshita, N. (1978). The polynomial eigenstrain problem for an anisotropic ellipsoidal inclusion, phys. stat. sol. (a) 48, 447-450.
(9) Kröner, E. (1958). Berechnung der elastischen Konstanten des Vielkristalls aus den Konstanten des Einkristalls, Z. Phys. 151, 504-518.
(10) Kröner, E. (1961). Zur plastischen Verformung des Vielkristalls, Acta Metall., 9, 155-161.
(11) Budiansky, B. & Wu, T. T. (1962). Theoretical prediction of plastic strains of polycrystals, Proc. 4th U.S. Nat. Congr. Appl. Mech., 1175-1185.
(12) Tanaka, K. & Mori, T. (1970). The hardening of crystals by non-deforming particles and fibres, Acta Metall. 18, 931-941.
(13) Mori, T. & Osawa, T. (1979). Softening of work-hardened Al-Si crystals and low-temperature creep, Phil. Mag. A40, 445-457.

(14) Okabe, M., Mochizuki, T. & Mori, T. (1980). Low-temperature recovery creep of a Cu-SiO_2 alloy, Phil. Mag. A41, 615-618.
(15) Okabe, M., Mori, T. & Mura, T. (1981). Internal friction caused by diffusion around a second phase particle Al-Si alloy, Phil. Mag. 44, 1-12.
(16) Mura, T. & Furuhashi, R. The inelastic inclusion with a sliding interface, J. Appl. Mech., in press.
(17) Oda, M. (1972). Deformation mechanism of sand in triaxial compression tests, Soils and Foundations, 12, 45-63.
(18) Padmanabhan, K. A. & Davis, G. J. (1980). Superplasticity, Springer-Verlag, Berlin.
(19) Langdon, T. G. (1982). The mechanical properties of superplastic materials, Metall. Trans. 13A, 689-701.
(20) Ghahremani, F. (1980). Effect of grain boundary sliding on anelasticity of polycrystals, Int. J. Solids and Structures, 16, 825-845.
(21) Jasiuk, I., Tsuchida, E. & Mura, T. The stress field of a sliding inclusion. In preparation.
(22) Walpole, L. J. (1967). The elastic field of an inclusion in an anisotropic medium, Proc. Roy. Soc. A300, 270-289.
(23) Mura, T. & Cheng, P. C. (1977). The elastic field outside an ellipsoidal inclusion, J. Appl. Mech. 44, 591-594.
(24) Yang, H. C. & Chou, Y. T. (1976). Generalized plane problems of elastic inclusions in anisotropic solids, J. Appl. Mech. 98, 424-430.
(25) Moschovidis, Z. A. & Mura, T. (1975). Two-ellipsoidal inhomogeneities by the equivalent inclusion method, J. Appl. Mech. 42, 847-852.
(26) Mindlin, R. D. (1953). Force at a point in the interior of a semi-infinite solid, Midwestern Conf. Solid Mech., 56-59.
(27) Seo, K. & Mura, T. (1979). The elastic field in a half-space due to ellipsoidal inclusions with uniform dilational eigenstrains, J. Appl. Mech., 46, 568-572.
(28) Tsutsui, S. & Saito, K. (1973). On the effect of a free surface and a spherical inhomogeneity on stress fields of a semi-infinite medium under axisymmetric tension, Proc. 23rd Japan Nat. Congr. Appl. Mech., 23, 547-560.
(29) Tsuchida, E., Saito, Y., Nakahara, I., & Kodama, M. (1982). Stress concentration around a prolate spherical cavity in a semi-infinite elastic body under all-round tension, Bulletin of the JSME, 25, 493-500.
(30) Tsuchida, E. & Mura, T. The stress field in an elastic half-space having a spheroidal inhomogeneity under all-around tension parallel to the plane boundary, J. Appl. Mech., in press.
(31) Sankaran, R. & Laird, C. (1974). Interfacial structure of plate-like precipitates, Phil. Mag. 29, 179-215.
(32) Lee, J. K. & Johnson, W. C. (1978). Calculation of the elastic strain field of a cuboidal precipitate in an anisotropic matrix, phys. stat. sol. (a) 46, 267-272.
(33) Chiu, Y. P. (1977). On the stress field due to initial strains in cuboid surrounded by an infinite elastic space, J. Appl. Mech., 44, 587-590.

(34) Mura, T. (1972). A variational method for micromechanics of composite materials, Proc. Int. Conf. Mech. Bahavior of Materials, ed. by The Soci. Materials Science Japan, 5, 12-18.
(35) Willis, J. R. (1980). A polarization approach to the scattering of elastic waves, I. Scattering by a single inclusion, J. Mech. Phys. Solids, 28, 287-306.
(36) Fu, L. S. & Mura, T. (1983). The determination of the elastodynamic fields of an ellipsoidal inhomogeneity, J. Appl. Mech., 50, 390-396.
(37) Furuhashi, R. & Mura, T. (1979). On the equivalent inclusion method and impotent eigenstrains, J. Elasticity, 9, 263-270.
(38) Furuhashi, R., Kinoshita, N. & Mura, T. (1981). Periodic distributions of inclusions, Int. J. Engineering Sci., 19, 231-236.
(39) Asimov, I. (1974). Asimov on Astronomy.
(40) Mikata, Y. & Taya, M. (1983). Stress field in and around a coated short fiber in an infinite matrix subjected to uniaxial and bi-axial loadings, submitted to J. Appl. Mech.
(41) Sendeckyj, G. P. (1967). Ellipsoidal inhomogeneity problem, Ph.D. Dissertation, Northwestern University, Evanston, Illinois.

THE ANALYSIS OF THE OVERALL ELASTIC PROPERTIES OF COMPOSITE MATERIALS

L. J. WALPOLE

School of Mathematics and Physics, University of East Anglia,
Norwich NR4 7TJ, England

ABSTRACT

We survey the literature and comment on the analysis of the 'overall' (or 'effective') elastic properties of composite materials, from a point of view which emphasises the early and enduring influence here of the papers (1,2,3) of Eshelby.

1. INTRODUCTION

A composite solid medium is an aggregate of a number of distinct homogeneous phases which remain bonded together firmly in the present circumstances when an equilibrated state of infinitesimal elastic strain is introduced. In every region of it we may identify a finite 'representative volume' whose structure is typical of the whole. But especially as this structure may not be prescribed in the fullest detail, the precise determination of the elastic state cannot be contemplated. For a wide range of practical purposes it is convenient and sufficient to treat the whole composite medium as if it were a particular ('equivalent') homogeneous one whose elastic moduli can be calculated to an acceptable accuracy once and for all by means of some mathematical expression of the equivalence. From among several ultimately coincident procedures, we may choose what amounts to the ('more severe') test which Eshelby (3) prescribes in particular for an aggregate of cubic crystals. Let the representative volume be embedded firmly, as a heterogeneous inclusion, in an infinite matrix of the equivalent homogeneous medium and let the accompanying elastic state be the one which reduces to a field of uniform strain e^A_{ij} and uniform stress p^A_{ij} in all remote regions, where the connection $p^A_{ij} = L_{ijk\ell} e^A_{k\ell}$ is made in terms of the fourth-rank tensor of elastic moduli assigned in advance to the unknown equivalent medium. The condition of equivalence is then expressed by either of the forms

$$\bar{p}_{ij} = L_{ijk\ell}\bar{e}_{k\ell}, \qquad \bar{e}_{ij} = M_{ijk\ell}\bar{p}_{k\ell}, \tag{1}$$

of the relation between the average stress $\bar{p}_{ij}$ and the average strain $\bar{e}_{ij}$ in the embedded representative volume, in order to extract an explicit formula for the stiffness tensor L of overall elastic moduli and for its inverse tensor M of elastic compliances. If the representative volume is given an ellipsoidal boundary then this same final formula would still be reached when any convenient homogeneous material is chosen for the outer matrix. Such a freedom of choice is referred to by Brown (4,5) in analogous contexts; it is allowed here by the property of Eshelby (1) that a homogeneous ellipsoidal inclusion is strained uniformly. The equivalence of the various, intuitively appealing, mathematical models for the calculations of overall properties can be demonstrated by showing that they all give rise to the same governing integral equations (6). In a persuasive review of what Hill (7) named and appraised later as the 'self-consistent method', Eshelby (3) employed a no longer fully representative sample of the aggregate of cubic crystals to achieve an analytical approximation of Eqns (1), by which the overall bulk modulus is equated correctly to the bulk modulus of the crystals, while the overall shear modulus is left as the only positive root of a certain cubic equation, to fall then (8,9) between the universal upper and lower bounds made available later by Hashin & Shtrikman (10). In general, the elastic moduli of a polycrystal cannot be specified precisely in terms only of those of its single crystal. The upper and lower bounds offered by Hashin & Shtrikman (10,11,12,13) are the best available to date but it is not known if they are the best possible in those terms. It is found that they do confine the overall moduli of many actual polycrystals quite narrowly, but that they would not do so if the single crystal were to be strongly anisotropic, in which event the self-consistent calculations would remain ready for consideration. The special example of a polycrystal will not enter our further discussion. We turn instead to multiphase media which each have a finite number of different constituents whose individual elastic properties and volume concentrations are prescribed, but whose geometrical arrangement is left largely unspecified or in any case too intricate or too special to allow for in full. The overall moduli are to be evaluated as precisely as possible, or desirable, between upper and lower limits, perhaps after imposing assigned structural features, which may serve to reduce the uncertainty.

2. MULTIPHASE MATERIALS

Let the composite medium be one of N phases, for which c_r is the concentration by volume of the rth phase, so that $\Sigma c_r = 1$, where the summation is from 1 to N . If it happens to be isotropic in the overall sense, its Eqns (1) can be written as

$$\bar{p}_{ij} = (\varkappa - 2\mu/3)\bar{e}_{kk}\,\delta_{ij} + 2\mu\,\bar{e}_{ij} \quad , \tag{2}$$

in terms of the overall bulk and shear moduli, $\varkappa$ and μ respectively, and of the Kronecker delta.

It is worth recalling straightaway that, according to Hill (14), an overall isotropy will prevail always if the phases are all made isotropic and are all given the same shear modulus μ , which is coincident then with the overall shear modulus. Moreover the overall bulk modulus is given always by the elegant exact expression

$$\varkappa = [\Sigma c_r \varkappa_r/(\varkappa_r + 4\mu/3)]/[\Sigma c_r/(\varkappa_r + 4\mu/3)] \qquad (3)$$

in terms only of the concentrations and moduli of the phases, $\varkappa_r$ being the bulk modulus of the rth phase, without any dependence at all on the finer ingredients of the possibly quite biased and anisotropic, geometry. The derivation finds its origins in the expression of Eshelby (3), anticipated also by Nabarro (15) and Robinson (16), for the elastic field of a uniformly expanding, misfitting inclusion in an infinite matrix of the same shear modulus but differing bulk modulus. Thus for Eqns (1) the strain field

$$e_{ij} = e^A_{ij} + e^A_{kk}\,\Sigma\,[(\varkappa_r - \varkappa)/4\pi(\varkappa_r + 4\mu/3)]\varphi_{r,ij}$$

is constructed with all its necessary properties, from the potential φ_r of matter of unit density imagined to fill the whole volume of the rth phase. These potentials need not be calculated at all fortunately since the harmonic property, that $\varphi_{r,ii}$ is equal to -4π at points in the rth phase and to zero elsewhere, suffices at once to bring about always the isotropic relation of Eqn (2), in which $\varkappa$ is evaluated by Eqn (3). This remarkable isotropic outcome is the one that Hill (14) derived (for just two phases) from an alternative standpoint. It has not apparently been taken note of anywhere in the later literature. Future calculations, for instance, for aligned or absolutely regular arrays of inclusions, must conform to it.

In the general case of N isotropic phases with differing shear moduli, that is, with bulk modulus $\varkappa_r$ and shear modulus μ_r assigned to the rth phase, an arbitrary anisotropy may persist at the overall level. Elementary bounds of the Voigt and Reuss type can be placed on the various moduli of interest. For instance, the inequalities

$$\Sigma c_r/\varkappa_r \geq M_{iijj} \geq 1/\Sigma c_r \varkappa_r$$

are set by the Reuss and Voigt (over- and under-) estimates of the 'volume compressibility', M_{iijj}. However these particular ones can be improved at once now on recalling a theorem of Hill (17) which implies in particular that if the phases are all 'strengthened' by replacing every shear modulus by the largest one, μ say, then M_{iijj} will be decreased to $1/\varkappa$, where $\varkappa$ is defined by Eqn (3). Thus this particular $\varkappa$ is specified as an upper or lower bound on $1/M_{iijj}$, and for similar reasons also on $L_{iijj}/9$, according as μ is equated to the largest or smallest of the $\mu_1, \mu_2, \ldots, \mu_N$. If an elastic isotropy or cubic symmetry is known or presumed to prevail overall then these two bounded quantities both coincide with its overall bulk modulus $\varkappa$. Specialising then to a composite of two phases only, the definition

$$\bar{\varkappa} = c_1\varkappa_1 + c_2\varkappa_2 - c_1c_2(\varkappa_1-\varkappa_2)^2/(c_1\varkappa_2 + c_2\varkappa_1 + 4\mu_o/3) \tag{4}$$

makes for a statement equivalent to that which Hill (17) derived in the present manner, namely that the difference $\bar{\varkappa} - \varkappa$ is positive or negative according as μ_o is equated to the larger or smaller of μ_1 and μ_2. These bounds on $\varkappa$ are known to be the best possible in that Hashin & Shtrikman (18) have found a particular two-phase medium which happens to make a coincidence with the upper bound and (when the numbering of the phases is reversed) with the lower bound, for all phase moduli and concentrations. Corresponding bounds do not appear to be available yet for any other modulus of an overall anisotropy, except for the shear modulus μ of an overall isotropy (18,19). When an isotropic composite has just two phases, and the definition

$$\bar{\mu} = c_1\mu_1 + c_2\mu_2 - c_1c_2(\mu_1-\mu_2)^2/(c_1\mu_2 + c_2\mu_1 + \mu_o^*), \tag{5}$$

where $$\mu_o^* = 3/2[1/\mu_o + 10/(9\varkappa_o + 8\mu_o)], \tag{6}$$

is introduced, we have that the difference $\bar{\mu} - \mu$ is positive or negative according as $\varkappa_o$ is equated to the larger or smaller of $\varkappa_1$ and $\varkappa_2$, and μ_o to the larger or smaller of μ_1 and μ_2. Improved bounds have been offered recently by Milton & Phan-Thien (20) in the exceptional case where $(\varkappa_1 - \varkappa_2)(\mu_1 - \mu_2) < 0$. Experimental comparisons with the present two-phase bounds on $\varkappa$ and μ have been made by Watt & O'Connell (21) and by the earlier authors to whom they refer. We turn next to two examples where a particular geometrical bias allows further development, and leaves a transverse elastic isotropy about a single preferred direction of symmetry.

For an overall transverse isotropy about say the x_3-axis, the Eqns (1) can be expressed in terms of a set of five independent overall moduli as

$$\bar{p}_{11} = (k+m)\bar{e}_{11} + (k-m)\bar{e}_{22} + \ell\bar{e}_{33},$$
$$\bar{p}_{22} = (k-m)\bar{e}_{11} + (k+m)\bar{e}_{22} + \ell\bar{e}_{33},$$
$$\bar{p}_{33} = \ell(\bar{e}_{11} + \bar{e}_{22}) + n\bar{e}_{33},$$
$$\bar{p}_{12} = \bar{p}_{21} = 2m\bar{e}_{12}, \quad \bar{p}_{23} = \bar{p}_{32} = 2p\bar{e}_{23}, \quad \bar{p}_{31} = \bar{p}_{13} = 2p\bar{e}_{13},$$

where in particular k is designated as the plane-strain bulk modulus, while m and p are the rigidity moduli for shearing over transverse and axial planes respectively. Among other dependent moduli, we may define the Young's modulus and Poisson's ratio for the axial direction by

$$E = n - \ell^2/k, \qquad \nu = \ell/2k,$$

and for a transverse direction by

$$4/E' = 1/m + 1/(k - \ell^2/n), \qquad \nu' = (E'/2m) - 1,$$

respectively. The phases may be taken as likewise transversely isotropic about the x_3-axis. A subscript r will distinguish then the corresponding elastic moduli of the rth phase. The complete overall isotropy of Eqn (2) is reverted to by means of all the connections

$$k = \varkappa + \mu/3, \quad \ell = \varkappa - 2\mu/3, \quad n = \varkappa + 4\mu/3, \quad m = p = \mu. \tag{7}$$

If the subscript r is attached throughout them they stipulate instead that the rth phase is an isotropic one with bulk modulus $\varkappa_r$ and shear modulus μ_r, and with dependent Young's modulus $E_r (= E'_r)$ and Poisson's ratio ν_r $(= \nu'_r)$.

The transverse isotropy is exhibited first by a completely stratified composite medium, in which the phases are separated by infinite parallel planes, all set perpendicular to the x_3-axis of symmetry. The five basic moduli are readily found (22) to be given by

$$k = \Sigma c_r k_r + \{\Sigma c_r \ell_r [\Sigma c_s(\ell_s - \ell_r)/n_s]/n_r\}/(\Sigma c_r/n_r),$$

$$\ell = (\Sigma c_r \ell_r/n_r)/(\Sigma c_r/n_r), \qquad n = 1/(\Sigma c_r/n_r),$$

$$m = \Sigma c_r m_r, \qquad p = 1/(\Sigma c_r/p_r).$$

Among the simpler, more palatable calculations that can be made when there are only two phases, we have

$$k = c_1 k_1 + c_2 k_2 - c_1 c_2 (\ell_1 - \ell_2)^2/(c_1 n_2 + c_2 n_1),$$

$$\ell/n = c_1 \ell_1/n_1 + c_2 \ell_2/n_2,$$

$$k - \ell^2/n = c_1(k_1 - \ell_1^2/n_1) + c_2(k_2 - \ell_2^2/n_2),$$

$$\nu/E = (c_1 k_1 \nu_1/n_1 + c_2 k_2 \nu_2/n_2)/(c_1 k_1 E_1/n_1 + c_2 k_2 E_2/n_2),$$

$$1/E = c_1/E_1 + c_2/E_2 - 4c_1 c_2(\nu_1/E_1 - \nu_2/E_2)^2/(c_1 n_2/k_2 E_2 + c_2 n_1/k_1 E_1). \tag{8}$$

We may verify that the overall isotropy of Eqns (7) is indeed regained, with $\varkappa$ defined by Eqn (3), when all the N phases are made to be isotropic and to have the same shear modulus μ. Postma (23) evaluated the overall moduli (the components $L_{ijk\ell}$) for a periodically stratified two-phase medium and discovered this particular isotropic outcome.

The second example of the overall transverse isotropy is one in which the phases are separated by cylindrical interfaces of arbitrary cross-section and infinite length, all aligned (with generators) parallel to the x_3-axis but otherwise dispersed randomly. In a 'fibre-strengthened material' one phase is a matrix containing continuous aligned cylinders of the other phases. We require the definitions of five barred moduli

$$\bar{k} = 1/[\Sigma c_r/(k_r+m_o)] - m_o, \quad \bar{\ell} = [\Sigma c_r \ell_r/(k_r+m_o)]/[\Sigma c_r/(k_r+m_o)],$$

$$\bar{n} = \Sigma c_r n_r + \{\Sigma c_r \ell_r[\Sigma c_s(\ell_s-\ell_r)/(k_s+m_o)]/(k_r+m_o)\}/[\Sigma c_r/(k_r+m_o)],$$

$$\bar{m} = 1/\{\Sigma c_r/[m_r + m_o k_o/(k_o+2m_o)]\} - m_o k_o/(k_o+2m_o),$$

$$\bar{p} = 1/[\Sigma c_r/(p_r+p_o)] - p_o ,$$

in terms of three new ones k_o, m_o and p_o which are to be specified. In the absence of a more detailed knowledge of the geometrical structure, we can (22) at least place upper and lower bounds on the overall moduli to the extent first that the six differences $\bar{k} - k$, $\bar{m} - m$, $\bar{n} - n$, $\bar{p} - p$, $(\bar{n} - \bar{\ell}^2/\bar{k}) - E$ and $(\bar{k} - \bar{\ell}^2/\bar{n}) - (k - \ell^2/n)$ are all positive (or all negative) when k_o is equated to the largest or smallest of $k_1,\dots,k_N$, and m_o to the largest (or smallest) of $m_1,\dots,m_N$, and p_o to the largest (or smallest) of $p_1,\dots,p_N$. It may be verified that this statement reduces precisely to the anticipated isotropic one of Eqns (7), with $\varkappa$ defined by Eqn (3), when all the phases are made to be isotropic and to have the same shear modulus μ. Bounds on the transverse Young's modulus E' are implied forthwith by those on m and on $(k - \ell^2/n)$. When there are just two phases, we may call upon the simplified definitions

$$\bar{k} = c_1k_1 + c_2k_2 - c_1c_2(k_1-k_2)^2/(c_1k_2+c_2k_1+ m_o), \tag{9}$$

$$\bar{m} = c_1m_1 + c_2m_2 - c_1c_2(m_1-m_2)^2/[c_1m_2+c_2m_1+m_ok_o/(k_o+2m_o)],$$

$$\bar{n} = c_1n_1 + c_2n_2 - c_1c_2(\ell_1-\ell_2)^2/(c_1k_2+c_2k_1+m_o),$$

$$\bar{p} = c_1p_1 + c_2p_2 - c_1c_2(p_1-p_2)^2/(c_1p_2+c_2p_1+p_o), \tag{10}$$

$$\bar{n} - \bar{\ell}^2/\bar{k} = c_1E_1 + c_2E_2 + 4c_1c_2(\nu_1-\nu_2)^2/(c_1/k_2+c_2/k_1+1/m_o), \tag{11}$$

which verify then that the present bounds on k, E and n are those given first by Hill (14), along with bounds directly on ℓ and ν, in consequence of the two universal relations that he found between k, ℓ and n, in terms only of the concentrations. For N isotropic phases, the present bounds on k, m and p are those given by Hashin (24) under the restriction that the largest (and also the smallest) bulk and shear moduli both belong to the same phase. More recently (25,26) he has indicated the extension to transversely isotropic phases and has also made some practical calculations of the margin between the various upper and lower bounds. It can be noted (27) particularly that, by Eqn (11), the margin between the bounds on E (the axial Young's modulus) is proportional to $(\nu_1 - \nu_2)^2$ and so is liable to be relatively small at least for isotropic phases, in which case E is left quite well approximated simply by $c_1E_1 + c_2E_2$. Moreover Eqn (8) reveals for the stratified two-phase medium that E is equal to $c_1E_1 + c_2E_2$ when both phases are isotropic and incompressible ($\nu_1 = \nu_2 = 1/2$). It would seem therefore that the simple weighting formula, $c_1E_1 + c_2E_2$, can be supported as a good first approximation to the axial Young's modulus E in practice whenever two isotropic (not too dissimilar) phases give rise to an overall transverse isotropy, for instance when a fibre strengthening is due to discontinuous ('short') rather than continuous fibres. Lastly, we recall that it is well known (26) that calculations for the overall transverse (thermal or electrical) conductivity transfer at once, because of a coincidence of the underlying mathematics, to the corresponding ones for the present 'longitudinal' shear modulus p. Thus the elegant two-phase relation of Keller (28), with its validity extended by Mendelson (29), is

$$p(p_1, p_2)\, p(p_2, p_1) = p_1p_2 \tag{12}$$

here. It enables a bound on one side of p to be changed to one on the other side, simply by forming its reciprocal and either by interchanging the arguments p_1 and p_2 and multiplying by the product p_1p_2, or by replacing each argument p_1 and p_2 by its own reciprocal. Each of the Voigt (upper) and Reuss (lower) bounds, $c_1p_1 + c_2p_2$ and $1/(c_1/p_1 + c_2/p_2)$ respectively, is changed to the other, and is shown to be an unattainable (not best possible) bound, by not satisfying Eqn (12). The superior bounds provided by Eqn (10) are similarly interchangeable. Moreover each satisfies Eqn (12) and is indeed found to be attained by a particular two-phase medium (24). [This best possible status can be assigned (24) also to the bounds on k, provided by Eqn (9) and to the related bounds on E, n, ℓ and ν presented by the connections of Hill (14).] For any two-phase medium which is oblivious to the interchange $(p_1 - p_2)$ of its phase moduli, Mendelson (29) notes that Eqn (12) implies the exact result

$$p = (p_1p_2)^{1/2} \tag{13}$$

which was found otherwise by Dykhne (30), and which as a geometric mean falls duly below the Voigt arithmetic mean (for $c_1 = c_2 = 1/2$) and above the Reuss harmonic mean. Dykhne (30) and Hashin (26) observe that Eqn (13) will apply when the cross-sectional geometry is the periodic one of a chess-board; the early derivation of this result by Lees (31) has apparently escaped notice in the modern literature. Extensive calculations have been reported recently (32,33) for the periodic geometry of a regular square or hexagonal array of circular cylindrical inclusions, in circumstances for which the bounds of Eqn (10) are too far apart to be useful. In particular, we note (32) that at the maximum concentration of touching cylinders (of say the first phase) the ratio p/p_2 is proportional to $p_1/p_2 \ell n(p_1/p_2)$ when p_1/p_2 is very large, where the square array has $\pi/2$ as its proportionality constant and approximately 6.0 as its next additive term, while the hexagonal array has $\sqrt{3}\pi/2$ and zero respectively as its corresponding numbers. It is to be expected that similar asymptotic and full numerical calculations will be made available in the near future for all the elastic moduli of such regular (and possibly of random) arrays.

3. PARTICULATE COMPOSITES

In our particulate composite of two phases, the first consists of inclusions dispersed throughout a matrix of the second. If the average strain tensor in the first phase is evaluated as $\bar{A}_{1ijk\ell}\bar{e}_{k\ell}$ in terms of the overall average strain in the representative volume then the tensor L of overall elastic moduli is expressed in turn by the connection

$$L = L_2 + c_1(L_1 - L_2)\bar{A}_1 \tag{14}$$

which is given by Hill (17) in this notation that suppresses the tensor suffixes in statements about only fourth-rank tensors. Inclusions that have a vanishing volume, as cracks for instance (34), can be allowed for by a limiting procedure which identifies the number density as the relevant parameter in place of the volume concentration. In many examples of great practical interest the inclusions are packed densely at a high concentration and their elastic moduli differ markedly from those of the matrix. Recent developments encourage the hope that they will succumb soon to accurate calculations. For, the analogous property of overall thermal or electrical conductivity, of random or regular arrays of spherical inclusions in an isotropic matrix, has yielded to calculations which are absolutely accurate in the limit of dense packing and of highly conducting inclusions, and which maintain a good accuracy away from this limit (35,36). Hitherto however more attention has been given to the other extreme where the inclusions have a much more moderate concentration and where, if they all have a given ellipsoidal shape, the analysis of Eshelby (1,3) for a single inclusion, and for many well separated inclusions, can be prevailed upon to furnish the ingredients for Eqn (14). Naturally as the exact distribution of the inclusions is not by any means prescribed, each overall modulus can be confined still only between upper and lower bounds. But in case the relevant bounds cannot be brought very close together, a well conceived and well tested approximate calculation may deserve consideration.

A particular, 'self-consistent', procedure has been introduced by Hill (37) and Budiansky (38) and has proved to be popular. It allows the equivalent medium (of stiffness tensor L) to extend right up to an inclusion, in place of the intervening remainder of the truly representative volume, in order to evaluate then the average strain tensor within the inclusion as $A_{1ijk\ell}e^{A}_{k\ell}$ say, for substitution in Eqn (1) or in the succeeding Eqn (14). According to Eshelby (1), an ellipsoidal inclusion would suffer then a uniform strain which can be calculated from the fourth-rank tensor S, which he introduces firstly for a preliminary 'transformation strain' of the inclusion, and evaluates explicitly for a general ellipsoid in an isotropic matrix. The connection is expressed by Hill (37) as

$$A_1 = [I + P(L_1 - L)]^{-1} , \qquad P = SM ,$$

where I is the appropriate unit tensor, where the superscript -1 denotes a tensor inversion, and where the tensor P offers to simplify the calculations by having the diagonal symmetry ($P_{ijk\ell} = P_{k\ell ij}$) not generally found in S. The strain field of each inclusion is evaluated in the same approximate manner for another, 'differential' self-consistent, procedure which has been introduced to the present context by Roscoe (39) and before to an analogous one by Bruggeman (40). However in place of Eqn (14), it employs the approximate differential relation

$$dL/dc_1 = (L_1 - L)\bar{A}_1/(1 - c_1) \tag{15}$$

where L is to coincide with L_2 when c_1 is zero.

When the inclusions are all indistinguishable in their shape and orientation, $\bar{A}_1$ is identified with A_1, to replace Eqn (14) by a number of algebraic equations, or Eqn (15) by differential ones, in order to solve uniquely and realistically (it is to be hoped) for all the components of L. When L is anisotropic, for instance, when the inclusions are aligned as short fibres (41,42) or cracks (34), the calculation of A_1 numerically for such an anisotropic matrix is demanded. Its calculation instead for the same (possibly isotropic) matrix material as that of the second phase is all that is required if c_1, the total concentration by volume of the inclusions, is small compared to unity, to leave Eqn (14) then as an explicit expression for L. For such dilute concentrations of aligned prolate spheroidal and other 'slender' inclusions, Russel (43) has given a thorough analysis particularly of the effect of varying the aspect (diameter/length) ratio. Moreover accurate calculations to the next order c_1^2 are available analytically when spherical (44) or aligned ellipsoidal (45) inclusions are regularly and widely spaced, and numerically (to adjust only the coefficient of c_1^2) when spherical inclusions are dispersed randomly (45,46).

The inclusions are presumed henceforth to be all of the same (not necessarily ellipsoidal) shape, variously oriented at random in an isotropic matrix of bulk modulus $\varkappa_2$ and shear modulus μ_2, and to be all made of an isotropic material of bulk modulus $\varkappa_1$ and shear modulus μ_1 or in particular to be all completely rigid or empty. Eqn (2) is then the appropriate overall relation by which the tensor L is given its isotropic expression

$$L_{ijk\ell} = \varkappa\delta_{ij}\delta_{k\ell} + \mu(\delta_{ik}\delta_{j\ell} + \delta_{i\ell}\delta_{jk} - 2\delta_{ij}\delta_{k\ell}/3),$$

which will be compressed to the briefer symbolic one

$$L = (3\varkappa, 2\mu) \tag{16}$$

of Hill (7) for each subsequent fourth-rank isotropic tensor in order to indicate similarly the two coefficients assigned to the elementary tensors formed from the Kronecker deltas. The numerical factors in Eqn (16) allow the neatest calculations of tensor products and inverses. The components of A_1 (and of P and S) vary now with orientation from one inclusion to another. For any such tensor X we define, by enclosure within curly brackets, an isotropic tensor

$$\{X\} = (X_{iijj}/3, \; X_{ijij}/5 - X_{iijj}/15)$$

derived from invariable combinations of the components. It is appropriate now to identify $\bar{A}_1$ with $\{A_1\}$ so that Eqn (14) is broken down as a pair of algebraic equations, and Eqn (15) as differential ones, to be solved simultaneously for the overall bulk and shear moduli, $\varkappa$ and μ.

In particular, when the preliminary calculations are taken far enough for ellipsoidal inclusions to determine that

$$\{P^{-1}\} = (3\varkappa', 2\mu'), \; \{(M - P)^{-1}\} = (3\varkappa'', 2\mu'') \tag{17}$$

say, we arrive then at the two pairs

$$\varkappa = \varkappa_2 + c_1\varkappa', \qquad \mu = \mu_2 + c_1\mu', \tag{18}$$

$$d\varkappa/dc_1 = \varkappa'/(1 - c_1), \qquad d\mu/dc_1 = \mu'/(1 - c_1), \tag{19}$$

respectively for rigid inclusions (for which M_1 vanishes) and the two pairs

$$\varkappa = \varkappa_2 - c_1\varkappa_2\varkappa''/\varkappa, \qquad \mu = \mu_2 - c_1\mu_2\mu''/\mu, \tag{20}$$

$$d\varkappa/dc_1 = -\varkappa''/(1 - c_1), \qquad d\mu/dc_1 = -\mu''/(1 - c_1), \tag{21}$$

respectively for empty inclusions (for which L_1 vanishes). All these pairs of equations are accurate when c_1, the total volume concentration of the inclusions, is sufficiently small compared to unity, and they can then be made into explicit formulae by replacing

κ and μ wherever they occur on the right-hand sides by κ_2 and μ_2 respectively. However, as illustrated by fuller analyses (37, 38) for spherical inclusions, the self-consistent Eqns (18) and Eqns (20) bring the overall moduli prematurely to infinite and zero values respectively well before a maximum c_1 is reached. There are indications (39, 47, 48) that the differential Eqns (19) and (21) do not suffer from such exaggerated effects, but instead give the extreme values to the moduli only when c_1 reaches its absolute maximum value of unity. It may be noted further that if the inclusion assumes a degenerate shape of vanishing volume, the primed moduli of Eqns (17) may tend to infinite values which however can be merged with the vanishing of c_1 to leave finite overall moduli in the end. For instance, if the original shape is that of a prolate or oblate spheroid, having two axes of length 2a and the third larger or smaller axis of length 2b, to make up the volume $4\pi a^2 b/3$, it is appropriate to express c_1 as

$$c_1 = 4\pi\varepsilon^2 b^3 n_1/3, \qquad c_1 = 4\pi\eta a^3 n_1/3,$$

respectively, where $\varepsilon = a/b$, $\eta = b/a$, and where n_1 is the number of inclusions per unit volume, in order to prepare for the extremely prolate limit of vanishing ε and the extremely oblate limit of vanishing η. We may calculate for the former ('needle') limit on retaining only the leading terms that

$$9\kappa' = \mu/\varepsilon^2[\ell n(2/\varepsilon) - 3(\kappa+\mu)/2(\kappa+4\mu/3)] = 15\mu',$$

$$\kappa'' = \kappa(\kappa+\mu)/\mu, \qquad \mu'' = 28\mu(\kappa+16\mu/21)/15(\kappa+\mu/3),$$

and similarly that

$$\kappa' = 16\mu(\kappa+4\mu/3)/9\pi\eta(\kappa+7\mu/3) = 4\mu'(\kappa+5\mu/3)/3(\kappa+11\mu/5),$$

$$\kappa'' = \kappa^2(\kappa+4\mu/3)/\mu\pi\eta(\kappa+\mu/3),$$

$$\mu'' = 4\mu(\kappa+4\mu/3)(\kappa+4\mu/9)/5\pi\eta(\kappa+\mu/3)(\kappa+2\mu/3),$$

for the other limit in which the shape becomes a circular disc, either rigid or empty (a "crack") respectively. In particular, for this last case of circular cracks, we can express the Eqns (20) as

$$\kappa/\kappa_2 = 1 - 16n_1a^3(1-\nu^2)/9(1-2\nu),$$
$$\mu/\mu_2 = 1 - 32n_1a^3(1-\nu)(5-\nu)/45(2-\nu), \tag{22}$$

in terms of $\nu\,[= (3\kappa - 2\mu)/2(3\kappa + \mu)]$, the overall Poisson's ratio, in order to identify them as those obtained (from self-consistent energy arguments) by O'Connell & Budiansky (49, 50), who proceed to a detailed application to cracked solids, not however without attracting some criticism (48, 51, 52, 53, 54). Bruner (48) prefers to employ instead what we may recognise as the differential Eqns (21). If ν is replaced by ν_2 (the Poisson's ratio of the matrix), Eqns (22) are then the explicit ones for $\varkappa$ and μ given by Bristow (55) when $n_1 a^3$ is very much smaller than unity, except that the factor $(2-\nu_2)$ is missing from his expression for μ.

More generally when the inclusions have the arbitrary shape and the (finite and positive) bulk modulus $\varkappa_1$ and shear modulus μ_1, we (22) employ the expressions

$$A_1 = I + P_1(L - L_1), \quad \{ P_1 \} = (p_1/3, q_1/2),$$

say, in terms of a diagonally symmetric tensor P_1, in order to extract from the 'algebraic' Eqns (14) the pair

$$\begin{aligned} \varkappa &= c_1\varkappa_1 + c_2\varkappa_2 - c_1c_2(\varkappa_1-\varkappa_2)^2/(c_1\varkappa_2 + c_2\varkappa_1 - \varkappa_1 + 1/p_1), \\ \mu &= c_1\mu_1 + c_2\mu_2 - c_1c_2(\mu_1-\mu_2)^2/(c_1\mu_2 + c_2\mu_1 - \mu_1 + 1/q_1), \end{aligned} \tag{23}$$

and from the 'differential' Eqns (15) the pair

$$\begin{aligned} d\varkappa/dc_1 &= (\varkappa_1 - \varkappa)[1 + p_1(\varkappa - \varkappa_1)]/(1 - c_1), \\ d\mu/dc_1 &= (\mu_1 - \mu)[1 + q_1(\mu - \mu_1)]/(1 - c_1), \end{aligned} \tag{24}$$

for which $\varkappa$ and μ are to coincide with $\varkappa_2$ and μ_2 respectively when c_1 vanishes. For the ellipsoidal shape there is the immediate connection

$$P_1 = (P^{-1} - L + L_1)^{-1}$$

with P. In particular, it is found for a spherical shape that

$$p_1 = 1/(\varkappa_1 + 4\mu/3), \quad q_1 = 1/(\mu_1 + \mu^*), \tag{25}$$

where μ^* is defined by Eqn (6) with the subscript o entirely omitted, to make Eqns (23) equivalent to those analysed further by Hill (37) and Budiansky (38), and Eqns (24) to those of Roscoe (39), Boucher (56) and McLaughlin (57). Hill (37) showed that there were unique, positive solutions for $\varkappa$ and μ, which moreover fell within the Hashin-Shtrikman bounds of our Eqns (4, 5), while McLaughlin (57) pointed to a similar outcome for the other 'differential' case. We shall not refer here to the rather lengthy expressions needed for a general ellipsoidal shape with semi-axes a, b and c. It will suffice to record the limit

$$p_1 = 1/(\varkappa_1 + \mu_1/3 + \mu),$$
$$5q_1 = 1/3(\varkappa_1 + \mu_1/3 + \mu) + 2/(\mu_1 + \mu) + 2/[\mu_1 + \mu(\varkappa + \mu/3)/(\varkappa + 7\mu/3)], \tag{26}$$

for the extremely prolate spheroid (for $a/c = b/c \to 0$: a 'needle') and the limit

$$p_1 = 1/(\varkappa_1 + 4\mu_1/3), \quad q_1 = 1/(\mu_1 + \mu_1^*), \tag{27}$$

for the extremely oblate spheroid (for $c/a = c/b \to 0$: a 'disc'), where μ_1^* is defined by Eqn (6) with the subscript o replaced entirely by 1, to leave then no dependence on $\varkappa$ and μ. For each of these substitutions, the Eqns (23) are equivalent to those of Wu (58), Walpole (22) and Boucher (56, 59) and the Eqns (24) to those of Boucher (56). It should be emphasised though that accuracy at (or nearly at) the limit of rigid or empty inclusions can no longer be presumed, unless we restore the next order of omitted terms in Eqns (26, 27), in order to reconcile any conflict between the two extreme limits (of shape and elastic moduli). When the Eqns (27) are employed for the disc shape, it is found (56) that Eqns (24) yield on integration the same explicit expressions as Eqns (23) do for $\varkappa$ and μ. Moreover that for $\varkappa$ coincides with either the upper or lower bound provided by Eqn (4), according as $\mu_1 > $ or $< \mu_2$, while that for μ coincides correspondingly with the upper or lower bound of Eqn (5) when $(\varkappa_1 - \varkappa_2)(\mu_1 - \mu_2) > 0$. More generally when the inclusions have any given shape, it can still be shown implicitly, as in an analogous context (47), that the 'algebraic' and 'differential' self-consistent methods emerge closely related algebraically to each other and to the Hashin-Shtrikman bounds, in an unanticipated manner that must await some future elucidation. Thus, the 'algebraic' calculation of $\varkappa$ (or of μ) is always larger or smaller than the 'differential' one according as $\varkappa_1$ (or μ_1) is larger than $\varkappa_2$ (or μ_2). Moreover each such calculation always falls faithfully somewhere between the relevant upper and lower bounds. We might speculate that it does so either because it is in fact an exact calculation for a particular unidentified species of composite, or, more significantly, because it is a general bound which is valid for the given shape of inclusions and which is better (or, for the aforementioned 'disc' shape, at least no worse) than one or other of the Hashin-Shtrikman upper and lower bounds. In the latter event furthermore, to decide whether the new bound is an upper or lower one, we could observe the effect of modifying the self-consistent calculation by leaving behind an infinitesimally thin layer of the second (matrix) phase between the inclusion and the matrix of equivalent material, provided it is supposed then that the calculation would be always improved. When the inclusions have a spherical shape, it is feasible to proceed albeit very laboriously (60, 61, 62) in terms of a finite spherical layer, only to find however that its thickness is liable to remain as an indeterminate parameter (63). Here we need only indulge in the very much simpler analysis for an infinitesimal thickness in order to predict that the self-consistent calculations of $\varkappa$ and μ, by Eqns (23, 25), are actually upper bounds on the overall bulk and shear moduli respectively when $\varkappa_1 > \varkappa_2$ and $\mu_1 > \mu_2$ (inclusions 'stronger' than matrix), and lower bounds when both these inequalities are reversed. In order to exemplify the foregoing remarks in an extreme case where the calculations can be easily completed, suppose that both phases are incompressible and that the first phase, of spherical inclusions, is completely rigid. Then the Hashin-Shtrikman bounds of Eqns (4, 5) give an infinite value to the overall bulk modulus and leave the overall shear modulus μ subjected to the inequalities

$$\mu_2(1 + 5c_1/2c_2) < \mu < \infty \, .$$

On the other hand, the self-consistent Eqns (23, 25) calculate that

$$\begin{aligned} \mu &= \mu_2/(1 - 5c_1/2) \quad \text{for } 0 \le c_1 \le 2/5 \, , \\ &= \infty \quad \text{for } 2/5 \le c_1 \le 1 \, , \end{aligned} \tag{28}$$

to comply therefore with the inequalities, while however reaching the infinite value when the inclusion concentration is only two-fifths. At all the lower concentrations, our present (not conclusively proven) understanding does offer now to improve the second, trivial, inequality, namely by converting Eqn (28) to an inequality, instead of the precise equality. At the lowest concentrations, the inequalities may be expanded then as

$$1 + 5c_1/2 + 5c_1^2/2 + \ldots < \mu/\mu_2 < 1 + 5c_1/2 + 25c_1^2/4 + \ldots \, ,$$

to correctly give the Einstein value 5/2 to the coefficient of c_1. The coefficient of c_1^2 is given its least value of 5/2 when the spherical inclusions are 'well separated' (44, 45, 46, 64) whereas when they are 'well stirred' its value is put at 5.2 ($\pm$ 0.3) by Batchelor and Green (64), 5.01 by Chen & Acrivos (46) and approximately 155/32 by Willis & Acton (45).

REFERENCES

(1) Eshelby, J. D. (1957). The determination of the elastic field of an ellipsoidal inclusion, and related problems. Proc. R. Soc. Lond. A 241, 376-396.

(2) Eshelby, J. D. (1959). The elastic field outside an ellipsoidal inclusion. Proc. R. Soc. Lond. A 252, 561-569.

(3) Eshelby, J. D. (1961). Elastic inclusions and inhomogeneities. In: Progress in Solid Mechanics, eds. I. N. Sneddon & R. Hill, Vol. II, pp. 87-140. Amsterdam: North-Holland Publ.

(4) Brown, W. F. (1955). Solid mixture permittivities. J. Chem. Phys. 8, 1514-1517.

(5) Brown, W. F. (1965). Dielectric constants, permeabilities, and conductivities of random media. Trans. Soc. Rheol. 9, 357-380.
(6) Walpole, L. J. (1981). Elastic behaviour of composite materials: theoretical foundations. In: Advances in Applied Mechanics, ed. C. S. Yih, Vol. 21, pp. 169-242. New York: Academic Press.
(7) Hill, R. (1965). Continuum micro-mechanics of elastoplastic polycrystals. J. Mech. Phys. Solids 13, 89-101.
(8) Kroner, E. et al. (1966). On the bounds of the shear modulus of macroscopically isotropic aggregates of cubic crystals. J. Mech. Phys. Solids 14, 21-24.
(9) Walpole, L. J. (1966). On bounds for the overall elastic moduli of inhomogeneous systems. II. J. Mech. Phys. Solids 14, 289-301.
(10) Hashin, Z. & Shtrikman, S. (1962). A variational approach to the theory of the elastic behaviour of polycrystals. J. Mech. Phys. Solids 10, 343-352.
(11) Watt, P. J. & Peselnick, L. (1980). Clarification of the Hashin-Shtrikman bounds on the effective elastic moduli of polycrystals with hexagonal, trigonal and tetragonal symmetries. J. Appl. Phys. 51, 1525-1531.
(12) Watt, P. J. (1979). Hashin-Shtrikman bounds on the effective elastic moduli of polycrystals with orthorhombic symmetry. J. Appl. Phys. 50, 6290-6295.
(13) Watt, P. J. (1980). Hashin-Shtrikman bounds on the effective elastic moduli of polycrystals with monoclinic symmetry. J. Appl. Phys. 51, 1520-1524.
(14) Hill, R. (1964). Theory of mechanical properties of fibre-strengthened materials: I. Elastic behaviour. J. Mech. Phys. Solids 12, 199-212.
(15) Nabarro, F. R. N. (1940). The strains produced by precipitation in alloys. Proc. R. Soc. Lond. A 175, 519-538.
(16) Robinson, K. (1951). Elastic energy of an ellipsoidal inclusion in an infinite solid. J. Appl. Phys. 22, 1045-1054.
(17) Hill, R. (1963). Elastic properties of reinforced solids: some theoretical principles. J. Mech. Phys. Solids 11, 357-372.
(18) Hashin, Z. & Shtrikman, S. (1963). A variational approach to the theory of the elastic behaviour of multiphase materials. J. Mech. Phys. Solids 11, 127-140.
(19) Walpole, L. J. (1966). On bounds for the overall elastic moduli of inhomgeneous systems. I. J. Mech. Phys. Solids 14, 151-162.
(20) Milton, G. W. & Phan-Thien, N. (1982). New bounds on effective elastic moduli of two-component materials. Proc. R. Soc. Lond. A 380, 305-331.
(21) Watt, P. J. & O'Connell, R. J. (1980). An experimental investigation of the Hashin-Shtrikman bounds on two-phase aggregate elastic properties. Phys. Earth Planet. Inter. 21, 359-370.
(22) Walpole, L. J. (1969). On the overall elastic moduli of composite materials. J. Mech. Phys. Solids 17, 235-251.
(23) Postma, G. W. (1955). Wave propagation in a stratified medium. Geophysics 20, 780-806.

(24) Hashin, Z. (1965). On elastic behaviour of fibre reinforced materials of arbitrary transverse phase geometry. J. Mech. Phys. Solids 13, 119-134.
(25) Hashin, Z. (1979). Analysis of properties of fiber composites with anisotropic constituents. J. Appl. Mech. 46, 543-550.
(26) Hashin, Z. (1983). Analysis of composite materials: a survey. J. Appl. Mech. 50, 481-505.
(27) Kelly, A. (1973). Strong solids, 2nd ed. Oxford: Clarendon Press.
(28) Keller, J. B. (1964). A theorem on the conductivity of a composite medium. J. Math. Phys. 5, 548-549.
(29) Mendelson, K. S. (1975). Effective conductivity of two-phase material with cylindrical phase boundaries. J. Appl. Phys. 46, 917-918.
(30) Dykhne, A. M. (1971). Conductivity of a two-dimensional two-phase system. Sov. Phys. JETP 32, 63-65.
(31) Lees, C. H. (1900). On the conductivities of certain heterogeneous media for a steady flux having a potential. Phil. Mag. 49, 221-226.
(32) Perrins, W. T. et al. (1979). Transport properties of regular arrays of cylinders. Proc. R. Soc. Lond. A 369, 207-225.
(33) Perrins, W. T. et al. (1979). Optical properties of dense regular cermets with relevance to selective solar absorbers. Thin Solid Films 57, 321-326.
(34) Hoenig, A. (1979). Elastic moduli of a non-randomly cracked body. Int. J. Solids Struct. 15, 137-154.
(35) Batchelor, G. K. & O'Brien, R. W. (1977). Thermal or electrical conduction through a granular material. Proc. R. Soc. Lond. A 355, 313-333.
(36) Sangani, A. S. & Acrivos, A. (1983). The effective conductivity of a periodic array of spheres. Proc. R. Soc. Lond. A 386, 263-275.
(37) Hill, R. (1965). A self-consistent mechanics of composite materials. J. Mech. Phys. Solids 13, 213-222.
(38) Budiansky, B. (1965). On the elastic moduli of some heterogeneous materials. J. Mech. Phys. Solids 13, 223-227.
(39) Roscoe, R. (1973). Isotropic composites with elastic or viscoelastic phases: general bounds for the moduli and solutions for special geometries. Rheol. Acta 12, 404-411.
(40) Bruggeman, D. A. G. (1935). Berechnung verschiedener physikalischer Konstanten von heterogenen Substanzen. I. Ann. Phys. (Leipzig) [5] 24, 636-664.
(41) Laws, N. & McLaughlin, R. (1979). The effect of fibrelength on the overall moduli of composite materials. J. Mech. Phys. Solids 27, 1-13.
(42) Chou, T. W. et al. (1980). A self-consistent approach to the elastic stiffness of short-fiber composites. J. Compos. Mater. 14, 178-188.
(43) Russel, W. B. (1973). On the effective moduli of composite materials: effect of fiber length and geometry at dilute concentrations. Z. Angew. Math. Phys. 24, 581-600.

(44) Walpole, L. J. (1972). The elastic behaviour of a suspension of spherical particles. Q. J. Mech. Appl. Math. 25, 153-160.
(45) Willis, J. R. & Acton, J. R. (1976). The overall elastic moduli of a dilute suspension of spheres. Q.J.Mech. Appl. Math. 29, 163-177.
(46) Chen, H. S. & Acrivos, A. (1978). The effective elastic moduli of composite materials containing spherical inclusions at non-dilute concentrations. Int. J. Solids Struct. 14, 349-364.
(47) Walpole, L. J. (1980). Evaluation of overall reaction of a composite solid with inclusions of any shape. In: Continuum Models of Discrete Systems (3), eds. E. Kroner & K. H. Anthony, pp. 455-464. Waterloo, Canada: University of Waterloo Press.
(48) Bruner, W. M. (1976). Comment. J. Geophys. Res. 81, 2573-2576.
(49) O'Connell, R. J. & Budiansky, B. (1974). Seismic velocities in dry and saturated cracked solids. J. Geophys. Res. 79, 5412-5426.
(50) Budiansky, B. & O'Connell, R. J. (1976). Elastic moduli of a cracked solid. Int. J. Solids Struct. 12, 81-97.
(51) O'Connell,R.J. & Budiansky,B. (1976).Reply. J.Geophys.Res.81, 2577-2578.
(52) Chatterjee, A.K. et al. (1978). Elastic moduli of two-component systems. J. Geophys. Res. 83B, 1785-1792.
(53) Budiansky, B. & O'Connell, R.J. (1979). Comment. J.Geophys. Res. 84B, 5687-5688.
(54) Chatterjee, A.K. et al. (1979). Reply. J.Geophys.Res. 84B, 5689-5690.
(55) Bristow, J.R. (1960). Microcracks, and the static and dynamic elastic constants of annealed and heavily cold-worked metals. Brit. J. Appl. Phys. 11, 81-85.
(56) Boucher, S. (1975). Modules effectifs de materiaux composites quasi homogenes et quasi isotropes constitues d'une matrice elastique et d'inclusions elastiques. II. Rev. M. 22, No. 1.
(57) McLaughlin, R. (1977). A study of the differential scheme for composite materials. Int. J. Engng Sci. 15, 237-244.
(58) Wu, T. T. (1966). The effect of inclusion shape on the elastic moduli of a two-phase material. Int. J. Solids Struct. 3, 1-8.
(59) Boucher, S. (1974). On the effective moduli of isotropic two-phase elastic composites. J. Compos. Mater. 8, 82-89.
(60) Smith, J. C. (1974). Correction and extension of van der Poel's method for calculating the shear modulus of a particulate composite. J. Res. Natl. Bur. Stand. A 78, 355-361.
(61) Smith, J. C. (1975). Simplification of van der Poel's formula for the shear modulus of a particulate composite. J. Res. Natl. Bur. Stand. A 79, 419-423.
(62) Christensen, R. M. & Lo, K. H. (1979). Solutions for effective shear properties in three phase sphere and cylinder models. J. Mech. Phys. Solids 27, 315-330.
(63) Hashin, Z. (1970). Theory of Composite Materials. In: Mechanics of Composite Materials, eds. F.W. Wendt et al., pp. 201-242. Oxford: Pergamon Press.
(64) Batchelor, G. K. & Green, J. T. (1972). The determination of the bulk stress in a suspension of spherical particles to order c^2. J. Fluid Mech. 56, 401-427.

THEORETICAL ASPECTS OF REINFORCEMENT AND TOUGHENING

L.R.F. ROSE

Aeronautical Research Laboratories, Melbourne, Australia

ABSTRACT

It is shown that the reduction in stress intensity factor due to a bonded reinforcing patch can be estimated analytically. It is also shown that the two approaches which can be used to explain the increased fracture toughness due to stress-induced phase transformations do not in general lead to equivalent measures of toughness. An expression for the crack extension force is derived which distinguishes the contribution due to the external load from that due to the change in the configuration of sources of internal stress. This distinction has important implications for the interpretation of experimental measurements.

1. INTRODUCTION

Starting from the premise that a crack will grow when the stress intensity factor K exceeds the fracture toughness K_C, or when the crack extension force $\mathcal{G}$ exceeds the incremental work of fracture R, one can attempt to prevent fracture by either reducing K, i.e. by reinforcement, or by increasing K_C, i.e. by toughening. We shall consider here an example of each, for mode I fracture, with a special emphasis on those aspects where Eshelby's work (1,2) has proved most useful. These include: the stress in and around an elliptical inhomogeneity (Section 2); the interaction energy associated with a transformed inclusion (Section 3); the configurational force on certain two-dimensional (2D) elastic singularities (Section 4).

2. CRACK PATCHING

Consider a centre-cracked plate which, after being repaired, will be subjected to a remote biaxial stress. The crack is taken to lie along the line segment ($|x| \le a$, $y = 0$), at right angles to the principal normal stress σ_{yy}^{∞} ($\ge \sigma_{xx}^{\infty}$). The repair consists in bonding to one face of the plate a reinforcing patch which covers an elliptical region $D = \{(x,y): x^2/A^2 + y^2/B^2 \le 1\}$. It will be shown that an explicit estimate can be derived for the stress intensity factor after the repair.

The analysis proceeds in two stages. First we shall consider the redistribution of stress which would be caused by the reinforcement if it were bonded to an *uncracked plate*. The quantity of interest is the normal stress σ_o in the plate, along the line segment ($|x| \le a$, $y = 0$). At the second stage, we shall make a cut in the plate along that line segment and allow the stress σ_o to relax to zero. The main problem will be to determine the resulting stress intensity factor. We shall ignore here the effects of out-of-plane bending due to one-sided reinforcement. Strictly speaking, therefore, the analysis is more appropriate to the case of two-sided reinforcement.

The value of dividing the analysis into these two stages is that different simplifying assumptions are appropriate for each stage. Thus, for stage I it is assumed that the bond does not allow any relative displacement between the plate and the reinforcement. This rigid-bond assumption is appropriate when the actual width of the load-transfer zone around the boundary of the reinforced region D is small compared with the in-plane dimensions (A, B) of the reinforcement. The width of the load-transfer zone can be estimated from the classical (one-dimensional) theory of bonded joints, which shows that the shear stress in the adhesive layer decays exponentially from the ends of a joint. That decay length is given by β^{-1},

$$\beta^2 = (G_A/t_A)\,\{(E_P\,t_P)^{-1} + (E_R\,t_R)^{-1}\}, \tag{1}$$

where G, E, t denote shear modulus, Young's modulus and thickness, and subscripts A, P, R serve to identify parameters pertaining to the adhesive layer, the plate and the reinforcement respectively. Thus the rigid-bond assumption is appropriate if $\beta^{-1} \ll A, B$. The reinforced region can then be treated as an inclusion of higher stiffness than the surrounding plate, and explicit formulae can be obtained for σ_o by proceeding in the following three steps.

(i) Determine the elastic constants of the equivalent inclusion in terms of those of the plate and the reinforcing patch.

(ii) Determine the stress in the equivalent inclusion.

(iii) Determine how the load which is transmitted through the inclusion is shared between the plate and reinforcement.

Step (ii) can be carried out analytically only for certain simple shapes, which is why we have chosen the reinforcing patch to be elliptical. For that shape, the stress is uniform within the equivalent inclusion, as could be anticipated from Eshelby's result (2) for the ellipsoid. This greatly simplifies the strategy for deriving an explicit solution (3). For an isotropic plate and reinforcement we find

$$\sigma_o = \sigma_{yy}^{\infty} (1 + q)/(1 + S), \tag{2}$$

$$q = (S/D)\ [(1 + S)\ \{(2B/A + 1)\ (1 - \nu\Sigma) + \Sigma - \nu\} - S(1 - \nu^2)],$$

$$D = 3(1 + S)^2 + 2(1 + S)\ (B/A + A/B + 2\nu) + 1 - \nu^2 S^2.$$

Thus σ_o is proportional to the principal applied stress σ_{yy}^{∞} and it depends in addition on the following three parameters: (i) the stiffness ratio $S = E_R t_R/E_P t_P$, (ii) the aspect ratio A/B, (iii) the applied-stress ratio $\Sigma = \sigma_{xx}^{\infty}/\sigma_{yy}^{\infty}$. The parameters characterizing the adhesive layer do not affect σ_o, but we recall that the idealization used to derive Eqn (2) relies on $\beta^{-1} << A, B$, and β^{-1} is of course dependent on the adhesive parameters.

The inclusion analogy also gives, as a natural by-product, the stress in the plate outside the reinforced region D. The stress at the point x = 0, y = B+ is of special interest, as it has been observed experimentally that cracking can initiate at that point. We find

$$\sigma_{yy}(0, B+) = (1 + S)\ \sigma_o. \tag{3}$$

For the second stage of the analysis, the assumption of a rigid bond must be discarded, as it would imply that the crack cannot open. It is now important to model more accurately the transfer of load through the adhesive layer in the immediate vicinity of the crack. But, if $\beta^{-1} << A, B$, the restraint on the crack opening will not depend sensitively on the precise shape or extent of the reinforcement, so that both the plate and the reinforcement can now be assumed to be of infinite extent. The crucial observation is that, for this composite structure, K does not increase indefinitely in proportion with $\sqrt{a}$, as it would for a single plate. Rather, K approaches asymptotically a limiting value, denoted here by K_∞, which is the value of the stress intensity factor for a semi-infinite crack. It can be determined by deriving first the corresponding crack extension force G_∞ as follows.

Consider a semi-infinite crack along x < 0, y = 0 in an infinite plate bonded to an infinite uncracked reinforcement, under the action of a uniform stress. If the semi-infinite crack extends by a distance δa, the stress and displacement fields are simply shifted to the right by δa. The change in energy δW (potential energy of loading mechanism plus strain energy) is the same as that involved in converting a strip of width δa from the state at $x \to \infty$ to that at $x \to -\infty$.

The crack extension force $\mathcal{G}_\infty$, by definition, is the ratio $\delta W/(t_P \delta a)$. To determine $\mathcal{G}_\infty$ therefore we only need to calculate δW. The fundamental simplification which has been achieved is that the configuration at $x \to -\infty$ is that of a simple overlap joint (4). Using the classical (one dimensional) theory of bonded joints one can derive an explicit formula for $\mathcal{G}_\infty$, and hence an upper bound K_∞ for K,

$$K_\infty = (E_P \mathcal{G}_\infty)^{\frac{1}{2}} = \sigma_o(\pi\Lambda)^{\frac{1}{2}}, \tag{4}$$

$$(\pi\Lambda)^2 = (E_P t_P t_A/G_A)\ (1 + E_P t_P/E_R t_R). \tag{5}$$

Another upper bound to K is given by

$$K_o = \sigma_o(\pi a)^{\frac{1}{2}}, \tag{6}$$

which is a more useful estimate when $(a/\Lambda) < 1$, while K_∞ is more useful for $(a/\Lambda) > 1$. Both estimates are conservative, which is a desirable feature for design purposes. The reduction of the analysis to that of an overlap joint also leads naturally to upper bounds for the maximum stress in the reinforcement and the adhesive layer (3,4).

3. TRANSFORMATION TOUGHENING

The increased fracture toughness due to stress-induced transformations can be explained from two distinct viewpoints. First, it can be argued that the net work of fracture is increased because a work of transformation must be supplied, in addition to the work of fracture for crack growth in the matrix. Secondly, the transformed inclusions can be regarded as sources of internal stress which alter the external load required for crack growth. This second viewpoint suggests that the toughness could be increased or decreased, depending on the particular distribution of internal stress, whereas the first suggests there should always be an increase in toughness. The aim of this section is to examine the relationship between these two viewpoints.

Consider a process of quasi-static crack growth in a material susceptible to transformation toughening, using a specimen of uniform thickness which is initially free from internal stresses. Suppose that (i) when the load is applied, inclusions around the crack tip begin to transform first, before the crack begins to grow; (ii) as the load is further increased, the crack grows in a stable manner, accompanied by further transformation; (iii) transformed inclusions do not revert during the course of further crack growth or on unloading, (iv) all deformations are accommodated elastically, matrix and inclusions having the same (isotropic) elastic constants; (v) the inclusions are distributed uniformly throughout the matrix.

It will be of fundamental importance for the present analysis to view the stress field σ as the superposition of (i) the nominal stress field σ^N which would prevail if no inclusions had transformed, and (ii) the perturbation σ^Z due to the zone of transformed inclusions. These distinguishing superscripts will also be used for all other field qualities for example,

$$K = K^N + K^Z, \tag{7}$$

where K denotes the conventional mode I stress intensity factor. Whereas the stress fields σ^N, σ^Z can be simply added to give the total stress, the total mechanical energy for the configuration contains an interaction energy E^{NZ}, corresponding to the change in potential energy of the external load due to the transformation of inclusions:

$$E = E^N + E^Z + E^{NZ}, \tag{8}$$

where E^N denotes the sum of the potential energy of the external load and the elastic energy of the specimen in the absence of transformed inclusions. E^Z denotes the elastic energy associated with the perturbation field,

$$E^Z = \int_V \tfrac{1}{2} \sigma^Z \varepsilon^Z \, dV,$$

with V denoting the volume of the specimen, ε the strain tensor and $\sigma \varepsilon$ the tensor product $\sigma_{ij} \varepsilon_{ij}$. The interaction energy E^{NZ} can be expressed as follows (2),

$$E^{NZ} = -\int_{V'} \sigma^N \varepsilon^T \, dV,$$

V' denoting the volume of all the transformed inclusions, and ε^T the stress-free transformation strain undergone by the inclusions.

On a 2D view, the transformed inclusions may be taken to lie within a zone of area A, so that a quasistatic crack extension δa is accompanied by a zone growth δA and results in an energy change δE. The conventional definition for G is

$$G = \lim_{\delta a \to 0} - t^{-1}(\delta E/\delta a), \tag{9}$$

$$= G^N + G^Z + G^{NZ}, \tag{10}$$

where t denotes the specimen thickness and Eqn (10) follows from the division of energy used in Eqn (8).

To derive an expression for G it will be convenient to consider first the energy change $\delta_1 E$ due to a hypothetical crack extension δa *without any change in the area* A. During this first stage we may regard the transformed inclusions as constituting a single composite source of internal stress which remains fixed, so that

$$\lim_{\delta a \to 0} - t^{-1}(\delta_1 E/\delta a) = \Gamma (K^N + K^Z)^2, \tag{11}$$

with $\Gamma = (1-\nu)/2\mu$ for plane strain.

Consider next the energy change $\delta_2 E$ due to an increase δA of zone area, with the crack length fixed, assuming that (i) the inclusions within δA transform when the interaction-energy density reaches a critical value $(\sigma \varepsilon^T)_c$; (iii) the average self-energy of the inclusions per unit volume of the transformation zone is a material constant which will be denoted by $\mathcal{E}$. Then we have

$$\delta_2 E = \{\mathcal{E} - V_f (\sigma \varepsilon^T)_c\} \, t \, \delta A + O(\delta A)^2, \tag{12}$$

where the term of order $(\delta A)^2$ is the contribution due to the image field of the inclusions within δA, which will not affect G. From Eqns (9-12) we can now derive

$$G = \Gamma(K^N + K^Z)^2 - \{E - V_f(\sigma\,\varepsilon^T)_c\}\,(\partial A/\partial a). \tag{13}$$

It is important to separate those contributions to G which depend on explicitly on the nominal field, and hence on the external load, from those which do not. To that end we first note while $(\sigma^N + \sigma^Z)\varepsilon^T$ can plausibly be viewed as a material constant, the relative contributions of σ^N and σ^Z will depend on position within the new zone area δA. A detailed stress analysis would be required to determine these relative contributions theoretically. Assuming this has been carried out, we shall use the following notation to distinguish between the averages of the two contributions:

$$(\sigma\,\varepsilon^T)_c = \langle \sigma^N\,\varepsilon^T\rangle + \langle \sigma^Z\,\varepsilon^T\rangle .$$

Then, recalling Eqn (10), we may write

$$\begin{aligned} G_{ext} &= G^N + G^{NZ}, \\ &= \Gamma\{(K^N)^2 + 2K^N K^Z\} + V_f\langle\sigma^N\,\varepsilon^T\rangle(\partial A/\partial a), \end{aligned} \tag{14}$$

The point of making this distinction is that the incremental work of fracture R measured experimentally is in fact the experimental measure of G_{ext} rather than of G. G includes an internal contribution G^Z which does not depend directly on the external load and which is not readily accessible to experimental measurement.

The other experimental measure of toughness is the nominal stress intensity factor during quasistatic crack growth, K_c^N. It is clear from the form of Eqn (14) that the two measures of toughness, $R(= G_{ext})$ and K_c^N, are not equivalent, in the sense that they do not, in general, satisfy the usual relation

$$R = \Gamma(K_c^N)^2 . \tag{15}$$

Rather surprisingly, this relation is not satisfied even for steady-state crack growth, as illustrated by the idealized model considered next.

4. A MODEL FOR TRANSFORMATION TOUGHENING

The two-dimensional (plane strain) point singularity described by the complex potentials

$$\Phi = C/(2\pi z) \,, \qquad \Psi = D/(2\pi z), \tag{16}$$

using conventional notation $(z = x + iy)$, has the following physical interpretations for particular choices of the real constants C, D :

(i) a half-line of 2D centres of dilatation (uniformly distributed along the negative x-axis in an uncracked infinite body) for $C = 0$; (ii) a half-line of 2D shear centres for $C = -D > 0$; (iii) an edge dislocation of Burgers vector $(0,b)$, with the extra half-plane along the negative x-axis, for $C=D=\mu b/2(1-\nu)$; (iv) a point force of strength F, pointing in the direction of the positive x-axis, for $C=-F/4(1-\nu)$, $D=(3-4\nu)F/4(1-\nu)$.

Now consider an idealized transformation zone represented by a continuous distribution of 2D centres of dilatation along two line segments $y = \pm w$, next to a semi-infinite crack along $y=0$, $x < 0$. Suppose that, as the crack grows, the points at the leading end of this zone retain fixed coordinates $z_1, \bar{z}_1$ relative to a coordinate system with the origin attached to the crack tip. The trailing end is fixed relative to the specimen and therefore has variable coordinates $z_2, \bar{z}_2$ relative to that coordinate system. The line segment between z_1 and z_2 can be viewed as the superposition of a half-line of dilatation of strength A with end point at z_1 and another of strength $-A$ with end point at z_2. The other line segment can be similarly represented by two point singularities, one at $\bar{z}_1$ and the other, of opposite sign, at $\bar{z}_2$. If we use Eqn (16), with a shift of origin, to describe these singularities, the same model can be used for transformation toughening due to shear centres with $C=-D > 0$, or, with $C=D > 0$, for dislocation shielding (5,6). The advantage of such models is that the crack tip together with the singularities at $z_1, \bar{z}_1$ can be treated as a single composite singularity, in Eshelby's sense (1). The appropriate crack extension force $\mathcal{G}$ can therefore be calculated from the Eshelby-Rice J-integral, expressed in terms of the complex potentials (7), taken around a contour $\mathcal{C}$ enclosing this composite singularity,

$$J(\mathcal{C}) = 2\Gamma\, \mathcal{I}m \int_{\mathcal{C}} (\Phi^2 + 2\,\Phi\Psi)\,dz. \tag{17}$$

The singularities of the integrand are poles at $z=0$, z_1, $\bar{z}_1$, so that, concentrating on the case of centres of dilatation, and using the assumption of symmetry (mode I) and the notation of Section 3, we can express J as a sum of residues as follows, with $\mathcal{C}_0$, $\mathcal{C}_1$ denoting contours enclosing only the poles at $z=0$ and $z=z_1$ respectively,

$$J(\mathcal{C}) = J(\mathcal{C}_0) + 2\,J(\mathcal{C}_1), \tag{18}$$

$$J(\mathcal{C}_0) = \Gamma(K^N + K^Z)^2, \tag{19}$$

$$J(\mathcal{C}_1) = 4\Gamma A \mathcal{R}e\,[\Phi^N(z_1) + \Phi^Z(z_1)]. \tag{20}$$

The division into two terms in Eqn (18) corresponds to the two stages used to derive an expression for $\mathcal{G}$ in Section 3. $J(\mathcal{C}_0)$ gives the energy release rate if we were to displace the crack tip only, the zone being left unchanged; $2J(\mathcal{C}_1)$ gives the energy release rate when extending the zone with the crack tip fixed. Grouping together those terms which depend explicitly on the nominal field, we may write, by analogy with Eqn (14)

$$J_{ext} = \Gamma\{(K^N)^2 + 2K^N K^Z\} + 8\Gamma A \mathcal{R}e[\Phi^N(z_1)]. \tag{21}$$

Now assume that (i) the specimen can be regarded as an infinite body with a semi-infinite crack for the purpose of calculating the image field due to the transformation zone, (ii) $\Phi^N(z_1)$ can be approximated by $K^N/2(2\pi z)^{\frac{1}{2}}$. We shall use a superscript H to label the perturbation due to the two singularities at the head of the transformation zone, so that

$$K^H = -A(2/\pi r_1)^{\frac{1}{2}} \cos(\theta_1/2), \tag{22}$$

where r_1, θ_1 are fixed and therefore K^H remains constant during crack growth. We shall use a superscript T to label the perturbation due to the two singularities at the trailing end of the zone, so that

$$K^T = A(2/\pi r_2)^{\frac{1}{2}} \cos(\theta_2/2), \tag{23}$$

where K^T is always positive, though it eventually decreases towards zero as the crack grows. The total perturbation is of course the sum of the contributions from these two sets of singularities. We can then readily verify that

$$4A\, \mathcal{R}e\, [\Phi^N(z_1)] = -K^N K^H, \tag{24}$$

so that Eqn (21) reduces to

$$J_{ext} = \Gamma\{(K^N)^2 + 2K^N K^T\}. \tag{25}$$

Thus, since $K^T > 0$, J_{ext} will always exceed $\Gamma(K^N)^2$, though it will approach that value in the steady-state limit as $K^T \to 0$. However, this result depends in an essential manner on the approximation used for $\Phi^N(z_1)$: a more complete representation

$$\Phi^N(z_1) = K^N/2(2\pi z_1)^{\frac{1}{2}} + T/4 + O(z_1^{\frac{1}{2}}), \tag{26}$$

would lead to

$$J_{ext} = \Gamma\{(K^N)^2 + 2K^N K^T + 2AT\}, \tag{27}$$

so that the two measures of toughness discussed in Section 3 would become equivalent only to the extent that 2AT can be neglected compared with $(K^N)^2$.

It is of interest to note that the half line of dilatation can be viewed as the superposition of an edge dislocation and a point force. If we had used a model with edge dislocations at $z_1, \bar{z}_1$, it would be found that the term T/4 in Eqn (26) makes no contribution to J_{ext}, a result which could have been anticipated from (8). Thus the T-stress effect in Eqn (27) could be associated with the presence of the point forces, a result which was known to Eshelby (private communication 1981), though he does not appear to have considered its application to transformation toughening.

I am deeply indebted to Dr A.A. Baker, who initiated an extensive research programme on crack patching (9), for kindling my interest in reinforcement problems, and to Dr M.V. Swain for many informative discussions on transformation toughening in partially stabilized zirconia (10).

REFERENCES

(1) Eshelby, J.D. (1956) The continuum theory of lattice defects. In: Solid State Physics, Vol. 3, ed. F. Seitz & D. Turnbull. New York: Academic Press.

(2) Eshelby, J.D. (1961) Elastic inclusions and inhomogeneities. In: Progress in Solid Mechanics, Vol. 2, ed. by R. Hill & I.N. Sneddon. Amsterdam: North Holland.

(3) Rose, L.R.F. (1981) Int. J. Solids Structures, 17, 827-838.

(4) Rose, L.R.F. (1982) Int. J. Fracture, 18, 135-144.

(5) Thomson, R.M. & Sinclair, J.E. (1982) Acta Met., 30, 1325-1334.

(6) Weertman, J. et al. (1983) Acta Met., 31, 473-482.

(7) Budiansky, B. & Rice J.R. (1973) J. App. Mech., 40, 201-203.

(8) Rice, J.R. (1974) J. Mech. Phys. Solids, 22, 17-26.

(9) Baker, A.A. (1981) Aircraft, 60, no. 12, 30-35.

(10) Hannink, R.H.J. & Swain, M.V. (1982) J. Aust. Ceram. Soc., 18, 53-62.

THE LOSS OF STIFFNESS OF CRACKED LAMINATES

N. LAWS

College of Aeronautics, Cranfield Institute of Technology,
Cranfield, Bedford, MK43 OAL, England.

G.J. DVORAK

Department of Civil Engineering, University of Utah,
Salt Lake City, UT 84112, United States of America.

ABSTRACT

The paper is concerned with the prediction of the reduction in stiffness of composite laminates due to extensive transverse matrix cracks. The required results are obtained via a self-consistent calculation. Specific numerical results are given for some typical cross ply and angle ply laminates.

1. INTRODUCTION

The study of composite laminates which contain extensive matrix crack distributions is a popular activity. The main reason for this interest is related to the experimental and theoretical analysis of damage. Certainly, matrix cracking is a frequently observed damage mechanism in fibrous composite structures. For example, in (0, 90, 0) graphite epoxy laminates which are subjected to monotonic loading, the cracking process usually starts in the 90° ply. The induced cracks extend in the (transverse) fibre direction across the entire laminate and across the thickness of the transverse ply. As the loading increases, additional cracks appear in the 90° ply until a certain saturation level is achieved, see Aveston and Kelly (1), Reifsnider and Talug (2), Bader et al (3), Bailey et al (4) and Reifsnider et al (5).

On the other hand, metal matrix laminates do not exhibit extensive matrix cracking under monotonic loading, but are quite susceptible to matrix fatigue cracking, see White and Wright (6) and Dvorak and Johnson (7).

Clearly other forms of damage occur in composite laminates, for example fibre breaks and delamination cracks, but this paper is concerned only with matrix cracking.

It is perhaps important to state at the outset that this contribution is concerned with the theoretical prediction of the loss of stiffness of cracked laminates. In particular we do not attempt to study the growth of cracks. Rather, we pose the question: Given an in situ crack distribution in a composite laminate, what is the corresponding loss in stiffness?

In this paper we are only concerned with the elastic response of composite laminates. As in so many studies of elastic composites, the essential idea is contained in the classic paper by Eshelby (8). Further developments, mainly concerned with cracked homogeneous materials, have been given by Bristow (9), Walsh (10), Budiansky and O'Connell (11), Hoenig (12), Willis (13), Gottesman et al (14), Horii and Nemat-Nasser (15). However, the work outlined here relies heavily on some earlier work by the present authors (16) which is specifically designed to analyse a matrix containing both fibres and cracks.

Three essential ideas underpin the work described below. First, we assume that the elastic moduli of a cracked ply are equal to those of a similarly cracked infinite fibrous material (i.e. with the same fibre concentration and crack density). Second, we use a self-consistent calculation to obtain the elastic stiffnesses and compliances of the infinite fibrous material containing cracks. Third, we determine the loss of stiffness of various cracked laminates from classical laminated plate theory. Finally we give some numerical results for typical E glass-epoxy, graphite-epoxy and boron-aluminium laminates.

2. TWO CAVITY PROBLEMS

Consider an orthotropic elastic solid with stiffness tensor L and compliance tensor M. The stress, $\underset{\sim}{\sigma}$, and strain, $\underset{\sim}{\varepsilon}$, in this material are thus related through

$$\underset{\sim}{\sigma} = L\underset{\sim}{\varepsilon}, \quad \underset{\sim}{\varepsilon} = M\underset{\sim}{\sigma}.$$

Suppose now that this solid contains a circular cylindrical cavity

$$x_1^2 + x_2^2 = a^2, \quad |x_3| < \infty,$$

with respect to Cartesian coordinates x_1, x_2, x_3. When this solid is loaded by a uniform stress $\underset{\sim}{\sigma}^A$ at infinity, and the walls of the cavity remain traction free, the form of the elastic field is known analytically, see (8, 17). In particular it may be shown that the interaction energy per unit length of the cavity is given by

$$\varepsilon_c = \tfrac{1}{2}\pi a^2 \underset{\sim}{\sigma}^A . Q^{-1} \underset{\sim}{\sigma}^A,$$

where

$$Q = L - LPL,$$

and where the components of P are obtained as follows: Let $\underset{\sim}{w}$ be the vector with components

$$\underset{\sim}{w} = (\cos\psi, \sin\psi, 0),$$

and let f_{ij} be the matrix inverse of $L_{ikjl}w_k w_l$. Also let

$$D_{ijkl} = \frac{1}{4}\left\{w_i f_{jk} w_l + w_j f_{ik} w_l + w_i f_{jl} w_k + w_j f_{il} w_k\right\},$$

then

$$P_{ijkl} = \frac{1}{2\pi}\int_0^{2\pi} D_{ijkl}\, d\psi.$$

We note that simple physical interpretations of both P and Q have been given by Hill (18).

Consider next the situation in which the infinite solid contains a slit crack:

$$|x_1| \leqslant a, \quad x_2 = 0, \quad |x_3| < \infty.$$

In this case the interaction energy, per unit length of the crack, is given by

$$\varepsilon_s = \frac{1}{2}\pi a^2 \underset{\sim}{\sigma}^A . \Lambda \underset{\sim}{\sigma}^A,$$

where the non-zero components of Λ are given by Laws (17) as

$$\Lambda_{22} = \frac{L_{11}(\alpha_1^{\frac{1}{2}} + \alpha_2^{\frac{1}{2}})}{L_{11}L_{22} - L_{12}^2},$$

$$\Lambda_{44} = \frac{1}{(L_{44}L_{55})^{\frac{1}{2}}},$$

$$\Lambda_{66} = \frac{(L_{11}L_{22})^{\frac{1}{2}}(\alpha_1^{\frac{1}{2}} + \alpha_2^{\frac{1}{2}})}{L_{11}L_{22} - L_{12}^2},$$

where α_1 and α_2 are the roots of

$$L_{11}L_{66}\alpha^2 - (L_{11}L_{22} - L_{12}^2 - 2L_{12}L_{66})\alpha + L_{22}L_{66} = 0.$$

3. A FIBROUS COMPOSITE CONTAINING LONGITUDINAL CRACKS

We now focus our attention on a composite consisting of a continuous matrix reinforced by a family of parallel fibres in the x_3- direction. In addition the composite contains a homogeneous distribution of parallel slit cracks which are aligned in the direction of the fibres and whose common normal is the x_2-axis, see Fig. 1. With the help of the results of the previous section we can easily write down the self-consistent estimate for the overall elastic moduli of the cracked composite.

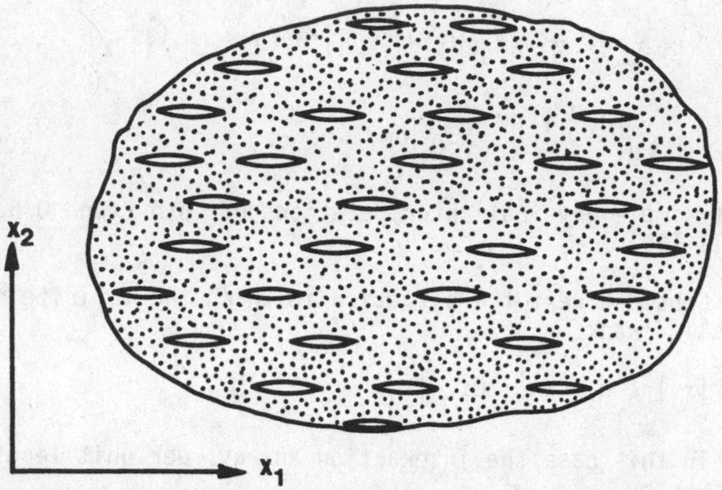

FIG.1. An Infinite Fibrous Medium Containing Cracks.

Let the fibres have stiffness tensor L_1 and occupy a volume fraction c_1 of the composite. The crack density parameter β is the average number of cracks of length 2a in a square of side 2a. In other words, if η is the number of cracks per unit area

$$\beta = 4\eta a^2 .$$

As is shown by Laws et al. (16), the overall stiffness of the cracked composite is given by

$$L = L_2+c_1(L_1-L_2)[I+P(L_1-L)]^{-1}-\frac{1}{4}\pi\beta L_2\Lambda L, \qquad (1)$$

where L_2 is the stiffness of the matrix. Since the matrix is isotropic and the fibres transversely isotropic (at worst), the overall composite is orthotropic. Thus the components of P and Q are defined above. In the terminology of Laws et al. (16), equation (1) refers to the three phase model of a cracked fibrous composite. We remark that here we do not consider the simpler two-phase model introduced in (16).

4. CRACKED LAMINATES

It is prudent to restrict our discussion of cracked laminates to relatively simple, but practically significant, configurations. We only consider symmetric cross ply laminates and balanced angle ply laminates in which the off-axis plies are equally cracked. This choice is made solely to guarantee that there is only one crack density parameter.

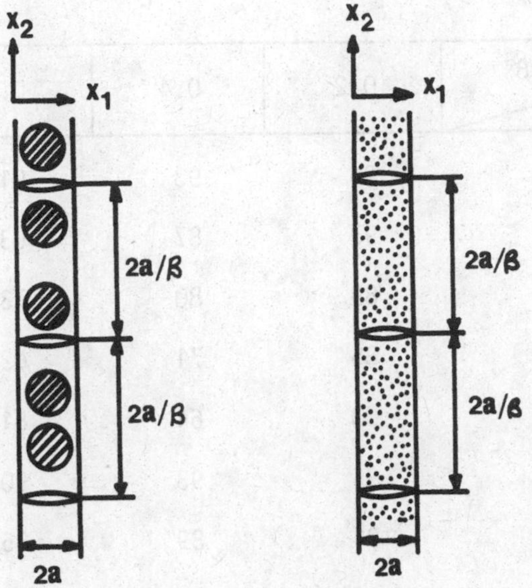

FIGS.2(a) A Filament Monolayer with Slit Cracks.
(b) A Lamina Containing Small Diameter Fibres and Slit Cracks.

Thus consider a cracked laminate in which a typical cracked inner layer is shown in Fig.2a or 2b. To a first approximation the properties of the cracked layer are equal to those of an infinite cracked medium with corresponding volume fraction of fibres and crack density. Clearly this transition poses some unresolved problems, but is similar to the approach that is followed in the evaluation of overall moduli of laminates without cracks. Once the stiffnesses of the cracked lamina have been calculated from Eqn.(1), the laminate stiffnesses can be computed by standard methods, see Tsai and Hahn (19). Thus it is possible to obtain numerical results for different matrices, fibres, lay-ups and crack densities. Typical results which indicate the predicted loss in longitudinal Young's modulus E_L are given below. The losses in stiffness of some 40% B/Al laminates are shown in Table 1. In addition, results for some angle ply E glass/epoxy laminates are given in Table 2. Finally, Table 3 contains some results for the common T300/5208 graphite/epoxy system.

TABLE 1: 40% Boron/Aluminium

% of uncracked modulus E_L

Lay-up \ β	0.2	0.4	0.6
$(0_4,90)$	96	93	91
$(0_2,90)$	93	87	83
$(0, 90)$	89	80	73
$(0, 90_2)$	85	71	62
$(0, 90_4)$	80	64	51
$(0_4,\pm45)$	96	93	90
$(0_2,\pm45)$	94	89	85
$(0, \pm45)$	91	84	78
$(0,(\pm45)_2)$	89	80	73

TABLE 2: 60% E Glass/Epoxy

% of uncracked modulus E_L

Lay-up \ β	0.2	0.6	1.0
$(0_2,\pm45)$	92	83	75
$(0, \pm45)$	88	75	63
$(0,(\pm45)_2)$	84	67	50
(±45)	77	51	25

TABLE 3: 40% T300/5208

% of uncracked modulus E_L

Lay-up \ β	0.2	0.6	1.0
$(0_4, 90)$	100	99	99
$(0_2, 90)$	99	98	97
$(0, 90)$	98	96	95
$(0, 90_2)$	97	92	90
$(0, 90_4)$	95	86	82
$(0_2, \pm 45)$	99	97	95
$(0, \pm 45)$	98	94	92
$(0, (\pm 45)_2)$	97	91	86
(± 45)	91	74	59
$(0_2, \pm 75)$	99	96	95
$(0_2, \pm 60)$	99	97	95
$(0_2, \pm 30)$	99	97	95
$(0_2, \pm 15)$	100	99	99

ACKNOWLEDGMENT

This work has been supported by a grant from the Air Force Office of Scientific Research.

REFERENCEES

(1) Aveston, J.& Kelly, A. (1973). Theory of Multiple Fracture in Fibrous Composites. J. Mater. Sci., 8, 352-362.
(2) Reifsnider, K.L. & Talug, A. (1978). Characteristic Damage States in Composite Laminates. Proc. ARO-NSF Research Workshop on Mechanics of Composite Materials (ed. G.J. Dvorak), pp.130-161, Duke University.
(3) Bader, M.G. et al. (1979). The Mechanisms of Initiation and Development of Damage in Multi-Axial Fibre-Reinforced Plastics Laminates. In: Proc. 3rd International Conf. on Mechanical Behaviour of Materials, Cambridge, pp.227-239. Oxford: Pergamon Press.
(4) Bailey, J.E. et al. (1979). On the Transverse Cracking and Longitudinal Splitting of Glass and Carbon Fibre Reinforced Epoxy Cross Ply Laminates and the Effect of Poisson and Thermally Generated Strain. Proc. R. Soc. Lond. A., 366, 599-623.
(5) Reifsnider, K.L. et al. (1983). Damage Mechanics and NDE of Composite Laminates. In: Mechanics of Composite Materials RECENT ADVANCES, ed. Z. Hashin & C.T. Herakovich, pp.399-420. Oxford: Pergamon Press.
(6) White, M.K. & White, M.A. (1979). The Fatigue Properties of Cross-Plied Boron 6061 Aluminium. J. Mater. Sci., 14, 653-662.
(7) Dvorak, G.J. & Johnson, W.S. (1980). Fatigue of Metal Matrix Composites. International J. Fracture, 16, 585-607.
(8) Eshelby, J.D. (1957). The Determination of the Elastic Field of an Ellipsoidal Inclusion and Related Problems. Proc. R. Soc. Lond. A., 241, 376-396.
(9) Bristow, J.R. (1960). Microcracks and the Static and Dynamic Elastic Constants of Annealed and Heavily Cold-Worked Metals. Brit. J. Appl. Phys., 11, 81-85.
(10) Walsh, J.B. (1969). New Analysis of Attenuation in Partially Melted Rocks. J. Geophys. Res., 74, 4333-4337.
(11) Budiansky, B. & O'Connell, R.J. (1976). Elastic Moduli of a Cracked Solid. Internat. J. Solids Structures, 12, 81-97.
(12) Hoenig, A. (1979). Elastic Moduli of a Nonrandomly Cracked Body. Internat. J. Solids Structures, 15, 137-154.
(13) Willis, J.R. (1977). Bounds and Self-Consistent Estimates for the Overall Properties of Anisotropic Composites. J. Mech. Phys. Solids, 25, 185-202.
(14) Gottesman, T. et al. (1980). Effective Elastic Moduli of Cracked Fibre Composites. In: Advances in Composite Materials, ed. A.R. Bunsell et al., pp.749-758. Oxford: Pergamon Press.
(15) Horii, H. & Nemat-Nasser, S. (1983). Overall Moduli of Solids with Microcracks: Load-Induced Anisotropy. J. Mech. Phys. Solids, 31, 155-171.
(16) Laws, N. et al. (1983). Stiffness Changes in Unidirectional Composites Caused by Crack Systems. Mechanics of Materials, 2, 123-137.

(17) Laws, N. (1977). A Note on Interaction Energies Associated With Cracks in Anisotropic Solids. Phil. Mag., 36, 367-372.
(18) Hill, R. (1965). A Self-Consistent Mechanics of Composite Materials. J. Mech. Phys. Solids, 13, 213-222.
(19) Tsai, S.W. & Hahn, T.W. (1980). Introduction to Composite Materials. Westport, CT., Technomic.

MEAN FIELD THEORY AND THE BAUSCHINGER EFFECT IN COMPOSITES

O.B. PEDERSEN

Metallurgy Department
Risø National Laboratory DK-4000 Roskilde, Denmark

ABSTRACT

The development of a rigorous mean field theory for dense thermoelastic composites from J.D. Eshelby's independent inclusion approximation for dilute elastic composites is outlined. The concept of an effectively uniform plastic strain in the matrix enables the mean field theory to give an accurate quantitative account of mean phase stresses measured by X-ray diffraction in metallic fibre composites. The mean in-situ friction stress in the matrix can be understood as an addition of plastic friction due to misfits equivalent with Orowan loops and elastic friction due to misfits arising from elastic heterogeneity. An analysis suggests that the plastic friction may be measured separately in a Bauschinger experiment if elastic friction adds linearly to plastic friction.

1. INTRODUCTION

The deformation and failure of real composite materials involves a variety of physical mechanisms of non-uniform plastic flow and cracking. These mechanisms are conveniently ignored in the linear theory of composites (LTC). The LTC provides mathematical approximations and rigorous bounds for the thermoelastic deformation of composites in which each phase undergoes perfectly uniform inelastic strains accommodated by non-uniform elastic interphase strains. Recent work on pure and dispersion strengthened metals suggests (1) that some of the physical mechanisms can be modelled realistically by judicious combinations of the LTC with continuum plasticity and dislocation mechanics.

In dispersion strengthened metals the particle volume fraction f is typically of the order of 0.01, so it is perhaps not unreasonable to represent the plastic strain in the matrix as a uniform strain. However, in fibre composites with large fibre volume fractions, say $f > 0.1$, it is much less obvious how such a crude picture of plastic flow in the matrix could possibly lead to a realistic theory. Here I shall therefore examine whether or not the concept of an EFFECTIVELY UNIFORM plastic strain in the matrix can indeed give a consistent account of experimental observations. A mean field approximation (2), based on Eshelby's equivalent transformation strain, will therefore be described in an attempt to make it clear exactly how the simple physical ideas developed for pure and dispersion strengthened metals might fit into the general framework of the LTC. Rigorous bounds, derived from the mean field approximation, will then be applied in an analysis of Cheskis and Heckel's measurement (3) of mean phase stresses by X-ray diffraction. An analysis of the Bauschinger effect will be attempted.

2. MEAN FIELD THEORY

In analogy with the Ampere-Lorentz principle of representing distributions of current and charge by magnetizations and polarizations Eshelby in 1957 introduced his "transformation" strain (4). He showed how this concept can be used in dealing with a large number of problems in the mechanics and physics of solids "in which the uniformity of an elastic medium is disturbed by a region within which it has changed its form or which has elastic constants different from those of the remainder" (4).

2.1 ESHELBY'S TRANSFORMATION STRAIN

The transformation strain is an arbitrary uniform inelastic strain ε^T_{ij} assigned to a region, the inclusion, in an infinite homogeneous linear elastic medium, the matrix. Eshelby showed that in analogy with the demagnetizing fields in ellipsoids the "inclusion stress" σ^I_{ij} generated by ε^T_{ij} inside an ellipsoidal inclusion is uniform. For convenience a second-order tensor f_{ij} may be divided up into its hydrostatic part $f = f_{kk}$ and deviatoric part $'f_{ij}$, as in

$$f_{ij} = \frac{1}{3} f \; \delta_{ij} + {'f_{ij}} \qquad (1)$$

where δ_{ij} is Kronecker's delta and summation from 1 to 3 is implied with respect to repeated indices, i, j, k ... The inclusion stress for the sphere is, in this notation,

$$\sigma^I_{ij} = \kappa(\varepsilon^\infty - \varepsilon^T)\delta_{ij} + 2\mu({'\varepsilon^\infty_{ij}} - {'\varepsilon^T_{ij}}) \qquad (2)$$

where

$$\varepsilon^\infty_{ij} = \frac{1}{3}\,\alpha\,\varepsilon^T\,\delta_{ij} + \beta\;{'\varepsilon^T_{ij}} \qquad (3)$$

and

$$\alpha = \frac{1}{3}\,\frac{1+\nu}{1-\nu} \qquad (4)$$

$$\beta = \frac{2}{15}\,\frac{4-5\nu}{1-\nu} = 1 - \gamma \tag{5}$$

Here γ is Eshelby's "accommodation" factor for a sphere, κ is the bulk modulus, μ is the shear modulus and ν is Poisson's ratio. The stresses outside the inclusion σ^{∞}_{ij} are of course not uniform (5); they fall off rapidly with distance from the inclusion, as implied by the principle of Saint Venant.

In general the elastic constants of inclusions differ from those of the matrix and the inclusions are then called elastic inhomogeneities. Eshelby showed that, in the case of a spherical inhomogeneity subject to a uniform surrounding stress σ^{U}_{ij} and a transformation strain ε^{T}_{ij}, the effect of elastic heterogeneity is simply to replace ε^{T}_{ij} by an "equivalent" transformation strain

$$\varepsilon^{T*}_{ij} = \frac{1}{3}\,(C_1\varepsilon^{T}+C_2\varepsilon^{U})\delta_{ij} + (D_1'\varepsilon^{T}_{ij}+D_2'\varepsilon^{U}_{ij}) \tag{6}$$

with

$$C_2 = \frac{\kappa_1-\kappa_2}{\kappa_2} \quad C_1 = \frac{\kappa_1-\kappa_2}{(\kappa_2-\kappa_1)\alpha_1+\kappa_1} \tag{7}$$

$$D_2 = \frac{\mu_1-\mu_2}{\mu_2} \quad D_1 = \frac{\mu_1-\mu_2}{(\mu_2-\mu_1)\beta_1+\mu_1} \tag{8}$$

The basis of this elegant result is the uniformity of the inclusion stress, so similar results can be obtained in the case of general ellipsoids (whose inclusion stresses are also uniform). It should perhaps be mentioned that Eshelby did not in his papers (4-6) explicitly provide C_1 and write ε^{T*}_{ij} by superposing the terms proportional respectively to ε^{T}_{ij} and ε^{U}_{ij}.

The effect of an external surface must also be dealt with. Eshelby (6) included it by writing the stress in a finite matrix as

$$\sigma_{ij} = \sigma^{\infty}_{ij} + \sigma^{im}_{ij} \tag{9}$$

where σ^{∞}_{ij} is the stress calculated outside the inclusion (assuming an infinite matrix) and σ^{im}_{ij} is an image stress required in order to satisfy whatever boundary conditions are imposed on the external surface. The stress inside the inclusion is written similarly as $\sigma^{I}_{ij} + \sigma^{im}_{ij}$.

2.2 INDEPENDENT INCLUSION APPROXIMATION

The changes of free energy and internal stress which occur when transformation strains are assigned to inclusions in a body at-

tached to a loading system can be dealt with in terms of simple energy
relations, supplied by Eshelby. He showed how his concept of the interaction energy of an inhomogeneity with an external stress leads to
expressions for the effective elastic constants (μ_c, κ_c) of a composite which consists of a very small volume fraction f of spherical
inhomogeneities in a matrix. Since f is small, each sphere makes its
contribution to the elastic energy almost independently of the other
spheres. Eshelby (4) could therefore use an independent inclusion approximation to obtain

$$\mu_c = \mu_1/(1+fD_2) \tag{10}$$

$$\kappa_c = \kappa_1/(1+fC_2) \tag{11}$$

In their study of work-hardening due to stored energy, Tanaka
and Mori (7) combined Eshelby's energy approach with Ashby's continuum
model (8) of plastic strain ε^P_{ij} in the matrix

$$\varepsilon^T_{ij} = -\varepsilon^P_{ij} \tag{12}$$

For large volume fractions the independent inclusion approximation breaks down; it is no longer realistic to ignore terms due to
the elastic interactions of the inclusions with each other and with the
surface of the material. Exact calculations of the elastic energy and
the internal stress as functions of the phase geometry therefore become
extraordinarily difficult. However, it is possible to make progress by
invoking Newton's second law of motion for a continuum: the density of
external forces, including inertial forces, is $-\sigma_{ik,k}$ and Gauss' theorem
provides the identity

$$\langle\sigma_{ij}\rangle = \frac{1}{V}\int_V \sigma_{ij}dV = \frac{1}{V}\int_S \sigma_{ik}x_jn_k dS - \frac{1}{V}\int_V x_j\sigma_{ik,k}dV \tag{13}$$

where S is the external surface, n_k is its outward normal and a comma
preceding an index k denotes the partial differentiation $\partial/\partial x_k$. In the
static case of interest the applied stress σ^A_{ij} is uniform on S and
$\sigma_{ik,k}$ vanishes. The stress equilibrium requirement is then simply

$$\langle\sigma_{ij}\rangle = \sigma^A_{ij} \tag{14}$$

Brown and Stobbs (9) introduced Eshelby's image stress into
Eq. 14 and pointed out that for stress-free external surfaces

$$\langle\sigma^\infty_{ij}\rangle_M + \langle\sigma^{im}_{ij}\rangle + f\,\sigma^I_{ij} = 0 \tag{15}$$

whatever the value of f, provided the material is elastically homogeneous. They also pointed out that if the inclusion and the external
surface are concentric spheres then $\langle\sigma^\infty_{ij}\rangle_M$ vanishes. This result is
easily generalised using the uniformity of the inclusion stress in el-
lipsoids, e.g. Tanaka and Mori (10). The net force on a rigid straight

dislocation would therefore be entirely due to the value of the quasi-uniform image stress. Mori and Tanaka (11) used the stress equilibrium requirement to modify their initial energy analysis. However, their modified analysis is not complete, and in particular it breaks down at large f when the elastic constants of the inclusions differ from those of the matrix. A simple solution to the large f problem is given by a second model based on Eshelby's equivalent inclusion, the mean field approximation.

2.3 MEAN FIELD APPROXIMATION AND BOUNDS

For elastically heterogeneous composites it must be taken into account that the inclusions respond as elastic inhomogeneities to the stress they feel. Following Eshelby's ideas, one might replace each inhomogeneity by its equivalent inclusion, thus converting the elastically heterogeneous composite into an "equivalent homogeneous composite" (EHC) whose elastic constants are those of the matrix.

Eshelby's equivalent inclusions must sample a perfectly uniform surrounding stress σ^U_{ij}. However, in an externally stressed composite whose inhomogeneities have a transformation strain ε^T_{ij} the surrounding stress is a non-uniform superposition σ^S_{ij} of σ^A_{ij} and the fields produced by the equivalent transformation strains. The surrounding stress is difficult to calculate; but according to Eq. (14) its mean value, for a uniform σ^I_{ij}, is simply

$$\langle\sigma^S_{ij}\rangle = \sigma^A_{ij} - f\,\sigma^I_{ij} \tag{16}$$

so it seemed worthwhile (2,14) to examine the consequences of the mean field approximation that $\sigma^U_{ij} = \langle\sigma^S_{ij}\rangle$. For simple shear with spherical inclusions, with $\sigma^A_{ij} = 0$ and $\varepsilon^T_{ij} = 0$ except for $e^T = 2\varepsilon^T_{13}$ and $\tau^A = \sigma^A_{13}$, one then finds using the above equations that the equivalent transformation strain is the sum

$$e^{T*} = \frac{D_1 e^T + D_2 \tau^A/\mu_1}{1 - f\gamma D_2} \tag{17}$$

of a "plastic" term proportional to the transformation strain and a new "elastic" term proportional to the applied stress.

The mean field approximation modifies Eshelby's analysis of μ_c and κ_c: the mean shear strain $\langle e\rangle$ of the EHC under τ^A with $e^T = 0$ can be written either as $\tau^A/\mu_1 + f\,e^{T*}$ or as τ^A/μ_c, which provides

$$\mu_c = \mu_1\Big/\left(1 + \frac{f\,D_2}{1 - f\gamma D_2}\right) \tag{18}$$

This expression, and the corresponding expression for κ_c (2,14) differ from Eshelby's independent inclusion approximation, Eqs. 10 and 11; but they coincide (14) with Hashin and Shtrikman's (13) expressions for bounds on the effective shear and bulk moduli of isotropic two phase composites with $(\kappa_1-\kappa_2)(\mu_1-\mu_2) > 0$. The use of an "effectively uniform" surrounding stress $\sigma^U_{ij} = \langle\sigma^S_{ij}\rangle$ then does not just lead to some arbitrary approximation, it leads to rigorous bounds.

Furthermore, a matrix formulation revealed (15) that the mean field approximation reproduces Levin's relation for thermoelastic two-phase composites, as presented by Laws (16). The reason why the mean field approximation coincides with rigorous bounds becomes clear in the light of Walpole's general theory of bounds for elastic composites, which displays (17) a crucial connection with Eshelby's equivalent inclusion. As it turns out (18), the EHC and Walpole's "comparison material" are closely related, and so are the mean surrounding field and Walpole's "image" field. Finally, the analogy between Eshelby's transformation strain and a uniform magnetisation, of course, suggests that the mean field approximation has a very wide range of physical interpretations. The present μ_c expression, for example, translates into an expression coinciding with bounds (12) for the composite's linearised magnetic permeability.

2.4 MATRIX MEAN STRESS

The mean field theory, written in (18) as a generalisation of the mean field approximation, is equivalent to a combination of the formulations (16,17) by Walpole and Laws. Consequently it deals quite rigorously with the thermoelasticity of a general class of composite materials which includes fibre composites. Let us consider fibre composites and assume that the fibres are parallel to the x_3 axis and randomly distributed in the matrix, see Fig. 1. We shall now examine the possibility of using the mean field theory as a rigorous theoretical framework for studies of the plasticity and failure of fibre composites. We shall see that it is essential in this application that the theory supplies the equivalent transformation strain ε^{T*}_{ij} of the fibres.

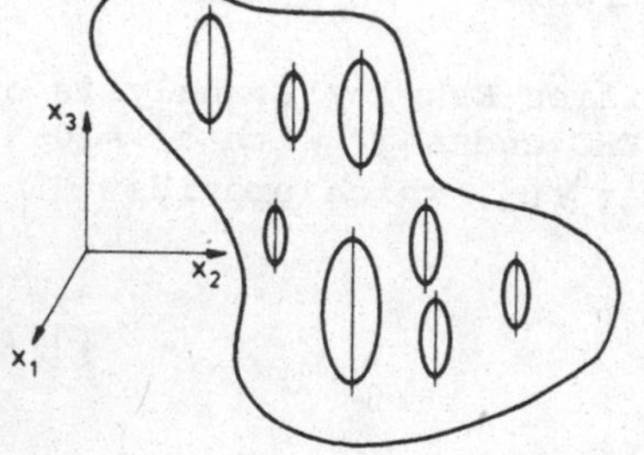

FIG. 1. The mean field theory features an arbitrarily shaped matrix phase. The inclusions are general ellipsoids of any size, but all of the same shape and orientation. From (18).

Plastic strain in the matrix is non-uniform at the level of discrete dislocations and also at the level of continuum plasticity. Let us nevertheless examine a model in which the plastic strain in the matrix is "effectively" uniform. The matrix may, for example, flow in single slip at 45° to the fibre axis, i.e. all $\varepsilon_{ij}^{P} = 0$ except $\varepsilon_{33}^{P} = -\varepsilon_{11}^{P} = \varepsilon$; or it may flow in multiple slip, i.e. all $\varepsilon_{ij}^{P} = 0$ except $\varepsilon_{33}^{P} = -2\varepsilon_{11}^{P} = -2\varepsilon_{22}^{P} = \varepsilon_{p}$. In the absence of stress relaxation ε_{ij}^{P} can in general be represented as the negative of the transformation strain ε_{ij}^{T} of the fibres. The mean field theory then supplies the plastic extension of the composite as

$$\varepsilon_{33}^{PC} = C\varepsilon_{p} \tag{19}$$

For elastically homogeneous composites $C = 1-f$ and the composite's volume is conserved, $\varepsilon_{kk}^{PC} = 0$. By contrast, $C \neq 1-f$ and $\varepsilon_{kk}^{PC} \neq 0$ in the general case of elastically heterogeneous composites, see Fig. 2.

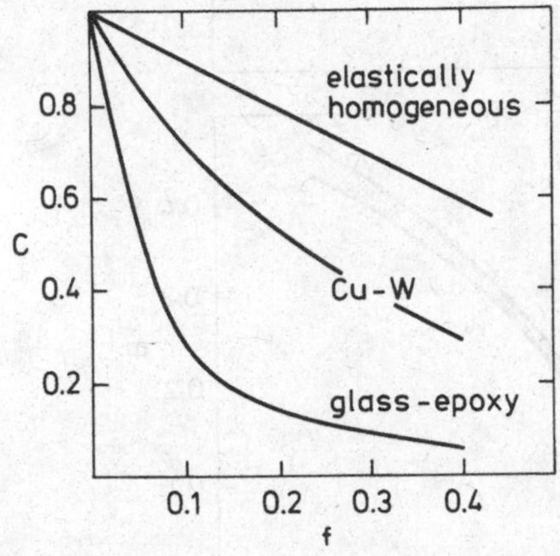

FIG. 2. The effect of elastic heterogeneity on the parameter C, Eq. (19), for the permanent inelastic extension ε_{33}^{PC} of unrelaxed fibre composites. The curves, calculated on the basis of the mean field theory, represent upper and lower bounds which are too closely spaced to be resolved in the graph.

For randomly arranged parallel fibres the mean matrix stress equals the $\langle\sigma_{ij}^{S}\rangle$ of Eq. (16), with σ_{ij}^{I} proportional to the equivalent transformation strain ε_{ij}^{T*}. The mean matrix stress must therefore (like ε_{ij}^{T*}) contain "elastic" terms, proportional to the applied stress, and "plastic" terms, proportional to the plastic strain in the matrix, see for example Eq. (17). For a longitudinal stress $\sigma_{33}^{A} = \sigma^{A}$ and all other $\sigma_{ij}^{A} = 0$ we can write

$$\langle\sigma_{11}\rangle_{M} = A_{1}\varepsilon_{p} + B_{1}\sigma^{A} \tag{20}$$

$$\langle\sigma_{22}\rangle_{M} = A_{2}\varepsilon_{p} + B_{2}\sigma^{A} \tag{21}$$

$$\langle\sigma_{33}\rangle_{M} = A_{3}\varepsilon_{p} + B_{3}\sigma^{A} + \sigma^{A} \tag{22}$$

where the general matrix expressions for A_i and B_i are supplied by the mean field theory (18). It is difficult to give an accurate estimate of the effect of the mean transverse constraint since bounds on transverse mean stresses tend to be widely spaced. Also, the transverse constraint appears to depend critically upon the assumed mode of flow. Figure 3 illustrates this dependence in the case of copper with tungsten fibres: the coefficient A_1 for the transverse plastic mean stress is reduced by by a factor of ~ 3 when multiple slip is assumed instead of single slip.

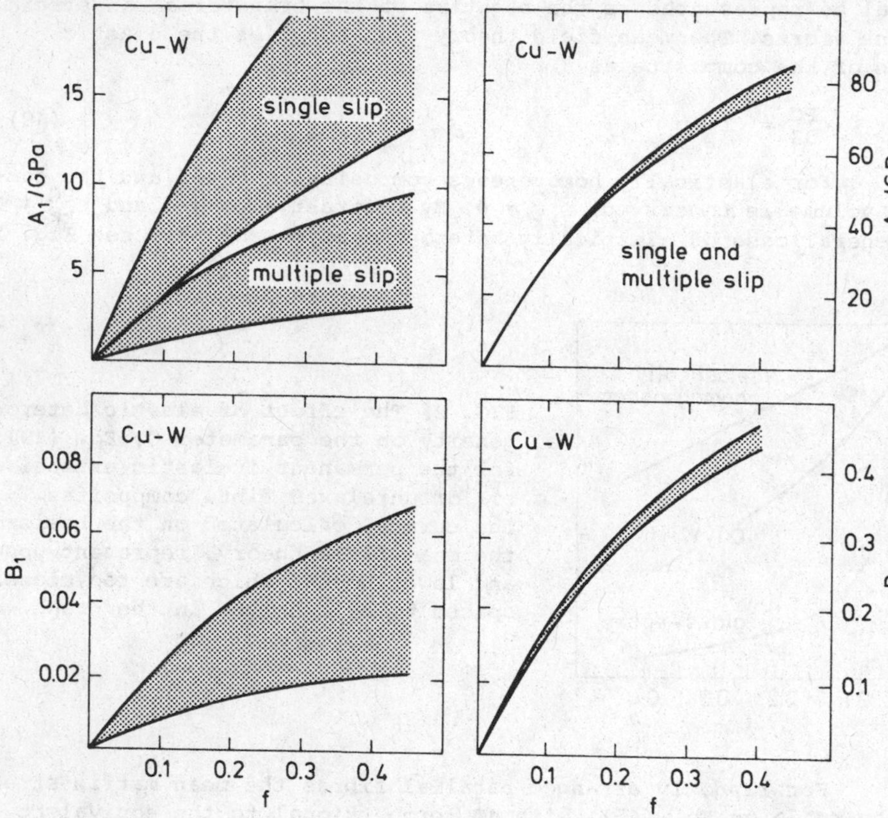

FIG. 3. Bounds on the parameters for the plastic mean stress (A_i) and the elastic mean stress (B_i) in copper-tungsten. The transverse parameter A_1 depends sensitively upon the deformation mode; but the difference between the single- and multiple slip values of the longitudinal parameter A_3 is too small to be shown in the graph.

It seems futile, at least at the present stage, to try to calculate the level of effectively uniform "unrelaxed" plastic strain ε_p^* in the matrix. Let us instead consider the following scheme: 1) derive the quantity that would be ε_p^* from direct X-ray measurements of mean lattice strains in a composite; 2) substitute this quantity into Eqs. (19) to (22) and compare with experiment to see if the concept of ε_p^* works.

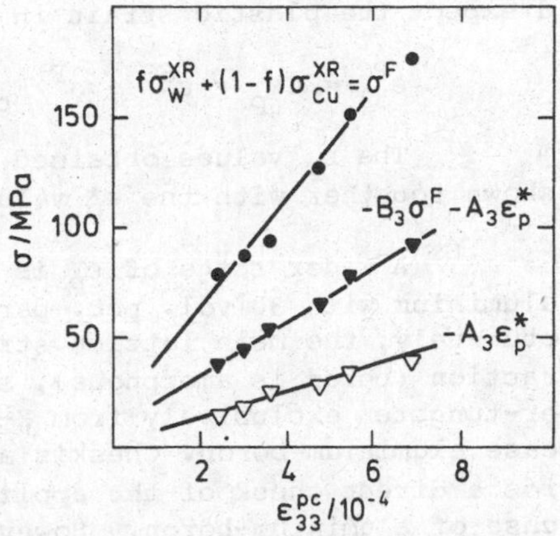

FIG. 4. X-ray diffraction data by Cheskis and Heckel (3) analysed in terms of an effectively uniform unrelaxed plastic strain ε_p^* in the matrix. Pairs of closely spaced bounds supplied by the mean field theory appear as single data points. The analysis shows that the elastic matrix mean stress is a bit higher than the plastic matrix mean stress.

The X-ray diffraction measurements by Cheskis and Heckel (3) on polycrystalline copper with 20 vol. pct. long parallel 25 μm diameter tungsten fibres can provide what would be ε_p^* in their composite; by subtracting the mean longitudinal lattice strain in the matrix from the mean longitudinal lattice strain in the (non-yielding) fibres one would obtain ε_p^*. The resulting longitudinal plastic mean matrix stress $A_3\varepsilon_p^*$ is compressive; its negative is shown in Fig. 4 as a function of the plastic extension of the composite, measured as

$$\varepsilon_{33}^{PC} = \varepsilon^{SG} - \sigma^F/E_c \tag{23}$$

where ε^{SG} is the composite's total longitudinal strain measured by a strain gauge (SG), E_c is its effective longitudinal Youngs modulus and σ^F is its forward flow stress, calculated from

$$\sigma^F = f\sigma_W^{XR} + (1-f)\sigma_{Cu}^{XR} \tag{24}$$

where σ_W^{XR} and σ_{Cu}^{XR} are the mean longitudinal stresses obtained from X-ray diffraction (XR). The σ^F values are shown in Fig. 4 together with (the negative of) the corresponding values of the longitudinal elastic and plastic mean matrix stresses. The total longitudinal mean matrix stress in the deforming composite is according to Eq. (22) the sum

$$\langle\sigma_{33}\rangle_M = A_3\varepsilon_p^* + B_3\sigma^F + \sigma^F \tag{25}$$

which is plotted in Fig. 5 together with the measured value σ_{Cu}^{XR}. The agreement between theory and experiments is good, so it seems that the concept of ε_p^* can be used successfully within the framework of the mean field theory, certainly in the description of longitudinal deformation.

Cheskis and Heckel's data suggest a linear relationship between the effective uniform plastic strains in the matrix before and after stress relaxation: if stress relaxation did not intervene one would expect the plastic strain in the matrix to be given as ε_p in

$$\varepsilon_{33}^{PC} = C\varepsilon_p = \varepsilon^{SG} - \sigma^F/E_c \qquad (26)$$

The ε_p values obtained for copper-tungsten using Eq. (26) are shown together with the ε_p^* values in Fig. 6.

A wider range of ε_p is covered by Cheskis and Heckel's data for aluminium with 30 vol. pct. parallel 95 µm diameter boron fibres. Unfortunately, the mean lattice strain in B cannot be measured by X-ray diffraction (boron is amorphous), so while ε_p^* can be obtained in copper-tungsten exclusively from X-ray data, this is not possible in the case aluminium-boron. Cheskis and Heckel's data therefore do not provide a direct check of the applicability of the concept of ε_p^* in the case of aluminium-boron. However, the outcome of assuming that the concept works, in other words of assuming that

$$\sigma_{Al}^{XR} = A_3\varepsilon_p^* + B_3\sigma^F + \sigma^F \qquad (27)$$

is not unacceptable: the ε_p^* derived using Eq. (27) suggests that the stress relaxation behaviour in aluminium-boron is similar to that in copper-tungsten, in both cases ε_p^* depends linearly upon ε_p, see Fig. 6.

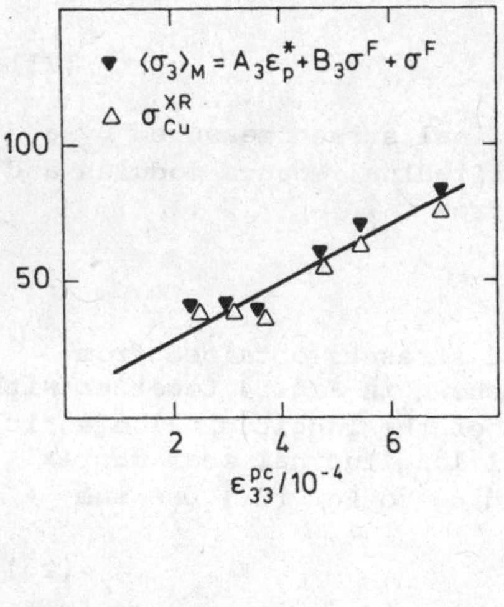

FIG. 5. The heuristic concept of an effectively uniform ε_p^* leads to accurate quantitative agreement between the longitudinal matrix mean stress σ_{Cu}^{XR} measured during flow and $\langle\sigma_{33}\rangle_M$, the overall flow stress σ^F off-set by the longitudinal elastic and plastic matrix mean stresses. The small systematic difference between σ_{Cu}^{XR} and $\langle\sigma_{33}\rangle_M$ lies within the uncertainty arising from the use of isotropic elasticity.

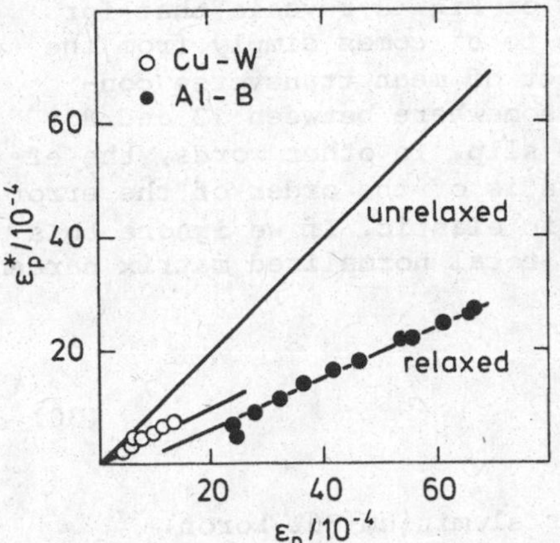

FIG. 6. A comparison of X-ray diffraction data and strain-gauge data shows that the unrelaxed and the relaxed plastic strains in the matrix, ε_p respectively ε_p^*, are approximately linearly related in Cheskis and Heckel's composites.

3. BAUSCHINGER EFFECT

We have seen in the last section that a rigorous mean field theory of bounds for thermoelastic composites can be combined with an "effectively" uniform plastic strain in the matrix. When this approach is applied to the analysis of X-ray diffraction data on the mean stresses in metallic composites it reveals that they consist of "elastic" and "plastic" terms of comparable magnitude. It is also found that the plastic mean stress is reduced considerably by stress relaxation. We shall now discuss how elastic heterogeneity affects the in-situ matrix flow stress and the Bauschinger effect of the composite.

3.1 MATRIX FRICTION STRESS

Is the concept of an effectively uniform plastic strain in the matrix also consistent with the matrix flow stress suggested by X-ray diffraction? An essential point is that the longitudinal mean matrix stress is not by itself a conventional flow stress; an appropriate yield criterion must be specified before a flow stress can be written. The criterion should refer to some "representative" region of the matrix, rather than to a single point in the matrix. The simplest approach is to use a Tresca type criterion: an increment of ε_p will occur when the maximum mean shear stress in the matrix reaches a critical value

$$\max\{\langle\sigma_{33}\rangle_M-\langle\sigma_{11}\rangle_M\ ,\ \langle\sigma_{33}\rangle_M-\langle\sigma_{22}\rangle_M\ ,\ \langle\sigma_{11}\rangle_M-\langle\sigma_{22}\rangle_M\} = 2\tau_M \tag{28}$$

The critical value τ_M is an "effectively uniform" friction stress in a representative region of matrix. By combining Eqs. (20)-(22) and (28) we can write the composite's forward flow stress as

$$\sigma^F = 2\tau_M + A\varepsilon_p^* + B\sigma^F = \frac{2\tau_M + A\varepsilon_p^*}{1-B} \tag{29}$$

where $A = A_1 - A_3$ and $B = B_1 - B_3$. Inspection of Fig. 3 reveals that for copper-tungsten the dominant contribution to σ^F comes simply from the longitudinal matrix mean stress: the effect of mean transverse constraint, measured as A_1/A and B_1/B, lies somewhere between 12 and 33% for single slip and 4 to 13% for multiple slip. In other words, the effect on σ^F of the transverse mean stresses is of the order of the error introduced by taking copper to be isotropic elastic. If we ignore this small effect we find from Fig. 7 that the total normalized matrix hardening rate converted to shear values

$$\theta_M/\mu = \frac{1}{4}\,\frac{\Delta\sigma^{XR}/\mu}{\Delta\varepsilon_p^*} \tag{30}$$

is 0.5 for copper-20% tungsten and 0.2 for aluminium-30% boron.

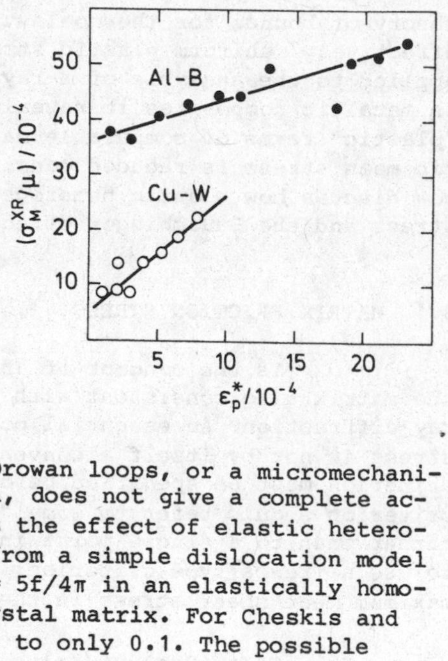

FIG. 7. The normalised longitudinal mean matrix stresses during flow σ_M^{XR}/μ shown as functions of the unrelaxed plastic strain in the matrix ε_p^*. A simple Tresca type criterion suggests that it is a good approximation to identify σ_M^{XR} with an "effectively uniform" friction stress in the matrix. The friction may be explained partly in terms of Orowan loops; but a major part of it has its origin in misfit strains directly due to elastic heterogeneity.

The frictional hardening by Orowan loops, or a micromechanically equivalent misfit of plastic strain, does not give a complete account of these θ_M values if one ignores the effect of elastic heterogeneity. Brown and Clarke (19) estimate from a simple dislocation model that the Orowan effect contributes $\theta_M/\mu = 5f/4\pi$ in an elastically homogeneous fibre composite with a single crystal matrix. For Cheskis and Heckel's data their estimate then amounts to only 0.1. The possible additional contributions from grain boundary strengthening or forest hardening in Cheskis and Heckel's composites would be too small to explain the observed θ_M/μ values of 0.2 and 0.5. The observation that the elastic and plastic mean stresses are comparable suggests that the discrepancy is an effect of elastic heterogeneity.

3.2 EFFECT OF ELASTIC HETEROGENEITY

We saw that the effectively uniform ε_p^* is consistent with measurements of the effectively uniform τ_M, provided that elastic heterogeneity makes a substantial contribution to τ_M. The origin of the "elastic" contribution to τ_M lies in the elastic term of ε_{ij}^{T*}, see Eq. (17); but ε_{ij}^{T*} does not conserve volume in an elastically heterogeneous composite, so one may at best represent the deviatoric components of ε_{ij}^{T*} by fictious Orowan loops. However, a simple and hopefully realistic approach is to assume linear addition and write

$$2\tau_M = \alpha\varepsilon_p^* + \beta\sigma^A + \delta \tag{31}$$

where we shall call $\alpha\varepsilon_p^*$ the "plastic" friction stress and $\beta\sigma^A$ the "elastic" friction stress. The term δ contains the various other friction stresses due to grain boundaries, forest dislocations, impurities etc. The assumed linear addition suggests a simple model of the composite's work hardening and Bauschinger effect which may be compared with experiments:

By combining Eqs. (29) and (31) with $\sigma^A = \sigma^F$ we can write the composite's forward flow stress as

$$\sigma^F = (A+\alpha)\varepsilon_p^* + (B+\beta)\sigma^F + \delta \tag{32}$$

For an unrelaxed composite with $\varepsilon_p^* = \varepsilon_p$ the overall monotonic hardening rate follows from Eqs. (19) and (32) as

$$\theta_F = \frac{\partial\sigma_F}{\partial\varepsilon_{33}^{PC}} = \frac{A+\alpha}{(1-B-\beta)C} \tag{33}$$

We see that it is impossible to separate the elastic and plastic terms on the basis of monotonic stress-strain data alone. However, further information on the reverse flow stress allows such a separation: in the Bauschinger experiment the flow direction is reversed after a forward prestrain of $\varepsilon_{33}^{PC} = C\varepsilon_p$. Since all the frictional terms and the elastic mean stress (for $B > 0$) oppose flow whatever its direction, while the plastic mean stress aids reverse flow, we can write

$$\sigma^R = (-A+\alpha)\varepsilon_p^* + (B+\beta)\sigma^R + \delta \tag{34}$$

The difference between σ_F and σ_R is called the "permanent softening" $\Delta\sigma$, so from Eqs. (32) and (34) we find that

$$\Delta\sigma = \frac{2A\varepsilon_p^*}{1-B-\beta} \tag{35}$$

whatever the level of stress relaxation in the composite. Note that a non-zero value of $\Delta\sigma$ implies kinematic matrix hardening as a result of the B+β term due to elastic heterogeneity. This contrasts with the isotropic matrix hardening expected from Orowan loops alone.

It is instructive to consider the quantity

$$\theta_\Delta = \frac{1}{2}\frac{\partial \Delta\sigma}{\partial \varepsilon_{33}^{PC}} \tag{36}$$

which may be measured in a Bauschinger experiment. The θ_Δ is closely related to the conventional monotonic hardening rate θ_F. This can be seen in the unrelaxed case where Eqs. 35 and 36 give

$$\theta_\Delta = \frac{A}{(1-B-\beta)C} \tag{37}$$

For an ideal matrix, flowing without relaxation and without in-situ hardening, α and β would both vanish so that θ_F and θ_Δ would simply coincide. However, this coincidence will be destroyed by in-situ matrix hardening because the plastic friction coefficient α appears in θ_F but not in θ_Δ. This is an interesting result, because it suggests that α can be measured separately as the ratio

$$\alpha = A\left(\frac{\theta_F}{\theta_\Delta} - 1\right) \tag{38}$$

The validity of this result for unrelaxed fibre composites is independent of the fibre volume fraction. It will also apply for relaxed composites if stress relaxation reduces θ_F and θ_Δ in the same proportion. The basis of the result is the hypothesis that elastic friction and plastic friction add linearly.

4. CONCLUSIONS

(1) It is possible to combine a simple mean field theory, equivalent with the theory of bounds for thermoelastic composites, with the concept of an effectively uniform plastic strain in the matrix. This approach seems to work when applied to X-ray diffraction data on the longitudinal deformation of metallic fibre composites. It shows very clearly that the mean matrix stress consists of an "elastic" term and a "plastic" term and that the plastic term is reduced by stress relaxation.

(2) When the approach is combined with the concepts of an effectively uniform in-situ friction stress in the matrix and a Tresca type yield criterion it suggests that the X-ray data can be understood partly in terms of a "plastic" friction stress, which is micromechanically equivalent with the hardening due to Orowan loops. However, for a full understanding of the data it is necessary to invoke an additional "elas-

tic" friction stress, which arises from the interaction of mobile dislocations with the local stresses set up purely as a result of the composite's elastic heterogeneity.

(3) The plastic friction is isotropic; but the elastic friction adds a kinematical hardening contribution to the in-situ matrix hardening whenever the composite as a whole displays a Bauschinger effect.

(4) A simple model of the Bauschinger effect emerges from the assumption that elastic and plastic matrix friction add linearly. This model suggests the possibility of measuring the plastic friction separately in a Bauschinger experiment.

ACKNOWLEDGEMENTS

It is a pleasure to acknowledge my indebtedness to L.M. Brown, T. Leffers, W.M. Stobbs and A.T. Winter for collaboration on related problems in the plasticity of metal crystals.

REFERENCES

(1) Pedersen, O.B. & Brown, L.M. (1983). The strength of heterogeneous materials - continuum models and discrete models. In: Deformation of Multi-phase and Particle-Containing Materials. Proc. 4th Risø International Symposium on Metallurgy and Materials Science. Roskilde: Risø National Laboratory, 83-102.

(2) Pedersen, O.B. (1976). The inhomogeneous PMS, and effective elastic constants. Manuscripts, Cavendish Laboratory.

(3) Cheskis, H.P. & Heckel, R.W. (1968). In-situ measurements of deformation behaviour of individual phases in composites by X-ray diffraction. ASTM STP 438, 76-91.

(4) Eshelby, J.D. (1957). The determination of the elastic field of an ellipsoidal inclusion, and related problems. Proc. Roy. Soc., A 241, 376-396.

(5) Eshelby, J.D. (1959). The elastic field outside an ellipsoidal inclusion. Proc. Roy. Soc., A 252, 561-569.

(6) Eshelby, J.D. (1961). Elastic inclusions and inhomogeneities. Prog. Solid Mech., 2, 89-140.

(7) Tanaka, K. & Mori, T. (1970). The hardening of crystals by non-deforming particles and fibres. Acta Met. 18, 931-941.

(8) Ashby, M.F. (1966). Work hardening of dispersion hardened crystals. Phil. Mag. 14, 1157-1178.

(9) Brown, L.M. & Stobbs, W.M. (1971). The work hardening of copper-silica I. A model based on internal stresses, with no plastic relaxation. Phil. Mag. 23, 1185-1199.

(10) Tanaka, K. & Mori, T. (1972). Note on volume integrals around an ellipsoidal inclusion. J. Elasticity, 2, 199-200.

(11) Mori, T. & Tanaka, K. (1973). Average stress in matrix and average elastic energy of materials with misfitting inclusions. Acta Met., 21, 571-574.
(12) Hashin, Z. & Shtrikman, S.A. (1962). A variational approach to the effective magnetic permeability of multiphase materials. J. Appl. Phys., 33, 3125-3131.
(13) Hashin, Z. & Shtrikman, S.A. (1963). A variational approach to the theory of the elastic behaviour of multiphase materials. J. Mech. Solids, 11, 127-140.
(14) Pedersen, O.B. (1978). Transformation theory for composites. ZAMM, 58, 227-228.
(15) Pedersen, O.B. (1979). Thermoelasticity and plasticity of composite materials. In: Mechanical Behaviour of Materials. Proc. ICM-3, Cambridge, Pergamon Press, Vol. 3, 263-273.
(16) Laws, N. (1973). On the thermostatics of composite materials. J. Mech. Phys. Solids, 21, 9-17.
(17) Walpole, L.J. (1966). On bounds for the overall elastic moduli of inhomogeneous systems - I. J. Mech. Phys. Solids, 14, 151-162.
(18) Pedersen, O.B. (1983). Thermoelasticity and plasticity of composites - I. Mean field theory. Acta Met., 31, 1795-1808.
(19) Brown, L.M. & Clarke, D.R. (1977). The work hardening of fibrous composites with particular reference to the copper-tungsten system. Acta Met., 25, 563-570.
(20) Foreman, A.J.E. (1955). Dislocation energies in anisotropic crystals. Acta Met., 3, 322-330.
(21) Lowrie, R. & Gonas, A.M. (1965). Dynamic elastic properties of polycrystalline tungsten, 24°-180°C. J. Appl. Phys., 36, 2189-2192.
(22) Kelly, A. (1973). Strong Solids. Clarendon Press, Oxford.

APPENDIX

The numerical calculations in the present and previous papers (15,18) were made with the following values for the isotropic elastic constants of individual phases:

Material	μ/GP_a	ν	Reference
Copper	41.2	0.45	(20)
Tungsten	158.9	0.28	(21)
Aluminium	27	0.3	(22)
Boron	144	(1/3)	(22)
Epoxy	1.1	0.37	(22)
Glass	29	0.25	(22)

ELASTIC INTERACTION AND ELASTOPLASTIC DEFORMATION OF INHOMOGENEITIES

Jong K. Lee
Michigan Technological University, Houghton, MI 49931
William C. Johnson
Carnegie-Mellon University, Pittsburgh, PA 15213

ABSTRACT

Eshelby's ingenious works solved the elasticity problems associated with an inclusion in an infinite matrix. However, the elastic interaction and elastoplastic deformation behavior of inhomogeneities, which are critical to materials performance, have yet to receive careful examination. In this paper an integral equation approach, developed to solve a general inclusion problem, is first outlined. We then address some numerical results of inclusion problems obtained via this approach: they are the elastic stress fields associated with two inclusions and the elastoplastic deformation behavior under external stress. Finally, elastic interaction energies between two spherical inclusions are reexamined.

1. INTRODUCTION

The elastic interaction of precipitates or second phase particles is of fundamental importance in connection with precipitate alignment during coarsening, stress induced phase transformations including autocatalytic and sympathetic nucleation, and void lattice formation (1-9). Interaction energies depend strongly on particle morphologies, separation and relative orientation. Differences in elastic constants between precipitate and matrix can affect the morphological evolution of a precipitate and the qualitative nature of the elastic interaction between precipitates. Recent work (10) has demonstrated that particle interactions can lead to situations of reverse coarsening; small particles can grow at the expense of larger particles.

Numerous approaches have been employed in the calculation of interaction energies. Many assume elastic homogeneity even though this assumption can result in qualitatively inaccurate energies. Other approaches are often restricted to particles of specified shape. One of the first general treatments concerning particle interactions is due to Knops (11) in his attempt to extend Eshelby's work on the isolated precipitate. He wrote the displacement field in terms of an integro-differential equation assuming system isotropy for several spherical particles. This approach proved to be fruitful in that elastic inhomogeneity and anisotropy can be accounted for in a rigorous way. Subsequent investigators (12-14) developed and applied the integral equation approach to a variety of problems in materials science. In this paper we present a derivation of the integral equation for several arbitrarily shaped precipitates in a matrix, discuss several solution techniques as applied to specific problems, demonstrate how the integral equation can be extended to some simple problems involving the plastic relaxation around inclusions, and then develop expressions for interaction energies between two spherical inclusions.

2. DEVELOPMENT OF THE INTEGRAL EQUATION

In the present paper we consider the displacement and strain field associated with an infinite, inhomogeneous system possessing a stress-free transformation strain (eigenstrain) or applied strain. Our discussion will be limited to small displacements and strains and to linear elasticity. By inhomogeneous it is meant the elastic constants of the materials are a function of position. First, a reference system from which the displacements and strains can be measured is defined. Then an integral equation for the displacements is derived and this reference state used to determine the homogeneous solution to the integral equation.

A reference state is established from which strains are measured and is chosen to be a strain-free configuration. The strain-free reference state can be established in the manner of Eshelby. If any transformation strains (eigenstrains) are to be imparted to the material, this region or a part thereof is cut out and is allowed to undergo the transformation as given by the stress-free transformation strain e_{ij}^{T*}. The region removed is taken small enough so as to be essentially homogeneous. It can be assumed to behave according to the elastic properties of homogeneous matter, obeying a stress-strain constitutive equation appropriate to the material of the particular region. A system of surface forces is applied to the transformed region such that the region is returned to its original configuration. The element is defined to be strain free although it is not stress-free since it is being constrained by forces to its original shape. The element can now be reinserted exactly into the hole left from where it was removed and reattached to the contiguous matter. The system of surface forces confining each small element has become a layer of body forces distributed throughout volumes of materials where either the elastic constants or the stress-free transformation strain is changing. This procedure can be performed for each element of the material experiencing a transformation strain. After each of these regions has been reinserted into the solid, there is still no displacement of the material as a whole, i.e., it still corresponds exactly with the initial material except that it is not stress-free.

The resultant displacement field on relaxation of the surface and body forces existing in the reference state must satisfy the following conditions of mechanical equilibrium

$$\partial\sigma_{ij}/\partial x_j = 0 = \partial_j[C_{ijkl}(\vec{x})u_{k,l}(\vec{x})] \tag{1}$$

where σ_{ij} is the Cauchy stress tensor, $C_{ijkl}(\vec{x})$ are the elastic constants of the material at the point in question and a comma or ∂ denotes differentiation with respect to x_j. At equilibrium, Eqn (1) must be satisfied everywhere along with any specified external boundary conditions. Eqn (1) can be rewritten in the form

$$C^m_{ijkl}u_{j,ik}(\vec{x}) = \partial_k\{[C^m_{ijkl} - C_{ijkl}(\vec{x})]u_{j,i}(\vec{x})\} \tag{2}$$

where C^m_{ijkl} are the elastic constants of some designated region, for example, the matrix phase. A solution to Eqn (2) can now be obtained using a Green's function technique. The left hand side of the equation serves to define the Green's function as

$$C^m_{ijkl}G_{km,lj}(\vec{x} - \vec{x}') + \delta_{im}\delta(\vec{x} - \vec{x}') = 0 \tag{3}$$

where $G_{km}(\vec{x} - \vec{x}')$ is the elastic Green's function and $\delta(\vec{x} - \vec{x}')$ is the Dirac delta function. The formal solution for the displacement is thus given by

$$u_j(\vec{x}) = u^h_j(\vec{x}) - \iiint_\infty G_{jl}(\vec{x} - \vec{x}')\partial'_k\{[C^m_{inkl} - C_{inkl}(\vec{x}')]\partial'_i u_n(\vec{x}')\}dV(\vec{x}') \tag{4}$$

where the integral is taken over all space and $u^h_j(\vec{x})$ is the homogeneous solution to Eqn (2), i.e., when the right hand side of Eqn (2) equals zero.

Integrating Eqn (4) by parts, invoking the divergence theorem, assuming that far from the region of interest the displacement and strain are zero or $C_{inkl}(\vec{x}) = C^m_{inkl}$ and differentiating to give the distortion yields

$$u_{j,n}(\vec{x}) = u^h_{j,n}(\vec{x}) + \iiint_\infty \Delta C_{imkl}(\vec{x}')G_{jl,kn}(\vec{x} - \vec{x}')\partial'_i u_m(\vec{x}')dV(\vec{x}'), \tag{5}$$

$$\Delta C_{inkl}(\vec{x}) = C_{inkl}(\vec{x}) - C^m_{inkl}.$$

Eqn (5) is a system of integral equations which must be solved simultaneously for the elastic distortions.

The homogeneous term $u^h_{j,n}(\vec{x})$ can be calculated for desired transformation strains and imposed displacement fields. To do so it must be remembered that the reference state is the state of zero strain but is not necessarily stress-free, the stresses arising from the process by which the transformation strains are imparted to the material. The magnitude of these initial stresses is related directly to the elastic constants of the small region to which the transformation strain has been given. It is these stresses that contribute to the straining of the material as measured from our reference state. The homogeneous solution of the differential equation corresponds to the relaxation of the real stresses engendered by the creation of the reference state in a system in which the elastic constants are everywhere taken as uniform. If no region of the system undergoes a transformation strain, the homogeneous solution becomes the applied displacement.

For the specific case of two precipitates with uniform elastic constants, C^*_{ijkl}, embedded in an infinite matrix, the integral over infinity appearing in Eqn (5) reduces to volume integrals over each of the precipitates. To calculate the homogeneous solution, refer once again to the reference system. Consider one of the precipitates removed from the matrix and the stress-free transformation strain imparted. The surface forces $-C^*_{ijkl}e^{T*}_{kl}n_j$ must be applied to the precipitate to return every element of the precipitate to its original shape. Inserting the precipitate back into the matrix and reattaching it leaves a distributed layer of body forces. The homogeneous solution is now obtained by assuming the elastic constants of the precipitate to be equal to those of the matrix (note that it is necessary to compute the body forces using the elastic constants of the actual precipitate). The distortion field arising from the relaxation (i.e., application of forces of equal magnitude but opposite sign) of the surface forces used to reinsert the transformed precipitate into the matrix is by definition of the Green's function

$$u^h_{j,n}(\vec{x}) = \int\int_{S_A+S_B} [C^*_{iklm}e^{T*}_{lm}n_k]G_{ji,n}(\vec{x} - \vec{x}')\,dS(\vec{x}') \qquad (6)$$

where S_A and S_B are over the surfaces of the precipitates A and B, respectively. Using the divergence theorem with Eqn (6) and combining with Eqn (5) gives the following integral equation for the displacement field associated with two arbitrarily shaped precipitates with uniform transformation strains in the presence of an applied stress field as

$$u_{m,n}(\vec{x}) = u^a_{m,n}(\vec{x}) - \sigma^{T*}_{kl}\int\int\int_{V_A+V_B} G_{ml,kn}(\vec{x} - \vec{x}')\,dV(\vec{x}') \qquad (7)$$
$$+ \Delta C_{ijkl}\int\int\int_{V_A+V_B} G_{ml,kn}(\vec{x} - \vec{x}')\partial'_j u_i(\vec{x}')\,dV(\vec{x}')$$

where $\sigma^{T*}_{kl} = C^*_{klij}e^{T*}_{ij}$. The integral equation can be written for any number of precipitates in a straightforward manner.

3. SOLUTION OF THE INTEGRAL EQUATION

In general, the solution of Eqn (7) is analytically intractable. An approximate, but analytic, approach is to assume that the difference in elastic constants between precipitate and matrix is not too large and then to use the Born approximation. This technique is especially suitable to precipitates that possess no misfit strains and are subjected to a uniformly applied stress field (15). Under such conditions, Eqn (7) may be approximated as

$$u_{m,n}(\vec{x}) = u^a_{m,n} + C_{ijkl}\, u^a_{i,j} \int\int\int_{V_A+V_B} G_{ml,kn}\,(\vec{x} - \vec{x}')\,dV(\vec{x}') \quad (8)$$

For an accurate solution one has to use a numerical method which usually employs an iterative scheme (12, 16). Since it is instructive to know the behavior of the stress field of a two-inclusion system, several numerical results are presented here. The results are based on a technique in which the integral equation is first discretized by setting up a number of grid points within the inclusions (17). The integral equation is then treated as a set of non-linear simultaneous equations whose roots are obtained through a standard iterative scheme, such as the Newton-Raphson method. Table 1 compares the present results with other methods for the case of two rigid spheres under both uniaxial and hydrostatic applied stress. For this problem, Shelly and Yu (18) employed the Boussinesq stress-function approach, while Moschovidis and Mura (19) introduced a modified equivalent inclusion method (MEIM). Fig. 1 shows the alignment of two spherical inclusions along the X_3-axis. The center of the first particle with the radius a_A is situated at (o, o, d) while the center of the other particle with radius a_B is at the origin. From Table 1, the integral equation results are found to agree quite well with the other methods, especially with the Boussinesq stress-function approach to within the three significant figures. The first order MEIM of Moschovidis and Mura (19) also provides reasonable numbers but with some significant error indicating the need to include the higher order terms.

Table 1

Comparisons of methods for two rigid spherical inclusions. For this system $\nu^* = \nu = 1/4$, $a_A = a_B = a$, $\vec{d} = (0, 0, 3.62a)$, and σ_{ij} is in units of σ_A.

	$\sigma^A_{ij} = \sigma_A\,\delta_{3i}\,\delta_{3j}$				$\sigma^A_{ij} = \sigma_A\,\delta_{ij}$			
	$\vec{X} = (0,0,-a)$		$\vec{X} = (0,0,a)$		$\vec{X} = (0,0,-a)$		$\vec{X} = (0,0,a)$	
	σ_{11}	σ_{33}	σ_{11}	σ_{33}	σ_{11}	σ_{33}	σ_{11}	σ_{33}
Ref. (18)	0.679	2.030	0.771	2.310	0.607	1.820	0.603	1.890
Ref. (19)	0.703	2.063	0.691	2.198	0.617	1.828	0.599	1.867
IEM	0.673	2.025	0.760	2.292	0.605	1.819	0.628	1.890

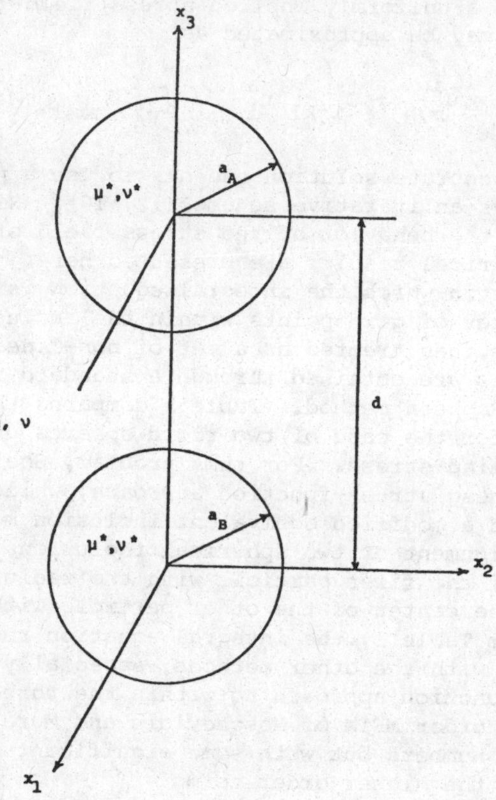

Fig. 1. Arrangement of two spherical inhomogeneities

Figure 2 shows the σ_{33} stress component along the X_3-axis within and without the spherical inclusion located at the origin under a uniaxial loading of $\sigma_{ij}^A = \sigma_A\delta_{3i}\delta_{3j}$. For this case, $a_A = a_B = a$, $\nu^* = \nu = 1/3$, $d = 4a$ and five different values of $\gamma = \mu^*/\mu = 0, 1/3, 1, 3, \infty$ are considered. In the absence of the second particle, the stress field within an isolated spherical particle is a constant (20). It is clearly shown that the constancy of the σ_{33} stress field is no longer held under the influence of the second particle, and the pertubation is stronger the harder the particles, and more pronounced in the regions near the second particle.

In order to see the influence of the second particle, Fig. 3 shows the σ_{33} stress component in an enlarged picture under the condition of $\gamma = 3$, $d = 4a$, $\sigma_{ij}^A = \sigma_A\delta_{3i}\delta_{3j}$. For comparison, the stress component obtained through superposition of an individual particle's stress field is given as a broken curve. Even for this large intercenter distance, $d = 4a$, we find that there is a significant interaction term, $\sigma_{33}^{int} = \sigma_{33}(\text{total}) - \sigma_{33}(\text{superposition})$. For a shorter intercenter distance,

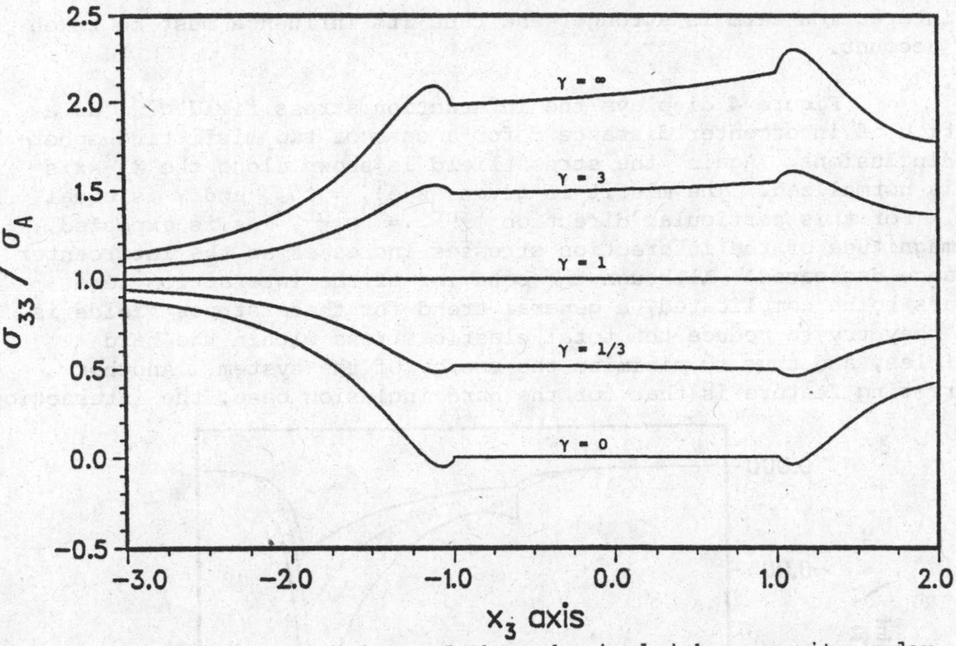

Fig. 2. The σ_{33} stress in/around the spherical inhomogeneity under $\sigma^A_{ij}=\sigma_A\delta_{3i}\delta_{3j}$ along the X_3-axis. $\nu^*=\nu=\frac{1}{3}$, $a_A=a_B=1$, and $d=4a_A$.

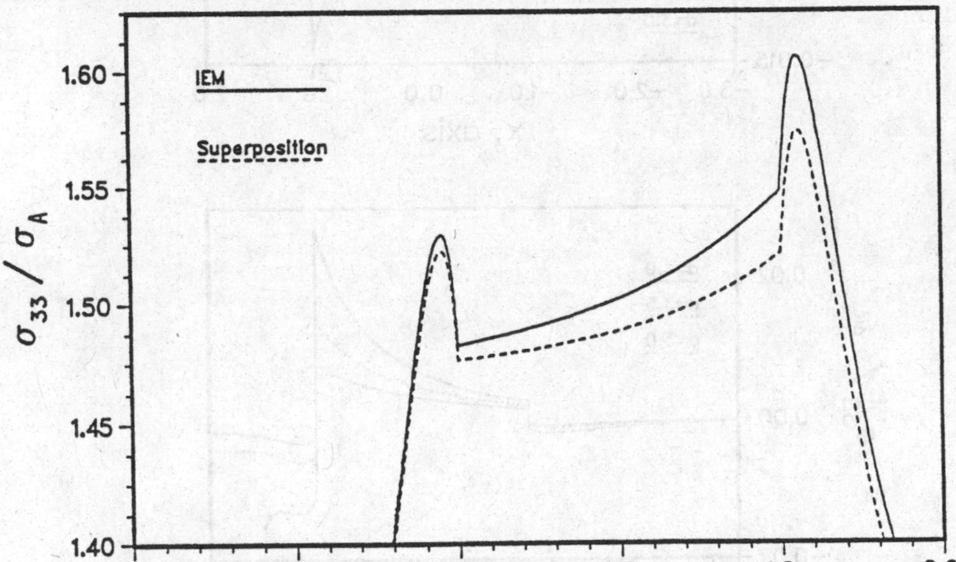

Fig. 3. The σ_{33} stress under $\sigma^A_{ij}=\sigma_A\delta_{3i}\delta_{3j}$ along the X_3-axis in the case of $\gamma=3$. The broken line indicates the stress produced through superposition. $\nu^*=\nu=\frac{1}{3}$, $a_A=a_B=1$, and $d=4a_A$.

the interaction term is stronger and thus its influence must be taken into account.

Figure 4 displays the interaction stress field σ_{ij}^{int} as a function of intercenter distance d for a case of two misfitting spherical inclusions. Again, the stress field is shown along the X_3-axis and is normalized. The misfit is given as $e_{ij}^{T*} = \varepsilon\delta_{ij}$ and γ is equal to 3. For this particular direction $\sigma_{11}^{int} = \sigma_{22}^{int}$. As is expected, the magnitude of the interaction stresses increases as the intercenter distance decreases. Although the behavior of the interaction terms appears to be complicated, a general trend for these stress fields is that they try to reduce the total elastic stress within the hard particles, and thus to minimize the energy of the system. Another interesting feature is that for the hard inclusion case, the interaction

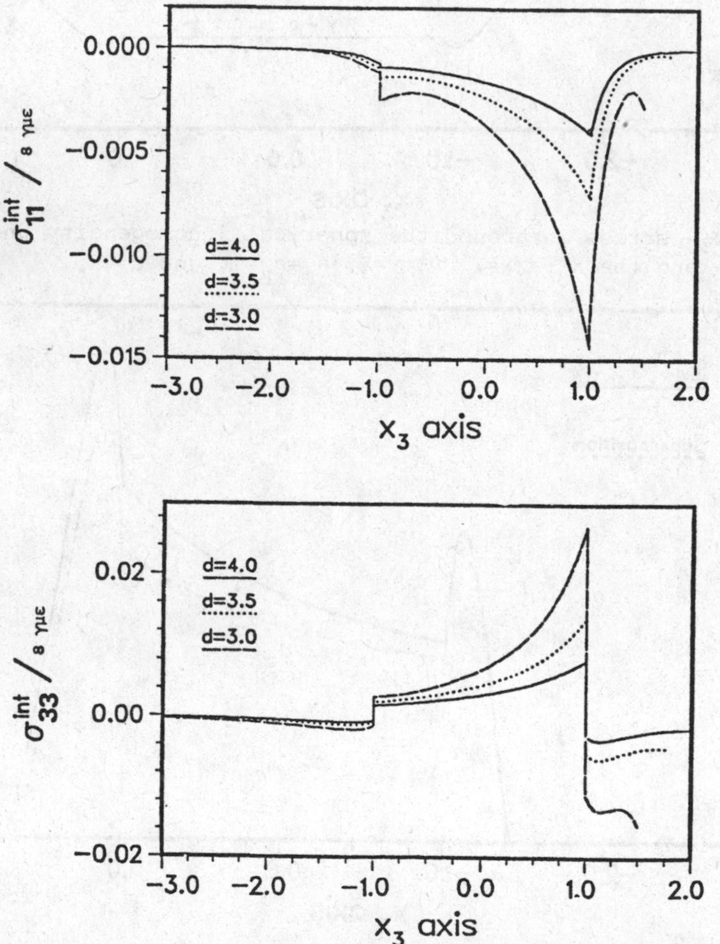

Fig. 4. The interaction stress σ_{ij}^{int} under the $e_{ij}^{T*} = \varepsilon\delta_{ij}$ condition along the X_3-axis for two misfitting spherical particles. $\nu^* = \nu = \frac{1}{3}$, $a_A = a_B = 1$ and d is in units of a_A.

stress in the region between the two inclusions is compressive (when the misfit is positive), whereas for the soft inclusion the stress in the ligament is tensile.

4. PLASTIC RELAXATION OF ELASTIC STRESS FIELD

Plastic relaxation of the transformation stress and elastic strain energy associated with the formation of misfitting precipitates has been often observed (21). Even if the applied stress is less than the plastic yield stress of the matrix phase, the induced stress field of an inclusion can cause the local stress field to exceed the yield stress and thus provide plastic deformation in the matrix phase as well as the inclusion. Solutions to inclusion problems in elastoplasticity are usually complex, especially for the three-dimensional problem. If radial symmetry exists, relatively simple expressions can be obtained for the stress, strain, and the strain energy as shown by recent works (22, 23). However, most cases do not exhibit radial symmetry and recourse must often be made to plane stress or plane strain assumptions.

By assuming that the inclusion remains elastic and thus plastic deformation occurs only in the matrix adjacent to the inclusion, Johnson and Lee (24) derived an integral equation for the calculation of the elastoplastic strain field:

$$\begin{aligned} u_{m,n}(\vec{x}) &= u^{a}_{m,n}(\vec{x}) - \sigma^{T*}_{kl} \iiint_V G_{ml,kn}(\vec{x} - \vec{x}')\,dV(\vec{x}') \\ &+ \Delta C_{ijkl} \iiint_V G_{ml,kn}(\vec{x} - \vec{x}')\,\partial'_j u_i(\vec{x})\,dV(\vec{x}') \\ &+ C_{ijkl} \iiint_M e^{P}_{ij}(\vec{x}')\,G_{ml,kn}(\vec{x} - \vec{x}')\,dV(\vec{x}'), \end{aligned} \qquad (9)$$

where e^{P}_{ij} is plastic strain, V is the inclusion volume and M indicates integration over the plastic zone. In the derivation, the plastic relaxation is assumed to be independent of strain rate and stress orientation. Fig. 5 shows the geometry of the plastic zone for a relatively hard precipitate in which $\gamma = 3$. The figures show one quadrant of the plane $X_2 = 0$. The solid line represents the precipitate-matrix interface and the broken line the extent of the plastic relaxation. Linear strain hardening is employed for ideal plastic behavior with zero hardening coefficient. The yield stress is taken as $\sigma_y = 10^{-3}\mu$, the left figure is for $\sigma_A = 0.80\ \sigma_y$ and the right for $\sigma_A = 0.85\ \sigma_y$. The majority of the plastic relaxation takes place in the direction of the applied stress field. What is interesting to note is the presence of an elastic region completely surrounded by a plastic zone situated on the precipitate-matrix interface and extending from $X_1/a = 0$ to $X_1/a \cong .38$ even for the case $\sigma_A = 0.85\ \sigma_y$.

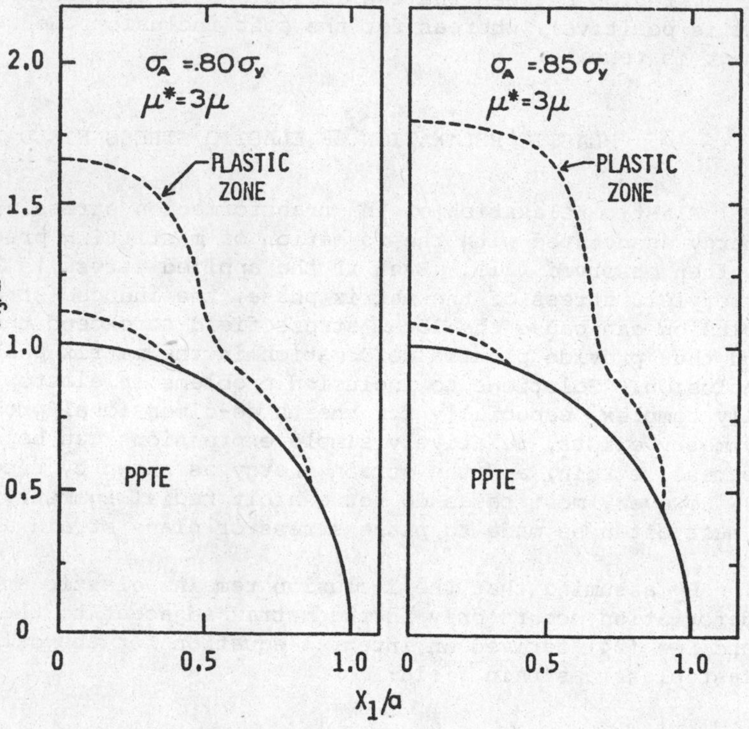

Fig. 5. The geometry of the plastic zone for a spherical particle with $\gamma = 3$ under $\sigma^A_{ij} = \sigma_A\delta_{3i}\delta_{3j}$. The solid line represents the particle-matrix interface while the broken line indicates the extent of the plastic relaxation (24).

5. INTERACTION ENERGIES BETWEEN TWO SPHERICAL INCLUSIONS

In principle, complete knowledge of the elastic strain field will allow calculation of the interaction energy between two inclusions. The absence of analytical solution for Eqn (7), however, permits no derivation of a general expression for the elastic interaction energy dependent on the relative positions of A and B. In view of this difficulty, we will consider the two spherical inclusions for which Eshelby's equivalency method is useful. Plastic relaxation will not be included. Let us consider first the interaction energy between the applied stress, σ^A_{ij} and an isolated spherical inclusion with a stress-free transformation strain e^T_{ij}. The interaction energy, which is the change in the free energy of the system, is given by (3, 20)

$$\Delta G = - \iiint_V \sigma^A_{ij}\, e^T_{ij}\, dV. \tag{10}$$

If the inclusion is an inhomogeneity, e^T_{ij} becomes the equivalent stress-free transformation strain obtainable from

$$C_{ijkl}(e^C_{kl} - e^T_{kl} + e^A_{kl}) = C^*_{ijkl}(e^C_{kl} + e^A_{kl}) \tag{11}$$

where e^C_{kl} is the constrained elastic strain which is given by

$$e^C_{kl} = \tfrac{1}{2}(u_{k,l} + u_{l,k}) - e^A_{kl}. \tag{12}$$

5.1 TWO SPHERICAL INHOMOGENEITIES UNDER $\sigma^A_{ij} = \sigma_A \delta_{3i}\delta_{3j}$ and $e^{T*}_{ij} = 0$

Eshelby (20) pointed out that in the case of two inclusions, Eqn (10) is still the interaction energy term if we refer σ^A_{ij} to the stress produced by one inclusion and e^T_{ij} to the stress-free strain associated with the other. Expanding this concept to the present case, we may consider a series of applied stresses acting on the particle A: σ^A_{ij}, $\sigma^C_{ij}(B)$, $\sigma^{C2}_{ij}(B)$, $\sigma^{C3}_{ij}(B)$, . . . where σ^A_{ij} is the externally applied stress, $\sigma^C_{ij}(B)$ the σ^A_{ij}-induced elastic stress associated with the particle A, $\sigma^{C2}_{ij}(B)$ is the secondary $\sigma^C_{ij}(A)$-induced stress associated with the particle B, but acting on the particle A, and so on. When σ^A_{ij} is a constant, $\sigma^C_{ij}(B)$ is proportional to d^{-3} and $\sigma^{C2}_{ij}(B)$ is proportional to d^{-6}. Similarly, there will be a series of stress-free transformation strains associated with particle A: e^T_{ij}, e^{T2}_{ij}, e^{T3}_{ij}, . . . where e^T_{ij} is the equivalent stress-free transformation strain from Eqn (11), e^{T2}_{ij} is the secondary equivalent stress-free strain induced by $\sigma^C_{ij}(B)$, e^{T3}_{ij} the ternary equivalent term due to $\sigma^{C2}_{ij}(B)$. Collecting cross terms among the applied stresses and the equivalent stress free strains, the total interaction energy may be given, in the order of the intercenter distance dependence,

$$\Delta G = -\iiint_{V_A} 2\sigma^A_{ij} e^T_{ij}\, dV - \iiint_{V_A} (\sigma^C_{ij}(B) e^T_{ij} + 2\sigma^A_{ij} e^{T2}_{ij})\, dV -$$

$$\iiint_{V_A} (\sigma^C_{ij}(B) e^{T2}_{ij} + \sigma^{C2}_{ij}(B) e^T_{ij} + 2\sigma^A_{ij} e^{T3}_{ij} + \sigma^{I2}_{ij} e^{T2}_{ij})\, dV + O(d^{-9}), \tag{13}$$

where the factor two in front of σ^A_{ij} reflects two identical inclusions and $\sigma^{I2}_{ij} e^{T2}_{ij}$ represents self inclusion energy dependent on the distance d. In Eqn (13), the first term is independent of d, the second term proportional to d^{-3} and the third term to d^{-6}. Defining E_{int} as the interaction energy depending upon the relative positions of the two particles, and neglecting terms of short-range nature, we get

$$E_{int} = -\iiint_{V_A} \sigma^C_{ij}(B) e^T_{ij}\, dV - 2\iiint_{V_A} \sigma^A_{ij} e^{T2}_{ij}\, dV. \tag{14}$$

Evaluation of the first term in Eqn (14) is straight forward (20). When $\nu^* = \nu = 1/3$ and $\gamma = \mu^*/\mu$, Eqn (11) yields

$$e^T_{ij} = \alpha\, e^A_{kk}\,\delta_{ij} + \beta\, e^A_{ij}, \tag{15}$$

where $\alpha = 3(1-\gamma)^2/(8+7\gamma)(1+2\gamma)$ and $\beta = 15(1-\gamma)/(8+7\gamma)$. If we introduce $\sigma^A_{ij} = \sigma_A\delta_{3i}\delta_{3j}$ into Eqn (15), we obtain, for σ^T_{ij},

$$\sigma^T_{ij} = \sigma_A(\alpha\,\delta_{ij} + \beta\,\delta_{3i}\delta_{3j}). \tag{16}$$

Since $e^C_{ij}(B) = \sigma^T_{kl}\,(3\,\psi_{,ijkl} - 4\delta_{ik}\phi_{,jl} - 4\delta_{jk}\phi_{,il})/32\pi\mu$, where ϕ is the harmonic potential and ψ is the biharmonic potential function, we get, for $|\vec{x}| = r > a_B$,

$$e^C_{ij}(B) = \frac{\sigma_A\, a_B^3}{24\mu}\Big[\frac{(2\alpha-3\beta)\delta_{ij} + 2\beta\delta_{i3}\delta_{j3}}{r^3} + \frac{6\beta(X_i\delta_{j3}+X_j\delta_{i3})X_3 + (9\beta-6\alpha)X_iX_j + 9\beta\delta_{ij}X_3^2}{r^5} - \frac{45\beta X_iX_jX_3^2}{r^7}\Big] + \frac{\sigma_A a_B^5\beta}{40\mu}\Big[\frac{6\delta_{i3}\delta_{j3} + 3\delta_{ij}}{r^5} - \frac{30(X_i\delta_{j3} + X_j\delta_{i3})X_3 + 15X_iX_j + 15\delta_{ij}X_3^2}{r^7} + \frac{105X_iX_jX_3^2}{r^9}\Big]. \tag{17}$$

Substitution Eqn (16) into the first term of Eqn (14) and integrating the result:

$$-\iiint_{V_A} \sigma^C_{ij}(B)e^T_{ij}dV = \frac{\pi\sigma_A^2 a_A^3 a_B^3}{18\mu d^3}\,[\beta^2 - 4\alpha\beta + (12\alpha\beta - 30\beta^2)\,X_p^2 + 45\beta^2X_p^4 + \frac{9}{5}\,\frac{(a_A^2 + a_B^2)\beta^2}{d^2}\,(-3 + 30X_p^2 - 35X_p^4)] \tag{18}$$

where X_p is the directional cosine of the intercenter distance vector $\vec{d}$ with respect to the orientation of the applied stress and is equal to $d_3/|\vec{d}|$. Starting with Eqn (8), Johnson (15) recently obtained Eqn (18) in the limiting case of $\gamma \simeq 1$. When $\gamma \simeq 1$, $\alpha << |\beta|$. Thus, if we neglect terms involving $\alpha\beta$, Eqn (18) becomes Johnson's Eqn (17).

The second term in Eqn (14) requires exact evaluation of the secondary equivalent stress-free strain e_{ij}^{T2} which is not, at present, analytically tractable. However, when $d>>a_A$, we may treat $e_{ij}^{C}(B)$ as a constant strain acting on A. Under this condition, an equivalent equation similar to Eqn (11) may be written for e_{ij}^{T2}:

$$C_{ijkl}(e_{kl}^{C2} - e_{kl}^{T2} + e_{kl}^{C}(B)) = C_{ijkl}^{*}(e_{kl}^{C2} + e_{kl}^{C}(B)). \qquad (19)$$

Substituting $e_{ij}^{C}(B)$ for e_{ij}^{A} into Eqn (15), we obtain, for e_{ij}^{T2},

$$e_{ij}^{T2} \simeq \alpha e_{kk}^{C}(B)\delta_{ij} + \beta e_{ij}^{C}(B). \qquad (20)$$

Use of Eqn (20) yields

$$\sigma_{ij}^{A} e_{ij}^{T2} = \sigma_A \delta_{3i}\delta_{3j}\,[\alpha e_{kk}^{C}(B)\delta_{ij} + \beta e_{ij}^{C}(B)]$$

$$= \sigma_A\,[\alpha e_{kk}^{C}(B) + \beta e_{33}^{C}(B)]$$

$$= \sigma_{ij}^{C}(B) e_{ij}^{T}. \qquad (21)$$

It is surprising to see that if $e_{ij}^{C}(B)$ is treated as a constant, $\sigma_{ij}^{A} e_{ij}^{T2} = \sigma_{ij}^{C}(B) e_{ij}^{T}$. Although it is assumed that $d>>a_A$, Eqn (20) should be a good approximation if we consider $e_{ij}^{C}(B)$ as the average value over the volume V_A. Therefore, the interaction energy between two identical spherical inhomogeneities becomes

$$E_{int} \simeq -3\iiint_{V_A} \sigma_{ij}^{C}(B) e_{ij}^{T} dV$$

$$= \frac{\pi\sigma_A^2 a^6}{6\mu d^3}\Big[\beta^2 - 4\alpha\beta + (12\alpha\beta - 30\beta^2)X_p^2 + 45\beta^2 X_p^4$$

$$+ \frac{18a^2\beta^2}{5d^2}(-3 + 30X_p^2 - 35X_p^4)\Big]. \qquad (22)$$

Figure 6 displays E_{int}, in units of $\pi\sigma_A^2 a^6/6\mu d^3$, as a function of the orientation angle $\theta_p = \cos^{-1}X_p$. To include the second term in the bracket of Eqn (22), d is taken as 3a and several different γ values are considered. We note that the behavior of two voids ($\gamma = 0$) is similar to that for two rigid spheres. Likewise, the interaction between two relatively-hard particles ($\gamma = 3$) is very similar to the case of two relatively-soft particles ($\gamma = 1/3$). We also note that the nature of the interaction depends not on the sense of applied stress (whether compressive or tensile) but on the inclusion-orientation relative to the direction of applied stress. It is also interesting to see that the minimum interaction energy occurs at the angle $\theta_p \simeq 55°$ which is the directional cosine angle in the bcc structure of a void lattice (9).

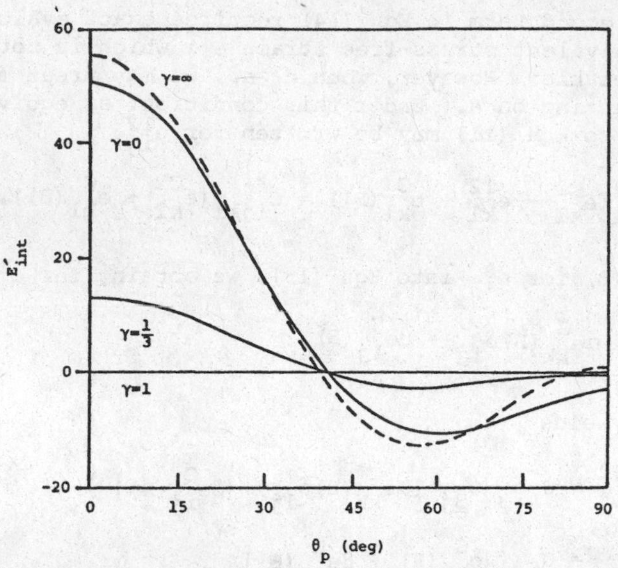

Fig. 6. Position-dependent interaction energy between two spherical inhomogeneities under $\sigma_{ij}^{A} = \sigma_A \delta_{3i}\delta_{3j}$. θ_p is the angle between the applied stress direction and the intercenter direction vector $\vec{d}$. $\nu^* = \nu = \frac{1}{3}$, $a_A = a_B$ and $d = 3a_A$.

5.2 TWO SPHERICAL PARTICLES WITH $e_{ij}^{T*} = \varepsilon(\delta_{ij} + t\delta_{3i}\delta_{3j})$

Since applied stress is zero, Eqn (13) yields

$$E_{int} \simeq -\iiint_{V_A} \sigma_{ij}^{C}(B) e_{ij}^{T} dV - \iiint_{V_A} (\sigma_{ij}^{C}(B) e_{ij}^{T2} + \sigma_{ij}^{C2}(B) e_{ij}^{T} + \sigma_{ij}^{I2} e_{ij}^{T2}) dV, \quad (23)$$

where e_{ij}^{T} is the equivalent stress-free strain obtained from (20)

$$C_{ijkl}(e_{kl}^{C} - e_{kl}^{T}) = C_{ijkl}^{*}(e_{kl}^{C} - e_{kl}^{T*}). \quad (24)$$

When $e_{ij}^{T*} = \varepsilon(\delta_{ij} + t\delta_{3i}\delta_{3j})$, Eqn (24) yields, for σ_{ij}^{T},

$$\sigma_{ij}^{T} = \mu\ (\alpha_1 \delta_{ij} + \beta_1 \delta_{3i}\delta_{3j}), \quad (25)$$

where $\alpha_1 = 6\gamma\varepsilon(32 + 28\gamma + 9t + 6\gamma t)/(8 + 7\gamma)(1 + 2\gamma)$ and $\beta_1 = 30\gamma\varepsilon t/(8 + 7\gamma)$. Since Eqn (25) is the same form as Eqn (16), the interaction energy between two misfitting spherical particles with a tetragonal defect ($t \neq 0$) is given by

$$E_{int} \simeq - \iiint_{V_A} \sigma^C_{ij}(B) e^T_{ij} dV$$

$$= \frac{\pi\mu a_A^3 a_B^3}{18d^3} [\beta_1^2 - 4\alpha_1\beta_1 + (12\alpha_1\beta_1 - 30\,\beta_1^2)X_p^2 + 45\beta_1^2 X_p^4$$

$$+ \frac{9}{5} \frac{(a_A^2 + a_B^2)\beta_1^2}{d^2} (-3 + 30X_p^2 - 35X_p^4)]. \quad (26)$$

Short range order terms are neglected. In Fig. 7, E_{int}', in units of $\pi\mu a^6\varepsilon^2/18d^3$, is plotted as a function of the orientation angle, θ_p. For this calculation, the tetragonal defect t is taken as -1 and $d = 3a$. In this interaction, the nature of the tetragonal defect t (whether negative or positive) determines the interaction behavior. For example, when $t = +1$, $E_{int} > 0$ for the range of $0° < \theta_p \simeq 50°$ which is nearly opposite to the behavior shown in Fig. 7.

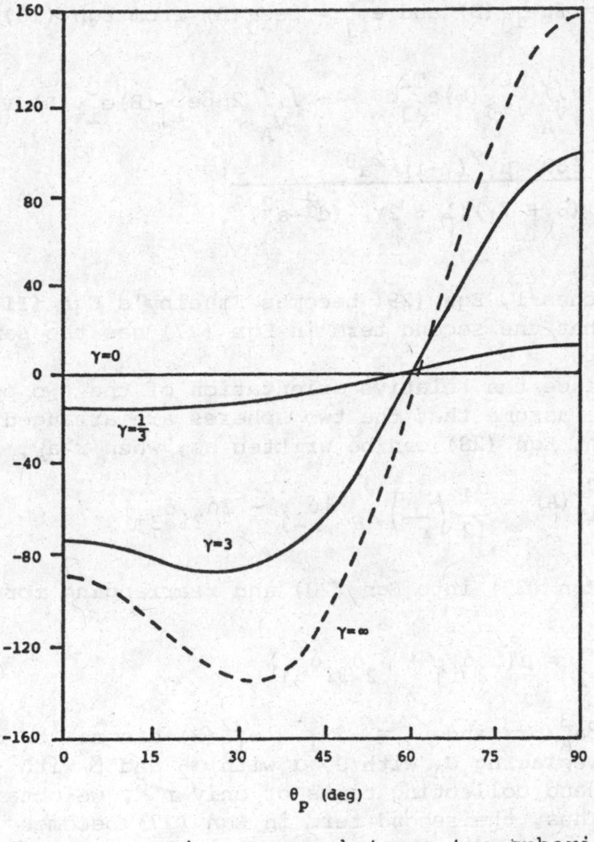

Fig. 7. Interaction energy between two spherical particles with $e^{T*}_{ij} = \varepsilon(\delta_{ij} + t\delta_{3i}\delta_{3j})$. θ_p is the angle between the X_3-axis and the intercenter distance vector $\vec{d}$. $\nu^* = \nu = \frac{1}{3}$, $a_A = a_B$, $d = 3a_A$ and $t = -1$.

When $t = 0$, $\beta_1 = 0$ and thus Eqn (26) has no contribution to the interaction between two misfitting spheres. This is a well-known result discussed by Eshelby (1). Therefore, when the defect is purely dilatational, the leading term is proportional to d^{-6} and given by, from Eqn (23),

$$E_{int} \simeq - \iiint_{V_A} (\sigma^{C}_{ij}(B)e^{T2}_{ij} + \sigma^{C2}_{ij}(B)e^{T}_{ij} + \sigma^{I2}_{ij}e^{T2}_{ij})\,dV. \tag{27}$$

To evaluate the first term in Eqn (27), use of Eqn (17) and (20) will be made. Replacing σ_A with μ, α with α_1, and β with 0 (because $\beta_1 = 0$) in Eqn (17):

$$e^{C}_{ij}(B) = \frac{\alpha_1 a_B^3}{12}\left[\frac{\delta_{ij}}{r^3} - \frac{3X_iX_j}{r^5}\right]. \tag{28}$$

Since $\sigma^{C}_{ij}(B) = 2\mu e^{C}_{ij}(B)$ and $e^{T2}_{ij} = \beta e^{C}_{ij}(B)$ from Eqn (20), we have

$$- \iiint_{V_A} \sigma^{C}_{ij}(B)e^{T2}_{ij}dV = - \iiint_{V_A} 2\mu\beta e^{C}_{ij}(B)e^{C}_{ij}(B)\,dV$$

$$= \frac{960\ \pi\mu\gamma^2(\gamma-1)\varepsilon^2a^9}{(8 + 7\gamma)(1 + 2\gamma)^2(d^2-a^2)^3}. \tag{29}$$

When γ approaches 1, Eqn (29) becomes Eshelby's Eqn (II 12). However, we will see that the second term in Eqn (27) has the same magnitude.

Since the relative orientation of the two particles is immaterial, we assume that the two spheres are arranged along the X_3-axis. Then, Eqn (28) can be written as, when $r>a_A$,

$$e^{C}_{ij}(A) = \frac{\alpha_1}{12}\left(\frac{a_A}{r}\right)^3 \{\delta_{ij} - 3\delta_{3i}\delta_{3j}\}. \tag{30}$$

Substituting Eqn (30) into Eqn (20) and rearranging for σ^{T2}_{ij}:

$$\sigma^{T2}_{ij} = \mu(\alpha_2\delta_{ij} + \beta_2\delta_{3i}\delta_{3j}) \tag{31}$$

where $\alpha_2 = \alpha_1\beta a_A^3/6r^3$ and $\beta_2 = -3\alpha_1$. $e^{C2}_{ij}(B)$ can now be obtained from Eqn (17), by replacing σ_A with μ, α with α_2 and β with $-3\alpha_2$. Making use of $X_3 = r$ and collecting terms of only r^{-6}, we obtain $e^{C2}_{kk}(B) = \alpha_1\beta a^6/12r^6$. Thus, the second term in Eqn (27) becomes

$$-\iiint_{V_A} \sigma^{C2}_{ij}(B)e^{T}_{ij}dV = - \iiint_{V_A} \mu\alpha_1\delta_{ij}e^{C2}_{ij}(B)\,dV =$$

$$- \mu\alpha_1 \iiint_{V_A} e^{C2}_{kk}(B)\,dV = \frac{960\pi\mu\gamma^2(\gamma-1)\varepsilon^2a^9}{(8+7\gamma)(1+2\gamma)^2(d^2-a^2)^3}. \tag{32}$$

Although d>>a is assumed in deriving Eqn (30), (31) and (32), it is quite surprising to see that the second term in Eqn (27) has a strength equal to the first term.

Finally, we consider the last term in Eqn (27). From Eqn (31), $e_{ij}^{T2} = (\alpha_2\delta_{ij} + \beta_2\delta_{3i}\delta_{3j})/2$. The equivalency relationship for an interior point (20) yields $e_{ij}^{I2} = -8e_{ij}^{T2}/15$. Therefore, we get

$$-\iiint_{V_A} \sigma_{ij}^{I2} e_{ij}^{T2} dV = -2\mu \iiint_{V_A} (e_{kk}^{I2}\delta_{ij} + e_{ij}^{I2}) e_{ij}^{T2} dV$$

$$= \frac{8(\gamma-1)}{(8+7\gamma)} \left[-\iiint_{V_A} \sigma_{ij}^{C}(B) e_{ij}^{T2} dV\right] . \qquad (33)$$

Eqn (33) is proportional to $(\gamma-1)^2$ while the first two terms are proportional to $(\gamma-1)$. Certainly, when $\gamma \simeq 1$, this second-order contribution is negligible compared with the first order terms as pointed out by Eshelby (1). Recently, using Eqn (8), Johnson (10) derived this second-order term in the limit of $\gamma \to 1$. However, because of different approximation, his expression shows a factor of 2/3 rather than 8/15 shown in Eqn (33). We note that when γ is very different from unity, the strength of the second order term is comparable to the first two terms of Eqn (29). Combining Eqn (29), (32) and (33), we find, for the interaction energy between two misfitting spherical particles,

$$E_{int} = \frac{1920\pi(4+11\gamma)\gamma^2(\gamma-1)\mu\varepsilon^2 a^9}{(8+7\gamma)^2(1+2\gamma)^2(d^2-a^2)^3} . \qquad (34)$$

ACKNOWLEDGMENT

This work was supported by the Divison of Materials Research of the National Science Foundation, under Grants DMR 81-16363 (contribution of J. K. L.) and DMR 82-19783 (contribution of W. C. J.) for which much appreciation is expressed.

REFERENCES

(1) Ardell, A. J., Nicholson, R. B. and Eshelby, J. D.: Acta Met., 14, 1295 (1966).
(2) Khachaturyan, A. G., Theory of Structural Transformations in Solids, Wiley, 1983, p. 198.
(3) Mura, T, Micromechanics of Defects in Solids, Martinus Nijhoff Publishers, 1982, p. 63.
(4) Seitz, E. and de Fontaine, D.: Acta Met., 26, 1671 (1978).
(5) Johnson, W. C. and Lee, J. K.: Metall. Trans., 10A, 1141 (1979).
(6) Perovic, V., Purdy, G. R. and Brown, L. M.: Acta Met., 27, 1075 (1979)

(7) Mihazaki, T., Imamura, H. and Kozakai, T.: Mat. Sci. & Eng., 54, 9 (1982).
(8) Russell, K. C., Barnett, D. M., Altstetter, C. J., Aaronson, H. I., and Lee, J. K.: Scripta Met., 11, 485 (1977).
(9) Evans, J. H., Bullough, R. and Stoneham, A. M.: Radiation Induced Voids in Metals, eds. Corbett, J. W. and Ianniello, L. C., U. S. Atomic Energy Commission, 1972, p. 522.
(10) Johnson, W. C.: To appear in Acta Metallurgica.
(11) Knops, R.J.: Proc. Edinburgh Math. Society, 14, 61 (1964)
(12) Chen, F. C. and Young, K.: J. Math. Physics, 18, 1412 (1977).
(13) Gubernatis, J. E. and Krumhansl, J. A.: J. Appl. Phys., 46, 1975 (1975).
(14) Johnson, W. C. and Lee, J. K., Proc. Internat. Conf. on Solid→Solid Phase Transformations, eds. Aaronson, H. I. et al, TMS-AIME, 1981, p. 151.
(15) Johnson, W. C.: Metall. Trans., 14A, 2219 (1983).
(16) Johnson, W. C., Earmme, Y. Y. and Lee, J. K.: J. Appl. Mech., 47, 775 (1980).
(17) Aikin, R. M., Jr. and Lee, J. K.: unpublished research.
(18) Shelley, J. F. and Yu, Y. Y.: J. Appl. Mech., 88, 68 (1966).
(19) Moschovidis, R. A. and Mura, T.: J. Appl. Mech., 42, 847 (1975).
(20) Eshelby, J. D.: Prog. in Solid Mech., 2, 89 (1961).
(21) Rigsbee, J. M. and VanderArend, P. J., Formable HSLA and Dual-Phase Steels, ed. A. T. Davenport, TMS-AIME, 1977, p. 58.
(22) Lee, J. K., Earmme, Y. Y., Aaronson, H. I. and Russell, K. C.: Metall. Trans., 11A, 1837 (1980).
(23) Earmme, Y. Y., Johnson, W. C. and Lee, J. K.: Metall. Trans., 12A, 1521 (1981).
(24) Johnson, W. C. and Lee, J. K.: J. Appl. Mech., 49, 312 (1982).

THE MECHANICS OF DYNAMIC CRACK GROWTH IN SOLIDS

L. B. FREUND

Brown University, Providence, RI 02912 USA

ABSTRACT

A general discussion of the mechanics of dynamic crack growth in materials is presented. First, a very simple one-dimensional model of dynamic crack growth is described; this model serves as a convenient vehicle for illustrating some of the main points to be considered. Then, focusing on elastodynamic crack growth, an expression for dynamic energy release rate is derived from a simple conservation law for elastodynamics. The dependence of dynamic stress intensity factor on crack tip motion is discussed for those two-dimensional situations for which mathematical solutions are available, and the results are next coupled with an energy balance crack growth condition to consider the implications of a particular crack tip equation of motion. Finally, examples are given of several contemporary research issues in the field, including illustrations of large scale computational approaches, some results obtained from taking crack tip plasticity into account in dynamic crack growth analysis, and observations on experimental methods in dynamic crack growth studies.

1. INTRODUCTION

A principal objective in the area of dynamic fracture mechanics is the prediction of the way in which a crack will grow in a particular material, given the geometrical configuration of the solid, a characterization of the fracture resistance of the material, the applied load distribution, and the initial state. In the interpretation of the laboratory data on rapid crack propagation, an objective of equal importance is the determination of the values of fracture characterizing parameters from measurements of the crack growth history and applied load distribution. In either case, a stress analysis for the body which is valid during the crack propagation event is required, including the effects of material inertia and of crack growth at nonuniform rates. The fruits of research in this area have found application in studies of rapid crack propagation and crack arrest in pressure vessels and piping systems, cleavage crack growth in crystalline materials, dynamic earth faulting viewed as a fracture process, stress wave emission from growing cracks as a diagnostic tool in material evaluation, and the erosion of material surfaces by high speed droplet or particle impact, among others. From the fundamental point of view, the area continues to be rich in challenging and potentially important problems.

2. A ONE-DIMENSIONAL CRACK GROWTH MODEL

Analytical models of dynamic crack growth involving a single spatial dimension have been developed in connection with seismic source modeling (1,2), dynamic fracture toughness testing (3,4,5), and peeling of a bonded layer (6). Some of these models have been remarkably successful in enhancing insight into the particular physical process, and the few models for which complete mathematical solutions exist provide the rare opportunity to see all aspects of a dynamic crack growth event in a common and relatively transparent framework. The discussion here will focus on the one-dimensional model which was developed by Freund (5) in order to illustrate the influence of reflected waves on crack growth in a double cantilever beam fracture specimen by means of a simple shear beam model of the specimen. The model may be rephrased in terms of the dynamics of an elastic string, and it has been analyzed with numerous variations from this point of view by Burridge and Keller (7).

Consider a stretched elastic string lying along the positive x-axis. The string has mass per unit length ρ and characteristic wave speed c. The time-dependent transverse deflection of the string is $w(x,t)$, and $q = \partial w/\partial x$, $v = \partial w/\partial t$. Initially, the string is free in the interval $0 < x < a_o$ and is bonded to a rigid surface for $x > a_o$. The end $x = 0$ is deflected an amount w_o so that the initial slope in the free portion is $-w_o/a_o$. At time $t = 0$, the string begins to peel away from the rigid surface, so that at some later time $t > 0$ the free length is $a(t) > a_o$. The field equations are to be satisfied for $0 < x < a(t)$, $t > 0$ subject to the end conditions $v = 0$ at $x = 0$, $qa' + v = 0$ at $x = a(t)$ where $a' = da/dt$, and the initial conditions $v = 0$, $q = q_o = -w_o/a_o$ at $t = 0$. Finally, it is observed that the field equations are hyperbolic, so that discontinuities in q and v can propagate along characteristic curves $dx/dt = \pm c$ in the (x,t)-plane subject to the familiar jump conditions. The length $a(t)$ is identified with crack length, and the mathematical problem can be solved in a straight-forward manner for arbitrary $a(t)$ by means of the method of characteristics. Details of the solution procedure are omitted, and only certain relevant results are included here. Attention is limited to crack speeds a' less than c. With reference to the (x,t)-plane shown in Fig. 1, the solution at the representative points P and R in the regions enclosed by the curves ABS and BCD, respectively, are given by

$$v_P = -a'_Q q_o/(1 + a'_Q/c), \quad q_P = q_o/(1 + a'_Q/c) \tag{2.1}$$

and

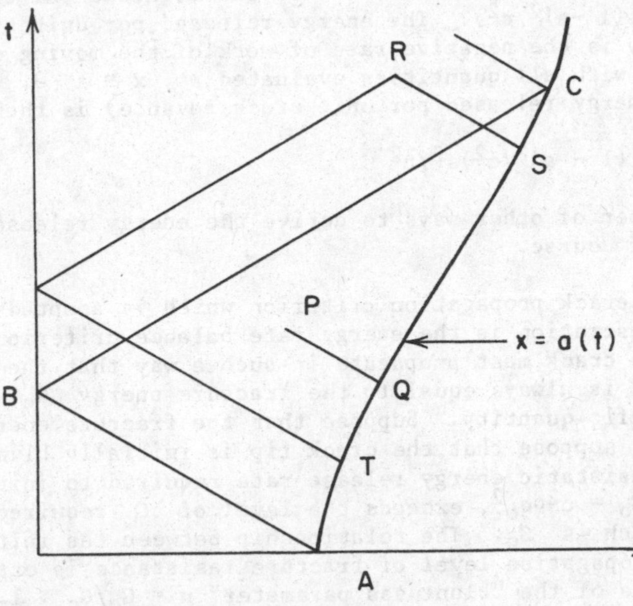

Figure 1 The (x,t) plane for the one-dimensional crack growth model.

$$v_R = q_o(a_T' - a_S')/(1 + a_T'/c)(1 + a_S'/c)$$
$$q_R = q_o(1 - a_S'a_T'/c^2)/(1 + a_T'/c)(1 + a_S'/c) \qquad (2.2)$$

and construction of the solution could be continued indefinitely for arbitrary $a(t)$. In order to determine the actual crack tip motion, it is necessary to introduce a physical crack propagation criterion.

For this simple model, the nature of the traction distribution on the prospective fracture plane can easily be determined. Everywhere ahead of the propagating tip, the stress $s = c^2\rho q$ and transverse particle velocity v are zero. Just behind the tip, these quantities are nonzero and are related through the boundary condition at $x = a(t) - 0$. Thus, discontinuities are carried along with the crack tip. Because these functions are governed by a hyperbolic system of equations, on the other hand, discontinuities may propagate only at the characteristic speeds $\pm c$ if linear momentum is conserved. The immediate inference is that linear momentum is not conserved at the crack tip, and that a source or sink of linear momentum is operative there. Because each point on the string has only a single degree of freedom, transverse displacement w, the momentum source must be a generalized force which is work-conjugate to w. It follows that there is a concentrated transverse force moving with the crack tip at $x = a(t)$ and tending to hold the crack closed for $x > a(t)$. Application of the impulse-momentum principle to the string segment

$a - 0 < x < a + 0$ yields the result that the magnitude of the cohesive force is $Q = -s(1 - a'^2/c^2)$. The energy released per unit time from the elastic body is the negative rate of work of the moving concentrated force, or Qva', with all quantities evaluated at $x = a - 0$. The energy release rate (energy released per unit crack advance) is then

$$G = (1 - a'^2/c^2)s^2/c^2 \tag{2.3}$$

There are a number of other ways to derive the energy release result in Eqn (2.3), of course.

The crack propagation criterion which is adopted here for purposes of illustration is the energy rate balance criterion, according to which the crack must propagate in such a way that the energy release rate G is always equal to the fracture energy G_c, which is a material specific quantity. Suppose that the fracture energy G_c is constant. Also, suppose that the crack tip is initially blunt, that is, that the quasistatic energy release rate required to initiate fracture, which is $G_o = c^2\rho q_o^2$, exceeds the level of G required to sustain growth, which is G_c. The relationship between the initiation level and the propagation level of fracture resistance is expressed through the value of the "bluntness parameter" $n = G_o/G_c > 1$. Given a value of n, the relation $n = G_o/G_c$ actually specifies the end displacement w_o which must be imposed to initiate dynamic crack growth. For subsequent times the crack must propagate so that $G = G_c$, or $a'(t) = 0$ if $G < G_c$.

From Eqn (2.1), it is evident that $q(a,t) = q_o/(1 + a'/c)$ at any point in the interval AC in Fig. 1. The crack propagation criterion requires that

$$G_c = G_o(1 - a'^2/c^2)/(1 + a'/c)^2 \tag{2.4}$$

If G_o is eliminated in favor of n in Eqn (2.4), it is seen that the crack speed has the constant value $(n - 1)/(n + 1)$ in the interval AC. If a critical crack tip force criterion had been assumed, rather than the critical energy release rate criterion, then the crack tip speed would have the constant value $n_f - 1/n_f$, where n_f is the ratio of the force required for initiation to the propagation value.

The two crack tip speeds a'_S and a'_T which appear in Eqn (2.2) are both equal to this constant speed, so that in the region BCD of Fig. 1

$$v = 0, \qquad q = (G_c/nc^2)^{\frac{1}{2}} \tag{2.5}$$

At time t_C, the body is entirely at rest and the crack tip strain is below the level required to meet the crack growth criterion (only positive crack speeds are admissible). Because Eqn (2.5) represents an equilibrium field, it is concluded that the crack arrests at t_C. The arrest length of the crack can be determined to be $a_C = na_o$, which is significantly larger than the equilibrium length of $n^{1/2}$ for the end displacement w_o. The energy release rate just before arrest is G_c and its value just after arrest is G_c/n, so that the energy release rate experiences an abrupt jump at the instant of arrest due to stress wave effects. Jumps in stress intensity factor showing this general trend have been observed experimentally by Kalthoff, Winkler and Beinert (8) in double cantilever beam specimens of a brittle plastic.

Because of the simplicity of the solution, the strain energy and kinetic energy as functions of crack length can be calculated easily, and a typical result is shown in Fig. 2 for the case $n = 4$. For this case, the crack speed up to arrest is $3c/5$, the arrest length is $4a_o$, and the crack length when the unloading wave reflects from the end $x = 0$ is $8a_o/5$. The interpretation of the energy variations shown in Fig. 2 in terms of the physical process is straightforward. At the instant of fracture initiation, q is reduced at the crack tip from q_o to the value necessary to satisfy the fracture criterion, and this reduction in strain implies a reduction in strain energy. Because of the rate of this reduction in q, however, the inertia of the material comes into play. The value of q is thus reduced behind the left-going wave front in AB in Fig. 1, and the strain energy decreases and the kinetic energy increases as the wave engulfs more and more of the string length. Because energy is being drawn from the body at the crack tip, the decrease in U is not balanced by the increase in T. When the stress wave reflects from the fixed end $x = 0$, the fixed displacement condition requires that it do so in just the right way to cancel the particle velocity. Thus, as the wave reflects back onto

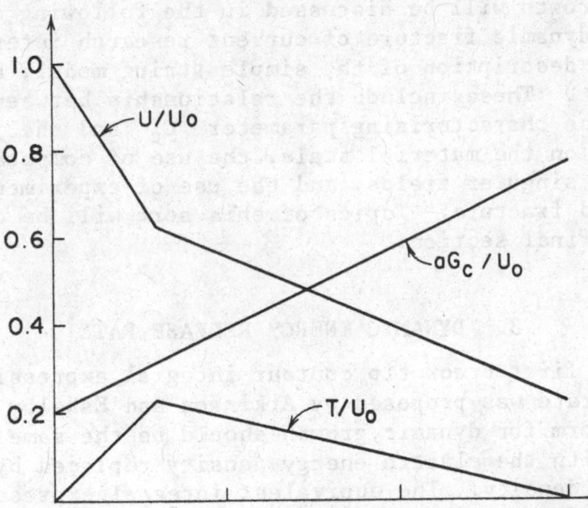

Figure 2 Energy variations with crack length for the one-dimensional crack growth model.

itself, the particle velocity v is reduced to zero and the slope q is further reduced behind the right-going wave front BC in Fig. 1. The kinetic energy decreases after wave reflection, and the strain energy also decreases, but at a slower rate than before. It is important to note that, until the reflected stress wave overtakes the crack tip, the crack considers itself to be propagating in an unbounded body, that is, the crack tip is unaware of the presence of the boundary at $x = 0$. Although the kinetic energy decreases continuously from the instant of wave reflection at $x = 0$ to the instant of crack arrest, this kinetic energy is not being recovered to assist in sustaining crack growth. Instead, the decrease is simply part of a wave reflection process and the crack tip is ignorant of the decrease. The crack tip first becomes aware of the presence of the end $x = 0$ when the reflected stress wave overtakes the propagating crack tip and, because the q level carried by this wave is below the minimum level required to sustain crack growth, arrest occurs simultaneously with the arrival of the reflected wave in this model. In effect, by the time that the stress wave communicates to the crack tip that the applied end displacement w_o is appropriate to maintain a certain equilibrium length for a sharp crack, the crack has already grown to a length greater than this equilibrium length. Thus, arrest is instantaneous with the arrival of the reflected wave and the post-arrest state is subcritical.

Again, the point of considering this simple string crack growth model is to illustrate the main features of dynamic crack growth analysis. Among the features which were given special attention were energy release rate, crack tip stress field, and crack tip equation of motion. Generalizations of these to the case of plane strain elastodynamic crack growth will be discussed in the following sections. Certain issues in dynamic fracture of current research interest were not included in the description of the simple string model, but their relevance is evident. These include the relationship between the crack growth resistance characterizing parameter G_c and the inelastic separation process on the material scale, the use of computational methods to describe the singular fields, and the use of experimental methods to observe rapid fracture. Topics of this sort will be considered briefly in the final section.

3. DYNAMIC ENERGY RELEASE RATE

The first crack tip contour integral expression for dynamic energy release rate was proposed by Atkinson and Eshelby (9), who argued that the form for dynamic growth should be the same as for quasi-static growth with the elastic energy density replaced by the total internal energy density. The equivalent integral expression for dynamic energy release rate in terms of crack tip stress and deformation fields was subsequently derived directly from the field equations of elastodynamics by Kostrov and Nikitin (10) and by Freund (11). They enforced an instantaneous energy rate balance for the time-dependent volume of material bounded by the outer boundary of the solid, the crack faces, and small loops surrounding each moving crack tip and translating with it. By application of Reynold's transport theorem and the divergence theorem, an expression for crack tip energy flux in the form of an integral of a field quantity along the crack tip loops was obtained. The same result can be obtained by means of the alternate derivation which is outlined here, with some observations made by Eshelby (12) as point of departure.

In the absence of body forces, the equation of motion in terms of Cartesian components of stress s_{ij} and displacement u_i is

$$s_{ij,j} - \rho u_{i,tt} = 0 \tag{3.1}$$

where ρ is mass density and subscripts following a comma denote partial differentiation. If the inner product of Eqn (3.1) with particle velocity $u_{i,t}$ is formed, and if the material is taken to be elastic (not necessarily linear) with strain energy function U, then the resulting equation may be written in the form

$$(u_{i,t}s_{ij})_{,j} - (U + T)_{,t} = 0 \tag{3.2}$$

where T is the kinetic energy density, $T = \rho u_{i,t}u_{i,t}/2$. Eqn (3.2) is a standard differential form of the energy balance principle [cf., Eqn (58) of the article by Eshelby (12)]. If the time coordinate is given the same status as a spatial coordinate, then Eqn (3.2) is a requirement that a certain vector expression is divergence free in space-time. If Eqn (3.2) holds over some "volume" of space-time, then direct application of the divergence theorem leads to the result that the integral over the bounding surface of this volume of the inner product of the divergence-free vector with the surface normal vector vanishes.

For definiteness, consider a two-dimensional body which is bounded by a curve C and which contains an extending crack. The plane of the body is the x_1,x_2-plane and crack growth is along the x_1-axis. The instantaneous crack tip speed is v. A small contour C^* begins on one traction free face of the crack, surrounds the tip, and ends on the opposite traction free face. Consider the volume in the x_1,x_2,t-space which is bounded by the planes t = 0 and t = t, by the right cylinder swept out by C between the initial time and t, by the planes swept out by the crack faces between the initial time and t, and by the tubular surface swept out by the small contour C^* as the crack tip advances in the x_1-direction with increasing time. For this particular choice of surface, the surface integral is

$$\int_0^t \int_C u_{i,t}s_{ij}n_j dC\, dt = \left[\int_{A(t)} (U + T)dA\right]_0^t - \int_0^t \int_{C^*} [u_{i,t}s_{ij}n_j + (U + T)v_n]\, dCdt \tag{3.3}$$

where A(t) is the cross-section of the volume at t = constant, n_i

is a unit vector which is normal to C or C* in any plane t = constant and which points away from the crack tip, and v_n is the instantaneous velocity of any point on C* in the direction of n_i at this point.

The term on the left side of Eqn (3.3) is the total work done on the body between the initial time and t, and the first term on the right side is the increase in internal energy during the same time. Because the time interval is arbitrary, it follows immediately that

$$F(C^*) = \int_C [s_{ij}n_j u_{i,t} + (U + T)vn_1]dC \qquad (3.4)$$

is the instantaneous rate of energy flow out of the body through C*. The interpretation of the terms in Eqn (3.5) is straightforward. The first term is the rate of work of the material outside of C* on the material inside of C*. If the curve were a material line, then this would be the only contribution to the energy flux. The curve C* is moving through the material, however, and the second term in Eqn (3.4) represents the contribution to the energy flux due to the flux of material.

The dynamic energy release rate G is defined as the limiting value of F(C*)/v as the contour C* is shrunk onto the crack tip. It was shown by Freund (11) that the value of G obtained in this limiting process does not depend on the shape of C* as it is shrunk onto the tip. It should be noted that F/v does reduce to the path-independent J-integral for quasistatic elastic crack growth. In general, however, the value of F will depend on the path used to evaluate it (9). However, under the special condition that the complete elastic field is time-independent as seen by an observer moving with the crack tip, the integral F is path-independent (9,13).

4. STRESS INTENSITY FACTOR FOR NONUNIFORM EXTENSION

Much of the early analytical work on dynamic fracture was based on the assumption that the crack tip moved with a constant speed. Some of the pioneering steps toward lifting this restriction were taken by Kostrov (14) and Eshelby (15) in their work on the nonuniform extension of cracks in the antiplane shear mode (that is, mode III). Kostrov considered an antiplane shear crack with the faces subjected to general equal but opposite time-dependent traction distributions. The complete elastic field was determined for arbitrary motion of the crack tips for times sufficiently small so that wave interactions between the crack tips does not occur. The analytical method, which was developed within the theory of supersonic fluid flow, makes use of Volterra's representation of the displacement solution of scalar wave equation in terms of boundary tractions, and it results in an expression for the dynamic stress intensity factor in terms of the crack face traction distribution. For example, it was found that if a uniform traction of magnitude p_o begins to act at time $t = 0$, and the crack tip simultaneously begins to move with speed $a'(t)$, then the stress intensity factor is

$$K_{III} = 2p_o(1 - a'/c_s)^{\frac{1}{2}}(c_s t)^{\frac{1}{2}} \tag{4.1}$$

where c_s is the elastic shear wave speed. Other special cases were considered by Kostrov (14), and the same mathematical technique was used by Achenbach (16,17) in a study of antiplane shear fracture initiation and crack propagation under stress wave loading.

Eshelby (15) considered the dynamic extension of an antiplane shear crack at nonuniform rate under general quasistatic loading conditions. He constructed a complete solution to the problem by making use of a result on the electromagnetic radiation from a nonuniformly moving line charge. It was found that the dynamic stress intensity factor had the form of a function of the instantaneous crack tip speed times the static stress intensity factor for the given loading and the instantaneous crack length. If the amount of crack growth is denoted by $a(t)$, then

$$K_{III}(a,a') = (1 - a'/c_s)^{\frac{1}{2}} K_{III}(a,0) \tag{4.2}$$

Note that the applied loading enters only through the equivalent static stress intensity factor. Eshelby argued that his result was different from Kostrov's because the latter assumed the condition of suddenly applied loading on a body initially stress free and at rest. It can be shown through an argument based on superposition of solutions, however, that Eshelby's particular result (4.2) can be derived by means of the analytical method used by Kostrov. On the other hand, the general discussion presented in section 5 of the paper by Eshelby (15) on the elements of dynamic fracture analysis and their interplay in a complete description of the process represents a major contribution to the field. The ideas contained in Eshelby's discussion (many of which were illustrated through the simple string model in section 2 above) were important in opening an exciting and productive era in dynamic fracture mechanics, and they had an important influence on subsequent research on nonuniform crack growth.

It was evident to both Kostrov and Eshelby that the analytical methods which they used to study nonuniform growth of antiplane shear cracks could not be extended to the case of the plane strain opening mode of crack growth, that is, to mode I. A close examination of their results provided clues, however, toward solution of the corresponding plane strain problems. First, it is noted that both of the special cases Eqn (4.1) and Eqn (4.2) have the form of a particular function of the instantaneous crack tip speed $(1 - a'/c_s)^{\frac{1}{2}}$ times the stress intensity factor which would exist if the crack tip had always been at its current position. The second clue was contained in the observation by Eshelby (15,18) that if the mode III crack extending nonuniformly under the action of time-independent loading suddenly stopped, then the static elastic field appropriate for the given applied loads and for the instantaneous crack tip position was fully established everywhere behind a shear wave front which radiated out from the stopping position of the tip. This was a truly remarkable result for a two-dimensional stress wave problem. In any case, it suggested the possibility of building up a solution for nonuniform crack growth as a sequence of many small constant speed segments. Because of the presence of free surface Rayleigh waves in plane strain wave fields, it was most unlikely that the same strong result on the radiation of the static solution would carry over to the plane strain case. It was noted that for the argument to go through in the case of plane strain, on the other hand, all that was required was the weaker result that the static stress distribution radiated out only on the prospective fracture plane ahead of the crack tip when the crack suddenly stopped (19). It was established that this is indeed the case, and the exact stress intensity factor for a half-plane crack extending nonuniformly in an elastic solid subjected to general time-independent loading was determined in this pair of papers. This provided the plane strain analog of Eshelby's result Eqn (4.2), and the plane strain analog of Kostrov's result (4.1) was also obtained by means of the constructive procedure by Freund (20,21) in subsequent work. The general result may be summarized as follows: The stress intensity factor for mode I extension of a half-plane crack is given by a universal function of instantaneous crack tip speed $k(a')$ times the stress intensity factor appropriate for a crack of fixed length, equal to the instantaneous length, subjected to the given applied loading, whether this loading is time-independent or time-dependent. That is, the stress intensity factor K is given by

$$K(t,a,a') = k(a')K(t,a,0) \tag{4.3}$$

An exact expression for $k(a')$ is given by Freund (19), from which it is concluded that k decreases monotonically from one at $a' = 0$ to zero at $a' = c_r$.

The result in Eqn (4.3) was verified and extended by means of more direct procedures by Kostrov (22) and Burridge (23). Kostrov used integral transform methods, while Burridge used similarity methods. He defined an influence function in such a way that the stress intensity factor for any particular problem was a linear convolution integral of the influence function with the particular applied crack face traction. He then formulated and solved a boundary value problem for the influence function. A constructive approach to solving nonuniformly growing crack problems based on superposition of solutions for moving elastic dislocations was introduced by Freund (24). This technique has been used to construct solutions for the sudden stopping of a steadily growing crack, and other similar situations.

With reference to the simple string model described in section 2, the present results on the dependence of stress intensity factor on the applied loads and on the crack motion correspond to the solution q in Eqn (2.1) or Eqn (2.2) when the observation point P or R coincides with the crack tip at $x = a(t) - 0$. The crack tip equation of motion is next determined and examined.

5. CRACK TIP EQUATION OF MOTION

In his 1969 article, Eshelby stated that "The main reason for calculating the elastic field for an arbitrarily moving tip is to be able to pick out the actual motion from the ensemble of all possible motions. The simplest way of doing this is to suppose, along with A.A. Griffith, that the energy released at the tip is all used to provide the surface energy of the faces of the crack." Of course, the term surface energy is used somewhat euphemistically here to denote the energy which is dissipated in the physical process of material separation. The specific fracture energy, which will be denoted by G_c, is assumed to completely characterize the resistance of the material to crack propagation. This quantity may be a constant, such as the true surface energy of a covalently bonded single crystal, or it may depend on crack tip motion in order to reflect the influence of strain rate sensitivity, material inertia or fracture mechanism transition in the physical separation process in plastics, metals or other materials. A detailed comparison between the Griffith energy criterion and the Barenblatt cohesive zone criterion was presented by Willis (25).

Just as in the case of the elastic string in section 2, the mathematical statement of the energy rate balance fracture criterion is $G = G_c$. For the nonuniform extension of a crack at sub-Rayleigh wave speeds, each term in the integrand of the energy flux expression Eqn (3.4) is proportional to the stress intensity factor to the power two, and because of the universal spatial dependence of the near crack tip stress and deformation fields, the integral in Eqn (3.4) can be evaluated to determine the velocity dependent proportionality factor (26). For a half-plane mode I crack, the stress intensity factor is given for arbitrary motion of the crack tip in Eqn (4.3) and an equation of motion follows immediately. Restricting attention for the time being to the case of time independent loading, the energy rate balance requires that

$$E\, G_c/[(1 - \nu^2)\, K(a,0)^2] = g(a') \qquad (5.1)$$

where E is the elastic modulus and ν is the Poisson ratio for the material. The function $g(a')$, which is a universal function of crack speed for this configuration and which is independent of the applied loading, is very nearly approximated by $(1 - a'/c_r)$ where c_r is the Rayleigh wave speed for the material. If it assumed that G_c is a bounded function of crack tip speed, then several general properties of crack tip motion are evident from Eqn (5.1): (i) The equation of motion is a first order differential equation for $a(t)$. This implies that the crack velocity, rather than the acceleration, changes in phase with the crack tip driving force [cf., Irwin's discussion following the article by Eshelby (14)] and that the crack tip speed may vary discontinuously. (ii) If the level of applied loading is such that $K(a,0)$ increases indefinitely with increasing a, then the crack tip speed increases indefinitely toward the Rayleigh wave speed c_r. In this sense, the Rayleigh wave speed is the highest attainable crack tip speed for elastodynamic crack growth. This result was anticipated on the basis of a clever physical argument by Stroh (27). (iii) If a cohesionless interface (i.e., an interface for which $G_c = 0$) is forced to separate, then the separation fronts will move with speed c_r [cf., Papadopoulos (28); Freund (24)].

As an illustration of the equation of motion Eqn (5.1), consider the following plane strain situation. A half-plane crack exists in an otherwise unbounded isotropic elastic solid on $x_2 = 0$, $x_1 < a_o$. The crack faces are traction free in the interval $0 < x_1 < a_o$, and the crack is propped open for $x_1 < 0$ with a uniform separation $2w_o$ between the faces. The specific fracture energy of the material is assumed to be a constant G_c for simplicity, and the initial crack tip is slightly blunted so that an energy release rate of nG_c is required to initiate crack growth. The bluntness parameter n is used here in precisely the same way in which it was used in the string analysis above. For an initial crack "length" a_o, the magnitude of the displacement w_o is gradually increased until the crack tip energy release rate is nG_c, whereupon a sharp crack begins to grow dynamically in the x_1-direction with no further increase in w_o. Thus, except for the way in which the elastic body is described, the problem is the same as that considered in section 2. The stress intensity factor for this elastic crack configuration is $K(a,0) = Ew_o/[(1 - \nu^2)(2\pi a)^{\frac{1}{2}}]$ for any crack "length" a. The displacement w_o may be eliminated in favor of the parameter n, and the solution of the crack tip equation of motion which satisfies the initial condition $a(0) = a_o$ is then $a(t) = na_o[n - (n - 1)\exp(-c_r t/na_o)]$. In writing this solution, it is assumed that the waves which are scattered from the point $x_1 = 0$, $x_2 = 0$ and subsequently overtake the moving crack tip do not significantly influence the crack growth process.

The differences between the crack tip motion predicted here and the motion found in section 2 point out the importance of the shape of the elastic body in which a crack is growing on the nature of the crack tip motion. For example, in the case of the string model, the crack grew at constant speed and then arrested with a sudden jump in speed. Here, the crack speed decreased gradually to zero. Furthermore, the final length of the crack in the string model was found to be much greater than the equilibrium length for the imposed end displacement, whereas in the plane strain case the final length is found to be identical with the equilibrium length. These differences are apparently due to the way in which waves are radiated from, and subsequently reflected back onto, the moving crack tip. In the one-dimensional structure, loads can be transferred over long distances without dilution of their effect and waves are guided along the length. In the plane strain case, on the other hand, the influence of an applied load becomes increasingly diffuse with increasing distance from its point of application and waves are continuously scattered rather than guided.

6. RECENT RESEARCH PROGRESS

The foregoing discussion provides a relatively complete description of the process of dynamic crack growth in solids, at least within the context of elastodynamics and brittle fracture. For the sake of simplicity, however, several issues of great importance toward understanding the process of dynamic crack growth in real materials were overlooked in this discussion. For example, how is the dynamic crack growth resistance of a material measured? What is the influence of material microstructure and physical separation mechanism on the observed fracture resistance? What methods are available for solving boundary value problems when the configuration and/or the material behavior of the physical system being analyzed preclude use of the simple models introduced above? The answers to these questions are not yet complete, but some recent progress toward resolving them is summarized in this section.

6.1 THE OPTICAL SHADOW SPOT MEASUREMENT METHOD

Consider a planar mode I fracture specimen in the single-edge-notch configuration. Suppose that, for points at a distance of about one-half to one plate thickness from the crack tip, the mechanical fields are adequately characterized by the theory of plane stress elasticity. Under the action of the applied loads, the specimen undergoes nonuniform thinning, with the through-the-thickness contraction being greatest near the crack tip. If the lateral face of the specimen was initially planar and of high optical reflectivity, then the deformed specimen face is a nonplanar reflecting surface. If parallel light rays are directed onto the deformed reflecting surface, then the reflected rays will deviate from parallelism. That is, the reflected light field will have some geometrical structure (when viewed on a plane parallel to the specimen plane, say), and the general idea is to relate a feature of this structure to the deformation of the specimen near the crack tip, which is completely characterized by the elastic stress intensity factor. If the reflected light rays are gathered by a camera which is focused on a suitable plane behind the specimen, then the photograph produced will show a dark ovoid (region devoid of light rays) surrounded by an illuminated region. The ovoid is the so-called shadow spot, and the level of stress intensity factor can be related to a transverse dimension of the spot. The optical shadow spot method has been applied in the measurement of stress intensities during fast fracture by Kalthoff and coworkers (19,29). Their work clearly established the importance of dynamic effects in rapid crack growth phenomena.

The shadow spot method has several features which make it very attractive for use in dynamic crack growth experiments: it is a direct crack tip measurement and thus avoids the complexities of analyzing the stress wave propagation and reflection in the entire specimen to interpret the data; the optical procedure does not interfere with the process being observed; and the response of the optical system is virtually instantaneous on the time scale of mechanical processes. The main disadvantage of the method is that it provides only a surface observation, and the extent to which the surface response of the specimen represents the internal response is somewhat uncertain.

The optical shadow spot method was used by Rosakis, et al. (30) to measure the crack propagation resistance of AISI 4340 steel for crack tip speeds up to about 1100 m/s, and the results are shown as the individual data points in Fig. 3. The double cantilever beam specimen configuration was used, and the specimens were loaded in the stiff wedge loading arrangement. The material was austenitized at 843°C for one hour, oil quenched and tempered at 316°C for one hour, which resulted in a Rockwell hardness number of $R_c = 45$. The data in Fig. 3 were normalized by $K_{Ic} = 30\ MPa(m)^{\frac{1}{2}}$ and $c_s = 3200$ m/s.

6.2 TRANSIENT CRACK-TIP STRESS HISTORIES

In the use of direct optical methods in fracture testing under very high rates of loading, such as the shadow spot method mentioned above, the question arises of how well the crack tip singular field describes the stress distribution over the portion of the specimen on which observations are made during the short observation times involved. That is, observations are made over a region near the crack tip of transverse dimensions from perhaps 3-4 mm up to 20 mm, depending on the technique employed, and the lowest characteristic wave speed for brittle plastics is about 1 mm per microsecond. Under the action of rapidly applied loads or due to rapid crack growth, how long does it take for the stress intensity factor-controlled field to become fully established within the region of observation? Consideration of this question was originally motivated by the experimental studies reported by Ravi-Chandar (31), and only a representative calculation is described below.

As the basis for a simple calculation, consider a half-plane crack in an unbounded elastic solid under plane strain conditions. The material is initially at rest and stress free. The crack faces are then subjected to a spatially uniform suddenly applied pressure at time $t = 0$. At some later time, say $t = t_o$, the crack begins to grow with crack tip speed v. As the crack grows, the pressure does not spread, that is, it continues to act only over the initial crack faces. An exact solution of this problem may be determined from the analysis due to Freund (19,20). Attention is focused on a moving point which is always a fixed distance, say β, ahead of the moving crack tip. Of particular interest is the time required for the stress level

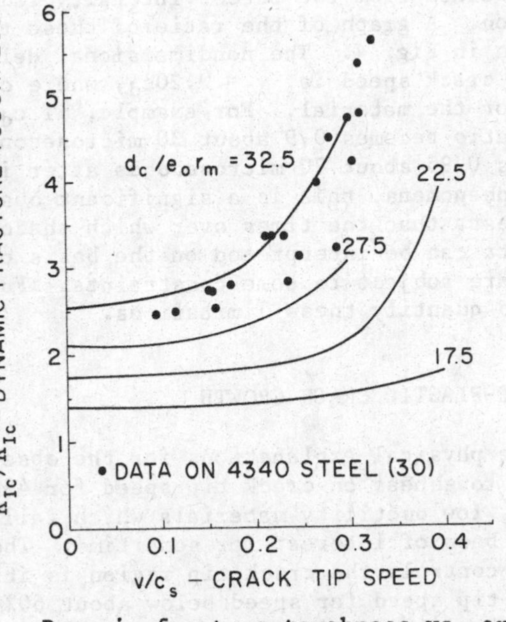

Figure 3 Dynamic fracture toughness vs. crack speed for critical crack opening displacement growth criterion.

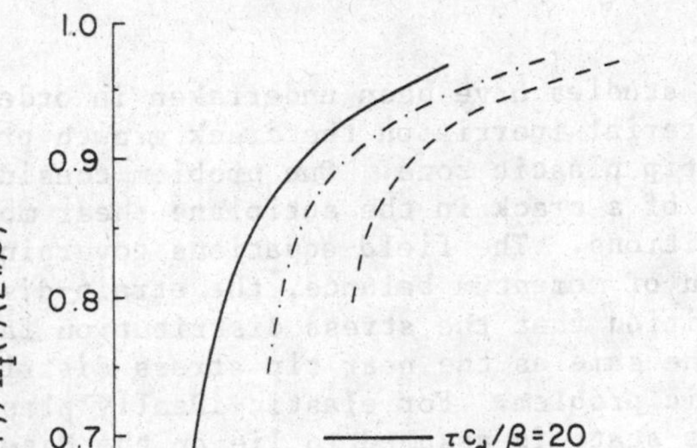

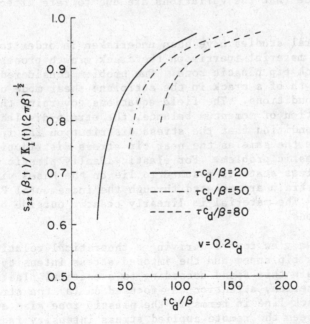

Figure 4 Ratio of total stress to singular stress term vs. time at fixed distance ahead of moving crack tip.

at this point obtained from a complete stress distribution to approach the stress level obtained from the stress intensity factor-controlled singular distribution. A graph of the ratio of these two stress levels versus time is shown in Fig. 4. The nondimensional delay time is $t_o c_d/\beta = 50$ and the crack speed is $v = 0.20c_d$, where c_d is the dilatational wave speed for the material. For example, if $c_d = 2000$ m/s and $\beta = 2$ mm then the ratio becomes 0.9 about 30 microseconds after initiation, and it becomes 0.95 about 70 microseconds after initiation. For high rate fracture phenomena, this is a significant observation time and the result suggests that the times over which shadow spot or photoelastic near tip data can be interpreted on the basis of a singular crack tip solution are subject to some constraints. Further work is required in order to quantify these limitations.

6.3 DYNAMIC ELASTIC-PLASTIC CRACK GROWTH

Possible physical explanations for the observed dependence of dynamic fracture toughness on crack tip speed for 4340 steel and other high strength, low ductility materials which fail in a locally ductile manner have been of interest for some time. The elastic field which surrounds and controls the crack tip region is insensitive to variations in crack tip speed for speed below about 60% of the shear wave speed of the material and, consequently, cannot account for the observed toughness variations. Furthermore, these materials are relatively strain rate insensitive in their bulk response, so that it is difficult to argue that the variations are due to rate effects on a local scale.

Several studies have been undertaken in order to determine the influence of material inertia on the crack growth process at the level of the crack tip plastic zone. One problem considered was the steady-state growth of a crack in the antiplane shear mode under small scale yielding conditions. The field equations governing this process include the equation of momentum balance, the strain-displacement relations, and the condition that the stress distribution far from the crack tip must be the same as the near tip stress distribution in a corresponding elastic problem. For elastic-ideally plastic material response, the stress state is assumed to lie on the Mises yield locus, and stress and strain are related through the incremental Prandtl-Reuss flow rule. The material is linearly elastic outside of plastically deformed regions.

With a view toward deriving a theoretical relationship between the crack tip speed and the imposed stress intensity factor required to sustain this speed according to a critical plastic strain crack growth criterion, attention was focused on (a) the strain distribution on the crack line in terms of the plastic zone size and (b) the relationship between the remote applied stress intensity factor and the plastic zone size. It was found that the strain distribution on the crack line could be determined exactly, and results are presented by Freund and Douglas (32) and Dunayevsky and Achenbach (33). The main observation on the strain distribution is that the level of plastic strain is significantly reduced from its corresponding slow crack growth levels due to material inertia. This implies that, if a fixed level of plastic strain is to be maintained at a given distance ahead of the crack tip for growth to occur, then the level of applied stress intensity factor would necessarily increase with increasing crack tip speed. To quantify this idea, the remote applied stress intensity factor was related to the plastic zone strain distribution through a full-field numerical solution (32,34). A similar antiplane analysis has been carried out for an elastic-viscoplastic material model by Freund and Douglas (35) and Lo (36). The same numerical approach has recently been followed in the case of steady dynamic crack growth in an elastic-plastic material for the plane strain tensile mode of deformation, i.e. mode I, by Lam and Freund (37). The general features of the model were the same as those in the antiplane shear case analyzed previously by Freund and Douglas (32). The material was characterized by the Mises yield condition and J_2 flow theory of plasticity. It was found that the influence of material inertia on the response in the crack tip region was as large as that observed for the case of antiplane shear. Of central interest was the relationship between the applied crack driving force or dynamic fracture toughness, represented by a remote stress intensity factor, and the crack tip speed on the basis of a particular ductile crack growth criterion. The criterion adopted was that which required growth to occur with a specified crack face opening displacement d_c maintained at a characteristic distance r_m behind the crack tip [see the article by Rice and Sorenson (38), for example]. For purposes of comparison with the data on 4340 steel, the results for four values of the characterizing parameter $d_c/e_o r_m$ are shown in Fig. 3, where e_o is the tensile yield strain of the material. The general features of this relationship are consistent with the limited data on dynamic growth in relatively rate insensitive steels as presented by Dahlberg, et al. (39), Kobayashi and Dally (40), and Rosakis, et al. (30). With the normalization factors used, the calculated toughness versus speed variation for $d_c/e_o r_m$ in the range from 30 to 35 is in very good agreement with the data for 4340 steel.

6.4 DYNAMIC CRACK GROWTH BY DUCTILE HOLE GROWTH

For crack propagation in a rate-independent material which separates by the locally ductile mechanism of hole growth under small scale yielding conditions, as is the case for the experiments on 4340 steel reported by Rosakis, et al. (30), a quantitative estimate can be made of the influence of material inertia on the relationship between K^d_{Ic} and v. Glennie (41) modified the void growth formulation of Rice and Tracey (42) to include the effect of inertia, and he obtained an approximate solution by means of a variational method. The behavior which he found is qualitatively consistent with the data in Fig. 4.

The influence of metallurgical factors on rapid crack growth in 4340 steel is discussed by Hahn, et al. (43).

6.5 NUMERICAL METHODS

A mathematical analysis of dynamic crack growth in solids can be performed for only the simplest geometrical configurations and loadings. Such analyses have been crucial to resolving certain fundamental issues in the area. On the other hand, the level of detail required to properly distinguish between geometrical effects and material effects in crack propagation experiments or to predict crack motion in a structure of a given material is beyond the means of methods of mathematical analysis. Thus, it has become essential that the study of dynamic crack growth in materials be pursued by means of computer simulation studies involving large scale numerical analysis. The actual model developed to simulate dynamic crack growth depends to a large extent on the objective of the calculation, and discrete systems based on finite difference methods, finite element methods, and atomistic models have been profitably studied.

Numerical solutions based on the finite element method have been obtained for a number of elastodynamic crack growth problems. For purposes of general discussion, the approaches may be classified according to the way in which the crack tip region is handled. The most common ways to deal with the crack tip region are to simulate crack growth through gradual release of element nodal forces, imbedding a moving element at the crack tip in the mesh, in which the interpolation functions are determined by the continuum near tip fields, and energy path integral considerations. Following spatial discretization, the system of differential equations in time for the node point variables must be integrated. Because the dynamic fields associated with rapid crack growth are rich in high frequency content, short time steps are required for accuracy. Because of this natural restriction to short time steps, conditionally stable explicit time integration schemes based on a suitable diagonalization of the mass matrix have been found to be both accurate and efficient in many cases.

The main idea of the gradual node release method is that a finite element node on the prospective crack plane is viewed as being two nodes which are kinematically constrained to move together. When the crack tip passes the point occupied by the node pair, the nodes are allowed to separate dynamically under the action of the equal but opposite resisting forces. The most common assumption about the magnitude of the restraining force is that it is proportional to the nodal constraint force which existed when the crack tip first reached the position of the node pair, and that thereafter the proportionality factor depends instantaneously on the fraction of the element side which has been traversed by the moving tip. For example, suppose that the nodal force had magnitude Q_o when the crack tip passed a particular node pair, that the spacing between nodes on the prospective fracture plane is H, and that the tip is instantaneously at a distance h beyond the particular node. Then, it is commonly assumed that the restraining force Q has the form

$$Q = Q_o F(h/H)$$

where F is a specified function of its argument. Some particular forms of the function F which have been used are $F(x) = (1 - x)^{\frac{1}{2}}$ by Rydholm, et al. (44), $F(x) = (1 - x)^{3/2}$ by Malluck and King (45), and $F(x) = (1 - x)$ by Kobayashi, Emery and Mall (46). Numerical experiments were described by Malluck and King (45) in which the transient crack motion was specified and the work done by the nodal forces was computed as a function along the crack path for different functions F. The results showed some sensitivity to the form of F.

A complete computational and experimental study of rapid fracture and arrest was described in a series of papers by Brickstad and Nilsson (47), Dahlberg, et al. (39) and Brickstad (48,49). A finite element approach based on the node release idea and on explicit time integration (at least in the later work) was used to analyze crack arrest experiments done with edge notched specimens of a high strength steel. First, measured crack tip motions were prescribed in the calculations to infer values of dynamic fracture toughness during crack growth. Then, the inverse problem was studied. That is, the inferred relationship between dynamic fracture toughness and crack speed was used as input, and the crack propagation history was calculated. The computed crack motions showed good agreement with the original experimental results.

The nodal force release technique has the virtue of simplicity, and it can be added quite easily to almost any versatile finite element code. The singularity element approach, while less commonly used, has been studied in some detail by King (50), Nishioka and Atluri (51), and Aoki, et al. (52). The energy integral approach has also been introduced recently by Burgers and Freund (53) and Kishimoto, et al. (54), among others.

ACKNOWLEDGEMENT

This report was prepared as part of an ongoing research program supported by the U.S. National Science Foundation (Solid Mechanics Program), the Office of Naval Research (Structural Mechanics Program), and the NSF Materials Research Laboratory at Brown University. This research support is gratefully acknowledged.

REFERENCES

(1) Knopoff, L. et al. (1973). The dynamics of a one-dimensional fault in the presence of friction. Geophysical Journal of the Royal Astronomical Society, 35, 169-184.

(2) Landoni, J.A. & Knopoff, L. (1981). Dynamics of one-dimensional crack with variable friction. Geophysical Journal of the Royal Astronomical Society, 64, 151-161.
(3) Kanninen, M.F. (1974). A dynamic analysis of unstable crack propagation and arrest in the dcb test specimen. International Journal of Fracture, 10, 415-430.
(4) Bilek, Z.J. & Burns, S.J. (1974). Crack propagation in wedged double cantilever beam specimens. Journal of the Mechanics and Physics of Solids, 22, 85-95.
(5) Freund, L.B. (1977). A simple model of the double cantilever beam crack propagation specimen. Journal of the Mechanics and Physics of Solids, 25, 69-79; see also (1979) A one-dimensional dynamic crack propagation model. In: Mathematical Problems in Fracture, ed. R. Burridge, pp. 21-37. American Mathematical Society.
(6) Hellan, K. (1978). Debond dynamics of an elastic strip - I. Timoshenko beam properties and steady motion. International Journal of Fracture, 14, 91-100; (1978). II. Simple transient motion. International Journal of Fracture, 14, 173-184.
(7) Burridge, R. & Keller, J.B. (1978). Peeling, slipping and cracking - some one-dimensional free boundary problems in mechanics. SIAM Review, 20, 31-61.
(8) Kalthoff, J.F. et al. (1977). Measurements of dynamic stress intensity factors for fast running and arresting cracks in double cantilever beam specimens. In: Fast Fracture and Crack Arrest, ASTM STP 627, 161-176.
(9) Atkinson, C & Eshelby, J.D. (1968). The flow of energy into the tip of a moving crack. International Journal of Fracture Mechanics, 4, 3-8.
(10) Kostrov, B.V. & Nikitin, L.V. (1970). Some general problems of mechanics of brittle fracture. Archiwum Mechaniki Stosowanej, 22, 749-775.
(11) Freund, L.B. (1972). Energy flux into the tip of an extending crack in an elastic solid. Journal of Elasticity, 2, 341-349.
(12) Eshelby, J.D. (1970). Energy relations and the energy-momentum tensor in continuum mechanics. In: Inelastic Behavior of Solids, ed. M.F. Kanninen, et al., 77-115, McGraw-Hill.
(13) Sih, G.C. (1970). Dynamic aspects of crack propagation. In: Inelastic Behavior of Solids, ed. M.F. Kanninen, et al., 607-639. McGraw-Hill.
(14) Kostrov, B.V. (1966). Unsteady propagation of longitudinal shear cracks. Applied Mathematics and Mechanics (English translation of PMM), 30, 1241-1248.
(15) Eshelby, J.D. (1969). The elastic field of a crack extending non-uniformly under general anti-plane loading. Journal of the Mechanics and Physics of Solids, 17, 177-199.
(16) Achenbach, J.D. (1970). Crack propagation generated by a horizontally polarized shear wave. Journal of the Mechanics and Physics of Solids, 18, 245-259.

(17) Achenbach, J.D. (1970). Extension of a crack by a shear wave. Zeitschrift fur angewandte Mathematik und Physik, 21, 887-900: see also erratum in International Journal of Fracture Mechanics, 8, (1972).

(18) Eshelby, J.D. (1969). The starting of a crack. In: Physics of Strength and Plasticity, ed. A.S. Argon, 263-275. MIT Press.

(19) Freund, L.B. (1972). Crack propagation in an elastic solid subjected to general loading. I. Constant rate of extension. Journal of the Mechanics and Physics of Solids, 20, 129-140; II. Nonuniform rate of extension. Journal of the Mechanics and Physics of Solids, 20, 141-152.

(20) Freund, L.B. (1973). Crack propagation in an elastic solid subjected to a general loading. III. Stress wave loading. Journal of the Mechanics and Physics of solids, 21, 47-61.

(21) Freund, L.B. (1974). Crack Propagation in an elastic solid subjected to general loading. IV. Obliquely incident stress pulse. Journal of the Mechanics and Physics of Solids, 22, 137-146.

(22) Kostrov, B.V. (1975). On the crack propagation with variable velocity. International Journal of Fracture, 11, 47-56; see also Applied Mathematics and Mechanics (English translation of PMM), (1974), 511-519

(23) Burridge, R. (1976). An influence function for the intensity factor in tensile fracture. International Journal of Engineering Science, 14, 725-734.

(24) Freund, L.B. (1974). The stress intensity factor due to normal impact loading of the faces of a crack. International Journal of Engineering Science, 12, 179-189.

(25) Willis, J.R. (1967). A comparison of the fracture criteria of Griffith and Barenblatt. Journal of the Mechanics and Physics of Solids, 15, 151-162.

(26) Freund, L.B. & Clifton, R.J. (1974). On the uniqueness of elasto-dynamic solutions for running cracks. Journal of Elasticity, 4, 293-299.

(27) Stroh, A.N. (1957). A theory of the fracture of metals. Advances in Physics, 6. From: Philosophical Magazine Supplement, 418-465.

(28) Papadopoulos, M. (1963). Diffraction of plane elastic waves by a crack, with applicationn to a problem of brittle fracture. Journal of the Australian Mathematical Society, 3, 325-339.

(29) Kalthoff, J.F. et al. (1980). Experimental analysis of dynamic effects in different crack arrest test specimens. In: Crack Arrest Methodology and Applications, ASTM STP 711, 109-127.

(30) Rosakis, A.J. et al. (1982). The determination of dynamic fracture toughness of AISI 4340 steel by relected caustics. Brown University Report, September.

(31) Ravi-Chandar, K. (1982). An experimental investigation into the mechanics of dynamic fracture. Ph.D. Thesis, California Institute of Technology.
(32) Freund, L.B. & Douglas, A.S. (1982). The influence of inertia on elastic-plastic antiplane shear crack growth. Journal of the Mechanics and Physics of Solids, 30, 59-74.
(33) Dunayevsky, V. & Achenbach, J.D. (1982). Boundary layer phenomenon in the plastic zone near a rapidly propagating crack tip. International Journal of Solids and Structures, 18, 1-12.
(34) Douglas, A.S. et al. (1981). Dynamic steady antiplane shear crack growth in an elastic-plastic material. In: Progress in Fracture Mechanics, Ed. D. Francois, 2233-2240. Pergamon.
(35) Freund, L.B. & Douglas, A.S. (1983). Dynamic growth of an antiplane shear crack in a rate-sensitive elastic-plastic material. In: Elastic-Plastic Fracture Mechanics, ASTM STP 803, Ed. C.F. Shih & J. Gudas, ASTM, 5-20.
(36) Lo, K.K. (1983). Dynamic crack-tip fields in rate sensitive solids. Journal of the Mechanics and Physics of Solids, 31, 287-305.
(37) Lam, P.S. & Freund, L.B. (1983). Elastic-plastic finite element analysis of steady state dynamic crack growth in plane strain tension. Brown University Technical Report.
(38) Rice, J.R. & Sorensen, E.P. (1978). Continuing crack tip deformation and fracture for plane strain crack growth in elastic-plastic solids. Journal of the Mechanics and Physics of Solids, 26, 163-186.
(39) Dahlberg, L. & Nilsson, F. (1976). Some aspects of testing crack propagation toughness. International Conference on Dynamic Fracture Toughness, London, 281-291.
(40) Kobayashi, T. & Dally, J.W. (1979). Dynamic photoelastic determination of the a' - K relation for 4340 steel. In: Crack Arrest Methodology and Applications, ed. G.T. Hahn and M.F. Kanninen, ASTM STP 711, 189-210.
(41) Glennie, E.B. (1972). The dynamic growth of a void in a plastic material and an application to fracture. Journal of the Mechanics and Physics of Solids, 20, 415-429.
(42) Rice, J.R. & Tracey, D.M. (1969). On the ductile enlargement of voids in triaxial stress fields. Journal of the Mechanics and Physics of Solids, 17, 201-217
(43) Hahn, G.T. et al. (1976). Influence of metallurgical factors on the fast fracture energy absorption rates. Metallurgical Transactions, 7, 49-54.
(44) Rydholm, G. et al. (1978). Numerical investigations of rapid crack propagation. Numerical Methods in Fracture Mechanics, Swansea, 660-672.
(45) Malluck, J.F. & King, W.W. (1977). Simulations of fast fracture in the dcb specimen using Kanninen's model. International Journal of Fracture, 13, 655-665.

(46) Kobayashi, A.S. et al. (1976). Dynamic finite element and dynamic photoelastic analyses of two fracturing Homalite-100 plates. Experimmental Mechanics, 16, 321-328.

(47) Brickstad, B. & Nilsson, F. (1980). Numerical evaluation by FEM of crack propagation experiments. International Journal of Fracture, 16, 71-84.

(48) Brickstad, B. (1983). A FEM analysis of crack arrest experiments. International Journal of Fracture, 21, 177-194.

(49) Brickstad, B. (1983). A viscoplastic analysis of rapid crack propagation experiments in steel. Journal of the Mechanics and Physics of Solids, 31.

(50) King, W.W. (1978). Toward a singular element for propagating cracks. International Journal of Fracture, 14, R7-R10.

(51) Nishioka, T. & Atluri, S.N. (1980). Numerical modeling of dynamic crack propagation in finite bodies, by moving singular elements - Part I: Formulation. Journal of Applied Mechanics, 47, 570-576; Part II: Results. Journal of Applied Mechanics, 47, 577-582.

(52) Aoki, S. et al. (1978). Elastodynamic analysis of crack by finite element method using singular element. International Journal of Fracture, 14, 59-68.

(53) Burgers, P. & Freund, L.B. (1979). Numerical analysis of dynamic linear elastic fracture mechanics problems by finite elements. Brown University Report, February.

(54) Kishimoto, K. et al. (1980). Dynamic stress intensity factors using J-integral and finite element method. Engineering Fracture Mechanics, 13, 387-394.

ADDITIONAL REFERENCES

(55) Krafft, J.M. & Irwin, G.R. (1965). Crack velocity considerations. In: Fracture Toughness Testing and Its Applications, ASTM STP 381, 114-132.

(56) Erdogan, F. (1968). Crack propagation theories. In: Fracture, 2, ed. H. Liebowitz, 498-586. Academic Press.

(57) Field, J.E. (1971). Brittle fracture: its study and application. Contemporary Physics, 12, 1-31.

(58) Achenbach, J.D. (1974). Dynamic effects in brittle fracture. In: Mechanics Today, ed. S. Nemat-Nasser, 1-57. Pergamon.

(59) Freund, L.B. (1976). The analysis of elastodynamic crack tip fields. In: Mechanics Today, ed. by S. Nemat-Nasser, 55-91. Pergamon.

(60) Freund, L.B. (1976). Dynamic crack propagation. In: The Mechanics of Fracture, AMD 19, ed. F. Erdogan, 105-134. ASME.

DISLOCATION DAMPING AND PHONON SCATTERING

A. V. Granato
University of Illinois at Urbana-Champaign, Urbana, IL 61801 U.S.A.

ABSTRACT

The dynamic aspects of the response of dislocations to applied stress form the basis for the description and understanding of many phenomena, most notably plastic flow, internal friction and thermal resistance. These can be represented by phenomenological coefficients for the inertia A and viscosity B in an equation of motion for the dislocation displacement y given by $A\ddot{y} + B\dot{y} + Cy = b\sigma$, where C is a restoring force constant, b is the Burgers vector and σ is the applied shear stress. This relation, sometimes called the rigid rod model, was first given and applied by Eshelby, who also calculated A and B. The viscosity coefficient B is given by $B = B_p + B_r + B_d + B_e$, where B_p is the drag provided by phonon scattering, B_r that from dislocation radiation, B_d that from dislocation interaction with dipolar point defects, and B_e that from electron scattering. With the exception of B_e, definitive calculations and discussions of the viscosity were given by Eshelby. For the most part, these calculations preceded the experiments in which they were measured. The dynamic aspects of dislocation response have become increasingly recognized in recent years. For example, low temperature thermal resistance is now known to arise from dislocation scattering by re-radiation, rather than from scattering by the static dislocation strain field. Also, measurements of the flow stress at low temperatures, particularly in superconductors, have shown that inertia and viscosity must be taken into account in a general description of plastic flow by use of a Brownian motion type of theory, rather than transition state theory. These theories have applicability to many other kinds of systems, but are best tested for dislocations because dislocation inertia and viscosity are so well understood today.

1. INTRODUCTION

An alternate title for this review might be Dislocation Dynamics, if the term is understood in the sense that I think Eshelby might have used it, rather than in the way it is most commonly used. I believe that Eshelby would have agreed that the term Dynamics should properly be used to describe properties determined by the instantaneous velocity, or acceleration of dislocations rather than by the average velocity of dislocations which are at rest most of the time.

1.1 PHONON SCATTERING

The basic process determining dislocation dynamics is phonon scattering by dislocations. There are two important aspects of this, a viscous drag force and thermal fluctuations in dislocation displacement. The scattering of a single phonon exerts a force on a dislocation, but there is no net force on a static dislocation because phonons are incident from all directions to give a zero net force. However, a moving dislocation scatters a higher flux of phonons from its leading edge, resulting in a viscous force. This is a mechanical effect.

In addition the phonons produce fluctuations in displacement, which is a thermal effect. This leads to dislocation diffusion, allowing the dislocation to overcome obstacles. In the presence of an external force, plastic flow occurs, and is described with the help of rate theory. Recently the question of the effect of mechanical viscosity on the thermal rates has become of special interest, not only for dislocations, but for rate theory generally. We shall see that dislocations provide a favorable testing ground for the general aspects of this question.

1.2 BROWNIAN MOTION OF DISLOCATIONS

The existence of two aspects to phonon scattering reminds us of Brownian Motion Theory, where the viscosity is associated with the systematic part of the scattering in an external force field, and the thermal fluctuations are connected with the random part of the scattering. For a particle falling in a fluid in a gravity field, the mechanical, systematic, viscous part of this motion is emphasized. For a particle which feels in addition a periodic potential, the thermal aspects may be emphasized. For example, a particle trapped in a well may escape by overcoming the barrier by thermal fluctuations. A dislocation subject to interactions with impurities and an external applied shear stress is a system of this type. Its motion is normally given by transition state theory (TST) as

$$v = d\nu \exp[-U(\sigma)/kT] \tag{1}$$

where v is the average dislocation velocity, d is the obstacle spacing, ν is the frequency of the particle in its potential well, U is the height of the barrier under the external applied stress σ, and kT has its usual meaning. It is noteworthy that there is no allowance for viscosity in this expression.

But it is clear from mechanical considerations that a particle, once started down a periodic stress hill with barriers, can overcome subsequent barriers by its inertia for sufficiently small viscosity. Thus the standard TST contains implicitly the assumptions that the viscosity is large enough to prevent inertial effects, and small enough so that the rate limiting process is the time taken to overcome the barriers rather than the time necessary to travel between the barriers. We will discuss effects arising in both of these limits.

1.3 EFFECTS OF PHONON SCATTERING

We will discuss three phenomena resulting from phonon scattering. The first is thermal resistance measurements, from which the mechanism of phonon scattering can be determined. The second is internal friction or ultrasonic attenuation from which the basic parameters for the mechanical aspects of the interaction can be measured (viscosity and frequency of the "particle"). The third is plastic flow, which is described as a diffusion of dislocations, and the effect of viscosity on rate theory will be of particular interest here.

In each of these cases, the displacement of a dislocation away from its equilibrium position in the presence of an external shear stress and an internal force field arising from impurities is required. A simple phenomenological equation of motion for this displacement was first given by Eshelby (1) in the form

$$A\ddot{y} + B\dot{y} + Cy = b\sigma \qquad (2)$$

where A represents the mass, B the viscous force per unit velocity, C the restoring force constant, and bσ the applied force per unit length, where b is the Burgers vector. The dots represent time derivatives. This equation, sometimes known as the rigid rod model (2), is a useful approximation to the string model where Cy is replaced by $T\,\partial^2 y/\partial x^2$, where T is the dislocation tension and x is a coordinate along the dislocation line. For this case, C is approximated by $12T/L^2$, where L is the length of dislocation line between pinning points. Eshelby calculated A and B, and estimated C to predict a maximum in ultrasonic attenuation years before it was observed (3). Eshelby found $A \sim \rho b^2$, with the value for an edge dislocation only slightly larger than that for a screw dislocation (4). The viscous drag constant B is given by

$$B = B_p + B_r + B_e, \qquad (3)$$

where B_p is the drag arising from phonons, B_r is a re-radiation drag owing to emission of elastic waves by an accelerating dislocation, and B_e is an electron drag. In addition, point defects can provide a drag of a somewhat different nature which can be additive under some circumstances (5).

The temperature dependence expected for the drag constant B is illustrated schematically in Fig. 1. The phonon drag depends primarily on the phonon density or thermal energy and is thus expected to be linear in temperature at high temperature. This drag disappears at low temperature in insulators, but in metals, a temperature independent drag remains. For superconductors, the electronic drag disappears below the transition temperature T_c, and at the lowest temperature only a re-radiation drag for oscillating dislocations remains. Point defect drag effects can be much larger, with opposite temperature dependence.

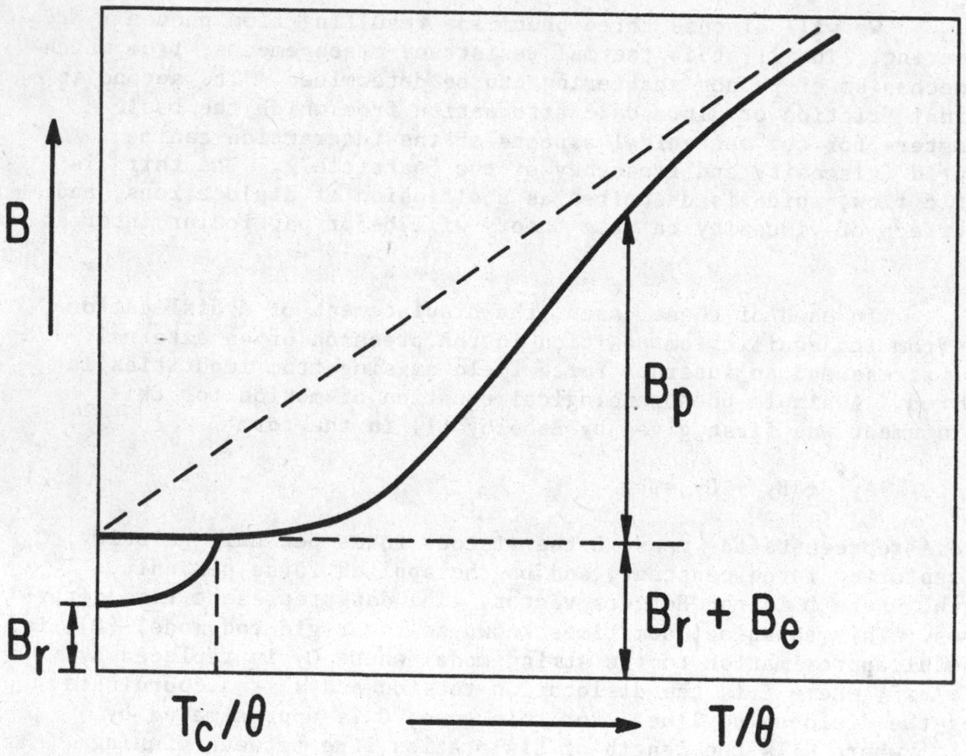

FIG. 1. Temperature dependence of the viscous damping parameter B (schematic). The temperature is normalized to the Debye temperature θ.

2. DRAG EFFECTS

2.1 THERMAL RESISTANCE

An old question in the literature of phonon scattering is whether the increase in thermal resistivity at low temperature after plastic deformation is primarily due to dynamic or static scattering of phonons by dislocations. Eshelby (1) and Nabarro (6) made the earliest important contributions to this subject by discussing the two mechanisms by which sound waves or phonons are scattered by dislocations. Eshelby proposed that if the sound wave induces the dislocation to vibrate, the incident energy will be dissipated as the dislocation radiates elastic waves. Taking into consideration the wavelength dependence of the effective mass of the dislocation, Nabarro showed that the scattering cross section for this process (reradiation scattering) increases with the wavelength λ, though not as fast as linearly. Even if the dislocation does not vibrate, phonons will be scattered due to the anharmonicity of the dislocation strain field. Nabarro suggested that the scattering cross section for this

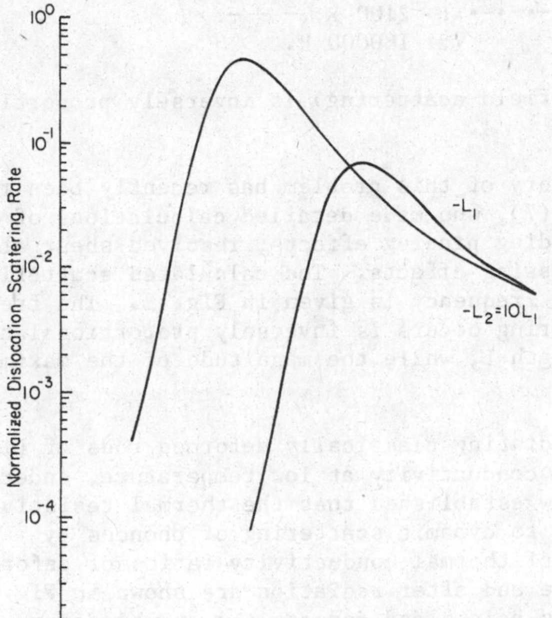

FIG. 2. The normalized scattering rate $\tau^{-1}/2\omega_1(L_2)\Omega\Lambda L_2^2$ vs the normalized frequency $\omega/\omega_1(L_2)$ for average dislocation lengths L_1 and $L_2 = 10L_1$. [After Kneezel and Granato (7).]

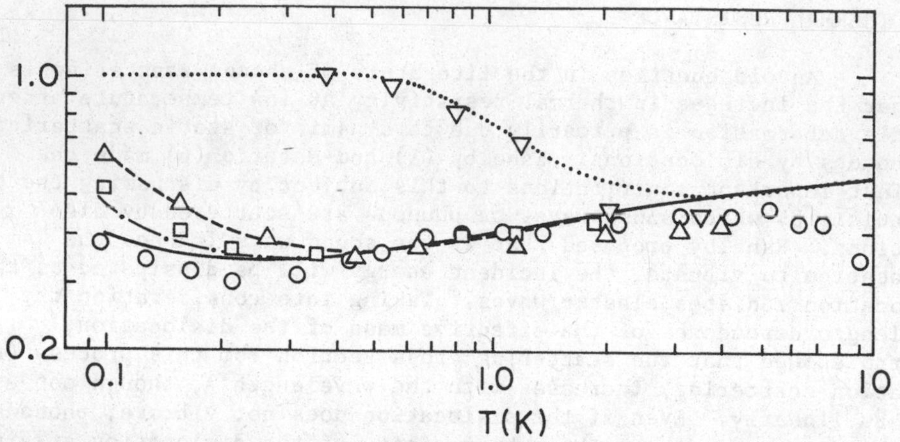

FIG. 3. Measured (Ref. 9) k/k_0 before and after irradiation.
○: Freshly bent sample; —:; After 720-Rγ;
-•-•-•:; 2100 R; - - -:
∇: 180000 R.

process (static strain-field scattering) is inversely proportional to λ.

The long history of this problem has recently been reviewed by Kneezel and Granato (7), who made detailed calculations of the dynamic mechanism including pinning effects, resolved shear stress factors and phonon focussing effects. The calculated scattering rate as a function of phonon frequency is given in Fig. 2. The frequency at which maximum scattering occurs is inversely proportional to the dislocation segment length L, while the magnitude of the maximum is proportional to L.

By gamma irradiating plastically deformed rods of LiF and remeasuring the thermal conductivity at low temperature, Anderson and coworkers (8,9) have now established that the thermal resistivity change is primarily due to dynamic scattering of phonons by dislocations. Results of thermal conductivity ratios of deformed to undeformed sample before and after radiation are shown in Fig. 3. The γ-radiation adds pinning points and reduces the average segment length L, thereby increasing the resonant frequencies of the dislocations. As a result, low frequency phonons are not scattered and the thermal conductivity at the lower temperatures returns to that of the undeformed sample. More recently, experiments using focussed ballistic phonons (10) show directly a resolved shear stress dependence for the scattering.

2.2 DISLOCATION DAMPING

Internal friction or ultrasonic attenuation measurements can be used to measure the phenomenological coefficients of Eq. 2. The theory is analogous to that of dielectric theory, with the dislocation displacement y playing the role of charge displacement. The elastic constant change $\Delta C/C$ is given by

$$\frac{\Delta C}{C} = \frac{\varepsilon_d}{\varepsilon_{el}} = \frac{\Lambda G b \bar{y}(\sigma)}{\sigma}$$

where ε_d is the dislocation strain, ε_{el} is the elastic strain, Λ is the dislocation density, G the shear modulus, and $\bar{y}(\sigma)$ is the average dislocation displacement for stress σ. The in-phase part of the response is the elastic constant change and the out out-of-phase part is the attenuation. The frequency dependence of the logarithmic decrement is shown in Fig. 4 for the different values of the drag constant B. In the Figure $d = B/2A$, and $\omega_o = \pi C/L$. For increasing B, the resonance disappears and the maximum moves to lower frequencies. For the overdamped case ($d \gg 1$) the displacement and decrement are tension limited at low frequencies, but drag limited at high frequencies. Measurements as a function of γ-radiation (3) show that the low frequency part of the curves are sensitive to radiation, while the high frequency response is not.

The resonance frequency is usually not observed in common crystals, but in recent studies on He^3 by Beamish and Franck (11), the resonance is clearly seen as an anomalous velocity effect. The

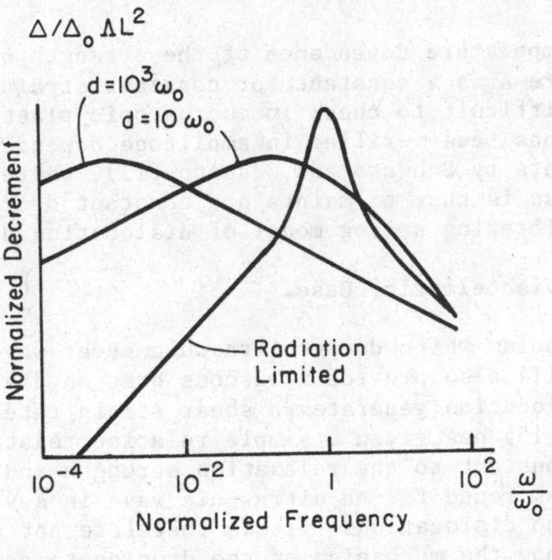

FIG. 4. Dislocation Decrement vs Frequency for different values of drag constant.

magnitude and temperature dependence of the drag constant obtained from such measurements are in reasonable agreement with those theoretically expected.

2.3 PLASTIC FLOW

The "Medium Drag" (Quasistatic) Case.

The plastic strain rate $\dot{\varepsilon}$ is given by

$$\dot{\varepsilon} = \Lambda b v \tag{4}$$

where Λ is the mobile dislocation density and v is the average velocity. The velocity v is then given by TST as Eq. 1. In Eq. (1) the effects of an internal stress field have been ignored.

It was shown by Granato and Lücke (12) that, for low temperatures and relatively pure materials,

$$U(\sigma) = U_o(1-\sigma/\sigma_m)^{3/2} \tag{5}$$

regardless of the form of the dislocation force-distance law, provided only that the force has a smooth maximum. In Eq. (5), U_o is determined by the barrier height and σ_m is the stress required to overcome the barrier in the absence of thermal fluctuations. Using Eq. (5) and (1) in (4), one thus obtains

$$\sigma = \sigma_m - AT^{2/3} \tag{6}$$

as the expected temperature dependence of the strength or flow stress on this model, where A is a constant for constant strain-rate tests. This relation is difficult to check in macroscopic plastic deformation measurements, but has been verified in amplitude dependent internal friction measurements by Schwarz and Granato (13), where the corresponding stress is that to maintain a constant decrement, according to the vibrating string model of dislocation damping (14).

The "Large Drag" (Viscoelastic) Case.

Any mechanism which damps ultrasonic shear waves viscoelastically will also provide a viscous drag on dislocations since a moving dislocation generates a shear strain rate in the crystal. Eshelby (15) has given a simple relation relating the dislocation drag constant to the relaxation strength and time constant for the damping peak found for an ultrasonic wave in a viscoelastic medium containing no dislocations. It is therefore not even necessary in such cases to know the mechanism of the damping to calculate the drag constant. Weertman (16) has interpreted the creep of ice in terms of proton motion using Eshelby's relation. When a drag constant B is calculated from ultrasonic data in ice by Schiller (17), a value about 7 orders of magnitude larger than typical room temperature phonon drag constants in metals is obtained. This drag, while linear in velocity, is so enormous that the travel time between obstacles is much longer than the time to overcome obstacles, so that obstacle effects can be neglected. In place of Eq. (1), one has now simply

$$v = b\sigma/B \tag{7}$$

so that the deformation should be described as a simple linear viscoelastic solid (18) of the type first described by Maxwell [1868] (19), who used as a model of a solid a simple spring and viscous dashpot in series. In fact the observed deformation behavior of predeformed ice single crystals is described to a good degree of approximation by such a model as shown by Joncich, Holder and Granato (18) in Fig. 5. Figure 5 shows the creep strain ε for two constant levels of stress in an ice single crystal at -10° C. The strain rate $\dot{\varepsilon}$ is the sum of an elastic strain rate $\dot{\sigma}/G$ and the plastic strain rate $\Lambda b\bar{v} = \Lambda b^2\sigma/B(\sigma)$. There is a small recoverable component which can be made negligibly small compared to the linear strain component for long stress application times. The dislocation component is given accurately by the drag coefficient using Eshelby's formula.

3. ANOMALOUS PLASTIC FLOW

3.1 MEASUREMENTS (The "Small Drag" or Inertial Case)

Recently, a quite unexpected loss of strength in superconducting materials entering the superconducting state has been found. Since the initial reports by Kojima and Suzuki (20) and Pustovalov, et. al (21), many studies have been made, which have been reviewed by Startsev (22). The results of Kojima and Suzuki for the stress-strain curve of a lead specimen changing between the normal and superconducting state are shown in Fig. 6.

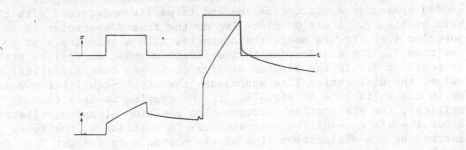

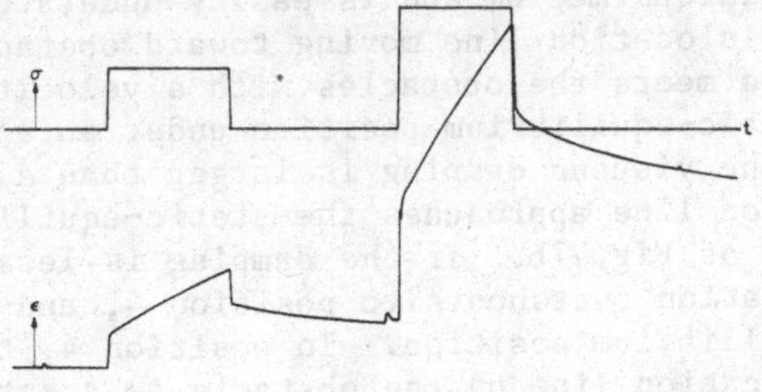

FIG. 5. Creep Strain ε for two stress levels in an ice single crystal at -10° C. [Ref. (18).]

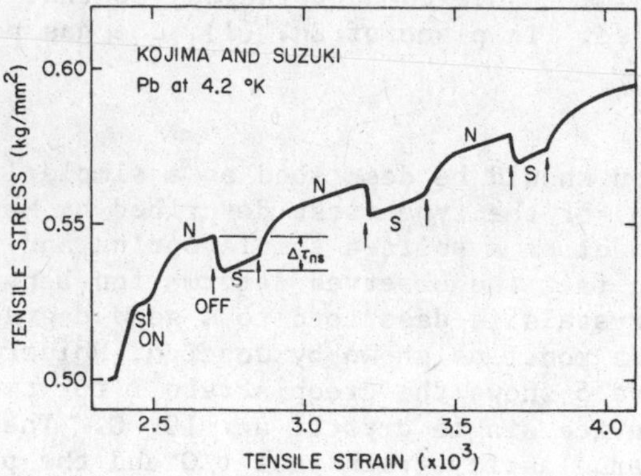

FIG. 6. Stress-strain curve in lead as specimen is switched between the normal and superconducting state. [After Kojima and Suzuki (20).]

Two types of theories have been offered to explain the effect. One by Natsik (23), modified by Suzuki (24), supposes that frequency factor ν in the rate theory expression in Eq. (1) decreases in the normal state. In a second theory (the mechanical inertial model) (25-28) it is supposed that the drag constant B at low temperatures is small enough so that dislocations can mechanically overshoot the obstacles inertially. A third theory which combines rate theory and mechanical inertial effects will be outlined in the next section.

3.2 MECHANICAL MODEL

The inertial mechanical model is entirely analogous to a loaded spring in a viscous medium and is easily understood with the help of Fig. 7. A dislocation line moving toward obstacles in position 1 of Fig. 7a meets the obstacles with a velocity v at position 2. The static-equilibrium position under an applied stress is position 3. If the viscous damping is larger than a critical value, the dislocation line approaches the static-equilibrium position as in the solid line of Fig. 7b. If the damping is less than critical, the dislocation overshoots to position 4, and oscillates about the static equilibrium position. In position 4, the force exerted by the dislocation line on the obstacle is greater than in the static-equilibrium case. Alternatively, one may say that a smaller stress is needed in the low-damping case to produce the same force on the obstacle.

From Eq. (2), one may deduce that the critical damping B_c to overdamp dislocations is given by

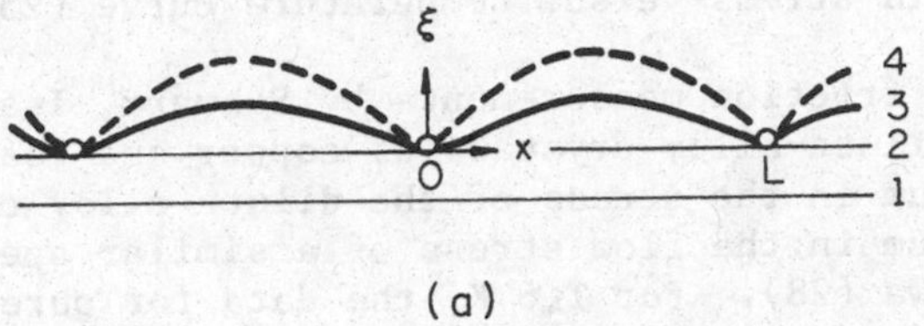

(a)

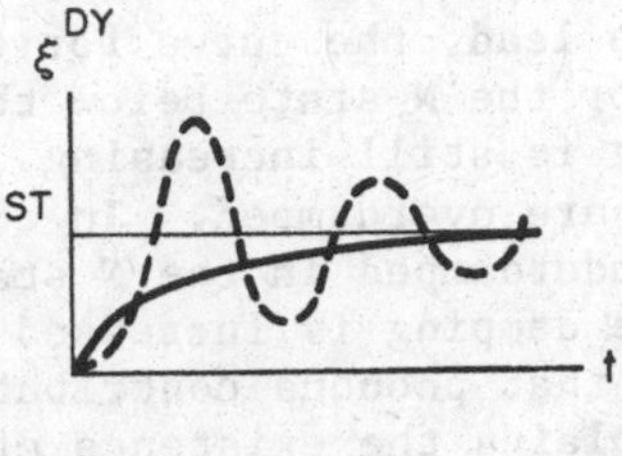

(b)

FIG. 7. (a) Schematic of the motion of the dislocation line.
1, dislocation line approaches pinning points;
2, dislocation line just touches pinning points;
3, static equilibrium position of dislocation line;
4, overshoot position of underdamped dislocation.
(b) Displacement as a function of time for an underdamped dislocation (solid line).

$$B_c = 2A\omega_o. \tag{8}$$

On the other hand, the smallest damping obtained at T=0 in the superconducting state is the re-radiation damping B_r, which may be found (29) from Eshelby's calculations (1) to be

$$B_r = A\omega_o/8. \tag{9}$$

This means that Br/Bc = 1/16, so that all dislocations are underdamped by a factor of 16 at T=0 in the superconducting state, independent of material or structure. This accounts for the universality of the effect (25,26).

In order to obtain the observed temperature dependence of the stress change, it is necessary to assume in the inertial model that the dislocations are also underdamped in the normal state, although by a lesser amount than in the superconducting state. If the dislocations are still underdamped in the normal state, then it should be expected that they may become overdamped at higher temperatures where the phonon density increases. If so, one would expect to see an anomaly in the yield stress versus temperature curve (25).

Internal friction measurements by Schwarz, Isaac and Granato (30) on pure and dilute-alloy crystals of copper and lead are shown in Fig. 8. The maximum in the stress of the dilute alloy of copper is similar to a maximum in the flow stress of a similar specimen seen by Kamada and Yoshizawa (28). For $T \lesssim 6$ K, the data for pure copper follow the strict $T^{2/3}$ dependence of Eq. (6), as was found earlier for pure aluminum (13) and which was explained in terms of a classical thermally activated process. The measurements in lead are similar to those in copper for the two purities, except that there are additional changes during the N-S transition. This change can be used as a key to determine whether the dislocations are over or underdamped in the normal state. In the pure lead, the curve for the S state joins smoothly onto the curve for the N state below the transition temperature where the drag is still increasing. This shows that the dislocations in pure lead are overdamped. In contrast, dislocations in the dilute alloy are underdamped in the N state so that they can become overdamped when the damping is increased when the temperature is raised sufficiently so that phonons contribute significantly to the viscous damping. This explains the existence of the anomalous maximum observed for the dilute lead alloy

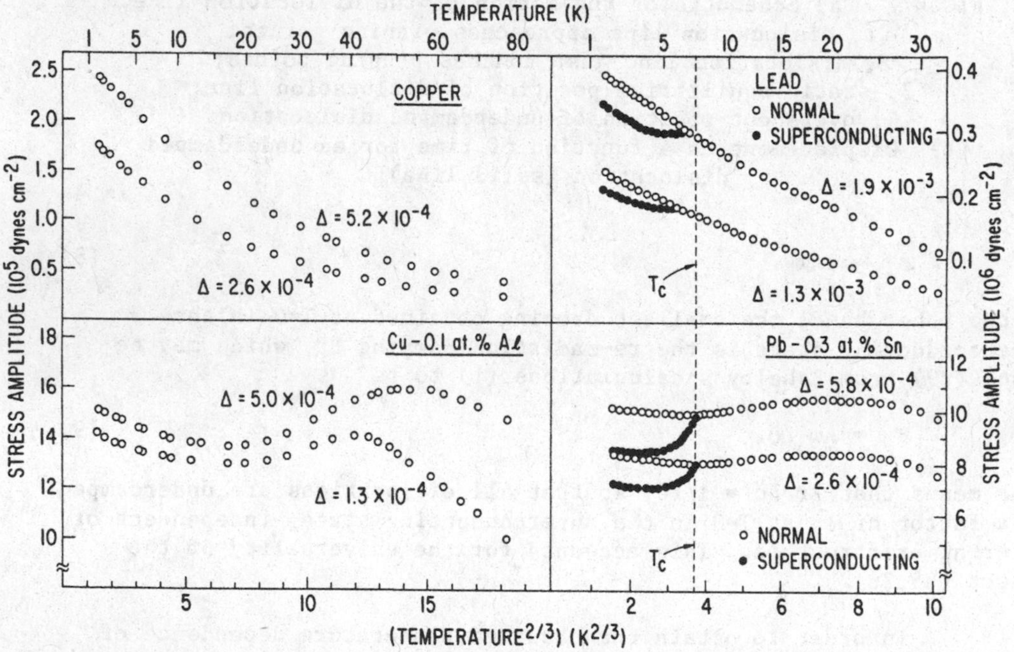

FIG. 8. Temperature dependence of the stress amplitude for constant decrement in internal friction measurements on pure and dilute alloys of copper (left column) and lead (right column). [After Schwarz, Isaac and Granato (30).]

at about 20 K. The fact that the temperature dependence of σ_o (Δ) in the N state of the dilute lead alloy is essentially the same as in the dilute copper alloy provides strong experimental evidence that the anomalous maximum in the dilute copper alloy is of inertial origin. Figure 8 also shows that below T = 3.5 K, the stress amplitude $\sigma_o(\Delta)$ for the S state resumes a linear $T^{2/3}$ dependence with approximately the same slope as in the N state. Thus the change in stress versus temperature is predicted well by a purely mechanical model. However, the temperature dependence of the absolute stress is characteristic of a rate process. This suggested to Schwarz, Isaac and Granato (3) that in the presence of inertial effects, dislocation motion is initiated by a thermally activated process and continues by the overcoming of additional obstacles inertially. The amount of plastic deformation in the latter process is determined by the degree of underdamping. A theory of this process is outlined in the next section.

4. A RATE THEORY INCORPORATING INERTIAL OVERSHOOT EFFECTS

Isaac and Granato (31) have recently developed a rate theory which incorporates inertial overshoot effects. It is found that the frequency factor ν in Eq. 1 is not dependent on the damping constant B for the range of B expected, but that the advance distance d per thermally activated event in Eq. 1 is strongly dependent on B. The theory is an extension to the many degrees of freedom corresponding to dislocation motion of some earlier results found for Brownian motion of a particle in a force field by Kramers (32), Chandrasekhar (33), and Wang and Uhlenbeck (34).

A Langevin equation for a dislocation in a stress and impurity field is solved for a distribution function $\rho(x, v, t)$ which gives the probability of finding a dislocation at position x with velocity v at time t. The form of this distribution at once explains one of the principal puzzles about many observations of the temperature dependence of the flow stress and its change at the superconducting transition. This is that a purely mechanical inertial theory (25-28) gives a satisfactory account of the changes in flow stress at a given temperature, including the predictions, (25,26) subsequently verified (Schwarz, Isaac and Granato (31)) that the flow stress change will go to zero below the transition temperature for sufficiently pure materials and that an anomaly in the temperature dependence of the flow stress above the transition temperature will appear otherwise. On the other hand, this mechanical theory cannot account for the fact that the absolute value of the flow stress has a normal temperature dependence expected from thermal overcoming of barriers in either the normal or superconducting state. This mechanical-thermal duality is explained by the form of the distribution function, which tends to be a cloud centered about the purely mechanical path, which differs for two different viscosities. The path depends on the viscosity, while the width of the cloud depends upon the temperature. If one holds the temperature constant and changes the viscosity, the cloud moves, but overall the change is mechanical-like. On the other hand, if one holds the viscosity fixed, and varies the temperature, the cloud position stays fixed but the cloud width changes, thus emphasizing the fluctuation aspect of the thermal bath. The superconducting transition thus provides a unique opportunity to emphasize either the systematic part of the Brownian motion effect leading to viscosities, or the random part leading to thermal fluctuations.

In general, both ν and d depend upon B. In the theory, there are two critical viscosities. Below the lower of these inertial effects occur, making d a function of B. Above the higher of these the frequency factor ν is decreased. The possibility of a decreasing ν with increasing B forms the basis of Natsik's theory. However, we (31) estimate that the upper viscosity limit is not reached even at the melting temperature for phonon drag.

In fact, these considerations are general, applying for any diffusing "particle" in an external force field. The quantities needed in the theory are the viscosity B, and the resonant frequencies of the "particle." The advantages of plastic flow measurements are that dislocations are Brownian particles for which these (1) have been calculated, (by Eshelby), (2) have had mechanisms determined (from thermal resistivity measurements), (3) can be measured (from internal friction measurements), and (4) can be controlled (as in the normal to superconducting transition).

5. SUMMARY

Eshelby's work has had a major impact on Dislocation Dynamics.

His calculations often preceded any experiments and have proven accurate. Particularly notable is his calculation of re-radiation effects, from which one knows that all dislocations are underdamped by a factor of 16 in the superconducting state at T=0.

Dislocations provide a model system for detailed studies of the effect of viscosity on reaction rates.

The range of validity of Transition State theory is limited to intermediate viscosities.

ACKNOWLEDGMENT

This work has been supported by the National Science Foundation under grant DMR80-15707.

REFERENCES

(1) Eshelby, J. D., Proc. R. Soc. London A197, 396 (1949).
(2) Lücke, K., and Granato, A. V., Phys. Rev. B24, 6991-7000 (1981) J. de Phys. C5, 327 (1981).
(3) Stern, R. M. and Granato, A. V., Acta Met. 10, 92 (1982).
(4) Eshelby, J. D., Proc. Roy. Soc. A266, 222 (1962).
(5) Granato, A. V. and Lücke, K., Phys. Rev. B24, 7007-7017 (1981).
(6) Nabarro, F. R. N., Proc. R. Soc. London A209, 279 (1951).
(7) Kneezel, G. A. and Granato, A. V., Phys. Rev. B25, 2851 (1982).
(8) Anderson, A. C. and Malinowski, M. E., Phys. Rev. B5, 3199 (1972).
(9) Roth, E. P. and Anderson, A. C., Phys. Rev. B20, 768 (1979).
(10) Northrop, G. A., Cotts, E. J., Anderson, A. C. and Wolfe, J. P., Phys. Rev. B27, 6395 (1983).
(11) Beamish, J. R. and Franck, J. P., Phys. Rev. Lett. 47, 1736 (1981).
(12) Granato, A. V. and Lücke, K., J. Appl. Phys. 52, 7136 (1981).
(13) Schwarz, R. and Granato, A. V., Phys. Rev. Lett. 34 (1975) 1174.
(14) Granato, A. V. and Lücke, K., J. Appl. Phys. 27 (1956) 583.
(15) Eshelby, J. D., Philos. Mag. 6 (1961) 953.
(16) Weertman, J., in Physics and Chemistry of Ice, ed. by E. Whalley, S. J. Jones and L. W. Gold, Royal Society of Canada, Ottawa (1973).
(17) Schiller, P., Z. Physik 153 (1958) 1.
(18) Joncich, D., Holder, J., and Granato, A. V., to be published.
(19) Maxwell, J. C., Philos. Mag. (IV) 35 (1868) 135.
(20) Kojima, H. and Suzuki, T., Phys. Rev. Lett. 21 (1968) 896.
(21) Pustovalov, V. V., Startsev, V. I., Didenko, D. A. and Fomenko, V. S., Fiz. Metal. i Metalloved. 23 (1967) 312.
(22) Startsev, V. I., in Dislocations in Solids, ed. by F. R. N. Nabarro (North Holland, New York, 1983) Vol. 6, Ch. 28.
(23) Natsik, V. S., Soviet Physics JETP 34 (1972) 1359.
(24) Suzuki, T., in Rate Processes in Plastic Deformation (American Society for Metals, Metals Park, OH, 1976).
(25) Granato, A. V., Phys. Rev. Lett. 27 (1971) 660.
(26) Granato, A. V., Phys. Rev. B4 (1971) 2196.
(27) Suenaga, M. and Galligan, J. M., Scripta Met. 5 (1971) 63.
(28) Kamada, K. and Yoshizawa, I., J. Phys. Soc. Japan 31 (1971) 1056.
(29) Garber, J. A. and Granato, A. V., J. Phys. Chem. Solids 31, 1863 (1970).
(30) Schwarz, R. B., Isaac, R. D., and Granato, A. V., Phys. Rev. Lett. 38 (1977) 554.
(31) Isaac, R. D. and Granato, A. V., Proc. 5th Int. Conf. Strength of Metals and Alloys, ed. by P. Hansen, et al. (Pergammon Press, Oxford, p. 425-493 (1979); Isaac, R. D., Ph.D. Thesis, University of Illinois (1977).
(32) Kramers, H. A., Physica 1 (1940) 284.
(33) Chandrasekhar, S., Rev. Mod. Phys. 15 (1943) 1.

(34) Wang, M. C. and Uhlenbeck, G. E., Rev. Mod. Phys. 17 (1945) 323.

(35) Granato, A. V., Lücke, K., Schlipf, J. and Teutonico, L. J., J. Appl. Phys. 35 (1963) 2732.

INVARIANT INTEGRALS FOR SOME TIME DEPENDENT CRACK AND HEAT CONDUCTION PROBLEMS

C. ATKINSON

Dept. of Mathematics, Imperial College, London SW7 2BZ.

ABSTRACT

A formalism based on the energy momentum tensor is used to derive invariant integrals for steady state crack propagation, dynamically similar expanding cracks and other dynamically similar anti-plane strain problems. Invariant integrals of the 'reciprocal equation' type are also given for dynamically similar elasticity and certain heat conduction problems. It is suggested that these integrals can be useful as ancillary tools for special stress analysis problems.

1. INTRODUCTION

In a number of papers (e.g. Eshelby (1), (2), (3)) the use of the elastic energy momentum tensor and various associated invariant integrals has been advocated. For the most part these papers were concerned with the application of the energy momentum tensor to the calculation of the force on a defect in a time independent elastic field. An exception is the paper by Atkinson and Eshelby (4) in which the energy flow into a moving crack tip is considered. In that paper it was shown that the energy release rate for a moving crack tip could be evaluated in terms of a certain integral taken around an infinitesimally small contour surrounding the crack tip. For an elastic material with energy density W this integral was defined as

$$I(\Sigma) = \int_{\Sigma} (\sigma_{ij}\dot{u}_i + vE\delta_{1j})dS_j \tag{1.1}$$

where

$$E = W + \tfrac{1}{2}\rho\dot{u}_i\dot{u}_i \quad . \tag{1.2}$$

The superposed dot denotes the time derivative, u_i is the displacement field, σ_{ij} the stress field (the application in mind was to the plane problem of a straight crack propagating in the X_1 direction), δ_{ij} is the kronecker delta and Σ is regarded as a surface embracing the tip of the crack and moving with it in the X_1 direction with the instantaneous velocity of the tip v. The integral I gives the total flow of energy through the contour Σ, the second term accounting for the convective transfer of both elastic and kinetic energy.

It is shown in (4) that under favourable circumstances

$$vG = \lim_{\Sigma\to 0} \int_\Sigma (\sigma_{ij}\dot{u}_i + vE\delta_{1j})dS_j \tag{1.3}$$

where vG is the energy release rate into the crack tip. Thus knowing the dynamic stress and displacement field at the crack tip the energy release rate can be calculated. Other accounts of a result like that given in equation (1.3) can be found in Cherepanov (5) and Freund (6) and a more extensive review of fairly recent literature in Atkinson (7).

In the special case of steady motion when the stress and displacement field depends on time only though the variable $X_1 = (X_1'-vt)$ equation (1.3) reduces to the result

$$G = \lim_{\Sigma\to 0} \int_\Sigma (-\sigma_{ij}\frac{\partial u_i}{\partial X_1} + E_1\delta_{1j})dS_j \tag{1.4}$$

where

$$E_1 = W + \tfrac{1}{2}\rho v^2 \frac{\partial u_i}{\partial X_1} \cdot \frac{\partial u_i}{\partial X_1} \quad . \tag{1.5}$$

The divergence of the integrand in equation (1.4) is zero so the surface Σ can be deformed into another far from the crack tip if the integrand is zero on the crack faces. This 'path independent' property can be useful for the solution of certain special problems such as those considered in section 2. When the inertia terms are neglected in equation (1.4) the integrand is the P_{1j} component of the elastic energy momentum tensor discussed by Eshelby (2,3) and G is the corresponding crack extension force. Such integrals are to be discussed elsewhere in this conference; we merely note here that it is not possible to deform the integral (1.3) into a surface far from the crack tip because the integrand does not have zero divergence in the time dependent situation. However, in section 3 we will investigate the possibility of invariant integrals for time dependent problems for a variety of special situations.

To set up the machinery for our later discussions we derive here the energy momentum tensor given a Lagrangian L. Suppose

$$L = L(X_i, u_i, u_{i,j}, \phi_i, \phi_{i,j}) \tag{1.6}$$

so L may depend explicitly on position (cartesian co-ordinates X_i) and two independent vector fields u_i, ϕ_i and their gradients $u_{i,j} = \partial u_i/\partial X_j$ etc. The conditions of stationarity of the functional $\int L dV$ (where V is the volume of the body) lead to the Euler equations

$$\frac{\partial}{\partial X_j}\left(\frac{\partial L}{\partial u_{i,j}}\right) - \frac{\partial L}{\partial u_i} \quad \text{and} \quad \frac{\partial}{\partial X_j}\left(\frac{\partial L}{\partial \phi_{i,j}}\right) - \frac{\partial L}{\partial \phi_i} = 0 \tag{1.7}$$

where the summation convention with respect to repeated indices ($i = 1,2,3$) has been used. If the tensor $P_{j\ell}$ is defined as

$$P_{j\ell} = \frac{\partial L}{\partial u_{i,j}} u_{i,\ell} + \frac{\partial L}{\partial \phi_{i,j}} \phi_{i,\ell} - L\delta_{j\ell} \tag{1.8}$$

then it can be shown by direct calculation, using the Euler equations (1.7), that

$$\frac{\partial P_{j\ell}}{\partial X_j} = -\left(\frac{\partial L}{\partial X_\ell}\right)_{exp} \tag{1.9}$$

where $(\partial L/\partial X_\ell)_{exp}$ means that all variables are held constant except explicit dependence on X_ℓ.

The property (1.9) means that a number of path independent integrals can be written down in terms of the energy momentum tensor. We quote here those derived originally by Gunther (8) by an application of Noether's theorem (see also Knowles and Sternberg (9))

$$F_\ell = \int_S P_{\ell j} dS_j \tag{1.10}$$

$$L_{k\ell} = \int (X_k P_{\ell j} - X_\ell P_{kj} + u_k p_{\ell j} - u_\ell p_{kj}) dS_j \tag{1.11}$$

and

$$M = \int (X_\ell P_{\ell j} - \tfrac{1}{2} u_\ell p_{\ell j}) dS_j \tag{1.12}$$

where

$$p_{ij} = -\frac{\partial L}{\partial u_{i,j}} \quad , \quad \ell,i,j,k = 1,2,3 \quad . \tag{1.13}$$

It is necessary to check by direct calculation under what conditions the above integrals are zero when taken over a closed surface S ($dS_j = n_j dS$, n_j the outward normal to S). From (1.9) it is clear that F_ℓ is path independent if $P_{\ell j}$ does not depend explicitly on X_ℓ. The $L_{k\ell}$ are path independent for a non-linear homogeneous isotropic medium. The integral M as quoted in (1.12) is path-independent for a homogeneous linear anisotropic medium (in this case $L = -W = -\frac{1}{2} p_{ij} u_{k,j}$). It is possible to amend the integral M to make it path independent for power-law elastic materials, see Budiansky and Rice (10) but look for a misprint in their definition of M. A number of applications of these integrals to problems of equilibrium elasticity are given in (3) and (10) and references contained in these papers and in (7).

For our purposes it is also worth recording the two dimensional version of (1.12). This can be written

$$M = \int X_\ell P_{\ell j} dS_j \quad , \tag{1.14}$$

where now $\ell = 1,2$.

It is worth stressing that the integrals defined above depend only on the Lagrangian equation (1.6); indeed the energy momentum tensor $P_{\ell j}$ is defined in Morse and Feshbach ((11), equation 3.4.2) for a variety of vector fields including those of elasticity. The object of the present paper will be to apply these results to some special dynamic crack problems and to obtain some invariant integrals for certain dynamic problems which have not, perhaps, been noted before.

The plan of the paper is as follows:- in section 2 the above method is applied to steady state crack propagation problems; in section 3 a discussion is given of how the above formulation might be applied to certain dynamic (time dependent) problems after various co-ordinate transformations have been made. In addition to deriving a Lagrangian formulation for the resulting equations and hence leading to the possibility of integrals such as considered in this introduction, we also give (in section 3) reciprocal theorems for equations arising from elasticity or heat conduction after certain similarity transformations have been made. We believe such integrals are potentially useful; space limitations mean we cannot pursue this point here.

2. STEADY STATE CRACK PROPAGATION

For the situations considered in this section we assume a state of plane strain in the (x_1', x_2') plane and a crack lying on the plane $x_2' = 0$ moving steadily in the positive x_1' direction at a constant speed v . Referring to a co-ordinate system (x_1, x_2, x_3) moving with the crack tip where

$$x_1 = x_1' - vt \quad , \quad x_2 = x_2' \quad , \quad x_3 = x_3' \tag{2.1}$$

the steady state assumption implies that the elastic shear and displacement field depend on time only through the x_1 co-ordinate. Making the change of variable (2.1), the equations of motion become

$$\frac{\partial \sigma_{ij}}{\partial x_j} = \rho v^2 \frac{\partial^2 u_i}{\partial x_1^2} \tag{2.2}$$

where u_i $(i = 1,2,3)$ are the components of displacement, σ_{ij} the stress tensor, and ρ the density. In our applications the stress and displacement field are functions only of x_1 and x_2 because of our plane assumption, but for anisotropic media there may be three non-zero displacement components.

It is easy to see from equations (1.6) and (1.7) of the introduction that a Lagrangian for equations (2.2) can be written

$$L = \frac{\rho v^2}{2} \frac{\partial u_i}{\partial x_1} \frac{\partial u_i}{\partial x_1} - W \tag{2.3}$$

where W is an energy density depending on $u_{i,j} \equiv \partial u_i/\partial x_j$ such that

$$\sigma_{ij} = \partial W/\partial u_{i,j} \quad . \tag{2.4}$$

A useful configuration for the application of the integral F_1 (defined in equation (1.10)), with L defined as in equation (2.3), is that of a semi-infinite crack moving steadily in a displacement loaded strip such as shown in Figure 1. For this situation

$$F_1 = G = \int_\Gamma [(W+\tfrac{1}{2}\rho v^2 \frac{\partial u_i}{\partial x_1} \frac{\partial u_i}{\partial x_1}) - \sigma_{ij} n_j \frac{\partial u_i}{\partial x_1}]\, d\Gamma \tag{2.5}$$

where Γ is any path enclosing the crack tip. Since the integral is path independent, it may be possible to choose a path for which evaluation of G is simple. This has been done for the linear elastic case by choosing a path such as AGEDKJHA of Figure 1. With the

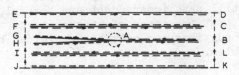

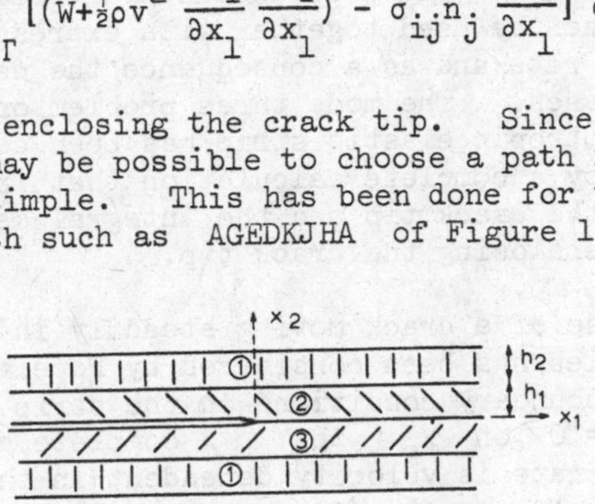

FIG. 1

crack stress free and displacement boundary conditions, for an isotropic medium, such as $u_1 = 0$, $u_2 = \pm u_{20}$ (a constant) on $x_2 = \pm h$ $(h = h_1 + h_2)$ for all x_1 the only non-zero contribution to the integral is that on the vertical ligament $x_1 = +\infty$, $-h \le x_2 \le h$. Thus G, the value of equation (2.5) taken around an infinitesimal contour at the crack tip, is equal to the energy stored in the ligament at $x_1 = +\infty$.

The energy release rate turns out to be velocity independent in this case. This follows if it is assumed that $\partial u_i/\partial x_1$ tends to zero as $x_1 \to \pm\infty$. A local eigenvalue analysis of the crack tip gives the form of the near crack tip stress and displacement field and the integral G, when evaluated in terms of this, equals (in symmetric situations) K^2 times a known velocity factor where K is the stress intensity factor. Thus the stress intensity factor can also be determined. This method has been used for crack propagation in elastic media with spatially varying moduli and the results checked by more complete calculations of the stress field (Atkinson (12)). A review of the problem of the infinite strip with a semi-infinite crack has been given by Nilsson (13) and some other applications of this method can be found in (14). Note that if media 2 and 3 of the figure have different elastic moduli and we consider the case of plane strain then the above argument works for G but not for K.

The simple argument given above is valid for steady crack propagation at subsonic velocities if the medium is elastic and homogeneous. In general a more complete analysis must be made to obtain accurate values of stress and displacement far from the crack tip. These values can then be used together with expression (2.5) to obtain the energy release rate and as a consequence the near crack tip stress field in certain cases. The mode three problem of a crack moving supersonically in an isotropic elastic strip has been considered by Nilsson (13) who verified by a complete calculation that in this case there is no energy flow to the crack tip and the integral defined in (2.5) is zero for any path enclosing the crack tip.

The case of a crack moving steadily in an elastic strip with shear-free boundaries has been considered by Popelar and Atkinson (15). For this case the boundary conditions on the strip sides are $u_2 = \pm u_{20}$ on $x_2 = \pm h$, $\sigma_{12} = 0$ on $x_2 = \pm h$. A complete analysis shows that the energy release rate is velocity dependent in this case and moreover the σ_{11} stress field and $\partial u_1/\partial x_1$ are not zero as $x_1 \to -\infty$ in the strip (they are in fact velocity dependent there). As $x_1 \to +\infty$ σ_{22} and $\partial u_1/\partial x_1$ tend to constant (velocity independent) values. The integral (2.5) is still path independent and zero along the crack flank and the strip sides, but the state of stress at $x_1 = \pm\infty$ consistent with the boundary conditions can only be determined by a fairly complete analysis of the problem.

The integral (2.5) still retains its path independent property for materials with non-linear stress strain laws provided they possess an energy density W. The stress analysis is complicated in this case because the equations of motion (2.2) will be of mixed type i.e. the equations will be hyperbolic or elliptic depending upon the value of the strains. Work is in progress on the case of a power law elastic solid $\sigma \sim \varepsilon^n$ $(n > 1$ say) in mode three deformation. In this situation for a crack propagating in a displacement loaded infinite strip a cursory analysis shows that the singular stress field at the crack tip should render the equations elliptic there whatever the velocity but presumably they become hyperbolic near the strip sides. For the case $v = 0$ the integral (2.5) can be used to give a simple solution to this problem for the near crack tip stress field. Recently Atkinson and Champion (16) have given a complete stress analysis solution of this problem and our hope is to rationalise the steady motion problem by using the solution given in (16) as an appropriate approximation. The difficulty is, of course, that the character of the differential equations changes for $v > 0$, however small.

3. TIME DEPENDENT PROBLEMS

We have seen in the last section that certain steady state problems can be solved by use of an invariant integral. In that section the integral F_1 had the physical significance of also being the energy release rate. However, even if such an integral had little physical significance and did not lead to a simple solution to a particular problem, it could still be useful either as a means of obtaining near crack tip information from the far field stress distribution or as a check on a numerical solution.

We consider below two types of dynamic problem. The first kind considered in section 3.1 refers to stationary cracks in a time dependent stress field. In such a case an application of the Laplace transform may lead to a set of equations for which a Lagrangian can be found and the methods of the introduction applied. We illustrate this method by an application to the equations of coupled time dependent thermoviscoelasticity following Atkinson and Smelser (17) where other related equations are also tackled. Earlier applications to homogeneous and variable moduli elastic or viscoelastic media can be found in (13) and (14).

The second type of dynamic situation considered is that in which the crack is moving but the stress or displacement field has the property of dynamic similarity. In this case the number of independent variables is reduced by defining new variables $\eta_i = x_i/t$. A Lagrangian can be found for the equation in the new variables or in a transformed form; for each case different invariant integrals can be found. These are discussed in section 3.2 where a reciprocal theorem is also given in terms of these new independent variables. The philosophy we adopt here is a little different from that usually followed, in that we simply try to construct invariant integrals for whatever form our equations take in the new variables. As discussed above they can always be used as an auxiliary tool to a full analysis of a problem. In section 3.3 a similar consideration is given for the heat equation.

3.1 COUPLED, TIME DEPENDENT, LINEAR THERMOELASTICITY AND THERMOVISCOELASTICITY

Consider the equations of coupled time dependent thermoviscoelasticity under conditions applicable to stationary cracks disturbing time dependent temperature and stress fields. Initial conditions are considered in which

$$\theta(t) = u_i(t) = \sigma_{ij}(t) = 0 \quad \text{for} \quad t < 0 \tag{3.1}$$

u_i and σ_{ij} are the usual displacement vector and stress tensor, $\theta(t)$ denotes the infinitesimal temperature deviation from the base temperature T_0. The formulation begins by Laplace-transforming the equations of motion etc. which become

$$\begin{aligned} \bar{\sigma}_{ij,j} &= \rho p^2 \bar{u}_i \\ \bar{\varepsilon}_{ij} &= \tfrac{1}{2}(\bar{u}_{i,j} + \bar{u}_{j,i}) \\ \bar{\sigma}_{ij} &= p\,\bar{G}_{ijk\ell}\,\bar{\varepsilon}_{k\ell} - p\,\bar{\phi}_{ij}\,\bar{\theta} \qquad \text{(anisotropic)} \end{aligned} \tag{3.2}$$

and for the temperature

$$(k_{ij}/T_0)\bar{\theta}_{,ij} = p^2 \bar{m}\,\bar{\theta} + p^2 \bar{\phi}_{ij}\,\bar{\varepsilon}_{ij} \tag{3.3}$$

where the Laplace transform is defined by

$$\bar{f}(p) = \int_0^\infty e^{-pt} f(t)\,dt \quad . \tag{3.4}$$

In general for viscoelastic media the coefficients $\bar{G}_{ijk\ell}$ etc. will be functions of p the transform variable.

The above field equations can be generated from a Lagrangian defined as

$$L = -\tfrac{1}{2}\bar{t}_{ij}\,\bar{\varepsilon}_{ij} - \frac{\rho p^2}{2}\bar{u}_i \bar{u}_i + p\bar{\phi}_{ij}\,\bar{\theta}\,\bar{u}_{i,j}$$
$$+ \frac{1}{2T_0 p} k_{ij}\,\bar{\theta}_{,i}\,\bar{\theta}_{,j} + \tfrac{1}{2}\,p\,\bar{m}\,\bar{\theta}^2 \tag{3.5}$$

where

$$\bar{t}_{ij} = p\,\bar{G}_{ijk\ell}\bar{\varepsilon}_{k\ell} \tag{3.6}$$

A 'pseudo' energy momentum tensor is then defined as

$$P_{\ell j} = \frac{\partial L}{\partial \bar{u}_{i,j}}\bar{u}_{i,\ell} + \frac{\partial L}{\partial \bar{\theta}_{,j}}\bar{\theta}_{,\ell} - L\delta_{\ell j} \tag{3.7}$$

so that we deduce that

$$P_{\ell j} = -(\overline{t}_{ij}-p\overline{\theta}\ \overline{\phi}_{ij})\overline{u}_{i,\ell} + \frac{k_{ij}}{T_0 p}\ \overline{\theta}_{,i}\ \overline{\theta}_{,\ell} - L\delta_{\ell j} \tag{3.8}$$

and the integrals

$$F_\ell = \int_S P_{\ell j}\ dS_j \tag{3.9}$$

follow as described in the introduction. A useful property of the integral F_1 is that, provided either the temperature θ is constant or the flux $k_{i2}\overline{\theta}_{,i}$ is zero on the crack faces, then P_{12} is zero on the crack face for a stress free crack. Thus in this case the integral can be deformed into the far field as discussed earlier. Also the near field integral, evaluated from a small contour round the crack tip can be explicitly obtained in terms of the coefficients of the singular transformed stresses $\overline{\sigma}_{ij}$ and temperature gradients $\overline{\theta}_{,i}$. In favourable circumstances explicit determination of these coefficients can be made: see Atkinson and Smelser (17) for some applications of these results.

3.2 ANTI-PLANE STRAIN ELASTICITY (THE WAVE EQUATION)

In this section we consider dynamic problems of anti-plane strain elasticity which possess the property of dynamic similarity. Although this restricts the range of situations covered it includes such model problems as uniformly expanding cracks and crack branching (see e.g. (18) for a recent account of this method applied to mode 3 crack kinking and (14) for earlier work).

For conditions of anti-plane strain (longitudinal shear) there is only one non-zero displacement u_3 in the x_3 direction and two non-zero stresses σ_{13} and σ_{23}. Since

$$\sigma_{i3} = \mu\frac{\partial u_3}{\partial x_i} \quad \text{and} \quad \frac{\partial \sigma_{i3}}{\partial x_i} = \rho\frac{\partial^2 u_3}{\partial t^2}\ , \quad i = 1,2 \tag{3.10}$$

are respectively the stress strain relations and the equation of motion, taking them together leads to the wave equation

$$\frac{\partial}{\partial x_i}\left(\mu\frac{\partial u_3}{\partial x_i}\right) = \rho\frac{\partial^2 u_3}{\partial t^2}\ . \tag{3.11}$$

If both the density ρ and modulus μ are constant this becomes

$$\frac{\partial^2 u_3}{\partial x_i \partial x_i} = \frac{\partial^2 u_3}{\partial t_1^2} \tag{3.12}$$

where $t_1 = ct$ with $\rho c^2 = \mu$.

We now consider situations for which u_3 depends upon x_i and t_1 only through $\eta_i = x_i/t_1$, i.e.

$$u_3 = t_1^s \phi(x_1/t_1, x_2/t_1) \equiv t_1^s \phi(w,\theta) \tag{3.13}$$

where

$$w = r/t_1 \quad , \quad r^2 = x_1^2 + x_2^2 \quad .$$

Substituting from equation (3.13) into equation (3.12) gives the equation

$$\frac{\partial^2\phi}{\partial\eta_i\partial\eta_i} = s(s-1)\phi - 2(s-1)\eta_i \frac{\partial\phi}{\partial\eta_i} + \eta_i\eta_j \frac{\partial^2\phi}{\partial\eta_i\partial\eta_j} \tag{3.14}$$

or in terms of w and θ the equation

$$\frac{1}{w}\frac{\partial}{\partial w}\left(w\frac{\partial\phi}{\partial w}\right) + \frac{1}{w^2}\frac{\partial^2\phi}{\partial\theta^2} = s(s-1)\phi - 2(s-1)w\frac{\partial\phi}{\partial w} + w^2\frac{\partial^2\phi}{\partial w^2} \ . \tag{3.15}$$

Multiplying equation (3.14) by $(1-w^2)^{(-s-0.5)}$, provided $w^2 < 1$, gives

$$\frac{\partial}{\partial\eta_1}\left[(1-w^2)^{(-s-0.5)}\left((1-\eta_1^2)\frac{\partial\phi}{\partial\eta_1} - \eta_1\eta_2\frac{\partial\phi}{\partial\eta_2}\right)\right]$$

$$+ \frac{\partial}{\partial\eta_2}\left[(1-w^2)^{(-s-0.5)}\left((1-\eta_2^2)\frac{\partial\phi}{\partial\eta_2} - \eta_1\eta_2\frac{\partial\phi}{\partial\eta_1}\right)\right]$$

$$= s(s-1)(1-w^2)^{(-s-0.5)}\phi \ . \tag{3.16}$$

This equation has Lagrangian

$$L = \frac{1}{2}(1-w^2)^{(-s-0.5)}\left[(1-\eta_1^2)\left(\frac{\partial\phi}{\partial\eta_1}\right)^2 + (1-\eta_2^2)\left(\frac{\partial\phi}{\partial\eta_2}\right)^2 - \right.$$

$$\left. 2\eta_1\eta_2\frac{\partial\phi}{\partial\eta_1}\frac{\partial\phi}{\partial\eta_2}\right] + \frac{1}{2}s(s-1)(1-w^2)^{(-s-0.5)}\phi^2 \tag{3.17}$$

but the fact that it is inhomogeneous, i.e. depends explicitly on η_1 and η_2, limits its usefulness. Perhaps a more useful result is the reciprocal theorem to be considered next.

Rewriting equation (3.16) in the form

$$\frac{\partial}{\partial \eta_i}(t_{i3}) = s(s-1)(1-w^2)^{(-s-0.5)}\phi \qquad (3.18)$$

where

$$t_{13} = (1-w^2)^{(-s-0.5)}\left((1-\eta_1^2)\frac{\partial\phi}{\partial\eta_1} - \eta_1\eta_2\frac{\partial\phi}{\partial\eta_2}\right)$$
$$t_{23} = (1-w^2)^{(-s-0.5)}\left((1-\eta_2^2)\frac{\partial\phi}{\partial\eta_2} - \eta_1\eta_2\frac{\partial\phi}{\partial\eta_1}\right) \qquad (3.19)$$

we can write down the reciprocal theorem

$$\int_S \hat{\phi}\, t_{i3} dS_i = \int_S \phi\, \hat{t}_{i3} dS_i \qquad (3.20)$$

where $\hat{\phi}$ and $\hat{t}_{i3}$ satisfy the same equations, (3.18) and (3.19), as ϕ and t_{i3}, and S is any closed contour in $(\eta_1\eta_2)$ space not including a singularity. In principle the identity (3.20) can be used to extract coefficients of stress singularities at crack and notch tips from an integral far from the crack tip provided suitable auxiliary functions $\hat{\phi}$ and $\hat{t}_{i3}$ are used.

APPLICATIONS OF CHAPLYGIN'S TRANSFORMATION. In the last section an attempt was made to obtain integrals for equation (3.16) as a function of the variables $\eta_i = x_i/t_1$. However, under some circumstances an easier equation is obtained by a transformation of variables. Thus, for example, if the similarity is such that $\partial u_3/\partial x_i$ depends only on w and θ one procedure is to differentiate equation (3.12) with respect to t_1 define

$$\phi_1 = \frac{\partial u_3}{\partial t_1}$$

and change the variables to $w = r/t_1$ and θ to obtain an equation similar to equation (3.15) but with $s = 0$ i.e.

$$w^2(1-w^2)\frac{\partial^2\phi_1}{\partial w^2} + w(1-2w^2)\frac{\partial\phi_1}{\partial w} + \frac{\partial^2\phi_1}{\partial\theta^2} = 0 \quad . \qquad (3.21)$$

For $w < 1$ Chaplygin's transformation $\beta = \mathrm{Cosh}^{-1}(1/w)$ reduces equation (3.21) to

$$\frac{\partial^2\phi_1}{\partial\beta^2} + \frac{\partial^2\phi_1}{\partial\theta^2} = 0 \quad , \quad \beta = \mathrm{Cosh}^{-1}(1/w) \quad . \tag{3.22}$$

If $w > 1$ the substitution $w = \sec\alpha$ gives the equation

$$\frac{\partial^2\phi_1}{\partial\alpha^2} - \frac{\partial^2\phi_1}{\partial\theta^2} = 0 \quad . \tag{3.23}$$

The real transformation (3.22) maps the interior of the domain $(-\pi \le \theta \le \pi\ ,\ w \le 1)$ into a semi-infinite strip $(-\pi \le \theta \le \pi\ ,\ 0 \le \beta < \infty)$. Within this strip a number of invariant integrals can be generated by the procedure outlined in the introduction simply by taking $2L = (\partial\phi_1/\partial\beta)^2 + (\partial\phi_1/\partial\theta)^2$. Furthermore, since stress free radial cracks in the original co-ordinate system will have $\partial\phi_1/\partial\theta = 0$ as a boundary condition, this is still a boundary condition in the mapped plane and similar singularities should be at the crack tip in the mapped plane. Thus the invariant integrals derived from L should be useful in the (β,θ) plane and similar integrals useful in subsequent planes derived by conformal mapping. A reciprocal theorem can also, of course, be written down in the (β,θ) plane and this should be useful for problems involving other singularities such as wedges.

3.3 THE HEAT EQUATION

If for the heat (or diffusion) equation (with thermal diffusivity unity for convenience)

$$\frac{\partial c}{\partial t} = \nabla^2 c, \tag{3.24}$$

situations are considered where there is no fundamental length (such as occur in the sudden heating of a semi-infinite plate, sudden applied temperatures to a wedge or cone, or certain moving boundary problems) then solutions with the property

$$c = t^s\, c(\eta_1,\eta_2,\eta_3) \quad \text{where} \quad \eta_i = x_i/t^{\frac{1}{2}} \tag{3.25}$$

are relevant. Making the substitution (3.25) into (3.24) the resulting equation can be rearranged as

$$\frac{\partial}{\partial\eta_i}\left(\exp(R^2/4)\,\frac{\partial c}{\partial\eta_i}\right) = s\,\exp(R^2/4)c \tag{3.26}$$

where $R^2 = \eta_1^2 + \eta_2^2 + \eta_3^2$. (See (19) for this result with $s = 0$ and various other results and applications.)

A Lagrangian, L , for equation (3.26) can be written as

$$2L = \exp(R^2/4)\left(\frac{\partial c}{\partial \eta_i}\right)\left(\frac{\partial c}{\partial \eta_i}\right) + s\,\exp(R^2/4)c^2 \qquad (3.27)$$

but perhaps more useful is a reciprocal theorem which can be written as

$$\int_S \exp(R^2/4)\,\hat{c}\,\frac{\partial c}{\partial \eta_i}\,dS_i = \int_S \exp(R^2/4)\,c\,\frac{\partial \hat{c}}{\partial \eta_i}\,dS_i \qquad (3.28)$$

where the auxiliary field $\hat{c}$ satisfies equation (3.26) and the integral S is a closed surface enclosing no singularities.

It is clear from section 3.1 that in problems for which the Laplace transformation is useful invariant integrals can be obtained for the heat equation by the approach outlined there.

4. CONCLUDING REMARKS

We have discussed in this paper how certain invariants can be derived for special problems by using the formalism of the introduction or by deducing directly theorems of reciprocal type for the set of differential equations appropriate to a given problem. Thus if the problem has the property of dynamic similarity, as discussed in section 3.2, invariants can be found in terms of the transformed variables and these are potentially useful. Of course, we have discussed here only the simplest equations. More complicated systems can be dealt with but more work is required to extend the results and make explicit applications of them.

REFERENCES

(1) Eshelby, J. D. (1951). The force on an elastic singularity. Phil. Trans. Roy. Soc. Ser. A, 244, 87-112.
(2) Eshelby, J. D. (1970). Energy relations and the energy-momentum tensor in continuum mechanics. In: Inelastic Behavior of Solids, ed. M. F. Kanninen et al., pp. 77-115. New York: McGraw-Hill.
(3) Eshelby, J. D. (1975). The elastic energy-momentum tensor. Journal of Elasticity, 5, 321-335.
(4) Atkinson, C. and Eshelby, J. D. (1968). The energy flow into the tip of a moving crack. Int. J. Frac. Mechs, 4, 1-8.
(5) Cherepanov, G. P. (1967). Crack propagation in continuous media. Appl. Maths and Mech., 31, 476-488.
(6) Freund, L. B. (1972). Energy flux into the tip of an extending crack in an elastic solid. Journal of Elasticity, 2, 341-349.

(7) Atkinson, C. (1979). Stress singularities and fracture mechanics. Appl. Mech. Reviews, 32, 123-135.
(8) Gunther, W. (1962). Uber einige Randintegrale der Elastomechanik. Abh. Braunschw. wiss. Ges., 14, 54-63.
(9) Knowles, J. K. and Sternberg, E. (1972). On a class of conservation laws in linearised and finite elastostatics. Arch. Rat. Mech. Anal., 44, 187-211.
(10) Budiansky, B. and Rice, J. R. (1973). Conservation laws and energy release rates. J. Appl. Mech., 40, 201-203.
(11) Morse, P. M. and Feshbach, H. (1953). Methods of theoretical physics, Vol. 1, McGraw-Hill.
(12) Atkinson, C. (1975). Some results on crack propagation in media with spatially varying elastic moduli. Int. J. Fracture, 11, 619-628.
(13) Nilsson, F. (1977). The infinite strip with a semi-infinite crack, static and dynamic considerations. S.M. Archives, 2, 205-261.
(14) Atkinson, C. (1977). Crack propagation in dissimilar media. In: Elastodynamic crack problems. Mechanics of Fracture, Vol. IV. ed. G. C. Sih, pp. 213-248. Leyden: Noordhoff.
(15) Popelar, C. H. and Atkinson, C. (1980). Dynamic crack propagation in a viscoelastic strip. J. Mech. Phys. Solids, 28, 79-93.
(16) Atkinson, C. and Champion, C. R. (1984). Some boundary value problems for the equation $\nabla \cdot (|\nabla\phi|^n \nabla\phi) = 0$. Quart. J. Math: Appl. Mech. (In press.)
(17) Atkinson, C. and Smelser, R. E. (1982). Invariant integrals of thermoviscoelasticity. Int. J. Solids Structures, 18, 533-549.
(18) Dempsey, J. P., Kuo, M. K. and Achenbach, J. D. (1982). Mode III crack kinking under stress wave loading. Wave Motion, 4, 181-190.
(19) Atkinson, C. (1981). Some elementary solutions of the heat equation and related equations. Int. J. Engng. Sci., 19, 713-728.

PLASTIC DEFORMATIONS NEAR A RAPIDLY PROPAGATING CRACK TIP

J. D. ACHENBACH
Northwestern University, Evanston, IL., 60201, USA

Z.-L. LI

Northwestern University, Evanston, IL., 60201, USA

ABSTRACT

For rapid crack propagation in an elastic perfectly-plastic material, explicit expressions have been obtained for the dynamic strains on the crack line, from the moving crack tip to the moving elastic-plastic boundary. The method of solution uses power series in the distance to the crack line, with coefficients which depend on the distance to the crack tip. Substitution of the expansions in the equations of motion, the yield condition (Huber-Mises) and the stress-strain relations, yields a system of nonlinear ordinary differential equations for the coefficients. These equations are exactly solvable for Mode-III, and they have been solved in an approximate manner for Mode-I plane stress. The crack-line fields have been matched to appropriate elastic fields at the elastic-plastic boundary. For both Mode-III and Mode-I plane stress, the plastic strains, which depend on the elastodynamic stress intensity factor and the crack-tip speed, have been used in conjunction with the crack growth criterion of critical plastic strain, to determine the relation between the far-field stress level and the crack-tip speed.

1. INTRODUCTION

At high crack-tip speeds the mass density of a material affects the fields of stress and deformation in the vicinity of a propagating crack tip. For essentially brittle fracture, near-tip dynamic effects have been investigated extensively on the basis of linear elastic fracture mechanics. By now, several papers have reviewed the computation of elastodynamic stress intensity factors, and they have discussed dynamic effects on the fracture criterion of the balance of rates of energies, see Achenbach (1), Freund (2) and Kanninen (3). The combined effect of plastic deformation and mass density on near-tip fields has not yet received that much attention. This is not surprising, considering the difficulties that are encountered in the quasi-static analysis of fields near a growing crack in an elastic-plastic material.

For quasi-statically growing cracks the asymptotic structure of near-tip fields in elastic perfectly-plastic solids has been analyzed in considerable detail. A recent review by Rice (4) includes a general formulation, and it presents detailed results for isotropic materials of the Huber-Mises type. In general, the analytical near-tip results must, however, be supplemented by numerical calculations to determine certain arbitrary functions that appear in the asymptotically valid near-tip results.

In recent papers, Achenbach and Dunayevsky (5) and Achenbach and Li (6) have constructed quasi-static solutions that are valid on the crack line, from the moving crack tip up to the moving elastic-plastic boundary. These solutions were obtained for an elastic perfectly-plastic material of the Huber-Mises type by expanding all fields in powers of the distance, y, to the crack line. Substitution of the expansions in the equilibrium equations, the yield condition and the constitutive equations yields a system of simple ordinary differential equations for the coefficients of the expansions. As shown in (6), the resulting equations are exactly solvable for the Mode-III case, and they are solvable for the Mode-I plane-stress case if it is assumed that the cleavage stress is uniform on the crack line. By matching the relevant stress components and particle velocities to the dominant terms of appropriate elastic fields at the elastic-plastic boundary, the plastic strains on the crack line were computed in terms of the elastic stress intensity factor.

The literature on dynamic effects in the presence of elastic-plastic constitutive behavior is growing. Investigations of the asymptotic structure of the dynamic near-tip fields were presented by Slepyan (7) and Achenbach and Dunayevsky (8). Dynamic near-tip effects for a strain-hardening material were investigated by Achenbach and Kanninen (9) and Achenbach, Kanninen and Popelar (10) on the basis of J_2-flow theory and a bilinear effective stress-strain relation. For Mode-III crack propagation in an elastic perfectly-plastic material, exact crack-line solutions were obtained by Achenbach and Dunayevsky (11) and Freund and Douglas (12).

In the present paper the expansion technique of Achenbach and Li (6) is extended to the dynamic formulation, for rapid crack growth in Mode-III and in Mode-I plane stress. Systems of nonlinear ordinary differential equations are established which are valid for the transient case. Solutions have, however, been obtained only for the steady-state dynamic crack line fields. The equations for the Mode-III case can be solved rigorously in implicit form. An approximate approach which gives excellent results for the Mode-III case has, however, also been developed. The equations for Mode-I plane stress cannot be solved rigorously, but the approximate approach can be used to yield the steady-state dynamic cleavage strain on the crack line. The plastic strains on the crack line have been used in conjunction with the crack growth criterion of critical plastic strain to determine the relation between the far-field stress level and the crack-tip speed.

The geometry is shown in Fig. 1. The x_3-axis of a stationary coordinate system is parallel to the crack front, and x_1 points in the direction of crack growth. The position of the crack tip is defined by $x_1 = a(t)$. A moving coordinate system (x,y,z) is centered at the crack tip, with its axes parallel to the x_1,x_2 and x_3 axes.

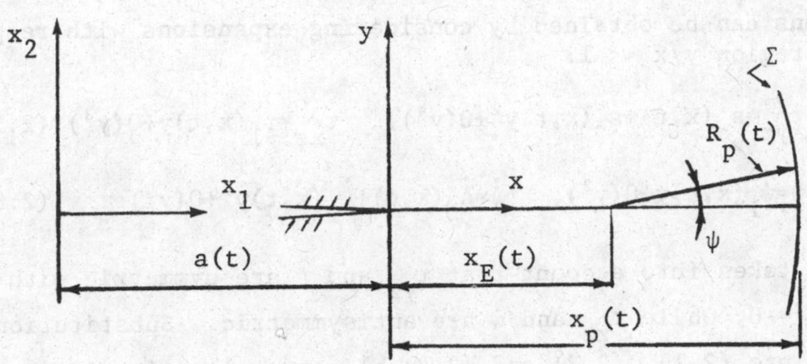

FIG. 1. Geometry for a propagating crack tip, with center of elastic field E, and elastic-plastic boundary Σ.

Because of space limitations it was not possible to include as many details as might have been desirable. Most of the details that were omitted can, however, be found in Ref.(13).

2. MODE-III CRACK PROPAGATION

In the moving coordinate system the equation of motion is

$$\frac{\partial \tau_{xz}}{\partial x} + \frac{\partial \tau_{yz}}{\partial y} = \rho \ddot{w} \tag{2.1}$$

where w(x,y,t) is the anti-plane displacement, and the material time derivative is defined as

$$(\dot{\ }) = (\partial/\partial t) - \dot{a}\ (\partial/\partial x) \tag{2.2}$$

Here $\dot{a} = da/dt$ is the speed of the crack tip. The Huber-Mises yield condition requires

$$\tau_{xz}^2 + \tau_{yz}^2 = k^2 \ , \tag{2.3}$$

where k is the yield stress in pure shear. The strain rates are related to the stresses and the stress rates by

$$\frac{1}{2}\frac{\partial \dot{w}}{\partial x} = \frac{\dot{\tau}_{xz}}{2\mu} + \dot{\Lambda}\tau_{xz} \ , \quad \frac{1}{2}\frac{\partial \dot{w}}{\partial y} = \frac{\dot{\tau}_{yz}}{2\mu} + \dot{\Lambda}\tau_{yz} \qquad (2.4a,b)$$

In (2.4a,b), μ is the shear modulus and $\dot{\Lambda}$ is a positive function of time and the spatial coordinates.

We are interested in solutions along the crack line $y = 0$, $0 < x \leq x_p$, where $x = x_p$ defines the elastic-plastic boundary. Such solutions can be obtained by considering expansions with respect to y in the region $y/x \ll 1$:

$$\tau_{yz}=s_o(x,t)+s_2(x,t)y^2+O(y^4), \quad \tau_{xz}=\tau_1(x,t)y+O(y^3) \qquad (2.5a,b)$$

$$\dot{w}=\dot{w}_1(x,t)y+O(y^3), \quad \dot{\Lambda}=\dot{\Lambda}_o(x,t)+\dot{\Lambda}_2(x,t)y^2+O(y^4) \qquad (2.6a,b)$$

Here we have taken into account that τ_{yz} and $\dot{\Lambda}$ are symmetric with respect to $y = 0$, while τ_{xz} and $\dot{w}$ are antisymmetric. Substitution of (2.5)-(2.6) into (2.1), (2.3) and (2.4a,b), and collecting terms of the lowest orders in y yields

$$\frac{\partial \tau_1}{\partial x} + 2s_2 = \rho \ddot{w}_1 \qquad (2.7)$$

$$s_o^2 = k^2 \ , \quad 2s_o s_2 + \tau_1^2 = 0 \qquad (2.8a,b)$$

$$\frac{1}{2}\frac{\partial \dot{w}_1}{\partial x} = \frac{\dot{\tau}_1}{2\mu} + \dot{\Lambda}_o\tau_1, \quad \frac{1}{2}\dot{w}_1 = \dot{\Lambda}_o s_o \qquad (2.9a,b)$$

It follows from (2.8a) that $s_o = k$. Elimination of s_2 from (2.7) and (2.8b) gives

$$\frac{\partial \tau_1}{\partial x} - \frac{\tau_1^2}{k} - \rho \ddot{w}_1 = 0 \qquad (2.10)$$

Similarly, $\dot{\Lambda}_o$ can be eliminated from (2.9a) and (2.9b) to yield

$$\frac{\partial \dot{w}_1}{\partial x} - \frac{\dot{\tau}_1}{\mu} - \frac{1}{k}\dot{w}_1\tau_1 = 0 \qquad (2.11)$$

Equations (2.10) and (2.11) define two coupled nonlinear partial differential equations. Analytical solutions to these equations, which would give the transient fields on the crack line, have not yet been obtained.

Equations (2.10) and (2.11) must be supplemented by conditions at the elastic-plastic boundary Σ. These have been discussed in some detail in Appendix A of Ref.(13), where it was shown that for conditions which may be assumed to hold ahead of a propagating crack tip, the stresses and the particle velocity are continuous at Σ.

The governing equations for the quasi-static case follow by setting $\rho \equiv 0$. The resulting system of coupled nonlinear ordinary differential equations can be solved. The quasi-static solution for $\dot{w}_1$ has been given in Ref.(5).

For the steady-state case the material time derivative (2.2) reduces to

$$(\dot{\ }) = -\dot{a}\,(d/dx)\,, \tag{2.12}$$

where $\dot{a}$ is now a constant crack tip speed. We define

$$\gamma_1 = dw_1/dx\,, \quad \text{and hence } \dot{w}_1 = -\dot{a}\gamma_1, \tag{2.13a,b}$$

and we note that (2.11) and (2.10) then may be written as

$$\frac{d\gamma_1}{dx} - \frac{1}{\mu}\frac{d\tau_1}{dx} - \frac{1}{k}\gamma_1\tau_1 = 0 \tag{2.14}$$

$$\frac{d\tau_1}{dx} - \frac{\tau_1^2}{k} - \mu M^2\frac{d\gamma_1}{dx} = 0\,, \tag{2.15}$$

where the Mach number M is defined as

$$M = \dot{a}/(\mu/\rho)^{\frac{1}{2}}. \tag{2.16}$$

For $y/x \ll 1$ the plastic fields in the loading zone will be matched to the dominant terms of the elastic fields. In polar coordinates R,ψ centered at point E, and for small values of the angle ψ, the dominant terms of the solution on the elastic side of the elastic-plastic boundary are taken as

$$w \cong \left(\frac{R}{2\pi}\right)^{\frac{1}{2}}\frac{2}{\mu}K_{III}\,\frac{1}{2}\psi\,, \quad \mu = \text{shear modulus} \tag{2.17}$$

$$\tau_{Rz} \cong \left(\frac{1}{2\pi R}\right)^{\frac{1}{2}}K_{III}\,\frac{1}{2}\psi, \qquad \tau_{\psi z} \cong \left(\frac{1}{2\pi R}\right)^{\frac{1}{2}}K_{III} \tag{2.18a,b}$$

Here the elastic stress-intensity factor K_{III} depends on M. The angular dependence on M enters in higher order terms of ψ. It should be noted that the center of the elastic field is not taken to coincide with the crack tip. The center is located at a moving point E. The geometry is shown in Fig. 1.

Since τ_{yz} is continuous at $y = 0$, $x = x_p$, we find

$$\left(\frac{1}{2\pi R_p}\right)^{\frac{1}{2}} K_{III} = k \ , \quad \text{or } R_p = (K_{III})^2/2\pi k^2 \qquad (2.19a,b)$$

where $R = R_p$ defines the radius of curvature of the elastic-plastic boundary, at least for small values of ψ. Other conditions are that τ_{Rz} (i.e., the shear stress in the R,ψ system) and γ_1 should be continuous at the elastic-plastic boundary. These conditions yield

$$\tau_1 = -k/2R_p \ , \qquad \gamma_1 = -k/2\mu R_p \qquad (2.20a,b)$$

For completeness we list the condition on the strain $\partial w/\partial y$ at $x = x_p$

$$\gamma_y = \frac{\partial w}{\partial y} = w_1 = \frac{k}{\mu} \qquad (2.21)$$

Equations (2.14) and (2.15) can be solved rigorously, as shown in Appendix B of Ref.(13). It is, however, of interest to note that an asymptotic solution for small values of x can be obtained with minimal effort in the form

$$\tau_1 \cong -\frac{k(1-M)}{x} \ , \quad \gamma_1 \cong -\frac{k}{\mu}\frac{1-M}{M}\frac{1}{x} \qquad (2.22a,b)$$

Since $\gamma_1 = \partial\gamma_y/\partial x$, we also have

$$\gamma_y \cong -\frac{k}{\mu}\frac{1-M}{M}\ \ell n(x/x_p) \qquad (2.23)$$

This solution is the same as the one derived earlier by Slepyan (7), see also Achenbach and Dunayevsky (8). Note that τ_1 reduces to the quasi-static solution as $M \to 0$. The strain γ_y has, however, not only the wrong behavior in x, but actually becomes singular in M.

As shown in Appendix B of Ref.(13) the solution to Eqs. (2.14) and (2.15) which satisfies the boundary conditions (2.20a) and (2.20b) at $x = x_p$ is defined by the following equations:

$$\frac{\tau_1}{k/2R_p} = -\left(\frac{1-M}{1+M}\right)^{1/2M} \frac{M}{(1-M^2)^{\frac{1}{2}}} \frac{\left(-2F - \frac{1}{1+M}\right)^{(1-M)/2M}}{\left(2F + \frac{1}{1-M}\right)^{(1+M)/2M}} \qquad (2.24)$$

where

$$F(\tau_1) = -\frac{\frac{1}{2}k}{(\tau_1)^2}\frac{d\tau_1}{dx}, \quad \gamma_1 = -\frac{\tau_1}{\mu M^2}\left[-k(1-M^2)\frac{1}{\tau_1^2}\frac{d\tau_1}{dx} + 1\right] \quad (2.25a,b)$$

Equation (2.24) gives F as a function of τ_1. Integration of (2.25a) then yields

$$x = -\frac{1}{2}k\int_{-k/2R_p}^{\tau_1}\frac{d\xi}{\xi^2 F(\xi)} + x_p \qquad (2.26)$$

Equation (2.26) yields τ_1 as a function of x and x_p. Substitution of the result in Eq.(2.25b) yields γ_1. By letting $\tau_1 \to \infty$ in (2.26) we obtain a relation between x_p and R_p. From $\gamma_1 = \partial\gamma_y/\partial x$, we finally find

$$\gamma_y = \frac{k}{\mu} + \int_{x_p}^{x}\gamma_1\, dx = \frac{k}{\mu} + \gamma_y^p \qquad (2.27)$$

The strain γ_y obtained from (2.27) is the exact solution on the crack line. This solution is equivalent to the one obtained earlier by Dunayevsky and Achenbach (8), and Freund and Douglas (12). It can be shown that for small x, Eq.(2.27) reduces to (2.23). In the limit $M \to 0$, (2.27) reduces to the quasi-static solution

$$\frac{\mu}{k}\gamma_y = 1 - \ell n(x/x_p) + \frac{1}{2}[\ell n(x/x_p)]^2 \qquad (2.28)$$

An explicit analytical expression for γ_y, albeit an approximate one, would be very useful for applications in conjunction with the crack growth criterion of a critical plastic strain. Another reason for an approximate approach to the Mode-III case is that the results can be tested by comparison with exact results. The same approach can then be used for the Mode-I plane-stress case, which is not amenable to an exact solution.

An approximate approach is suggested by the structure of Eqs.(2.14) and (2.15). If an acceptable approximation to τ_1 were available a-priori, then (2.14) would simply be a linear ordinary differential equation for γ_1. A first approximation to τ_1 is suggested by (2.22a), namely, $\tau_1 = -k(1-M)/x$. This expression has the correct limits at M = 0 and M = 1. A better result is obtained by adding a constant term

$$\tau_1 = -\ k(1-M)[\frac{1}{x} + \frac{M}{2x_p}] \tag{2.29}$$

The second term is chosen so that (2.15) is satisfied up to order O(M) near $x = x_p$. Equation (2.29) can now be substituted in (2.14), and the resulting equation can be solved rigorously for γ_1. The strain $\gamma_y = k/\mu + \gamma_y^p$ then follows from (2.27). In anticipation of difficulties with the Mode-I case, we elect, however, to solve γ_1 by using a perturbation solution which ignores terms of order $O(M^2)$. The corresponding expression for γ_y is obtained as

$$\frac{\mu}{k}\gamma_y^p(\frac{x}{x_p}) = \frac{(1-M)(2-3M-M^2)}{M(2-M+M^2)}\left[\frac{1}{M}(\frac{x}{x_p})^M - \frac{1}{M} - \frac{M(1-M)}{2(1+M)}(\frac{x}{x_p})^{1+M} + \frac{M(1-M)}{2(1+M)}\right]$$
$$- \frac{1-M}{M}\ \ell n(\frac{x}{x_p}) + \frac{1}{2}(1-M)(\frac{x}{x_p} - 1) \tag{2.30}$$

It is of interest that (2.30) yields (2.23) in the limit $x \to 0$, while it yields the quasi-static solution (2.28) as $M \to 0$, provided that $|M\ \ell n(x/x_p)| << 1$. A comparison of (2.30) with the exact result is shown in Fig. 3 of Ref.(13).

Application of the boundary condition (2.20a) to the approximate expression for τ_1, yields a relation between x_p and R_p. Subsequent use of (2.19b) gives

$$x_p = (1-M)(1+2M)\frac{1}{2\pi}(K_{III}/k)^2 \tag{2.31}$$

Finally, following Freund and Douglas (12) we apply the crack growth criterion of critical plastic strain to determine the value of K_{III} that would be required for crack growth at a given value of M. The crack-growth criterion, originally proposed by McClintock and Irwin (14), states that the crack will grow with (normalized) plastic strain $(\mu/k)\gamma_y^p = \gamma_f$ at $x = x_f$ on $y = 0$. For plastic strain below γ_f at $x = x_f$ the crack cannot grow. As discussed by Rice (15) the characteristic length x_f is related to K_c, the value of the Mode III stress intensity factor which is required to satisfy the fracture criterion for a stationary crack, by the relation

$$\pi x_f(\gamma_f+1)=(K_c/k)^2 \quad \text{, i.e.,} \quad x_f=(K_c/k)^2/\pi(\gamma_f+1) \tag{2.32a,b}$$

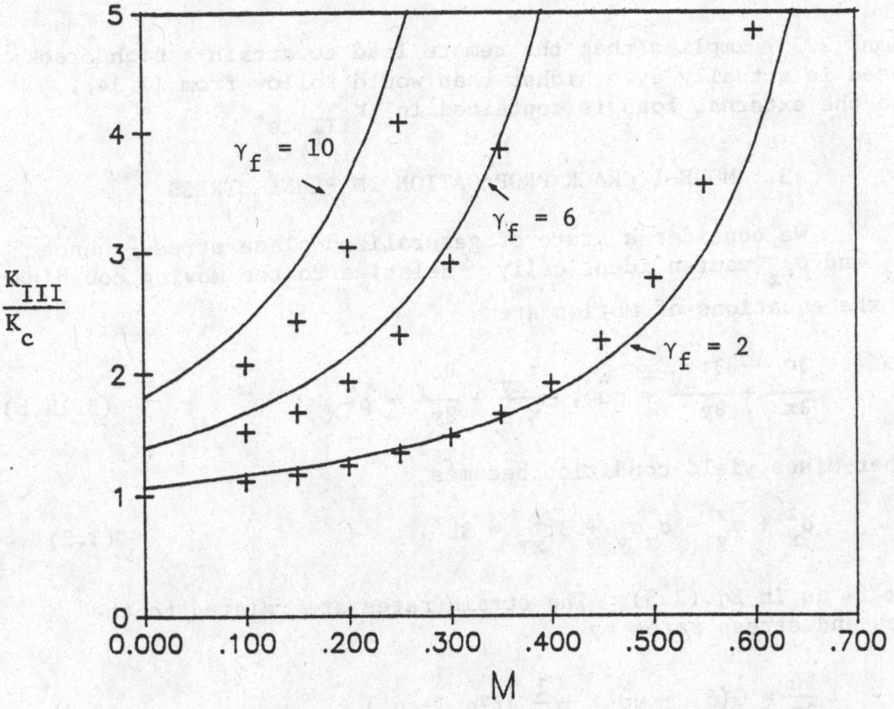

FIG. 2. Comparison between exact (+) and approximate (——) relation between K_{III}/K_c and M.

Next we compute γ_y^p from (2.30) at $\xi = x_f/x_p$. Since x_p is defined by (2.31) we have

$$\xi = 2(K_c/K_{III})^2/[(\gamma_f+1)(1-M)(2+M)] \tag{2.33}$$

The crack growth criterion now yields

$$\gamma_f = \frac{\mu}{k}\gamma_y^p(\xi) \tag{2.34}$$

where the functional form of γ_y^p is given by (2.30). For three values of γ_f, the relation between K_{III}/K_c and M given by (2.34) has been plotted in Fig. 2, and compared with the exact relation.

The elastodynamic stress intensity factor K_{III} is the dynamic factor. It is related to the corresponding quasi-static factor, see Ref. (1,p.35) by the relation

$$K_{III} = (1-M\ )^{\frac{1}{2}}(K_{III})_{qs} \tag{2.35}$$

Equation (2.35) implies that the remote load to attain a high crack tip speed is actually even higher than would follow from (2.34), because the external load is contained in $(K_{III})_{qs}$.

3. MODE-I CRACK PROPAGATION IN PLANE STRESS

We consider a state of generalized plane stress; hence σ_z, σ_{xz} and σ_{yz} vanish identically. Relative to the moving coordinate system the equations of motion are

$$\frac{\partial\sigma_x}{\partial x} + \frac{\partial\tau_{xy}}{\partial y} = \rho\ddot{u}\ , \quad \frac{\partial\tau_{xy}}{\partial x} + \frac{\partial\sigma_y}{\partial y} = \rho\ddot{v} \tag{3.1a,b}$$

The Huber-Mises yield condition becomes

$$\sigma_x^2 + \sigma_y^2 - \sigma_x\sigma_y + 3\tau_{xy}^2 = 3k^2, \tag{3.2}$$

where k is as in Eq.(2.3). The strain rates are related to the stresses and stress rates by

$$\frac{\partial\dot{u}}{\partial x} = \frac{1}{E}(\dot{\sigma}_x - \nu\dot{\sigma}_y) + \frac{1}{3}\dot{\Lambda}(2\sigma_x - \sigma_y) \tag{3.3}$$

$$\frac{\partial\dot{v}}{\partial y} = \frac{1}{E}(\dot{\sigma}_y - \nu\dot{\sigma}_x) + \frac{1}{3}\dot{\Lambda}(2\sigma_y - \sigma_x) \tag{3.4}$$

$$\frac{1}{2}\left(\frac{\partial\dot{u}}{\partial y} + \frac{\partial\dot{v}}{\partial x}\right) = \frac{1+\nu}{E}\dot{\tau}_{xy} + \dot{\Lambda}\tau_{xy}\ , \tag{3.5}$$

where E and ν are Young's modulus and Poisson's ratio, respectively, and $\dot{\Lambda}$ is a positive function of time and the spatial coordinates.

Analogously to (2.5)-(2.6) we consider

$$\sigma_x = p_o(x,t) + p_2(x,t)y^2 + p_4(x,t)y^4 + O(y^6) \tag{3.6}$$

$$\sigma_y = q_o(x,t) + q_2(x,t)y^2 + q_4(x,t)y^4 + O(y^6) \tag{3.7}$$

$$\tau_{xy} = s_1(x,t)y + s_3(x,t)y^3 + O(y^5) \tag{3.8}$$

$$\dot{u} = \dot{u}_o(x,t) + \dot{u}_2(x,t)y^2 + O(y^4) \tag{3.9}$$

$$\dot{v} = \dot{v}_1(x,t)y + \dot{v}_3(x,t)y^3 + O(y^5) \tag{3.10}$$

$$\dot{\Lambda} = \dot{\Lambda}_o(x,t) + \dot{\Lambda}_2(x,t)y^2 + O(y^4) \qquad (3.11)$$

Here we have taken into account that σ_x, σ_y, u and $\dot{\Lambda}$ are symmetric with respect to y = 0, while τ_{xy} and $\dot{v}$ are antisymmetric. Substitution of (3.6)-(3.8) into (3.1a,b) and collecting terms of the same order in y yields

$$\frac{\partial p_o}{\partial x} + s_1 = \rho\ddot{u}_o \,, \quad \frac{\partial p_2}{\partial x} + 3s_3 = \rho\ddot{u}_2 \qquad (3.12a,b)$$

$$\frac{\partial s_1}{\partial x} + 2q_2 = \rho\ddot{v}_1 \,, \quad \frac{\partial s_3}{\partial x} + 4q_4 = \rho\ddot{v}_3 \qquad (3.13a,b)$$

Substitution of (3.6)-(3.8) into the yield condition (3.2) yields by the same procedure

$$p_o^2 + q_o^2 - p_o q_o = 3k^2 \qquad (3.14)$$

$$(2p_o - q_o)p_2 + (2q_o - p_o)q_2 + 3s_1^2 = 0 \qquad (3.15)$$

$$p_2^2 + (2p_o - q_o)p_4 + q_2^2 + (2q_o - p_o)q_4 - p_2 q_2 + 6s_1 s_3 = 0 \quad (3.16)$$

Another 5 equations are obtained by using (2.12) and (3.6)-(3.11) in (3.3)-(3.5). These equations have been listed as Eqs.(21)-(25) by Achenbach and Li (6), and they are not reproduced here.

At this stage we have 14 unknowns and 12 equations. Clearly, the system cannot be solved without further simplifying assumptions. For the quasi-static problem (i.e., $\rho \equiv 0$), one assumption, namely that q_o = constant, suffices to produce a solvable system of equations, as shown by Achenbach and Li (6). For the dynamic case considerable additional argument is required to reduce the system of equations to a solvable one. For the steady-state dynamic case the necessary discussion has been given in Ref.(13). The resulting equations which will be used to analyze the Mode-I plane-stress fields are

$$\frac{d^2 p_2}{dx^2} + \frac{4}{k} p_2^2 + \frac{1}{2} EM^2 \frac{d^3 v_{1x}}{dx^3} = 0 \qquad (3.17)$$

$$\frac{1}{2}\frac{d^2 v_{1x}}{dx^2} + \frac{2}{3k} p_2 v_{1x} + \frac{1}{E}\frac{dp_2}{dx} = 0 \,, \qquad (3.18)$$

where

$$v_{1x} = dv_1/dx \tag{3.19}$$

The solutions for v_{1x} and p_2 must satisfy certain conditions at the elastic plastic boundary Σ. In Appendix A of Ref.(13) it was shown that for conditions which may be assumed to hold ahead of a propagating crack tip, the stresses and the components of the particle velocity are continuous at Σ. Matching at the elastic-plastic boundary of the plastic fields in the loading zone to the dominant terms of an appropriate elastic field yields

$$(1/2\pi R_p)^{\frac{1}{2}} K_I = \frac{3}{2} k, \qquad \text{or } R_p = (4/9)(K_I/k)^2/2\pi, \tag{3.20a,b}$$

and

$$\text{at } x = x_p: \quad p_2 = -\, 3k/4R_p^2 \tag{3.21}$$

$$v_{1x} = -3k/2ER_p, \; dv_{1x}/dx = 3(2+\nu)k/4ER_p^2 \tag{3.22a,b}$$

Details can be found in Ref.(13).

It appears to be difficult to solve (3.17) and (3.18) rigorously, Just as for the Mode-III case, an asymptotic solution for small values of x can, however, easily be obtained as

$$p_2 \cong -\,(3k/2)(1-M)/x^2, \; v_{1x} \cong -(3k/E)(1-M)/Mx \tag{3.23a,b}$$

The corresponding strain ε_y is

$$\varepsilon_y \cong -\,3\frac{k}{E}\,\frac{1-M}{M}\,\ln\left(\frac{x}{x_p}\right) \tag{3.24}$$

In the limit $M \to 0$, p_2 reduces to the quasi-static solution, but ε_y becomes singular.

The similarities in the structure of the equations for Mode-III and Mode-I plane stress suggest an approximate approach to (3.17)-(3.18) similar to the one used for solving (2.14) and (2.15). Thus, if an acceptable approximation to p_2 were available a-priori, then (3.18) would be a linear ordinary differential equation for v_{1x}. A first approximation to p_2 is provided by the asymptotic expression (3.23a). This expression has the correct limits at M = 0 (quasi-static case) and M = 1. It may, however, be expected that a better approximation will be obtained by adding a constant term, and use of

$$p_2 = -\frac{3}{2}k(1-M)\left[\frac{1}{x^2} + \frac{M}{2x_p^2}\right] \tag{3.25}$$

The second term is chosen so that (3.17) is satisfied up to order $O(M)$ near $x = x_p$. It is noted that (3.25) is completely analogous to (2.29). By enforcing the condition (3.21) on p_2 we obtain:

$$x_p = [(1-M)(2+M)]^{\frac{1}{2}}R_p \tag{3.26}$$

The equation for v_{1x} which is obtained by substituting (3.25) into (3.18), can be solved by means of a perturbation method which neglects terms of order $O(M^2)$. By integrating the result, the strain $\varepsilon_y = v_1$ is obtained as

$$\varepsilon_y = (\varepsilon_y)_{PB} + \varepsilon_y^p(x/x_p, M) \tag{3.27}$$

where

$$(\varepsilon_y)_{PB} = (k/E)(2-\nu) \tag{3.28}$$

The explicit expression for $\varepsilon_y^p(x/x_p, M)$ can be found in the Appendix of this paper.Equation (3.27) reduces to the quasi-static solution, which has been given by Achenbach and Li (6,Eq.(64)), provided that $|M\,\ell n(x/x_p)| \ll 1$. In the limit $x \to 0$, (3.27) reduces to (3.24). Numerical results for ε_y as given by (3.27) are shown in Fig. 6 of Ref.(13), for $\nu = 0.3$ and $M = 0.1, 0.3, 0.5$. The quasi-static solution has also been shown.

Finally, just as for the Mode-III case, we apply the crack growth criterion of critical plastic strain to determine the value of K_I that would be required for crack growth at a given value of M. For a stationary crack the quasi-static plastic strain follows from the results of Refs.(5) and (6) as

$$\varepsilon_y^p(x/x_p) = \frac{k}{E}\left\{B_2\left(\frac{x_p}{x}\right) - \frac{1}{2}C_2\left(\frac{x}{x_p}\right)^2 - \frac{3}{4}(2+\nu)\right\}, \tag{3.29}$$

where B_2 and C_2 are expressions in terms of the elastic constants which are given in Ref.(6).

Now suppose that the normalized critical strain

$$\varepsilon_f = (\varepsilon_y^p)_{cr}/(\varepsilon_y)_{PB} \tag{3.30}$$

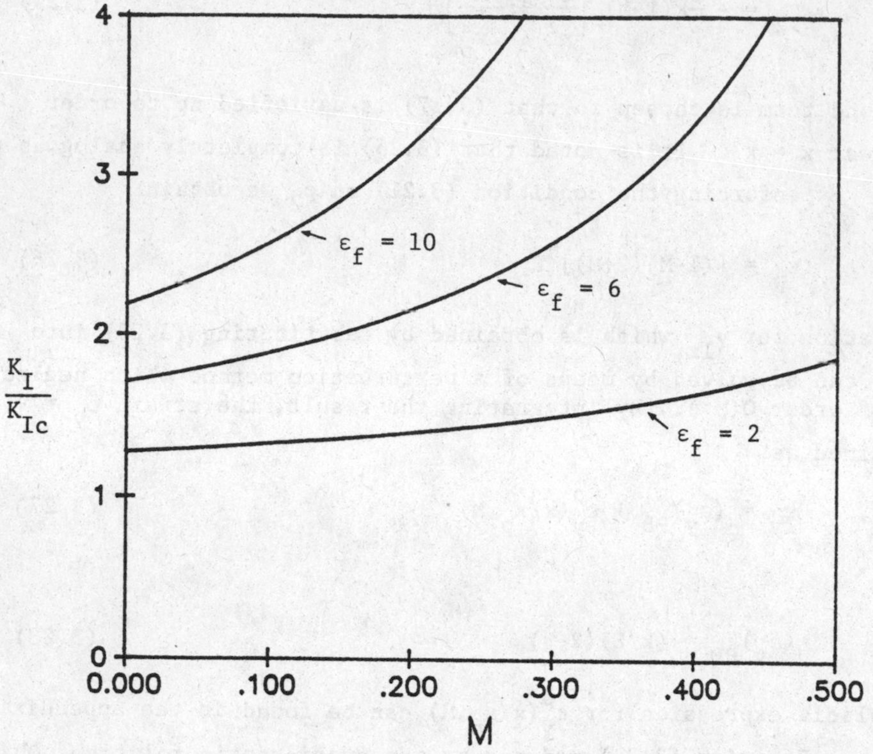

FIG. 3. K_I/K_{Ic} versus M for Mode-I plane stress.

is reached at $x = x_f$, for a value of $K_I = K_{Ic}$. The corresponding value of x_p is given by Ref.(6,Eq.(57)) as

$$x_{pc} = \frac{2\sqrt{2}}{9}\frac{1}{\pi}(K_{Ic}/k)^2 \tag{3.31}$$

A cubic equation for x_f/x_{pc} follows from (3.29). The relevant real-valued root is

$$x_f/x_{pc} = S \tag{3.32}$$

where S depends on the elastic constants and ε_f. Next we compute ε_y^p at $\xi = x_f/x_p$. Since x_p is given by (3.26) we have

$$\xi = \frac{S\sqrt{2}}{(1-M)^{\frac{1}{2}}(2+M)^{\frac{1}{2}}}(K_{Ic}/K_I)^2 , \tag{3.33}$$

where (3.20b) has also been used. The crack growth criterion now yields

$$\varepsilon_f = \varepsilon_y^p(\xi, M)/(\varepsilon_y)_{PB}, \tag{3.34}$$

where the functional form of ε_y^p is the same as in (3.27). Equation (3.34) has been used to plot K_I/K_{Ic} versus M for three values of ε_f in Fig. 3.

ACKNOWLEDGMENT

This paper was prepared in the course of research sponsored by the Office of Naval Research under Contract N00014-76-C-0063 with Northwestern University.

REFERENCES

(1) Achenbach, J.D. (1974). Dynamic effects in brittle fracture. In:Mechanics Today Vol. 1, ed. S. Nemat-Nasser, pp. 1-57. Oxford and New York: Pergamon Press.

(2) Freund, L.B. (1975). Dynamic crack propagation. In: The mechanics of fracture, ed. F. Erdogan, pp. 105-134. ASME AMD Vol. 19, New York: ASME.

(3) Kanninen, M.F. (1978). A critical appraisal of solution techniques in dynamic fracture mechanics. In: Numerical methods in fracture mechanics, eds. A.R. Luxmoore & D.R. Owen, pp. 612-633. Swansea: University College of Swansea Press.

(4) Rice, J.R. (1982). Elastic-plastic crack growth. In: Mechanics of Solids, eds. H.G. Hopkins and M.J. Sewell, pp.539-562. Oxford and New York: Pergamon Press.

(5) Achenbach, J.D. & Dunayevsky, V. (1984). Crack growth under plane stress conditions in an elastic perfectly-plastic material. J. Mech. Phys. Solids, 32, 89-100

(6) Achenbach, J.D. & Li, Z.L. (1984). Plane stress crack-line fields for crack growth in an elastic perfectly-plastic material. J. Eng. Fract. Mech., in press.

(7) Slepyan, L.I. (1976). Crack dynamics in an elastic-plastic body. Izv. Nauk SSSR, Mekhanika Tverdogo Tela, 12, No. 2, 144-153.

(8) Achenbach, J.D. & Dunayevsky, V. (1981). Fields near a rapidly propagating crack tip in an elastic perfectly-plastic material. J. Mech. Phys. Solids, 29, p. 283.

(9) Achenbach, J.D. & Kanninen, M.F. (1978). Crack tip plasticity in dynamic fracture mechanics. In: Fracture mechanics, eds. N. Perrone et al., pp. 649-670. University of Virginia Press.

(10) Achenbach, J.D. et al. (1981). Crack tip fields for fast fracture of an elastic-plastic material. J. Mech. Phys. Solids, 29, 211-225.

(11) Dunayevsky, V. & Achenbach, J.D. (1982). Boundary layer phenomenon in the plastic zone near a rapidly propagating crack tip. Int. J. Solids Structures, 18, No. 1, pp.1-12.

(12) Freund, L.B. & Douglas, A.S. (1982). The influence of inertia on elastic-plastic antiplane-shear crack growth. J. Mech. Phys. Solids, 30, No. 1/2, 59-74.

(13) Achenbach, J.D. & Li, Z.L. Plastic deformation near a rapidly propagating crack tip. Technical Report SU-SML-TR-No.84-1, Department of Civil Engineering, Northwestern University, Evanston, IL. 60201.

(14) McClintock, F.A. & Irwin, G.R. (1965). Plasticity aspects of fracture mechanics. In: Fracture toughness testing and its applications, ASTM STP 381, Philadelphia: ASTM.

(15) Rice, J.R. (1968). Mathematical analysis in the mechanics of fracture. In: Fracture, ed. H. Liebowitz, pp. 191-311. New York and London: Academic Press.

APPENDIX A: EXPRESSION FOR $\varepsilon_y^p(x/x_p, M)$

The expression given by Achenbach and Li (13) is

$$\varepsilon_y^p(x/x_p,M) = \frac{K}{E}\Bigg\{\frac{\Delta_1}{\Delta}\left[\frac{1}{1+\alpha_1}\left(\left(\frac{x}{x_p}\right)^{1+\alpha_1}-1\right)+\frac{\beta_1}{3+\alpha_1}\left(\left(\frac{x}{x_p}\right)^{3+\alpha_1}-1\right)\right]$$
$$+\frac{\Delta_2}{\Delta}\left[\frac{1}{1+\alpha_2}\left(\left(\frac{x}{x_p}\right)^{1+\alpha_2}-1\right)+\frac{\beta_2}{3+\alpha_2}\left(\left(\frac{x}{x_p}\right)^{3+\alpha_2}-1\right)\right]-\frac{3(1-M)}{M}\left[\ln\left(\frac{x}{x_p}\right)-\frac{M}{4}\left(\left(\frac{x}{x_p}\right)^2-1\right)\right]\Bigg\} \tag{A.1}$$

where

$$\alpha_1 = \tfrac{1}{2}[1 + (9-8M)^{\frac{1}{2}}], \quad \alpha_2 = \tfrac{1}{2}[1 - (9-8M)^{\frac{1}{2}}] \tag{A.2a,b}$$

$$\beta_1 = \frac{M(1-M)}{2[2+(9-8M)^{\frac{1}{2}}]}, \quad \beta_2 = \frac{M(1-M)}{2[2-(9-8M)^{\frac{1}{2}}]} \tag{A.3a,b}$$

$$\Delta_1 = \frac{3}{2}\left[\frac{(1-M)(2-M)}{M} - \frac{x_p}{R_p}\right](\alpha_2+\alpha_2\beta_2+2\beta_2) + \frac{3}{2}\left[\frac{(1-M)(2+M)}{M} - \frac{(2+\nu)}{2}\left(\frac{x_p}{R_p}\right)^2\right](1+\beta_2) \tag{A.4}$$

$$\Delta_2 = -\frac{3}{2}\left[\frac{(1-M)(2+M)}{M} - \frac{2+\nu}{2}\left(\frac{x_p}{R_p}\right)^2\right](1+\beta_1) - \frac{3}{2}\left[\frac{(1-M)(2+M)}{M} - \frac{x_p}{R_p}\right](\alpha_1+\alpha_1\beta_1+2\beta_1) \tag{A.5}$$

$$\Delta = (\alpha_2-\alpha_1)(1+\beta_1+\beta_2+\beta_1\beta_2) + 2(\beta_2-\beta_1) \tag{A.6}$$

Here x_p/R_p is given by (3.26).

WHAT HAPPENS AT FAST CRACK GROWTH?

K. B. BROBERG

Lund Institute of Technology, Box 725, S-220 07 Lund, Sweden

ABSTRACT

Crack growth is essentially determined by the properties of the process region. At fast crack growth, however, the process region seems to interact with the surrounding stress field in a way not yet understood. This phenomenon is discussed in the light of some recently reported experimental results. Mode II fast growth is discussed, including the behaviour at healing and the possibilities of edge velocities beyond the Rayleigh wave velocity. The conclusion is that the upper limit for the propagation velocity of a healing region is the velocity of irrotational waves.

1. CHARACTERISTICS OF THE PROCESS REGION

At the edges of a propagating crack or a slipping region there is a process region, in which material separation (mode I) or slip (modes II and III) is initiated. Although it is quite obvious that this region behaves in a very special way, different from surrounding regions, it is not easy to describe its behaviour in general terms. Perhaps the simplest way is to identify the process region with the region in which decohesion takes place. This is a rather unprecise specification, but the processes involved are so complex and varying that more precise specifications become problematical.

In static cases the process region would deform unstably under load controlled conditions. A simple, illustrative model of such behaviour is provided by the tensile test-piece. As soon as one neck is developed the force starts decreasing and no other neck is formed. In the dynamic case, however, more than one neck can be developed. The force on a cross-section varies along the test-piece at each instant of time and a multitude of cross-sectional states belonging to the same force is possible. Look, for instance, on the schematical expression for the cross-section,

$$P = (P_o + k\dot{\varepsilon})/(1 + \varepsilon) \quad (1)$$

where P is the force, ε the average strain and the dot denotes time derivative. Obviously, two different cross-sections may exhibit the same force P at the same instant of time even though their average strain ε is different. A more realistic expression for P would not change this conclusion. Necking at one site does not necessarily prevent necking at another one. In principle, necking can occur even at increasing cross-sectional force.

Assume now that the same tensile force is suddenly applied at both ends of a tensile test-piece and that three necks are formed,

FIG. 1. Three necks on a rod.

see Fig. 1. Due to inertia effects the force might decrease towards the middle part of the specimen. This implies that the development of the middle neck might become impeded. The two outer necks tend to "shelter" the middle part of the test-piece. On the other hand, if the applied force is very high, wave phenomena might promote development of the middle neck.

Even though necking phenomena in tensile test-pieces were introduced as simple illustrations of instabilities in process regions, they are obviously not very simple at dynamic loading. By analogy the following conclusions can be drawn for a process region at the edge of a crack (for simplicity specification is made to mode I):

1. In slow crack growth very few micro-separations (voids or micro-cracks) are opened. Generally not more than one micro-separation is found in each section across the crack front. Virtually all micro-separations coalesce to form the fracture surfaces. Intrinsic dimensions of the material (such as the distance between inclusions) specify the minimum number of micro-separations. This leads to the "minimum energy requirement" (1), i.e. a certain minimum amount of energy is to be consumed in the process region. This, in turn determines the size of the plastic region and the stress intensity factor at slow crack growth.

2. In rapid crack growth many more micro-separations are opened than those eventually forming fracture by coalescence. This might involve micro-separations in the middle of the cluster, but it

might as well involve peripheral micro-separations, especially when inertia effects tend to "shelter" the interior parts of the process region. No upper limit seems to exist for the size of the process region.

Further characteristics of the mode I process region is that the highest forces tend to work in the crack growth direction at rapid crack propagation. This would influence the morphology so that micro-separations would tend to grow in the direction normal to the main crack, see Fig. 2.

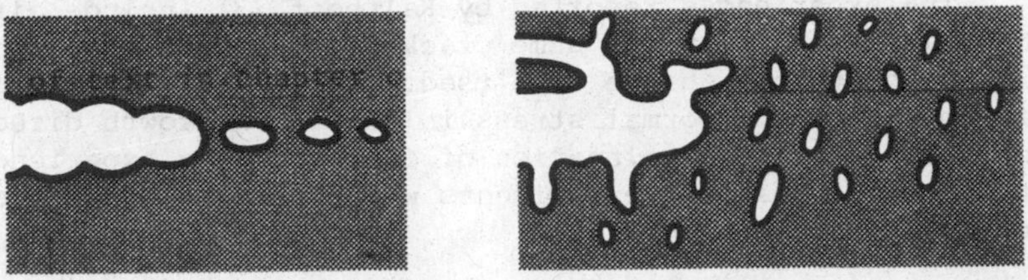

a) b)

FIG. 2. Distribution and shape of holes at a) slow crack growth and b) fast crack growth.

These points of view on mode I process regions were discussed earlier (1). Some recent experiments seem to confirm these viewpoints. Thus Ravi Chandar (2) comments on experiments with a plastic (Homalite-100): "At higher stress intensities, in the 'mist zone', the flaws are activated into growing and the crack propagates along different planes and voids grow before the arrival of the main crack front. The size of the fracture process zone, i.e. the zone in which the small flaws are activated, increases. In the 'hackle zone' all of the mist zone features exist, only more pronounced. There is also evidence of crack growth perpendicular to the main crack growth direction. The fracture process zone is much larger."

"Higher stress intensities" generally refer to higher crack growth velocities, but an interesting result from Ravi Chandar's (2) experiments is that no unique relation between the stress intensity factor and the crack tip velocity was found. The same experience has been reported by Kalthoff (3). In Ravi Chandar's case the crack tip velocity remained unchanged over several centimeters travel although the stress intensity factor varied substantially, in some cases by a factor of about 2.

As mentioned, the size of the process region increases with the stress intensity factor (or *vice versa*). Ravi Chandar points out that this is the case even at constant crack tip velocity: there seems to be a unique relation between stress intensity factory and process region size.

These results are in accordance with the conclusion in an earlier discussion (1) that the size of the process region at high velocities "will no longer be tied to intrinsic dimensions of the material but instead to the strength of the surrounding stress-strain field".

One could, of course, ask why the crack tip velocity turns out to remain constant. The answer is probably that this is a result of the specific experimental setup used by Ravi Chandar. Inertia effects may be sufficiently dominating to maintain a constant velocity, whereas the variation of energy expenditure in the crack tip vicinity plays a subordinate rôle.

The experiences reported by Kalthoff (3) include different stress intensity factors at the same crack tip velocity when test-pieces of different geometrical shapes were used. Such experiences might indicate an influence of the normal stress in the crack growth direction. This is probably also the explanation of corresponding experiences with different outcomes of static experiments when different test-piece geometries were used, cf. (4).

Most of the discussion of mode I cases could probably be carried over with only minor adjustments to modes II and III, even though experimental results concerning these modes are very scarce. Instead of voids or micro-cracks shear micro-failures appear, sometimes as shear bands. These micro-mechanisms may be the same in modes II and III. Therefore mixed mode II-III propagation seems to be possible, whereas mode I hardly mixes with the other two. On the other hand cracks may grow at alternating modes on the micro-scale. Thus zig-zag crack paths, indicating mode II micro-mechanisms in mode I macro-growth, have been observed. Likewise echelon cracks sometimes appear in mode II macro-growth. The echelons can be interpreted as successions of mode I cracks between regions of mode II growth, cf. (5).

Modes I and III are symmetry modes and, consequently their process regions are symmetrical (statistically seen) as regards physical quantities. Mode II is different. One side is subjected to dilatational, the other one to compressive forces. Furthermore, the shear symmetry depends on the friction properties of the "crack" surfaces. With, for instance, a Coulomb friction coefficient 0.6, the mode II growth direction forms the angle 15.5° with the direction of the largest remote shear stress (5). Hence a mixed mode II-III slip propagation radially outwards from some initiation site ought to create a non-plane (twisted) slipping surface.

2. THE HEALING REGION

In modes II and III the slipping region can propagate in essentially one direction, implying one leading and one trailing edge. Even at the trailing edge there is a process region, albeit the process is healing instead of tearing. Healing does not require energy, but can, in principle, generate energy. The classical Eshelby (6) treatment of the suddenly arrested mode III crack exhibits the feature of a healing edge, travelling with sound velocity and neither requiring nor generating energy. Through this treatment the first stepping stone was laid for a fruitful exploration of transient crack growth. The next stepping stone, laid by Freund (7), concerned mode I and, with almost only formal changes, also mode II. At arrest of a mode II crack, in the Freund treatment, a healing edge moves with the velocity of Rayleigh waves, neither requiring, nor generating energy.

Naturally the solutions by Eshelby (6) and Freund (7) do not involve any *physical* conditions at healing (or tearing) regions. The results obtained cannot be interpreted as proofs of energy-neutral processes at a healing region. One might recall the formal result that the energy flow to mode I edges, moving at the Rayleigh wave velocity at symmetrical crack extension is zero (8). However the energy given off from the healing region is probably negligible in comparison to the energy consumption at the matching tearing region. The safest estimate seems to be to assume that the healing process is energy-neutral.

3. MODE II GROWTH

A special feature of modes II and III consists of difficulties in demarcating the process region. The difficulties are small when it can be assumed that the shear stress on the slipping region (the "dynamic friction stress") is constant, see Fig. 3. However, if this shear stress decreases with increasing slip (for instance due to temperature rise) then it seems to be a matter of convenience to make a suitable division between the process region and the regular slipping region, see Fig. 4.

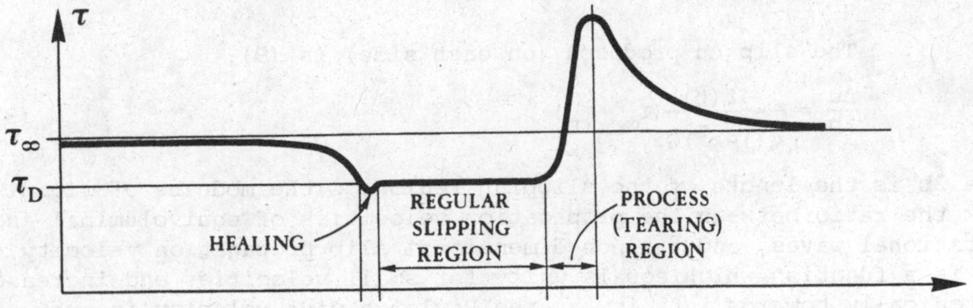

FIG. 3. Division between process region and regular slipping region when the shear stress is constant = τ_D on the latter region.

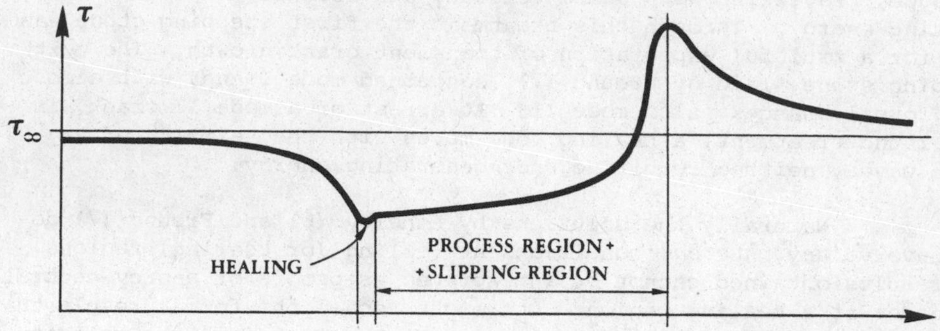

FIG. 4. The figure shows a difficulty in determining the boundary between process region and regular slipping region.

When the shear stress is constant (= τ_D) on a major part of the slipping region, then the energy dissipation in the process regions (the tearing and the healing region) in uni-directional mode II propagation at remote shear stress τ_∞ is the fraction

$$(\tau_\infty - \tau_D)/\tau_\infty \tag{2}$$

of the total energy dissipation (9). (Actually, due to erroneous sign of the healing contribution, a somewhat more complicated expression was given in (9)). Knowledge of the total energy dissipation and of the above fraction thus enables calculation of the energy dissipation in the tearing region (assuming the healing region to be energy-neutral) or *vice versa*.

The slip Δu produced (on each side) is (9):

$$\frac{\Delta u}{2b} = \frac{\pi f(\beta)}{4(1-k^2)G}(\tau_\infty - \tau_D) \tag{3}$$

where 2b is the length of the slipping region, G the modulus of rigidity, k the ratio between the propagation velocities of equivoluminal and irrotational waves, and β a non-dimensional slip propagation velocity. $f(\beta)$ is a function which equals unity for small velocities and increases monotonically towards infinity as the Rayleigh wave velocity is approached.

Assuming that the healing region is energy-neutral, the length of the slipping region is (9):

$$2b = \frac{1}{\pi^2} \cdot \frac{[T(\beta)]^2}{(\tau_\infty - \tau_D)^2} \tag{4}$$

where $T(\beta)$ is the modulus of tearing, connected with the stress intensity factor K_{IId} by the relation

$$T = (\pi/2)^{\frac{1}{2}} K_{IId} \tag{5}$$

(It is assumed here that there is a one-to-one relation between T or K_{IId} and β).

For a homogeneous infinite solid in a homogeneous stress state one of the three quantities Δu, 2b and β can be chosen arbitrarily (within certain limits). Then the other two follow from Eqns (3) and (4) provided that $T(\beta)$ and $(\tau_\infty - \tau_D)$ are known. A steady state solution exists.

Even though the steady state solution satisfies both equations and boundary conditions one cannot be certain that it represents a stable solution. Suppose, for instance, that either the leading or the trailing edge starts to accelerate. If the leading edge travels faster than the trailing one, then the slip deposited increases. It is not difficult to construct a solution for the case when the leading edge moves with a higher velocity than the trailing one, both velocities being constant, see (9). This solution shows a deposited slip that increases linearly with the distance from the origin.

If the trailing edge moves faster than the leading one, the slip deposited should decrease as the trailing edge moves on. If this process goes on until the trailing edge catches the leading one, the slip event collapses completely: it ceases to exist.

The questions posed would probably be answered if one could analyze the transient case, i.e. let the slip event start from rest and see whether it evolves towards a steady state, and, if so, which steady state (i.e. which combination of Δu, 2b and β) would result. The start conditions are obviously critical. The natural way would be to assume a non-homogeneous state at rest (residual stresses), for instance one having resulted from previous slip events of finite duration. These events may have imprinted the trace of a displacement pile up (10), similar to a dislocation pile up. Such a trace consists of one or more sets of local peaks and valleys along the fault. A new slip event would most probably be initiated at a stress peak, and it would start to grow radially from the initiation site. However, in the "backwards" direction (the "valley" side) the shear stresses are smaller and the front of the slipping region in this direction may eventually be arrested. This, of course, means that the propagation direction of the front is reversed, i.e. the front is converted from a tearing to a healing region, and a uni-directional slip propagation is created.

The initiation of slip is governed by local conditions. However, it cannot be taken for granted that an ensuing uni-directional slip event will be propagated indefinitely, even if homogeneous conditions prevail far away from the initiation site. It all depends on what these conditions are. If the slip produced when a uni-directional slipping region has been created is less than Δu_o, where, see (9):

$$\Delta u_o = \frac{T^2_{min}}{4\pi(1-k^2)G(\tau_\infty-\tau_D)} \tag{6}$$

then the slip event will come to rest. The minimum value of T, T_{min}, probably occurs for $\beta = 0$ or for a comparatively low β-value.

When a slip event comes to rest the sliding velocity (the mass velocity difference between the two mating surfaces) may decrease slowly towards zero, but the velocity of the healing region may be high. Not even intersonic velocities can be excluded.

4. SUPER-RAYLEIGH EDGE VELOCITIES

For modes I and II the subsonic case (edge propagation slower than equivoluminal waves) is well known. The energy flow to a leading mode II edge per unit of edge movement is

$$\frac{dW}{dS} = \frac{f(\beta)}{2\pi(1-k^2)G} T^2 \tag{7}$$

The function $f(\beta)$ is positive for $\beta < k_s$, where k_s is the (non-dimensional) Rayleigh wave velocity, and negative for $k_s < \beta < k$, where k is the (non-dimensional) propagation velocity of equivoluminal waves. The energy flow to a trailing mode II edge is (note a sign error in (9))

$$\frac{dW}{dS} = -\frac{f(\beta)}{2\pi(1-k^2)G} H^2 \tag{8}$$

where H is the modulus of healing (which probably is much smaller than T and thus can be set to zero).

The interesting thing is that there is a positive energy flow towards a healing region travelling with constant velocity in the subsonic but super-Rayleigh region. Since a healing region does not require energy this result indicates that it could accelerate if it entered the velocity region in question.

The intersonic case, $k < \beta < 1$, can be studied by assuming a line model of the process region, i.e. the material remains elastic outside the slipping region. The tearing region is assumed to occupy a portion of length d_T and the healing region a portion of length d_H along the slipping region. Then, cf. (11), the energy flow to a tearing region is found to be

$$\frac{dW}{dS} = -2\,\frac{f_o(\beta)}{\pi G}\sin^2\pi\gamma \int_0^{d_T}\int_0^{\infty} \frac{x^{1-\gamma}}{s^{1-\gamma}}\,\tau_o(-x)\,\frac{\tau_o(-s)-\tau_o(-x)}{s-x}\,ds\,dx \qquad (9)$$

where

$$f_o(\beta) = \beta^2(1-\beta^2)^{-\frac{1}{2}}/4k^2 \qquad k < \beta < 1$$

$$\gamma = \pi^{-1}\tan^{-1}\frac{4k^3(1-\beta^2)^{\frac{1}{2}}(\beta^2-k^2)^{\frac{1}{2}}}{(2k^2-\beta^2)^2} \qquad 0 < \gamma < 1/2$$

and $\tau_o(x)$ is the shear stress along the process region $-d_T < x < 0$. Stress continuity and smooth closing are assumed. For physical reasons it is assumed that $\tau_o(x)$ is a maximum at $x = 0$ and then dW/dS is positive, i.e. energy flows towards the leading edge

For a trailing edge (i.e. a healing region) the energy flow is given by Eqn (9) after interchange of d_T and d_H (the length of the healing region), interchange of γ and $1-\gamma$ and, finally, change of sign. Thus

$$[\gamma_{healing}]_{mode\ II} = 1 - [\gamma_{tearing}]_{mode\ II} \qquad (10)$$

$\tau_o(x)$ is the shear stress along the healing region $-d_H < x < 0$. Since $\tau_o(x)$ should be a minimum at $x = 0$ one arrives at the result that dW/dS is positive, i.e. energy flows also towards the trailing edge in the intersonic region.

It is somewhat surprising that the mode I results are different (11):

$$[\gamma_{opening}]_{mode\ I} = [\gamma_{tearing}]_{mode\ II} + 1/2 \qquad (11)$$

$$[\gamma_{closure}]_{mode\ I} = 1 - [\gamma_{opening}]_{mode\ I} \qquad (12)$$

dW/dS for a leading mode I edge is given by Eqn (9) after replacing γ by $1/2 + \gamma$ and after multiplication by the factor.

$$-(1-\beta^2)^{\frac{1}{2}}(\beta^2-k^2)^{-\frac{1}{2}} \qquad (13)$$

Energy flows away from a leading mode I edge at intersonic velocities and towards a trailing edge.

It seems hardly possible that a leading edge might bypass the "forbidden" part of the subsonic region. Even if this would happen — for instance after bridge formation — the energy requirement has to be met and, as seen from Eqn (9) the energy flow tends to be very small at low intersonic velocities. For a healing region, on the other hand, no such restrictions exist — no region is "forbidden" and acceleration towards the irrotational wave velocity seems to be possible.

REFERENCES

(1) Broberg, K. B. (1978). On the behaviour of the process region at a fast running crack tip. In: High Velocity Deformation of Solids, ed. K. Kawata and J. Shioiri, pp. 182-194. Berlin/Heidelberg: Springer-Verlag.

(2) Ravi Chandar, K. (1982). An Experimental Investigation into the Mechanics of Dynamic Fracture. Ph. D. Thesis. Pasadena, California: California Institute of Technology.

(3) Kalthoff, J. F. (1983). Presentation at Workshop on Dynamic Fracture, February 17 and 18, 1983. Pasadena, California: California Institute of Technology.

(4) Broberg, K. B. (1983). New approaches in fracture mechanics. In: Mechanical Behaviour of Materials - IV, ed. J. Carlsson and N. G. Ohlson, pp. 927-934. Oxford: Pergamon Press Ltd.

(5) Broberg, K. B. (1983). On crack paths. Paper presented at Workshop on Dynamic Fracture, February 17 and 18, 1983. Pasadena, California: California Institute of Technology.

(6) Eshelby, J. D. (1969). The elastic field of a crack extending non-uniformly under general anti-plane loading. J. Mech. Phys. Solids, 17, pp. 177-199.

(7) Freund, L. B. (1972). Crack propagation in an elastic solid subjected to general loading. — I. Constant rate of extension. J. Mech. Phys. Solids, 20, pp. 129-140.

(8) Broberg, K. B. (1964). On the speed of a brittle crack. J. Appl. Mech., 31, pp. 546-547.

(9) Broberg, K. B. (1978). On transient sliding motion. Geophys. J. R. astr. Soc., 52, pp. 397-432.

(10) Broberg, K. B. (1978). On modelling of friction. Paper presented at the IASPEI Symposium on Earthquake Source Physics at Bad Honnef, West Germany, 1978.

(11) Broberg, K. B. (1980). Velocity Peculiarities at Slip Propagation. Report from the Division of Engineering, Brown University, Providence, R.I. USA.

CREEP CRACK GROWTH IN DUCTILE ALLOYS

A.S. ARGON, C.W. LAU, B. OZMAT, AND D.M. PARKS

Massachusetts Institute of Technology, Cambridge, MA 02139, U.S.A.

ABSTRACT

Creep crack growth in ductile alloys involves considerable fragmentation of the crack tip region in its early phases of growth. This is a result of the defocusing action of crack tip blunting by both distortional and cavitational strains on the distribution of intergranular creep damage and is affected significantly by the initial sharpness of the crack. Specific models of intergranular damage combining non-steady creep flow, evolution and growth of grain boundary facet cracks in the inelastic deformation field leading to final fracture have been developed. When used in conjunction with finite element programs for crack tips, these damage models can explain such crack extension modes. The combination of mechanistic three dimensional damage models and large strain finite element codes, promise to be of wide-spread utility in predicting the development of creep damage under complex loading histories.

1. INTRODUCTION

Understanding the growth of cracks in creeping alloys has been of considerable interest over the past decade. The problem has been studied theoretically on the basis of singularity analysis; it has been studied experimentally in a wide variety of alloys ranging from rather brittle superalloys to very ductile stainless steels; and has been modeled by making use of extensive mechanistic developments on transgranular as well as intergranular cavitation. Mechanics analyses for sharp cracks permitting singular field representations have dominated much of the theoretical development. For example, Riedel & Rice (1) have considered an infinitely sharp stationary crack in a solid continuum combining both elasticity and non-linear viscous power-law behavior. They have obtained a solution for the crack tip singular field as the stresses relax from an initial elastic distribution characterized by an elastic stress intensity factor K_I to a steady state asymptotic distribution dominated by creep strains, in which stresses are distributed according to a lower grade singularity characterized by the steady state creep stress singularity factor C^*. Such singularity analyses have been extended by others by incorporating approximately other response modes such as power-law strain hardening plasticity, and primary creep (2,3). In these the initial elastic singularity at the sharp crack is first replaced by a Hutchinson (4) -- Rice & Rosengren (5)(HRR) singularity due to the power-law plastic response, which is replaced soon after by a lower grade singularity C_p due to primary creep, which in turn is eventually replaced by the C^* singularity of the steady state creep stress distribution. In a further extension of these singularity analyses Hutchinson (6) has considered the additional effect of cavitational strains due to growing facet cracks, and has found that at eventual asymptotic response at very long times the stress distribution is quantitatively altered by the cavitational component of strain but is still scaled by the steady state creep stress singularity factor C^*.

The primary lesson of these exercises has been that for times much shorter than a critical relaxation time the crack tip singular stress field is relatively unrelaxed and is characterized by an elastic stress intensity factor K_I, while for times much larger than this critical relaxation time the creep strains dominate everywhere and the stress singularity is characterized by the steady state creep singularity factor C^*.

Experiments (7,8) have demonstrated that in relatively brittle alloys in which the attainment of a small strain is sufficient to produce fracture, such strains can be reached at the crack tip during times much less than the critical relaxation time, making the cracks grow, primarily in an elastic surrounding, governed by the level of the stress intensity factor K_I. In very ductile alloys in which fracture requires substantial creep strains, their attainment at the crack tip can be accomplished only for times substantially larger than the critical relaxation time. For such cases it has been reported that crack growth rates can be correlated by the factor C^* (7,9). Although some of the mechanistic interpretations of these correlations are suspect, the engineering value of these distant field characterizing parameters in the correlation of data on crack growth rates is clear.

Based on the extensively studied mechanisms of transgranular and intergranular cavitation (for a review see Argon (10)), and using the singular stress distributions discussed above, many investigators have tried to model the growth of creep cracks in non-linearly creeping alloys (for a review of these see Wilkinson & Vitek (11) and Bassani & Vitek (12)). As in the cases of the experimental studies where relatively brittle alloys showing little crack blunting, were found to exhibit a correlation between crack velocity and K_I, the modelling studies of such alloys are most probably reliable. For ductile alloys, however, where the required large strains to fracture are accompanied by large crack tip blunting due to both distortional and cavitational strains, the crack tip stress distributions become radically altered from the singular field distributions obtained for asymptotic sharp crack behavior. As a result cracks in such alloys show large non-planar growth perturbations under the locally relaxed, and non-singular distributions of stress that strongly interact with both the blunting and the caviation process. The early growth of such cracks has been considered by Leckie & Hayhurst (13), and has been modelled by Hayhurst and coworkers (14,15) on the basis of phenomenological descriptions of creep damage, in combination with finite element (FE) techniques of local stress analysis. We will present here micromechanical models of creep damage, and will show how the early growth of creep cracks can be analysed by a combination of these models and FE stress analysis for large strains.

2. DESCRIPTION OF CRACK TIP PROCESSES

2.1 MODES OF CRACK EXTENSION IN DUCTILE ALLOYS

Figures 1a and 1b show the early growth phases of two cracks in annealed Type 304 stainless steel: the first after 150 hours at 750°C in a double edge notched (DEN) plane strain specimen, initially fatigue cracked, and subjected to a net section stress (NSS) of 80 MPa, while the second, after 140 hours at 775°C in a DEN plane strain specimen having an initially rounded crack tip, subjected to a NSS of also 80 MPa. Both figures show three pronged crack extensions consisting of some extension in the plane of the

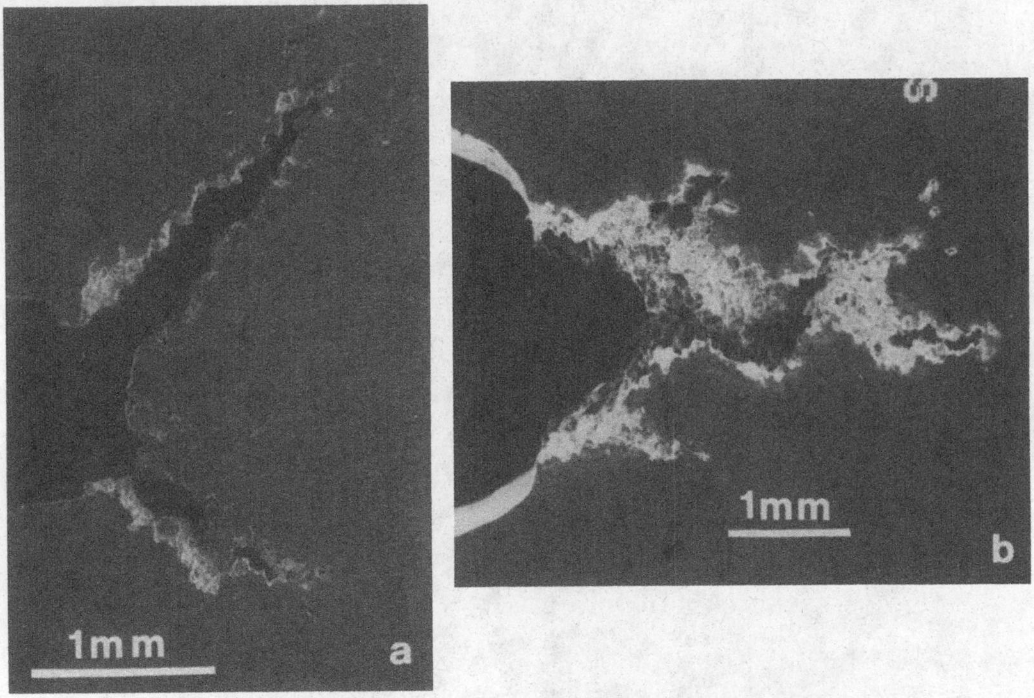

FIG. 1. a) Creep crack in initially fatigue cracked DEN plane strain specimen of 304 stainless steel, after 150 hours at 750°C, under NSS of 80 MPa, b) initially round creep crack (r = 0.2 mm) in DEN plane strain specimen of 304 SS, after 140 hours at 775°C, under NSS of 80 MPa.

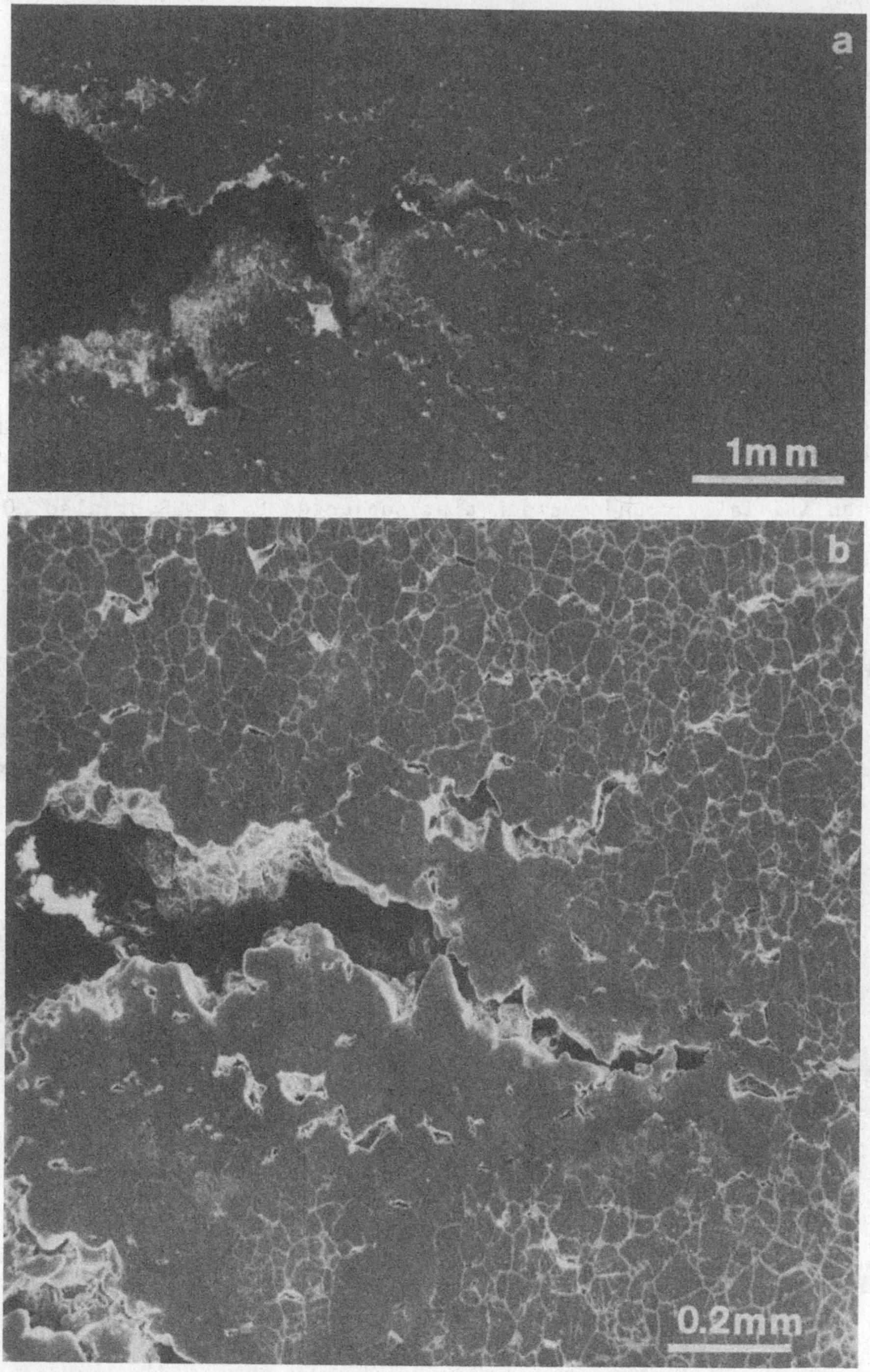

FIG. 2. Same creep crack as Fig. 1b; a) showing the fragmented quasi-planar growth of the crack, b) showing the crack tip region and surrounding diffuse intergranular facet cracks.

crack and some extension roughly at 45°-50° with the crack plane. While in Fig. 1a the inclined cracks have moved ahead and shielded the center crack, in the case of Fig. 1b the opposite has occurred. Figures 2a and 2b show two larger magnification views of the crack in the DEN specimen of Fig. 1b, with the initially round crack tip. Figure 2a shows the well dispersed nature of the diffuse extension process of the center branch crack while Fig. 2b shows a more magnified view of the tip region of the main crack in a field of a high concentration of readily recognizable grain boundary facet cracks. This crack front behavior between initially sharp and initially rounded cracks is typical of all plane strain loading geometries including also modified compact tension (CT) specimens. As all figures show, but particularly Figs. 2a and 2b, the cracks grow intergranularly by the development and eventual linkage of grain boundary facet cracks, and are associated with wide spread diffuse damage of the same kind remaining dormant in the flanks of the propagating cracks. At higher growth rates under higher NSS the crack growth mode is transgranular, and associated with little diffuse damage. We will concentrate here only on the slow intergranular mode of crack growth, and present specific models for the development of such damage, that are based on known details of intergranular cavitation.

2.2 MECHANISTIC CRACK TIP MODELS

Figures 2a and 2b show convincingly that singular field solutions for sharp cracks will not be sufficient in the modeling of these early phases of crack growth, but that on the basis of the very fine nature of the accompanying diffuse intergranular cracking, continuum descriptions of such damage will be necessary for such modeling. Although the pictures strongly suggest that the cavitation by facet cracking should have profound effects on stress relief, this is difficult to quantify in a strong strain gradient field capable of jacking up again the relieved stresses. Simple analysis gives that an increase of porosity dp in a quasi-spherical region will give rise to an elastic relief of mean normal stress $d\sigma_m$ that is

$$d\sigma_m = - [K/(\frac{3K}{4\mu} + \frac{1 + 2p}{1 - p})] dp \qquad (1)$$

where K, μ, and p are the bulk modulus and shear modulus of the fully dense material and p the current level of porosity in the spherical region. Clearly, the important question is the rate of change of mean normal stress imposed by the outer field, governed by the inelastic deformation of it, that accompanies this increase of porosity in the inner region. For a porosity level of 0.05, that is roughly what the field in Fig. 2b reflects, $d\sigma_m/dp \simeq - K/3$, and is clearly substantial. This is independent of whether the porosity increment is produced by facet cracks or grain boundary cavities. The important stress relieving role of the latter in diffusive cavity growth on grain boundaries has already been established by Rice (16) in his model of facet crack evolution by constrained cavitation.

In the following sections we deal in order: with the evolution of deformation resistance in diffusive primary creep leading to a general three dimensional constitutive relation for such creep; with the rates of evolution of grain boundary facet cracks by intergranular cavity growth starting from grain boundaries with a variability in the concentration of hard particles; with the growth and interaction of grain boundary facet cracks and its consequences on the acceleration of creep strain rates; with how these processes can be combined to account for the entire tensile creep response; and finally with how these damage mechanisms can be combined with large strain numerical programs to account for the above mentioned crack tip behavior.

3. EVOLUTION OF DEFORMATION RESISTANCE IN PRIMARY CREEP

In ductile alloys transient deformation can make large contributions to the overall creep strains and can significantly alter the rates of evolution of damage. This requires a sound description of primary creep based on appropriate internal parameters or state variables. Much mechanistic creep research over the past several decades has established that during primary creep the decrease of creep strain rate is accompanied by a monotonic increase in dislocation density. This evolving dislocation density prescribes a deformation resistance tensor that directly affects the current creep rate. Two measures of this resistance tensor are of direct consequence. The average value s, best measured by a reference hardness experiment at low temperature, characterizes the all important isotropic component of the deformation resistance governing the overall radial loading response of the alloy. The internal stress σ_i, usually measured by a so-called stress dip experiment gives a measure of the polarity of the resistance tensor opposed to the current principal deformation direction. Robinson (17) has shown that knowledge of this internal stress is particularly essential for stress reversals. We will demonstrate that for radial loading, and possibly for slowly varying direction changes in principal strain axes knowledge of the evolution of the isotropic deformation resistance s is sufficient for the description of all phases of diffusive power-law creep.

Mounting evidence suggests that the kinetic law for the inelastic strain rate even in diffusive creep is given by the well established Arrhenius expression for thermally activated glide, which is usually taken as (18)

$$\dot{\varepsilon} = \dot{A}_o \exp\left(-\frac{\Delta F}{kT}\left(1-\frac{\sigma}{s}\right)\right) \qquad (2)$$

where ΔF, s, and $\dot{A}_o$ are the Helmholtz free energy of the process to be activated, the athermal tensile deformation resistance, and a pre-exponential factor involving the frequency factor, the mobile dislocation density and the unit sweep of a dislocation segment after activation. The applied stress σ must be interpreted here as the Mises effective stress σ_e which we will abbreviate by σ for simplicity. In a narrow range of stress this kinetic law can be represented as a power law expression.

$$\dot{\varepsilon} = A(T)\left(\frac{\sigma}{s}\right)^n, \quad n = \frac{\Delta F}{kT}\frac{\sigma}{s} \qquad (3a,b)$$

where A(T) is a fitting factor depending on $\dot{A}_o$ and temperature, and n is a phenomenological exponent with typical values in the range of 30-40. We now assume that the evolution of the deformation resistance s, proportional to the square root of the dislocation density in the well known way, is governed by the Bailey (19), Orowan (20) equation, balancing strain hardening against recovery,

$$\frac{ds}{dt} = h(s,\sigma)\dot{\varepsilon} - B(T)\,\mu\left(\frac{s}{\mu}\right)^p \qquad (4)$$

where h(s,σ) is the athermal strain hardening rate that should in a more general development include dynamic recovery (21-23) but will be taken here as a constant; B(T) is a static recovery coefficient that has been developed in detail elsewhere (23), p (= 3 or 5 for volume diffusion, or core diffusion controlled recovery) is a specific recovery exponent, for which the value of 5 should be taken on the basis of recent experimental measurements (24). Incorporation of the kinetic law (3a) into (4) gives,

$$\frac{d(s/\mu)}{d(t/\tau_r)} = C\frac{(\sigma/\mu)^n}{(s/\mu)^n} - (s/\mu)^p, \therefore\ C = (h\dot{\varepsilon}_o\tau_r/\mu)^{(n+p)/n} \qquad (5a,b)$$

$$\dot{\varepsilon}_o = (\mu/h)^{n/(n+p)}B(T)[(A(T)/B(T))]^{p/(n+p)}, \therefore\ \tau_r = 1/B(T). \qquad (5c,d)$$

We note that Eqn (5a) leads to a stress dependent steady state deformation resistance

$$s_o = \mu[(h\,A(T)/\mu B(T))(\sigma/\mu)^n]^{1/(p+n)}, \qquad (6)$$

and through the kinetic law to a steady state creep strain rate of

$$\dot{\varepsilon}_s = \dot{\varepsilon}_o(\sigma/\mu)^m, \therefore\ m = np/(n+p), \qquad (7a,b)$$

where $\dot{\varepsilon}_o$ is given by Eqn (5c) above. Equation (7a) is of the well known Dorn form (25) with a creep stress exponent that asymptotically approaches p for very large n. For a typical n of 35 the effective creep exponent m will be 4.38, which is very close to what has been quoted for pure f.c.c. metals (25). We note further: that $(A(T)/B(T))^{p/(n+p)}$ should be a very weak function of temperature, and that the main temperature dependence of $\dot{\varepsilon}_o$ comes from B(T) which reflects the kinetics of diffusive recovery; that the processes of evolution go according to a dimensionless time B(T)t, already well established by Dorn and coworkers (26); and that when terms in the Dorn equation (7a) are regrouped, the strain rate is governed by a power function of stress divided by the plastic deformation resistance, as was observed by Brown & Ashby (27) and is our starting assumption of Eqn (3a).

The current inelastic creep rate at any time is then given by the deformation resistance s from the evolution law of Eqn (5a), used in the kinetic law of Eqn (3a), for any given level of current applied equivalent stress σ. Given any appropriate initial condition of s and any slowly varying stress history, Eqn (5a) can be solved and the current creep rate can be obtained from Eqn (3a). This procedure has been followed for many stress histories by Argon et al. (28) including transients associated with sudden stress increments and decrements, and has been found satisfactory in form. In the form used here, with a constant value for h, however, it is not possible to model steady state stress dependence characteristic of stress exponents $m > 4.4$ such as is required for Type 304 stainless steel with $m = 8$. When dynamic recovery is included, having hardening rates dependent on σ and s, as has been shown by several investigators (21-23), more realistic responses are obtained. Acceptable overall responses, however, can also be obtained, albeit less desirably, by formally choosing larger exponents p in Eqns (4) and (5a) purely on phenomenological grounds. Figs. 3a and 3b show the fits of the computed behavior to the data of Blackburn (29), at 750°C for 304 stainless steel for $h/\mu = 1.25 \times 10^{-2}$, $B(T) = 6 \times 10^{18}\ s^{-1}$, $n = 40$, $m = 8$, $p = 10$. Apart from n and p that are artificially elevated to compensate for the absence of dynamic recovery in the model, other assumed parameters are all compatible with the behavior of this material at this temperature. In more refined modeling studies the more desirable but more difficult approach of incorporating dynamic recovery into Eqns (4) and (5a) with a true static recovery exponent of $p = 5$ should be used. The results of this approach to obtain the uniaxial tensile creep response are described elsewhere (28).

4. EVOLUTION OF GRAIN BOUNDARY FACET CRACKS

Intergranular cavitation on grain boundaries to form facet cracks usually occurs non-uniformly even if all grain boundaries are stressed equally. This is a result of the non-uniform initial particle concentration along grain boundaries that give rise to cavities through a process of stochastic grain boundary sliding (30). On the basis of

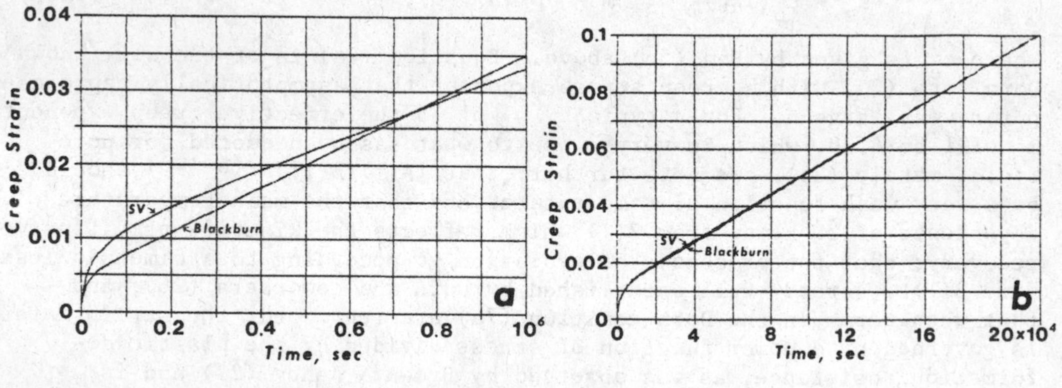

FIG. 3. Creep curves calculated from state variable (SV) model of primary creep for parameters given in text, compared with data of Blackburn (29): a) for 70 MPa; b) for 100 MPa.

the established random distribution of carbides in the volume of many steels (31), it can be assumed that this random coverage extends also to grain boundary carbides. Thus, it can be assumed that the measured overall cavitation response resulting from the constrained stochastic sliding of grain boundaries,

$$\dot{N} = \overline{\beta}\dot{\varepsilon}_e \quad \text{or} \quad N = \beta\left|\int\dot{\varepsilon}_e dt\right| , \tag{8a,b}$$

should also be randomly distributed among individual boundaries by a random distribution in the coefficients β that relate directly to the number density of carbides. Thus, if η is the fraction of all boundaries having a local nucleation coefficient β, with an overall population average for all boundaries being $\overline{\beta}$, then η should be given by

$$\eta = 1 - (1 + (\overline{\beta}/\beta)) \exp (-\overline{\beta}/\beta) \tag{9}$$

by direct extension of the measurements of Argon & Im (31) on carbide distributions in spheroidized steels. According to these measurements the coefficient β of any specific boundary should be inversely proportional to the area allocated to a specific carbide which is assumed constant for a given boundary but varies from boundary to boundary. The coefficient $\overline{\beta}$ then is the inverse area per particle for the overall population average. Equations (8a-b) can now be interpreted to apply to individual boundaries giving the local rate of cavity formation. The growth of these cavities by diffusional flow, in a constrained way at low stress and unconstrained way at high stress as described by Rice (16), will then lead to cracked grain boundary facets. In the presence of continued cavity formation the rate of change of cavitated area fraction A of any given boundary is given by (16,10),

$$\frac{dA}{dt} = \frac{(\pi\beta\varepsilon_e(t))^{3/2}}{h(\psi)\left[\frac{2\pi}{(0.9)}\left(\frac{A^{1/2}\beta\varepsilon_e(t)}{\dot{\varepsilon}_e d}\right) + \frac{kT}{K(a/b)D_b\delta_b\Omega\sigma_1(t)}\right]} \tag{10}$$

where $h(\psi)$ is a specific function of the dihedral half angle ψ of a grain boundary cavity having a typical value of about 0.33, d the grain size, $D_b\delta_b$ the grain boundary diffusional conductance, Ω the atomic volume, $\sigma_1(t)$ the maximum principal tensile stress acting across the boundary, and $K(a/b)$ ($\simeq$10) a slowly varying specific function of the cavity size to spacing ratio a/b that is described elsewhere (10). For any given local condition of stress, creep strain rate, and creep strain history the time τ to facet crack formation is obtained by integrating Eqn (10) from $A = 0$ to $A = A_{crit}$ ($\simeq$0.1) where rapid interaction between growing cavities produces final separation, i.e.

$$A_{crit} = \int_0^{\tau}\left(\frac{dA}{dt}\right)dt \ . \tag{11}$$

Thus, for different grain boundaries with different β, different dA/dt, different times $\tau(\beta)$ to facet crack formation will result under otherwise identical conditions. We write this symbolically as

$$\tau = G^{-1}(A_{crit},\beta), \tag{12}$$

where G stands for the integral of Eqn (11) and τ is recognized to be the inverse function of G for any given β, giving τ explicitly. The rate of change of facet cracking time with changing coefficient β in the cavitation history, then is

$$\frac{d\tau}{d\beta} = \frac{d}{d\beta}(G^{-1}(\beta)), \tag{13}$$

giving the time rate of increase of cracked fraction α of grain boundary facets,

$$\frac{d\alpha}{d\tau} = \frac{d\alpha}{d\beta}\frac{d\beta}{d\tau} = \frac{d\eta}{d\beta}\left[\frac{d}{d\beta}\{G^{-1}\}\right]^{-1} = -\frac{1}{\bar{\beta}}\left[\left(\frac{\bar{\beta}}{\beta}\right)^{3}\exp\left(-\frac{\bar{\beta}}{\beta}\right)\right]\left[\frac{d}{d\beta}\{G^{-1}\}\right]^{-1}. \tag{14}$$

Here we have made the observation in the second equality that the fraction $d\alpha$ of boundaries that are coming into the cracking range, because of the sampling by the process of the grain boundaries with coefficients between β and $\beta + d\beta$, is the same fraction $d\eta$ of boundaries having the initial particle coverage range pertaining to $d\beta$. The last equality is obtained by a direct substitution of the β derivative of Eqn (9). The right hand side of Eqn (14) is a function of β, which moreover through the equality of α and η is

$$\beta = \bar{\beta}\,\alpha^{-1}(\alpha) \quad . \tag{15}$$

Thus, substitution of Eqn (15) into (14) gives a differential equation of α in τ that can be numerically integrated to give for any local stress and strain rate history the evolution of the cracked boundary fraction α as a function of time τ. Figure 4 shows a computed evolution of cracked grain boundary fraction α for 304 stainless steel during tension creep at 750°C under a tensile stress of 86 MPa, utilizing the $\bar{\beta} = 1.74 \times 10^{10}\,m^{-2}$ measured by Chen and Argon (32), $A_{crit} = 0.1$, $h(\psi) = 0.33$, $K(a/b) = 10$, $d = 5 \times 10^{-5}$m, and all other values in Eqn (10) appropriate to γ-iron at the given temperature obtained from tabulated values (33). It is important to note here that we assume that at any point in the volume of a creeping solid facet cracks will develop only nearly normal to the local maximum principal tensile stress. Even in bi-axial tension where a second principal minor tensile stress might exist facet cracking should be only normal to the principal tensile stress. Any cracking across the minor principal stress should be retarded, or even suppressed by the additional compressive stresses that result from the block motion of grains induced by grain boundary sliding on the 45° boundaries. This is in agreement with experimental observations such as the facet crack orientations shown in Fig. 2b. In our continuum model of damage we make the further simplifying assumption that in the presence of moderate rotations of principal axes all increments of stress history are fully additive without any account being taken of actual boundaries remaining fixed in relation to the rotating

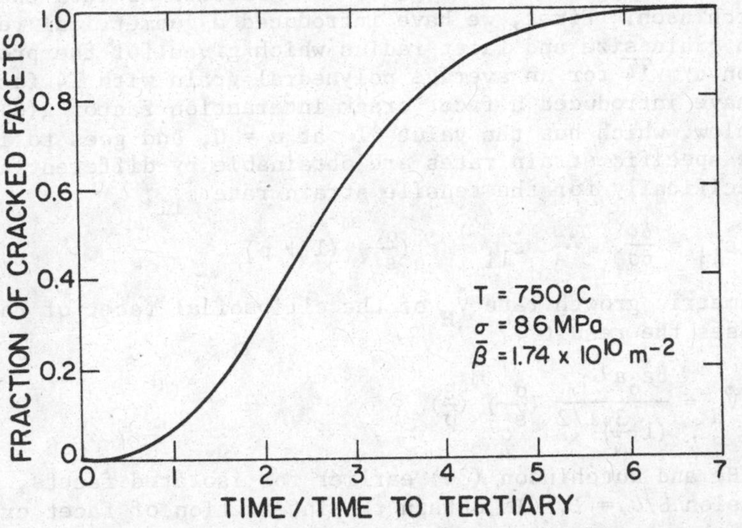

FIG. 4. Computed evolution of fraction of cracked grain boundary facets at 750°C under 86 MPa, and $\bar{\beta} = 1.74 \times 10^{10} m^{-2}$.

axes. Clearly, this gives an under estimate of the facet cracking time, and can lead to increasingly bad estimates when large systematic rotations of stress axes occur in the field as is the case in simple shear. Nevertheless, the results of Fig. 4 bring in an important element of variability in grain boundary strength which is present in all real alloys.

5. GROWTH AND INTERACTION OF FACET CRACKS

Hutchinson (6) has considered the effect of a constant volume density N of grain boundary facet cracks of radius a on the power law creep rate of a non-linearly viscous material with stress exponent m, by introducing a potential function Φ which when differentiated by the stresses σ_{ij} give the corresponding creep strain rates $\dot{\varepsilon}_{ij}$. This potential function is,

$$\Phi = \frac{\dot{\varepsilon}_o s_o}{(m+1)} \left(\frac{\sigma}{s_o}\right)^{m+1} \left[1 + \rho \left(\frac{S}{\sigma}\right)^2\right] \qquad (16)$$

where σ is the Mises equivalent stress, $\dot{\varepsilon}_o$ is the strain rate factor given by Eqn (5c), s_o is a constant steady state deformation resistance that is characteristic of the assumed non-linear viscous response chosen for the material, and S and ρ are defined as follows:

$$\frac{S}{\sigma} = \frac{\sigma_m}{\sigma} + \frac{2}{3}, \therefore \rho = \frac{4\alpha}{\pi\sqrt{14}} \frac{(1+m)}{(1+\frac{3}{m})^{1/2}} g(\alpha) \qquad (17a,b)$$

In Eqns (17a,b) σ_m is the negative pressure, and α is the same fraction of cavitated boundary facets nearly normal to the local maximum principal tensile stress that has been introduced in the preceding section.

Equation (17b) incorporates two important modifications into the form given by Hutchinson. First, we have introduced a geometrical relationship between grain size and facet radius which gives for the product a^3N of Hutchinson $\alpha/\pi\sqrt{14}$ for an average polyhedral grain with 14 facets. Second, we have introduced a facet crack interaction factor $g(\alpha)$ to be discussed below, which has the value 1 at $\alpha = 0$, and goes to infinity as $\alpha \to 1$. The specific strain rates are obtainable by differentiation, and give specifically for the tensile strain rate $\dot{\varepsilon}_{11}$,

$$\dot{\varepsilon}_{ij} = \frac{\delta\Phi}{\delta\sigma_{ij}}, \therefore \quad \dot{\varepsilon}_{11} = \dot{\varepsilon}_o\left(\frac{\sigma}{s_o}\right)^m (1 + \rho) . \qquad (18a,b)$$

For the volumetric growth rate $\dot{V}_H$ of the ellipsoidal facet of radius a Hutchinson uses the result

$$\dot{V}_H = \frac{8\dot{\varepsilon}_o a^3}{(1+\frac{3}{m})^{1/2}} \left(\frac{\sigma}{s_o}\right)^m \left(\frac{S}{\sigma}\right) \qquad (19)$$

obtained by He and Hutchinson (34) earlier for isolated facets, where for uniaxial tension $S/\sigma = 1$. To obtain the interaction of facet cracks accelerating the above growth rate and leading eventually to the rupture of ligaments between facet cracks we approximate the volume allocated to a typical facet crack by a center cracked tension specimen in a plane strain model of the phenomenon, as shown in Fig. 5. Since the medium is in fully developed creep flow, we use the plane strain results of Kumar et al. (35) for the analogous deformation problem for a power-law strain hardening material to model this interactive growth rate of the facet. Thus, from Kumar et al. we obtain directly,

$$\dot{V} \simeq \pi a^2 \dot{\delta} = \pi \dot{\varepsilon}_o a^2 h_2(a/b, m)(P/P_o)^m \therefore \; P_o = \frac{4as_o}{\sqrt{3}} \frac{1-(a/b)}{(a/b)} \qquad (20a,b)$$

where $\dot{\delta}$ is the crack flank separation rate at the center of the crack, $h_2(a/b, m)$ is a function calculated and tabulated by Kumar et al. (35) for the center cracked plate, and all other quantities are as defined in Fig. 5, or earlier above. Noting that $P/2b = \sigma$, $P_o/2b = 2s_o(1-a/b)/\sqrt{3}$, we extend these results into the required three dimensional problem by taking $a/b = \alpha$, and by assuming that when the plane strain blocks of Fig. 5 are transformed into space filling rectangular polyhedra the form of Eqns (20a,b) will not qualitatively alter. Thus, we obtain for the three dimensional case by formal extension,

$$\dot{V} = \frac{4}{3}\pi a^3 \frac{\dot{\varepsilon}_o}{(2/\sqrt{3})^m} \left(\frac{\sigma}{s_o}\right)^m \frac{1}{(1-\alpha)^m} h_2(\alpha,m) \quad . \qquad (21)$$

To associate this with Hutchinson's asymptotic result for $\alpha \to 0$ we factor out his growth rate given by Eqn (19) from the expression of Eqn (21). This defines immediately the facet crack acceleration factor $g(\alpha,m)$ as

$$\dot{V} = \dot{V}_H \; g(\alpha,m), \therefore \; g(\alpha,m) = \frac{(1+\frac{3}{m})^{1/2}}{(2/\sqrt{3})^m} \frac{\pi}{6} \frac{h_2(\alpha,m)}{(1-\alpha)^m} \quad . \qquad (22a,b)$$

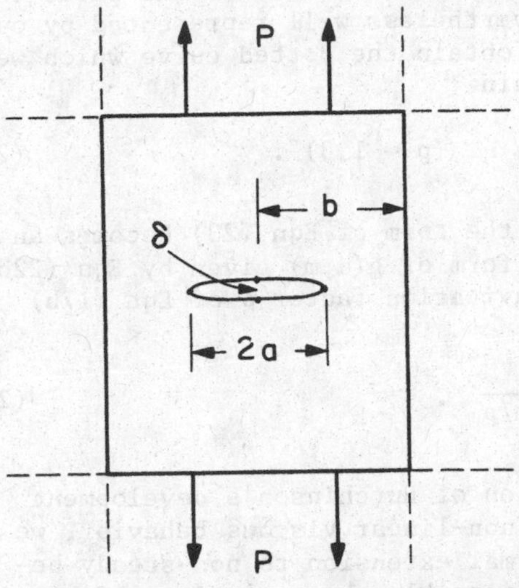

FIG. 5. Center cracked plate model of interaction of cracked grain boundary facets.

If the answer had been exact $g(\alpha,m) \to 1$ as $\alpha \to 0$. Since, however, considerable liberties were taken in the model, which ignores any induced or applied transverse stresses, and obtains the three dimensional solution by formal extension from an already approximate plane strain model we do not expect such precision from our model, and are prepared to make further slight ad hoc adjustments. To determine their nature we have evaluated $g(\alpha,m)$ of Eqn (22b) for several interesting values of m, from the tabulated $h_2(\alpha,m)$ values of Kumar et al. The result for m = 8 in Fig. 6 is typical, and shows that, as suspected, it does not

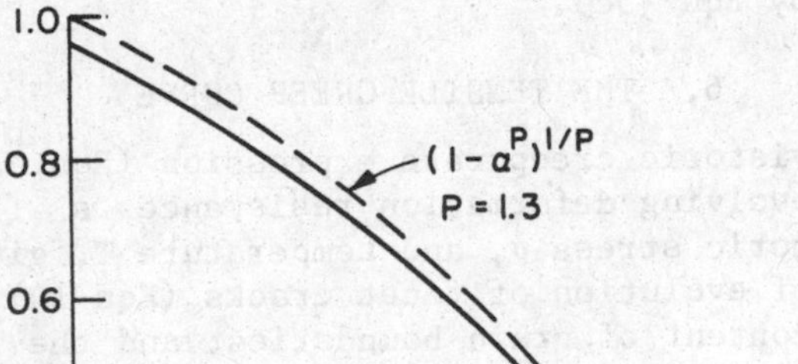

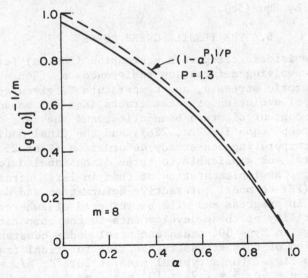

Fig. 6. Computed (solid) and parameterized (dotted) curves of the facet crack interaction factor $g(\alpha,m)$ for 304 stainless steel.

quite have the required asymptotic behavior at $\alpha = 0$. On the basis that the dependence of $g(\alpha,m)$ on α is nevertheless well represented by our model, we make a formal adjustment to obtain the dotted curve which we have empirically parameterized to obtain

$$g(\alpha,m) = \frac{1}{(1-\alpha^p)^{m/p}} \quad , \therefore \qquad (p = 1.3) \; . \tag{23}$$

We have found that for increasing m the form of Eqn (23) becomes an increasingly better fit to the actual form of $g(\alpha,m)$ given by Eqn (22b). With this modification we write the cavitation factor ρ of Eqn (17b) explicitly as

$$\rho = \frac{4}{\pi\sqrt{14}} \frac{1+m}{(1+\frac{3}{m})^{1/2}} \frac{\alpha}{(1-\alpha^p)^{m/p}} \; . \tag{24}$$

Although the above extension of Hutchinson's development should strictly be applicable only to non-linear viscous behavior, we note from Eqn (18b) that a further formal extension to non-steady behavior incorporating primary creep is possible, by replacing all steady state creep rates with non-steady creep rates given by the kinetic law of Eqn (3a), dependent upon the evolving current deformation resistance s while keeping all dependences of ρ on the m, characteristic of the steady state creep exponent, i.e. as given by Eqn (7b). This results in the following modified potential function Φ and uniaxial creep rate,

$$\Phi = \frac{\dot{\varepsilon}_o s}{(n+1)} \left(\frac{\sigma}{s}\right)^{n+1} \{1 + \rho(\alpha,m)\left(\frac{S}{\sigma}\right)^2\} \tag{25}$$

$$\dot{\varepsilon}_{11} = \dot{\varepsilon}_o \left(\frac{\sigma}{s}\right)^n \{1 + \rho(\alpha,m)\} \tag{26}$$

where $\dot{\varepsilon}_o$ is given by Eqn (5c).

6. THE TENSILE CREEP CURVE

The deviatoric creep rate expression (Eqn 3a) for non-steady flow based on the evolving deformation resistance s (Eqn 4), responding to the local deviatoric stress σ, and temperature T, given as a function of time; the rate of evolution of facet cracks (Eqn 14) based on the specific particle content of grain boundaries, and the rate of growth of facet cracks by creep (eqns 19, 22a, 22b); and the final modification of this growth incorporating non-steady deformation (Eqns 25, 18a) are all very general and applicable to three dimensional deformation and damage production. The implementation of them in large strain finite element programs (FE) to model interactive deformation and damage at crack tips is now in progress and will be reported elsewhere. Here we demonstrate the utility of these developments in the computation of the tensile creep curve in Type 304 stainless steel under constant stress, from the application of this stress to the time for final fracture. The result is shown in Fig. 7 for a typical case at 1023°K, $h/\mu = 0.0125$, $\sigma = 86$ MPa, $m = 8$, $n = 40$, $\bar{\beta} = 1.74 \times 10^{10} m^{-2}$, and $A_{crit} = 0.1$, and is compared with the experimental creep curves obtained by Swindeman (36). Since the experimental curves of Swindeman showed considerable scatter in the minimum creep rate strain among seemingly similar alloys, no doubt due to the variability in the facet cracking times of grain boundaries, we present the results in Fig. 7 on a normalized basis. The agreement is good.

7. DISTRIBUTION OF DAMAGE IN THE CRACK TIP FIELD

To understand the distribution of crack tip damage in the ductile Type 304 stainless steel alloys presented in Figs. 1 and 2 we examine the results of a large strain FE computation of the crack tip region in a DEN plane strain tension specimen of an elastic, power law creeping solid, with a stress exponent m = 8 that is typical for the alloy. The material parameters chosen for the computation were E = 133 GPa for the Young's modulus, $\nu = 0.33$ for the Poisson's ratio, and $\dot{\varepsilon}_o/s_o^{\ m} = 4.03 \times 10^{-23}\ MPa^{-8}s^{-1}$ for the coefficient of the dependence of the equivalent creep rate on the equivalent stress, that was obtained from the data of Blackburn (29). Figure 8a shows the outer (9-14) layer of elements in the FE problem of one quarter of the DEN specimen with the crack representing the left lower half of the box. The dotted figure of Fig. 8b shows the initial shape of the inner (1-3) layers of elements at the crack tip which was chosen to have an initial b/L ratio of 3.4×10^{-3}. The distorted figure on the right of Fig. 8b represents the same three inner elements for a later state of deformation of $b/L = 1.34 \times 10^{-2}$. The computation was continued to a final state of crack

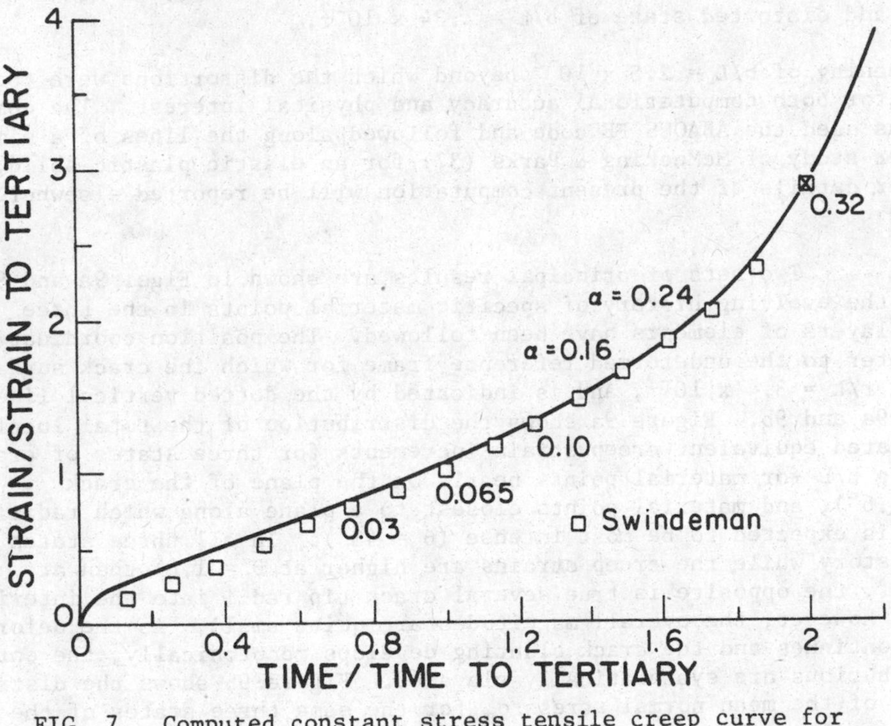

FIG. 7. Computed constant stress tensile creep curve for 304 stainless steel compared with data of Swindeman (36).

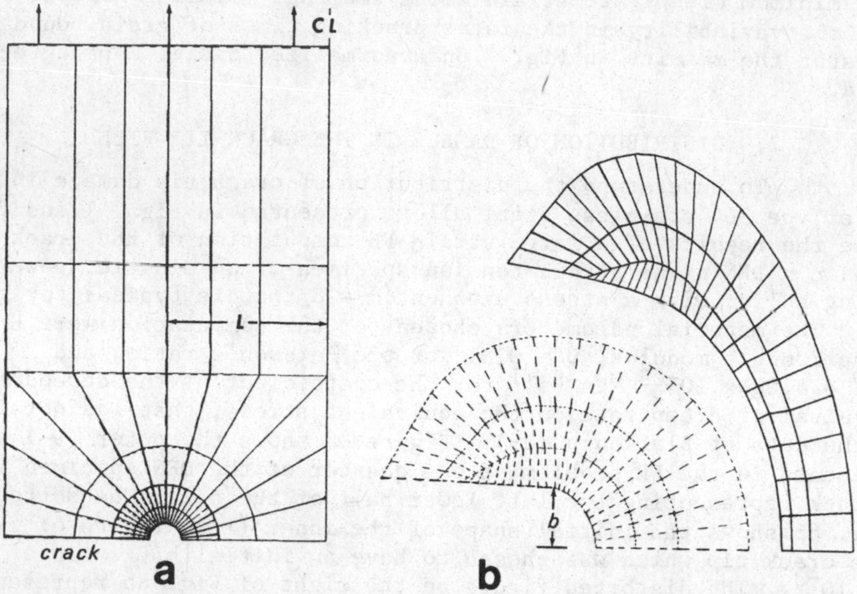

FIG. 8. Finite element net of DEN tension specimen showing outer layers (9-14) of elements, b) inner three layers (1-3) of finite elements in the initial state of $b/L = 3.4 \times 10^{-3}$ and distorted state of $b/L = 1.34 \times 10^{-2}$.

tip opening of $b/L = 2.5 \times 10^{-2}$ beyond which the distortions were too large for both computational accuracy and physical interest. The computations used the ABAQUS FE code and followed along the lines of a similar earlier study of McMeeking & Parks (37) for an elastic plastic solid. Further details of the present computation will be reported elsewhere later.

Two sets of principal results are shown in Figs. 9a and 9b where the evolving history of specific material points in the three inner layers of elements have been followed. The position coordinates r/L refer to the undeformed reference frame for which the crack surface was at $r/L = 3.4 \times 10^{-3}$, and is indicated by the dotted vertical line in Figs. 9a and 9b. Figure 9a shows the distribution of the total local integrated equivalent creep strain increments for three states of crack opening b/L for material points nearly on the plane of the crack ($\theta = 1.6°$), and material points closest to a plane along which radial shear is expected to be most intense ($\theta = 43°$). In all three states of the history while the creep strains are higher at $\theta = 1.6°$ than at $\theta = 43°$, the opposite is true several crack tip radii into the interior, where, however, the overall magnitudes are quite small. As the deformation continues and the crack blunting develops monotonically, the entire distributions are systematically elevated. Figure 9b shows the distribution of the mean normal stress σ_m for the same three states of the history. They show a clear and expected trend of monotonic decrease of crack tip mean normal stress with increasing blunting and an accompanying

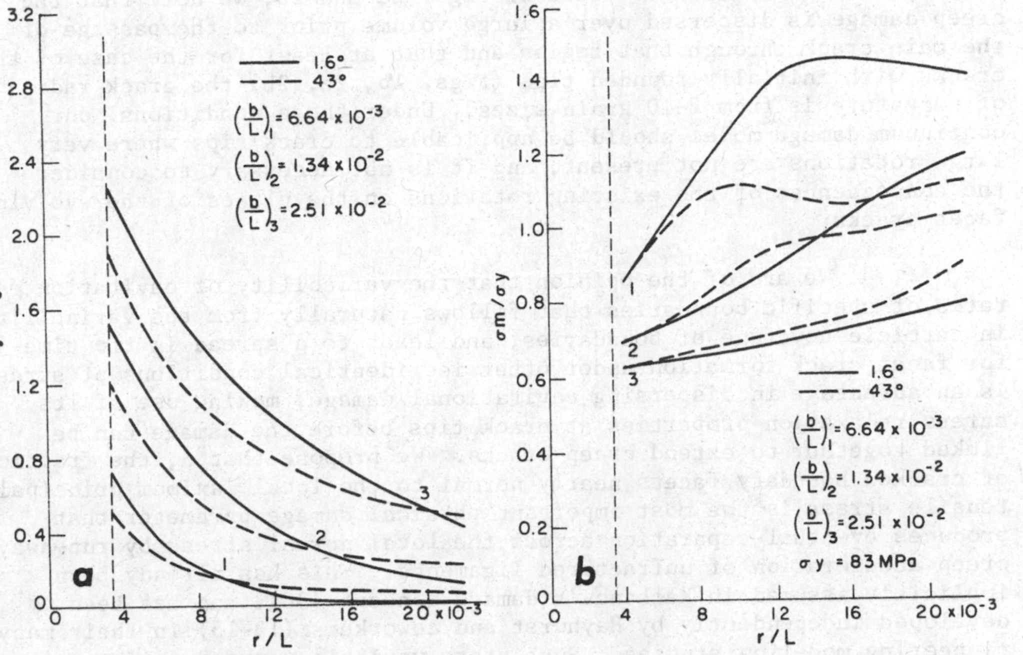

FIG. 9. a) Computed distribution of total creep strain around the creep of the specimen configuration introduced in Fig. 8 for three states of crack opening, b) computed distribution of mean normal stress for the same three states of crack opening.

systematic displacement outward of the peak in the distribution away from the crack surface. A further interesting trend is the systematic shift away from the plane of highest mean normal stress from $\theta = 1.6°$ to $\theta = 43°$. The trend of the local maximum principal tensile stress is qualitatively identical to that of the mean normal stress but the magnitudes are higher, and decrease less rapidly with blunting than the mean normal stress: for the three states of history they are roughly 1.4, 1.6 and 1.8 times the level of σ_m at $r/L = 1.2 \times 10^{-3}$. The distribution of the equivalent stress, as expected, remains relatively flat and orientation independent over this region of the crack tip; varying from approximately σ/σ_y of 1.15 at the crack surface to 0.95 at $r/L = 2.2 \times 10^{-2}$, and being affected little by crack blunting.

8. DISCUSSION

In our study we have used entirely mechanism based models for the development of grain boundary cavities, their growth to form grain boundary facet cracks, the growth of these in the prevailing stress field to interaction, and final run-away creep acceleration leading to local fracture. These models have been presented in three dimensional form based on two important state variables: of isotropic deformation resistance s, and the fraction α of separated grain boundary facets normal to the local maximum principal tensile stress. On the basis of the micro-structural evidence of Figs. 2a and 2b, we note that the creep damage is dispersed over a large volume prior to the passage of the main crack through that region and that at least for the case of the cracks with initially rounded tips (Figs. 1b, 2a, 2b) the crack radius of curvature is from 2-10 grain sizes. Under these conditions, our continuum damage model should be applicable to crack tips where very large rotations are not present, and it is not necessary to consider the consequences of the existing rotations on the planes of the evolving facet cracks.

We are of the opinion that the variability of cavitation rates of specific boundaries that follows naturally from the variability in particle coverage of boundaries, and leads to a spread in the time for facet crack formation under otherwise identical conditions of stress, is an advantage in dispersing cavitational damage, making use of its stress relaxation properties at crack tips before the damage can be linked together to extend creep cracks. We propose that α, the fraction of cracked boundary facets nearly normal to the local maximum principal tensile stress is the most important physical damage parameter that produces eventual separation across the local normal stress by run-away creep acceleration of unfractured ligaments. This has already been implicitly assumed in Kachanov's damage concept (38), and has been developed independently by Hayhurst and coworkers (13-15) in their many pioneering modeling studies. Such state variable representations of deformation resistance and damage evolution can now be used in fully interactive large strain FE models. Hayhurst and coworkers (14,15) have already used phenomenologically developed equivalent damage models in small strain FE programs and have demonstrated their predictive power.

The two different modes of crack tip processes shown: for sharp (initially fatigued) cracks in Fig. 1a; and initially rounded cracks in Fig. 1b, can now be explained on the basis of the sharp crack models investigated earlier by Bassani and McClintock (39) and the solution for rounded cracks with large blunting presented in Section 7 above. In the sharp crack models of Hutchinson (4) in strain hardening material and in the analogues models of Riedel and Rice (5) for creep, as Bassani and McClintock (39) clearly demonstrate, the levels of equivalent creep strain are nearly an order of magnitude higher along planes inclined about 50° with the crack plane than along the extension of the crack plane. Thus, in creeping alloys with very sharp cracks and little creep ductility the strain controlled damage processes such as intergranular cavity nucleation by constrained boundary siding and constrained cavity growth should develop more rapidly on these inclined planes. This should lead to crack extension along these planes, if the strain to fracture is small,to produce little blunting. Thus, even in Type 304 stainless steel, where the strains to fracture can be 0.2-0.4, sharp cracks will initially choose to propagate along the inclined planes at 45-50° as shown in Fig. 1a. This behavior is counteracted to some extent by the maximization along the extension of the main crack of both the mean normal stress and the maximum principal tensile stress that are instrumental in facet crack growth and cavity growth respectively. In fact examination of such initially sharp cracks has shown some incipient crack extension directly ahead of the initial crack as well, as Fig. 1a shows clearly. When the crack tip is initially slightly rounded, and when creep ductility is high, permitting additional blunting, as our results in Fig. 9a show, the intensity of the equivalent creep strain along the extension of the main crack is found to be higher than on any inclined plane at $\theta = 45\text{-}50°$. The mean normal stress and the maximum principal tensile stress are also higher along the $\theta \simeq 0°$ plane as Fig. 9b shows. Thus, initially rounded or blunted cracks will tend to grow in their plane. Here, however, further blunting reverses the order of relative intensities of the mean normal stress and maximum principal tensile stress between the $\theta \simeq 0$ and $\theta = 45\text{-}50°$ planes, away from the former, and toward the latter. This should counteract to some degree the favored nature of the $\theta = 0°$ plane and develops cracking conditions along the $\theta \simeq 45\text{-}50°$ planes. This tendency is shown also in Fig. 1b, and 2a as the two off-shoots of the main crack in the growth of this initially rounded crack.

ACKNOWLEDGMENT

This research has been supported by the U.S. Department of Energy under Contract No. DE-AC02-77ER04461.

REFERENCES

(1) Riedel, H. & Rice, J.R. (1980). In: Fracture Mechanics, ASTM STP 700, ed. P.C. Paris, p. 112. Philadelphia: ASTM.
(2) Riedel, H. (1979). Creep Deformation at Crack Tips in Elastic-Viscoplastic Solids. Div. Eng. Rept. MRL E-114. Providence, R.I.: Brown University.
(3) McClintock, F.A. & Bassani, J.L. (1981). In: Three Dimensional Constitutive Relations and Ductile Fracture, ed. S. Nemat-Nasser, p. 123. Amsterdam: North Holland.
(4) Hutchinson, J.W. (1968). J. Mech. Physics of Solids. 16, 13.
(5) Rice, J.R. & Rosengren, G.F. (1968). J. Mech. Physics of Solids. 16, 1.
(6) Hutchinson, J.W. (1983). Acta Met., 31, 1079.
(7) Riedel, H. (1984). In: Flow and Fracture at Elevated Temperature, ed. R. Raj. Metals Park, Ohio: ASM, in the press.
(8) Bensussan, P.L., Jablonski, D.A. & Pelloux, R.M. (1984). Met. Trans. 15A, 107.
(9) Saxena, A. (1980). In: Fracture Mechanics, ASTM STP 700, ed. P.C. Paris, p. 131. Philadelphia: ASTM.
(10) Argon, A.S. (1982). In: Recent Advances in Creep & Fracture of Engineering Materials and Structures, eds. B. Wilshire & D.R.J. Owen, p. 1. Swansea, U.K.: Pineridge Press.
(11) Wilkinson, D.S. & Vitek, V. (1982). Acta Met., 30, 1723.
(12) Bassani, J.L. & Vitek, V. (1982). In: Proc. 9th U.S. Natl. Congr. Appl. Mech., p. 127. New York: ASME.
(13) Leckie, F.A. & Hayhurst, D.R. (1980). Experiments in High Temperature Fracture. Report of Dept. Mech. & Ind. Eng. Urbana, Illinois: University of Illinois.

(14) Hayhurst, D.R., Morrison, C.J. & Brown, P.R. (1981). In: Creep in Structures, ed. A.R.S. Ponter, & D.R. Hayhurst, p. 564. Berlin: Springer.
(15) Hayhurst, D.R., Brown, P.R. & Morrison, C.J. (1983). The role of Continuum Damage in Creep Crack Growth. Report 83-4, Feb. 1983. Leicester, U.K.: University of Leicester (Dept. of Eng.)
(16) Rice, J.R. (1981). Acta Met., 29, 675.
(17) Robinson, D.N. (1978). A Unified Creep-Plasticity Model for Structural Metals at High Temperature. ORNL/TM-5969. Oak Ridge, Tenn.: Oak Ridge Natl. Lab.
(18) Kocks, U.F., Argon, A.S. & Ashby, M.F. (1975). In: Progress in Materials Science, eds. B. Chalmers, et al., vol. 19. Oxford: Pergamon Press.
(19) Bailey, R.W. (1926). J. Inst. Met., 35, 27.
(20) Orowan, E. (1946). J. West Scotl. Iron Steel Inst., 54, 45.
(21) Kocks, U.F. (1976). Laws for Work-Hardening and Low-Temperature Creep. J. Eng. Mater. & Tech., 98, 76.
(22) Anand, L. (1982). J. Eng. Mater. & Techn., 104, 12.
(23) Prinz, F. & Argon, A.S. (1984). Acta Met., in the press.
(24) Prinz, F., Argon, A.S. & Moffatt, W.C. (1982). Acta Met., 30, 821.
(25) Bird, J.E., Mukherjee, A.K. & Dorn, J.E. (1969). In: Quantitative Relations Between Properties and Microstructure, ed. D.G. Brandon & A. Rosen, p. 255. Jerusalem: Israel University Press.
(26) Bayce, A.E., Ludemann, W.D., Shepard, L.A. & Dorn, J.E. (1960). Trans. ASM, 52, 451.
(27) Brown, A.M. & Ashby, M.F. (1980). On the Power-Law Creep Equation. Scripta Met., 14, 1297.
(28) Argon, A.S., Bhattacharya, A. & Ozmat, B., to be published.
(29) Blackburn, L.D. (1972). In: The generation of Isochronous Stress-Strain Curves, ed. A.O. Schaefer, p. 15. New York: ASME.
(30) Argon, A.S., Chen, I-W. & Lau, C.W. (1980). In: Creep-Fatigue-Environment Interactions, ed. R.M. Pelloux & N.S. Stoloff, p. 46. New York: AIME.
(31) Argon, A.S. & Im, J. (1975). Met. Trans., 6A, 839.
(32) Chen, I-W. & Argon, A.S. (1981). Acta Met., 29, 1321.
(33) Needleman, A. & Rice, J.R. (1980). Acta Met., 28, 1315.
(34) He, M.Y. & Hutchinson, J.W. (1981). J. Appl. Mech., 48, 830.
(35) Kumar, V., German, M.D. & Shih, C.F. (1981). An Engineering Approach for Elastic-Plastic Fracture Analysis, EPRI Topical Report NP-1931. Palo Alto, Calif.: EPRI.
(36) Swindeman, R. (1983). Private communication.
(37) McMeeking, R.M. & Parks, D.M. (1979). In: Elastic Plastic Fracture, ASTM STP 668, eds. J.D. Landes, et al., p. 175. Philadelphia: ASTM.
(38) Kachanov, L.M. (1958). Izv. Akad. Nauk. SSSR, 8, 26.
(39) Bassani, J.L. & McClintock, F.A. (1981). Solids & Struct., 17, 479.

STRENGTH CONTRIBUTIONS DURING CREEP OF COMPOSITE MATERIALS

H. LILHOLT

Metallurgy Department

Risø National Laboratory, Roskilde, Denmark

ABSTRACT

The characteristics of creep behaviour of composite materials with aligned fibres are reviewed. A model for composites with rigid fibres is presented, and strengthening factors are worked out. Based on the analysis of internal stresses and strains by J.D. Eshelby several contributions to the creep strength are evaluated. A model for the behaviour of composites with creeping fibres is included, and the transition from rigid-fibre behaviour to creeping-fibre behaviour is demonstrated. The advantage of combined analyses of several sets of data is emphasised and demonstrated for a set of experimental data. These analyses show the possibility of deriving equations for the (in situ) properties of the fibres. It is also shown that a friction stress could possibly exist in certain composites with coarse microstructure; the nature of the friction stress can not be established. The analogous situations in tensile, plastic deformation and in creep deformation are illustrated for materials both in the unrelaxed state and in the relaxed state.

1. INTRODUCTION

The prediction of mechanical properties is of great importance for composite materials in practical engineering. The controllable variability of the properties of composites allows a design of the engineering material to a specific purpose, but it also makes it difficult to evaluate the many possible materials. Thus a predictive ability is significant in order to save time and avoid errors. This is of special interest for long-term behaviour, and we shall in this report consider the creep properties of fibrous composite materials.

A relatively large number of studies has been performed on many composites, and many creep data have been collected, often for purely practical reasons of testing potential materials for extreme applications. Various attempts have been made to rationalise the description of the macroscopic creep behaviour, mostly on the basis of purely mechanistic models. Recently experiments and analyses on eutectic composites have given a quite detailed description of the creep behaviour of these composites with a fine scale microstructure, demonstrating the existence of additional contributions to the creep strength of a composite material.

The initial elucidation of stress and strain distributions in two-phase materials was presented by J.D.Eshelby in his classical papers (1,2). On the basis of these fundamental ideas and analyses the general understanding of strength contributions in two-phase materials has increased substantially. Detailed descriptions for materials consisting of a (soft) matrix with (hard) inclusions have been given for tensile, plastic deformation by Brown and Stobbs (3,4), by Brown (5), and in previous, related papers by Brown and co-workers. Recently the creep behaviour has been considered by Brown (6) from a fundamental point of view. In a previous report (7) a first attempt was made to apply the ideas of Eshelby and subsequent workers to a series of experimental creep data for (model) composites to illustrate both mechanistic aspects and microstructural aspects of strength contributions during creep.

The present report extends the analyses to illustrate the relation between the matrix creep behaviour and the composite creep behaviour, and attempts an evaluation of strength contributions both for composites with rigid fibres and for composites with creeping fibres. The treatment tries to show that a combined analysis of many different data may give an improved understanding compared to individual analyses of various sets of data. The treatment is general and will be illustrated by metallic composites, but should also be applicable to polymeric composites.

2. COMPOSITE CREEP BEHAVIOUR

The macroscopic aspects of composite creep are based on the assumption that the matrix and the fibres have their usual properties, and that the composite behaviour results from a geometrical scaling or modification of the properties of matrix and fibres. This implies that the creep laws of the constituents are maintained in the composite. The microscopic aspects of creep are e.g. seen in dispersion hardened materials where the mechanism of deformation (e.g. climb of dislocations at particles) may be different from that of the pure matrix (e.g. diffusion), and thus the creep law may be different. Such effects of the scale of microstructure have been observed in fibrous composites when the microstructural dimensions (fibre sizes and spacings) are smaller than typically 10 μm (8), and have recently been studied in creep of eutectic composites by M.McLean and co-workers (9,10,11). We have here interactive effects between the constituents and at the interfaces, in addition to the macroscopic behaviour.

The previous report (7) considered composites with (short) rigid fibres, i.e. the matrix creep law governed the creep law of the composite. We shall here also include the possibility of creeping fibres, and thus the creep law of the fibres will exert its influence on the resulting composite creep behaviour. In (metallic) composites the fibres are typically refractory metals or compounds, and the temperatures encountered during creep are thus often low compared to the melting temperature of the fibres. In contrast, the matrix, metallic or polymeric, will often encounter temperatures near its melting point. It is thus of importance to include creep with the laws both at high and low temperatures. The creep of metals and other materials has traditionally been described by a power law at low and medium stresses (and often high temperatures, typically above 0.6 times the melting point), and by an exponential law at high stresses (and often low temperatures). A detailed description of these creep laws was given previously (7), and only the results will be summarised here.

CREEP BY POWER LAW. The equation governing power law creep is

$$\dot{\varepsilon} = A_o \frac{D\,G\,b}{k\,T} \left(\frac{\sigma}{G} \right)^n \tag{1}$$

where σ is the (tensile) stress and $\dot{\varepsilon}$ the (tensile) creep rate, G is the shear modulus, b Burgers vector, k Boltzmann's constant, T the temperature, A a constant and D the diffusion coefficient. In analysis of creep data it is convenient to simplify the form to

$$\dot{\varepsilon} = A_1\,\sigma^n$$

$$\log \dot{\varepsilon} = \log A_1 + n \log \sigma \tag{2}$$

$$\sigma = B\,\dot{\varepsilon}^{1/n} \tag{3}$$

CREEP BY EXPONENTIAL LAW. The equation for exponential law creep is

$$\dot{\varepsilon} = \dot{\varepsilon}_o \exp\left(- \frac{Q_o}{RT} \left(1 - \frac{\sigma}{\sigma_o} \right) \right) \tag{4}$$

where σ is the (tensile) stress and $\dot{\varepsilon}$ is the (tensile) creep rate, $\dot{\varepsilon}_o$ is a constant, Q_o is the activation energy for dislocation glide, and σ_o is the "strength" of the glide obstacles at 0 K. In analysis of experimental data it is often convenient to use it in the form (M=log e)

$$\log \dot{\varepsilon} = \log \dot{\varepsilon}_o - \frac{Q_o\,M}{RT} + \frac{Q_o\,M}{RT\sigma_o}\,\sigma$$

$$\log \dot{\varepsilon} = A + \alpha\,\sigma \tag{5}$$

$$\sigma = (1/\alpha) \log \dot{\varepsilon} - A/\alpha \tag{6}$$

3. COMPOSITES WITH RIGID FIBRES

We shall consider composites with short, aligned fibres, and treat the case of non-creeping fibres. The fibres thus act as inclusions and modify the creep rate and the stress level of the composite. The

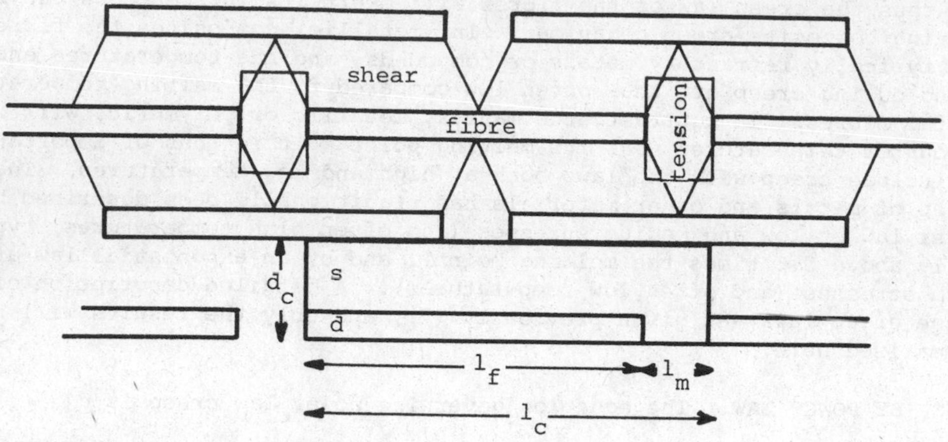

FIG. 1. Assemblage model: section of composite with shear regions and tensile regions, and a unit element with definition of parameters.

composite creep law therefore derives from the matrix creep law via suitable modifications. The matrix creep law is written

$$\dot{\varepsilon}_m = Z(\sigma_m) \tag{7}$$

We shall use the form where the functional relationship Z is retained and $\dot{\varepsilon}_m$ and σ_m are modified, see (7).

The model of the composite assumes that the matrix region between the fibres creep in shear and that the matrix region at the fibre ends creep in tension, Fig.1. The basic element of the model is similar to the element in the assemblage model used for composites (12). There is some small complication of overlaps of the shear regions and tensile regions; this is not considered to be serious, and the model offers the advantage that the shear regions have the length l_f and the tensile regions the length l_m. A further approximation is that we have ignored a possible tensile deformation in the shear regions, and possible multiaxial stress states in the tensile regions. Earlier models, notably the model of D.McLean (13), have considered only the shear regions between the fires. We shall present the results of D.McLean´s analysis as well as those of the present assemblage model to illustrate the effect of the details of the model on the resulting creep properties.

In all models it is implicitly assumed that the unit element is representative for the whole composite, i.e. that the element has dimensions much smaller than the dimensions of the actual test material or specimen used in experiments. Certain (model) composites used in laboratory experiments may not fulfil this requirement, e.g. specimens with long fibres may have rather few end regions, and thus may be described less unrealistically by D.McLean´s model than by the assemblage model.

3.1. MODIFICATION OF CREEP RATE

The resulting creep rate of the composite is

$$\dot{\varepsilon}_c = \dot{\varepsilon}_{c,s} (l_f/l_c) + \dot{\varepsilon}_{c,t} (l_m/l_c) \tag{8}$$

where $\dot{\varepsilon}_{c,s}$ is the tensile rate produced by shear, and $\dot{\varepsilon}_{c,t}$ is the simple tensile rate. The detailed derivation was given previously (7), and implies the assumption of constrained deformation in the tensile regions, and inclusion of the relevant geometry of the composite. From the definition

$$\dot{\varepsilon}_m = K_c \ \dot{\varepsilon}_c \tag{9}$$

we get

$$K_c = (4 d/l_f + \tfrac{1}{2})^{-1} (1 - V_f^{1/3})^{-1} \tag{10}$$

where K_c depends on the fibre aspect ratio l_f/d and on the volume fraction V_f. The effect on K_c of these parameters is less than in the model of D.McLean (13):

$$K_c \ (\mathrm{McL}) = 2/3 \ l_f/d \ V_f/V_m \tag{11}$$

In the limit of $V_f = 0$ we get $K_c = 2$ (the value 2 is caused by the constraint term). D.McLean´s value approaches zero; physically we expect $K_c = 1$.

3.2. MODIFICATION OF STRESS

Based on strain energies in the shear regions and the tensile regions, and on the definition

$$\sigma_c = X_c \ \sigma_m \tag{12}$$

we get

$$X_c = K_c V_m \tag{13}$$

The treatment by D.McLean (13) leads to

$$X_c = \tfrac{1}{2} \ l_f/d \ V_f \tag{14}$$

The effect on X_c of fibre aspect ratio and volume fraction is less in the assemblage model than in D.McLean´s model. In the limit of $V_f = 0$ we get $X_c = 2$, D.McLean´s model gives zero, and physically we expect $X_c = 1$. It is noteworthy that both models give a composite creep stress which is larger than the matrix creep stress by a factor and not by a additive term.

The general understanding of strength contributions in two-phase materials will be used here, but only in a simplified presentation:

$$\sigma_c = \sigma_m + \sigma_{fric} + \langle\sigma\rangle \tag{15}$$

The stress σ_{fric} is a frictional stress, and may be the (initial) Orowan stress σ_{OR} and possibly a source-shortening stress σ_{ss}, which refers to the decreasing free spacing between inclusions as deformation proceeds, $\sigma_{OR} + \sigma_{ss}$ is the resulting frictional stress. The stress $\langle\sigma\rangle$ is a mean stress (5) and is proportional to some function of the applied strain, and is thus an increasing function of the corresponding stress on the matrix. The assemblage model and the model by D. McLean are implicitly assumed to simulate macroscopic composites with relatively large volume fractions of coarse fibres. This gives

$$\sigma_{fric} \ll \langle\sigma\rangle \qquad \text{and} \qquad \sigma_m \ll \langle\sigma\rangle$$

$$\sigma_c \cong \langle\sigma\rangle \cong \text{const.}\ \sigma_m$$

This is the same form as equation (12) and the constant is established as X_c for this simple composite. This presentation does not attempt to evaluate the strength contributions in terms of the microstructure; this will probably be difficult for creep conditions. It has been done for short time loading of materials both in the unrelaxed and the relaxed condition, see e.g. (5). The presentation however seems to offer a possible way of analysing creep experiments in terms of strength contributions over and above the creep producing stress of the matrix itself.

3.3. COMPOSITE CREEP LAW

From the general creep law for the matrix Eqn (7) and the established modifications of creep rate and stress we get the general creep law for the composite:

$$\dot{\varepsilon}_c K_c = Z(\sigma_c - \sigma_{fric} - \langle\sigma\rangle) \tag{16}$$

$$\dot{\varepsilon}_c K_c = Z\left[(\sigma_c - \sigma_{fric})/(1 + X_c)\right] \tag{17}$$

We note that the factors K_c and X_c are large when they refer to a strong composite in the sense that the strength σ_c is high and/or the creep rate $\dot{\varepsilon}_c$ is low. The general relation between corresponding points of the matrix creep curve and the composite creep curve is shown in Fig.2.

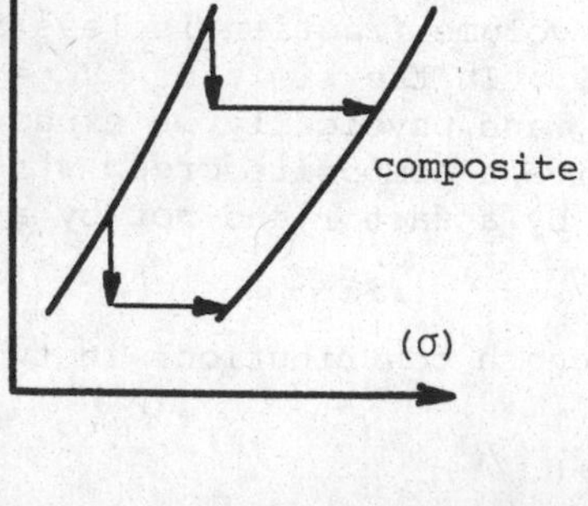

FIG. 2. Relation between matrix creep curve and composite creep curve.

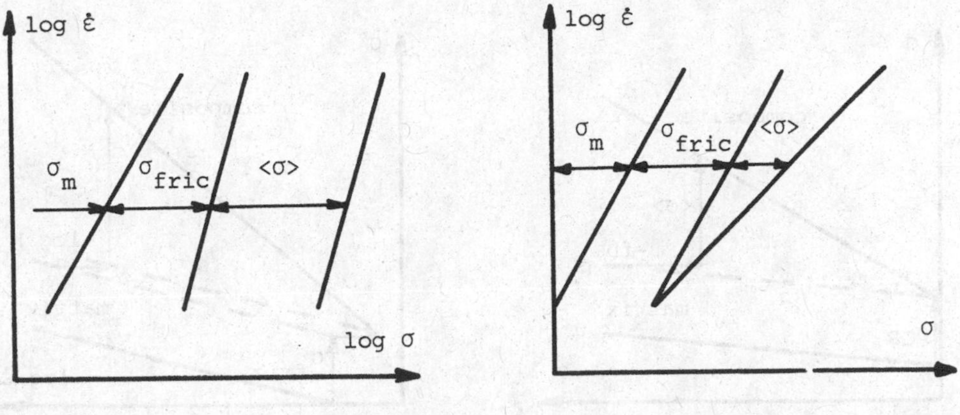

FIG. 3. Strength contributions in creep, (a) power law, (b) exponential law.

The creep law Z can have any form, and in practice it will often be possible to use the power law Eqn (1) or the exponential law Eqn (4). For these two creep laws the strength contributions are shown schematically in Fig. 3 a and b. The most commonly used double logarithmic diagram illustrates the effect of a (constant) friction stress in creep: the slope (stress exponent) increases relative to the slope of the (pure) matrix; this effect is quite often noted in plots of experimental data. It is, of course, possible to use any diagram to present experimental data, but the advantage of straight lines during analysis of the data indicates the convenience of the two diagrams for, respectively, the power law and the exponential law.

3.4. COMPARISON OF TENSILE AND CREEP BEHAVIOUR

It is interesting at this stage to compare the well established tensile behaviour with the creep behaviour. The relevant situation for tensile deformation is the unrelaxed state, i.e. no relaxation of dislocations in the matrix and/or no yield of the inclusions. This corresponds to the steady state creep of composites with non-creeping fibres, with tensile strains corresponding to creep strain rates (6). To facilitate the comparison we shall use the exponential law and present the diagram as σ vs $\log \dot{\varepsilon}$ rather than $\log \dot{\varepsilon}$ vs σ. From Eqn (9) and (15) we have

$$\log \dot{\varepsilon}_c = \log \dot{\varepsilon}_m - \log K_c \tag{18}$$

$$\sigma_c = \sigma_m + \sigma_{fric} + X_c \sigma_m \tag{19}$$

which means a shift along the x-axis by the constant amount $-\log K_c$, and along the y-axis by the constant amount $+\sigma_{fric}$ and the proportionally increasing amount $+X_c\sigma_m$. From the analysis of tensile curves (14) we have a shift along the x-axis by $-fD\varepsilon_p$, and along the y-axis by $+\langle\sigma\rangle = +2\gamma\mu fD\varepsilon_p$, both of which are proportionally increasing, and further along

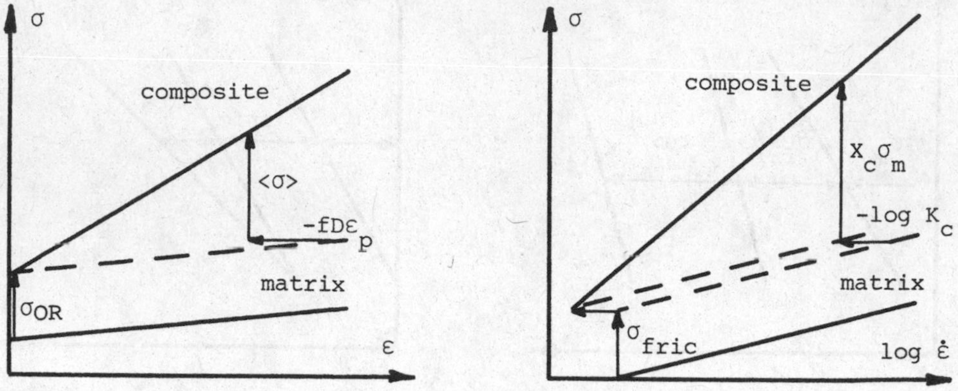

FIG. 4. Tensile and creep behaviour, unrelaxed state; (a) modification of strain and stress in tensile deformation; (b) modification of creep rate and stress in creep deformation.

the y-axis by the constant amount σ_{OR}. Fig. 4 a and b illustrate the two situations. We note the following features:

- σ_{fric} corresponds to σ_{OR};
- the high slope and the high stress level of the composite relative to the matrix is caused by two shifts along the two axes;
- in tensile deformation ε_p increases with "loading" and results in the proportionally increasing shift $-fD\varepsilon_p$; in creep deformation $\dot{\varepsilon}$ is constant with "time" and results in the constant shift $-\log K_c$.

4. COMPOSITES WITH CREEPING FIBRES

If the stress on the composite is increased it is likely that the stress in the fibres reach a level where the fibres will creep. Kelly and Street (15) have analysed this situation, and find that the fibres will creep in their middle region with a stress transfer region at the ends where the fibres remain rigid. The analysis further shows that with very long fibres the composite creep will approach the creep of a continuous composite, i.e. a composite with infinitely long fibres creeping along their full length. Both for short-creeping-fibre composites and for continuous composites it is assumed that the fibre creep rate is equal to the matrix creep rate and thus to the general composite creep rate.

We shall approximate the behaviour of real composites (of which none are truly continuous) with the simple description for continuous composites; we thus ignore the local creep in the matrix near the fibre ends, and the local creep situation of the stress transfer regions at the fibre ends. For continuous composites we take

$$\dot{\varepsilon}_c = \dot{\varepsilon}_m = \dot{\varepsilon}_f \tag{20}$$

and use the average stress (law of mixtures)

$$\sigma_c = V_f \sigma_f + V_m \sigma_m \tag{21}$$

From the creep laws for matrix and fibres

$$\dot{\varepsilon}_m = Z_m(\sigma_m) \quad \text{and} \quad \dot{\varepsilon}_f = Z_f(\sigma_f) \tag{22),(23}$$

the composite creep creep law can be worked out

$$\dot{\varepsilon}_c = Z_c(\sigma_c) \tag{24}$$

We shall consider two situations.

POWER LAWS. Both Z_m and Z_f are power laws according to Eqn (3), we get

$$\sigma_c = V_f B_f \dot{\varepsilon}_c^{1/n_f} + V_m B_m \dot{\varepsilon}_c^{1/n_m} \tag{25}$$

Although the individual power laws give straight lines in a log $\dot{\varepsilon}$ vs log σ diagram, this is not the case with the composite creep law. Furthermore, it is not very likely that the fibres creep in their power law region due to the relatively low temperature of the composite.

EXPONENTIAL LAWS. Both Z_m and Z_f are exponential laws according to Eqn(6):

$$\sigma_c = (V_f/\alpha_f + V_m/\alpha_m) \log \dot{\varepsilon}_c - (V_f A_f/\alpha_f + V_m A_m/\alpha_m) \tag{26}$$

In a log $\dot{\varepsilon}$ vs log σ diagram both matrix, fibre and composite give straight lines, this facilitates analysis of (limited) experimental data. The composite has a friction stress (at any given creep rate) in the sense that a larger stress (for a larger V_f) is needed to give the actual creep rate. But the friction stress is not caused by microstructural aspects, like the Orowan stress for dislocation bowing, because we have equal creep rates in matrix and fibres. It is proposed that the exponential law Eqn (26) and the log $\dot{\varepsilon}$ vs σ diagram are useful in analyses of data, and in particular in detection of possible creep in the fibres. It is often not easy to judge whether fibres do actually creep in composites, even with knowledge about the creep behaviour of the fibre-material itself, because often the structure of the fibres is changed during fabrication or during creep conditions (e.g. recrystallization; alloying of the fibres by the matrix-material).

5. TRANSITION BETWEEN RIGID AND CREEPING FIBRES

The transition between composites with rigid fibres and composites with creeping fibres is illustrated in a log $\dot{\varepsilon}$ vs σ diagram in Fig. 5. The transition is based on the idea that the case with the fastest creep rate will determine the type of composite creep. In Fig. 5 the composite shows rigid-fibre creep at low stresses and creeping-fibre creep at higher stresses. The creep of rigid-fibre composites is governed

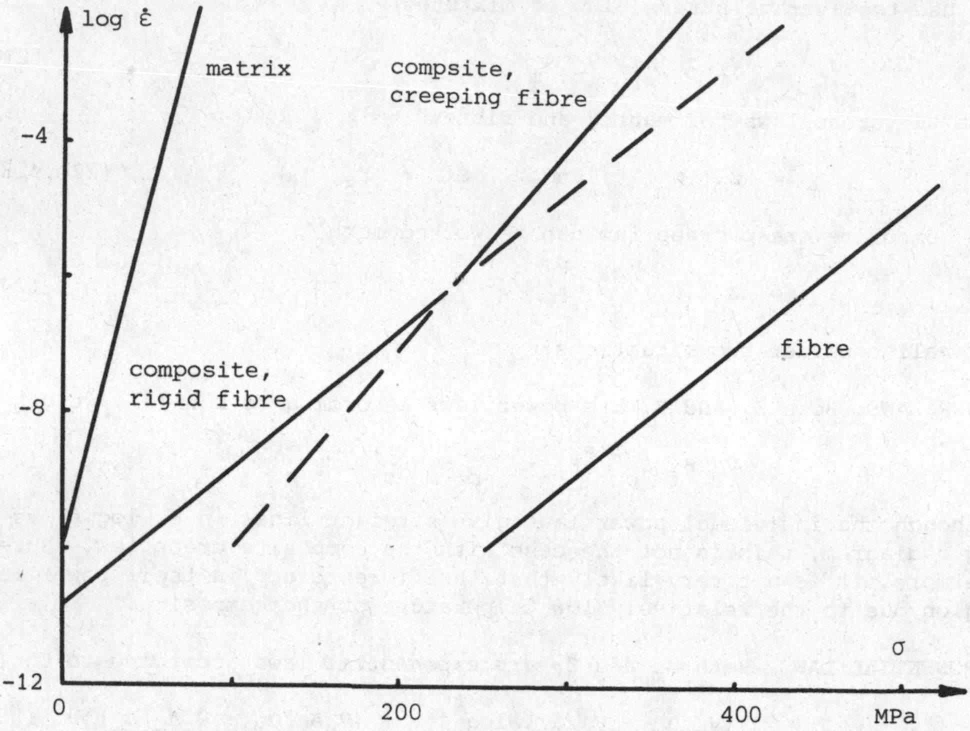

FIG. 5. Transition between rigid-fibre behaviour and creeping-fibre behaviour; sketch for volume fraction V_f=0.40 and fibre aspect ratio l_f/d=60.

by volume fraction and fibre aspect ratio but not by the fibre properties, while the creep of creeping-fibre composites is controlled by the (in situ) creep properties of the fibres, and their volume fraction. It is therefore possible for the position of the two types of composite creep curves to shift independently in the diagram, and thus change the extent of the two types of creep. It may also be possible for certain combinations of properties and parameters to fully suppress one type of creep.

6. ANALYSIS OF EXPERIMENTS

In the following we shall analyse a series of experiments on the basis of the equations and diagrams presented, and in particular evaluate the strength contributions in creep as well as the possibility of creeping fibres. In the previous report (7) some experiments were analysed primarily on the basis of rigid-fibre behaviour; it was clear that some of the discrepancies between experiments and models could originate from creeping fibres; it was also clear that some data indicated creep according to the exponential law rather than the power law. For these reasons we shall analyse the experiments on the basis of the exponential law and use the log $\dot{\varepsilon}$ vs σ diagram.

The following equations will be used:

Matrix: $$\log \dot{\varepsilon}_m = A_m + \alpha_m \sigma_m \tag{27}$$

Rigid-fibre composite:

$$\dot{\varepsilon}_c K_c = \dot{\varepsilon}_m \tag{28}$$

$$\sigma_c = (1 + X_c)\, \sigma_m + \sigma_{fric} \tag{29}$$

$$\log \dot{\varepsilon}_c = A_c + \alpha_c \sigma_c \tag{30}$$

where $$\alpha_c = \alpha_m/(1 + X_c) \tag{31}$$

$$A_c = A_m - \log K_c - \alpha_c \sigma_{fric} \tag{32}$$

Creeping-fibre composite:

Fibre: $$\log \dot{\varepsilon}_f = A_f + \alpha_f \sigma_f \tag{33}$$

$$\log \dot{\varepsilon}_c = A_c + \alpha_c \sigma_c \tag{34}$$

$$\sigma_c = (1/\alpha_c) \log \dot{\varepsilon}_c - A_c/\alpha_c \tag{35}$$

where $$\alpha_c = (V_f/\alpha_f + V_m/\alpha_m)^{-1} \tag{36}$$

$$A_c/\alpha_c = V_f A_f/\alpha_f + V_m A_m/\alpha_m \tag{37}$$

For rigid-fibre composite behaviour we can estimate X_c from Eqn (31), and thus K_c from Eqn (13), finally Eqn (32) can yield an estimate of a possible friction stress. The experimental values of X_c and K_c can be compared to the theoretical values of Eqn (10), (11), (13) and (14). For creeping-fibre composites we can estimate α_f and A_f and thus the governing equation for the (in situ) creep of the fibres; this behaviour can then be judged in relation to known creep properties of the fibre-material.

6.1. CREEP OF Ag-W

The creep data for the model composite of W-fibres in a Ag-matrix (16) are plotted in Fig. 6. The experimental equations are:

Ag-matrix: $$\log \dot{\varepsilon}_m = -10.0 + 0.1400\, \sigma_m \tag{38}$$

Rigid-fibre composite, $l_f/d = 30$:

$$\log \dot{\varepsilon}_c = -11.8 + 0.0255\, \sigma_c \tag{39}$$

From these we find

Exp.:	$X_c = 4.5$	$\log K_c = 0.9$	$\sigma_{fric} \cong 35$ MPa	
Calc.:	$X_c = 3.6$	$\log K_c = 0.8$		(Assembl.model)
	$X_c = 6.0$	$\log K_c = 1.1$		(McL. model) .

The agreement is acceptable, although it is not easy to judge the accuracy of the two models in relation to the structure of the real composite. It

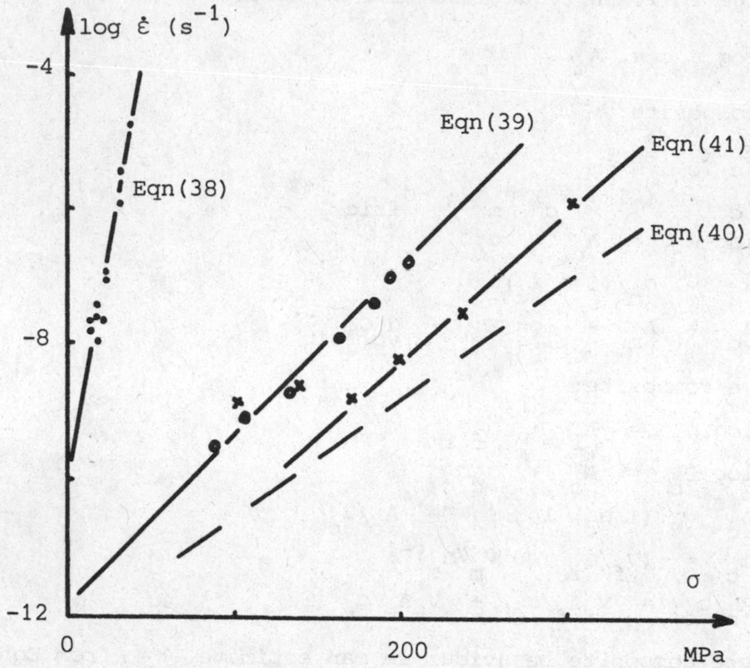

FIG. 6. Creep data at 400°C for Ag (·) and Ag-W (16); 40 vol % fibres, aspect ratios 30 (o) and 60 (x).

is not immediately clear that a relatively coarse macroscopic composite should have a friction stress.

An estimate of the expected parameters for the composite with $l_f/d = 60$ has been made on the basis of the relative agreement between models and experiments for the $l_f/d = 30$ composite, and on the basis of the trend for the friction stress from other analyses:

$$X_c \cong 7.0 \qquad \log K_c \cong 1.1 \qquad \sigma_{fric} \cong 70 \text{ MPa}$$

$$\log \dot{\varepsilon}_c = -12.3 + 0.0175\,\sigma_c \tag{40}$$

This is plotted in Fig. 6, and it seems reasonably clear that all experimental data for $l_f/d = 60$ are above this rigid-fibre composite curve. We then analyse these data on the basis of creeping-fibre behaviour:

Creeping-fibre composite, $l_f/d = 60$:

$$\log \dot{\varepsilon}_c = -12.6 + 0.0220\,\sigma_c \tag{41}$$

W-fibres: $$\log \dot{\varepsilon}_f = -12.9 + 0.0097\,\sigma_f \tag{42}$$

This equation for W-fibres can be evaluated together with results from other composites with W-fibres.

7. RELAXED BEHAVIOUR

During plastic tensile deformation the (initially) established internal stresses can be partly relaxed by various mechanisms, e.g. rearrangements of dislocations near inclusions, or yield of the inclusions. This relaxation is dependent on the strain and the temperature, and an example for Cu dispersion hardened with SiO_2-particles is shown in Fig. 7 a. During conditions of creep of composites there can be a transition from rigid fibres to creeping fibres; this change constitutes a relaxation and has been demonstrated in the analyses of the experiments. An example, which shows the similarity to the tensile deformation, is presented in the master diagram plot of all the data for rigid-fibre and creeping-fibre composites of Cu-W, Fig. 7 b. The diagram is the same as that of the previous report (7), but with the axes interchanged to facilitate the comparison with the tensile deformation in Fig. 7 a.

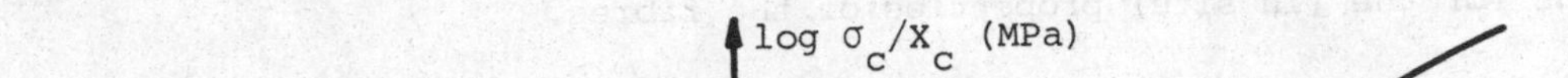

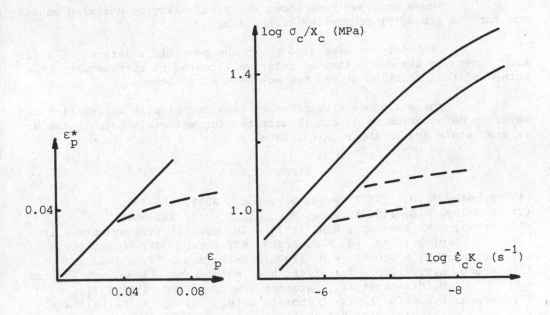

FIG. 7. Tensile and creep behaviour, relaxed state; (a) Cu with SiO_2-particles, tensile stress is proportional to ε_p^*, strain is equal to ε_p, from (3); (b) Cu with W-fibres, creep data are plotted in a master diagram, from (7). Full lines represent unrelaxed behaviour, dotted lines represent relaxed behaviour.

8. CONCLUSIONS

The characteristics of creep behaviour of composite materials with aligned fibres have been reviewed.

A model for the behaviour of composites with rigid fibres has been presented, and strengthening factors have been established. On the basis of Eshelby´s initial analysis of internal stresses and strains several contributions to the creep strength have been evaluated.

A model for the behaviour of composites with creeping fibres has been included, and the transition from rigid-fibre behaviour to creeping-fibre behaviour has been demonstrated.

The advantage of combined analysis of several sets of data has been emphasised, and illustrated for one set of experimental data by analyses according to the models for the different types of creep behaviour.

These analyses have shown the possibility of deriving equations for the (in situ) properties of the fibres.

The analyses have also shown the possible existence of a friction stress in the composites of relatively coarse microstructure; the nature of this friction stress has not been established.

The analogous situations in tensile, plastic deformation and in creep deformation have been illustrated for materials both in the unrelaxed state and in the relaxed state.

REFERENCES

(1) Eshelby, J.D. (1957). Proc. Roy. Soc., A241, 376-396.
(2) Eshelby, J.D. (1959). Proc. Roy. Soc., A252, 561-569.
(3) Brown, L.M. & Stobbs, W.M. (1975). In: Constitutive equations in plasticity, ed. A.S. Argon, MIT Press, USA, Chap. 10.
(4) Brown, L.M. & Stobbs, W.M. (1976). Phil. Mag., 34, 351-372.
(5) Brown, L.M. (1980). In: Strength of metals and alloys, vol. 3, ed. P. Haasen et al., Pergamon Press, Oxford, 1551-1571.
(6) Brown, L.M. (1982). In: Composite materials, ed. H. Lilholt & R. Talreja, Risø National Laboratory, Roskilde, 1-18.
(7) Lilholt, H. (1982). In: Composite materials, ed. H. Lilholt & R. Talreja, Risø National Laboratory, Roskilde, 63-76.
(8) Kelly, A. & Lilholt, H. (1969). Phil. Mag., 20, 311-328.
(9) Bullock, E., McLean, M. & Miles, D.E. (1977). Acta Met., 25, 333-344.
(10) McLean, M. (1980). Proc. Roy. Soc., A371, 279-294.
(11) McLean, M. (1980). Proc. Roy. Soc., A373, 93-109.
(12) Hashin, Z. & Rosen, B.W. (1964). J. Appl. Mech., 31, 223-232.
(13) McLean, D. (1972). J. Mat. Sci., 7, 98-104.
(14) Lilholt, H. (1977). Acta Met., 25, 587-593.
(15) Kelly, A. & Street, K.N. (1972). Proc. Roy. Soc., A328, 283-293.
(16) Kelly, A. & Tyson, W.R. (1966). J. Mech. Phys. Solids, 14, 177-186.

SUPPRESSION OF GRAIN BOUNDARY SLIDING BY SECOND PHASE PARTICLES

T. MORI

Tokyo Institute of Technology, Nagatsuta, Midori-ku, Yokohama 227, Japan

ABSTRACT

Second phase particles or inclusions on a grain boundary suppress grain boundary sliding. This particle blocking effect is formulated by applying the paper by Eshelby on ellipsoidal inclusions. A grain boundary is modelled as a flat ellipsoidal inclusion, the inelastic deformation of which is blocked by the disc-shaped inclusions that represent the intersection of the second phase particles with the boundary. The elastic state brought about by sliding on the boundary with the inclusions is analyzed in terms of the size and distribution of the inclusions and is detected by an internal friction measurement. The theoretical prediction has been confirmed by experiments on Cu alloys with Fe or SiO_2 particles on the boundaries. Two types of samples have been tested; normal polycrystalline samples with constraint from boundary edges and bamboo structured samples free from boundary edge constraint. The agreement between the theory and the experiment is satisfactory.

1. INTRODUCTION

The papers on ellipsoidal inclusions by J. D. Eshelby have been applied to many fields of mechanics and metallurgy (1-3). Since the early 1970's, these papers have been successfully used in the understanding of work hardening and related problems in dispersion strengthened materials (4,5). Interestingly, this subject is one which was not suggested in Eshelby's original papers as a possible problem for direct application of his theory. In work hardening of a dispersion strengthened alloy, inclusions act as blocking particles against plastic deformation of the matrix by producing internal stresses to oppose further plastic deformation. In the present paper, a subject similar to this will be dealt with in two-dimensional plastic deformation, that is, grain boundary sliding.

At a sufficiently high temperature, a grain boundary slides when a shear stress is applied along it. Apart from high temperature deformation such as creep where plastic deformation inside grains and diffusional accommodation occur besides boundary sliding, a simple way to detect grain boundary sliding is to observe the anelastic deformation measured by internal friction and anelastic creep, as demonstrated many years ago (6,7). It is intuitively conceivable that grain boundary sliding is obstructed by second phase particles or inclusions on boundaries and thus that the anelastic deformation associated with grain boundary sliding is modified in an inclusion bearing material. Although this point was noted previously (8), no quantitative or theoretical study has been performed on this subject except for a paper by Raj and Ashby (9). However, what concerns us in the present work on this subject differs from the situation Raj and Ashby considered. They discussed the case where the intrinsic boundary viscosity is vanishingly small and inclusions in the boundary offer resistance but at the same time diffusion around the inclusions relaxes the inclusion-blocking effect. Raj and Ashby formulated this problem by an electrostatic analog. Physically, the blocking effect of the inclusions on boundary sliding comes from the internal stress produced by the inclusions which stand against the penetration of the sliding. However, this internal stress was not specifically formulated by Raj and Ashby. This situation is felt to be unsatisfactory, considering that a fairly clear picture of work hardening in a dispersion strengthened alloy now exists because of the formulation of the internal stress state. As will be shown in the present work, the above unsatisfactory situation is solved by again applying the Eshelby paper on ellipsoidal inclusions, this time to the two-dimensional plastic deformation of grain boundary sliding (10).

2. INTERNAL STRESS CAUSED BY BOUNDARY SLIDING

2.1 CONSTRAINED BOUNDARY

Here we consider a grain boundary which has inclusions on it and is subjected to an external shear stress $\sigma_{31}^A=\sigma^A$, as depicted in Fig. 1. The grain boundary, Ω_0, is assumed to have a flat ellipsoidal shape described by

$$\Omega_0;\ (x_1^2 + x_2^2)/a^2 + x_3^2/c^2 \leq 1,\ c/a \ll 1 \tag{1}$$

where $2a$ is the diameter of the boundary and $2c$ the thickness of the boundary layer where boundary sliding occurs. The boundary sliding is not only restricted by the edges of the boundary (triple lines) but is blocked by the inclusions on the grain boundary. These effects are quantitatively treated by considering the following combination of two stage deformation. In stage 1, a uniform plastic distortion of $\beta_{31}^*=\beta^*$ is considered to occur throughout Ω_0. In stage 2, to account for the inclusion blocking effect on the boundary sliding, a plastic distortion of $\beta_{31}^*=-\beta^*$ is given in domains, each of which is described by Ω as

$$\Omega;\ (x_1^2 + x_2^2)/r^2 + x_3^2/c^2 \leq 1,\ c/r \ll 1. \tag{2}$$

Here, r represents the radius of the inclusions on the boundary. See Fig. 2. It is clear that the combination of stages 1 and 2 is identical to the supposition that a uniform plastic distortion of β^* occurs only in the inclusion-free domain of the boundary, $\Omega_0 - \Omega$.

By applying Eshelby's work, the internal stress, σ_{31}^{o}, present in Ω_0 in the stage 1 deformation is calculated. It is uniform in Ω_0 and is given by

$$\sigma_{31}^{o} = -A(c/a)\mu\beta^* \tag{3}$$

with

$$A = \pi(2 - \nu)/[4(1 - \nu)], \tag{4}$$

where μ is the shear modulus and ν the Poisson ratio. The elastic strain energy, E_0, and the external potential energy, V_0, induced by the stage 1 deformation are calculated, respectively, as

$$E_0 = (A/2)(c/a)\mu(\beta^*)^2(4\pi a^2 c/3) \tag{5}$$

and

$$V_0 = -\sigma^A\beta^*(4\pi a^2 c/3). \tag{6}$$

The stage 2 deformation introduces an additional internal stress, σ_{31}^{P}. In Ω it is uniform and is given by

$$\sigma_{31}^{P} = A(c/r)\mu\beta^*. \tag{7}$$

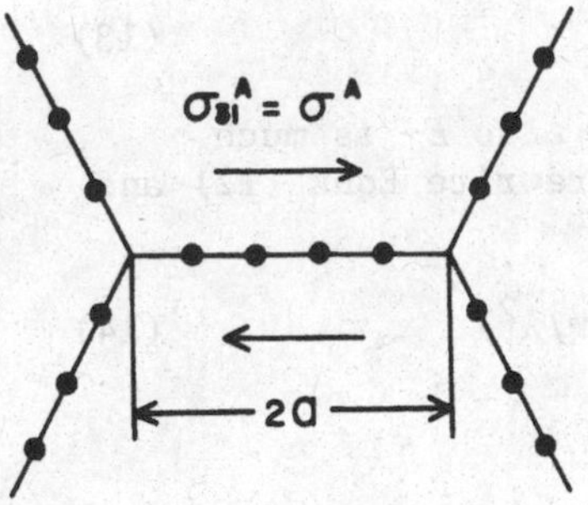

FIG. 1. Grain boundaries with slide-blocking inclusions.

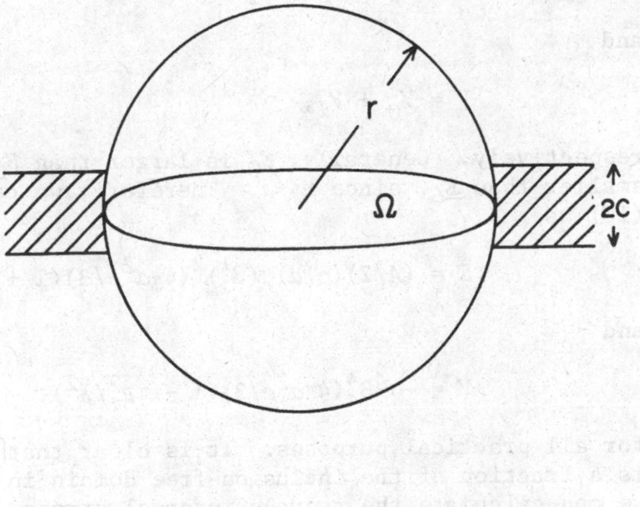

FIG. 2. Modelling of a slide-blocking inclusion on a boundary.

The associated (self) elastic strain energy, E_Ω, is

$$E_\Omega = (A/2)(c/r)\mu(\beta^*)^2 4\pi r^2 c/3 \tag{8}$$

per inclusion. When the inter-inclusion distance is λ, the elastic strain energy, E_S, introduced by the stage 2 deformation is calculated from Eqn (8) as

$$E_S = (A/2)(c/r)\mu(\beta^*)^2(4\pi r^2 c/3)(\pi a^2/\lambda^2) \tag{9}$$

per grain boundary. Here the interaction between the inclusions is ignored. Besides E_S, the stage 2 deformation introduces two energy terms, E_I and V_I. E_I is the elastic strain interaction energy between σ^o_{31} given by Eqn (3) and $-\beta^*$ in Ω and is calculated as

$$E_I = -A(c/a)(\beta^*)^2(4\pi r^2 c/3)(\pi a^2/\lambda^2) \tag{10}$$

per grain boundary. V_I is the interaction between the applied stress σ^A and $-\beta^*$ in Ω or equivalently the change in the external potential energy introduced by the stage 2 deformation. It is given by

$$V_I = \sigma^A\beta^*(4\pi r^2 c/3)(\pi a^2/\lambda^2) \tag{11}$$

per grain boundary.

In total, when only the inclusion-free domain of the boundary undergoes the plastic distortion of β^* under the external shear stress of σ^A, the elastic strain energy, E, and the potential energy of the loading system, V, per grain boundary are given by

$$E = E_0 + E_S + E_I \tag{12}$$

and

$$V = V_0 + V_I, \tag{13}$$

respectively. Generally, E_0 is larger than E_S. Also E_I is much smaller than E_S, since $r \ll a$. Therefore, we can rewrite Eqns (12) and (13) as

$$E = (A/2)(c/a)\mu(\beta^*)^2(4\pi a^2 c/3)(1 + \pi ar/\lambda^2) \tag{14}$$

and

$$V = -\sigma^A\beta^*(4\pi a^2 c/3)(1 - \pi r^2/\lambda^2) \tag{15}$$

for all practical purposes. It is clear that the term, $(1 - \pi r^2/\lambda^2)$, is a fraction of the inclusion-free domain in the boundary. From this, we can calculate the average internal stress, σ_{31}, which exists in the inclusion-free domain and is developed by the plastic distortion occurring only in this domain. The work done against the internal stress σ_{31} by the increment of $\delta\beta^*$ in the inclusion-free domain is $-\sigma_{31}\delta\beta^*(4\pi a^2c/3)(1 - \pi r^2/\lambda^2)$ from Eqn (15). This must be equal to the increment of the elastic strain energy, δE. δE is calculated from Eqn (14). Therefore, we obtain

$$\sigma_{31} = -A(c/a)\mu\beta^*(1 + \pi ar/\lambda^2)/(1 - \pi r^2/\lambda^2)$$
$$\fallingdotseq -A(c/a)\mu\beta^* - A\mu\beta^*(\pi rc/\lambda^2). \qquad (16)$$

The first term comes from the existence of the boundary edges (triple lines) and the second term from the inclusions blocking the boundary sliding.

2.2 UNCONSTRAINED BOUNDARY (BAMBOO STRUCTURE)

By putting $a \to \infty$ in Eqn (16) and thereby eliminating the boundary edges, we obtain

$$\sigma_{31} = -A\mu\beta^*(\pi rc/\lambda^2). \qquad (17)$$

This is the average internal stress in the inclusion-free domain when an unconstrained boundary slides β^* within this domain. The result may also be obtained by the following reasoning:

Consider an unconstrained flat boundary Ω_∞ containing a number of regions which, for simplicity, have an identical shape. Each region is called a 2-D inclusion, Ω_{2D}. The boundary is perpendicular to the X_3 axis. Suppose an internal stress, σ_{ij}, is developed when the boundary slides uniformly by $\beta^*_{31}=\beta^*$ only on $M=\Omega_\infty - \Omega_{2D}$. The following equation is required by the force balance.

$$(1 - f)\langle\sigma_{3j}\rangle_M + f\langle\sigma_{3j}\rangle_I = 0, \qquad (18)$$

where f is the plane fraction of the 2-D inclusion, $\langle\ \rangle_I$ the average defined over all the 2-D inclusions and $\langle\ \rangle_M$ the average defined in the domain unoccupied by the 2-D inclusions. Assuming a random distribution of the 2-D inclusions, we have

$$\langle\sigma_{3j}\rangle_I = \sigma_{3j}^\infty + \langle\sigma_{3j}\rangle_M, \qquad (19)$$

where σ_{3j}^∞ is the internal stress within a 2-D inclusion when only it exists on an infinitely extended boundary. From Eqns (18) and (19), we obtain

$$\langle\sigma_{3j}\rangle_M = -f\sigma_{3j}^\infty. \qquad (20)$$

Eqn (17) is just an example of this general expression. σ_{31}^∞ is the stress given by Eqn (7) and $f=\pi r^2/\lambda^2$. Therefore

$$\langle\sigma_{31}\rangle_M = -A\mu\beta^*(\pi rc/\lambda^2), \qquad (21)$$

which is identical to Eqn (17). There is an expression, similar to Eqn (20), in the three-dimensional plastic deformation of an inclusion bearing material (11).

3. ANELASTIC DEFORMATION BY BOUNDARY SLIDING

A simple way to detect grain boundary sliding is to use the internal friction where grain boundary sliding appears as an anelastic strain. In the following, we formulate the characteristic quantities of anelastic deformation caused by grain boundary sliding with and without slide-blocking inclusions.

3.1 CONSTRAINED BOUNDARIES (NORMAL POLYCRYSTALLINE SAMPLE)

The equilibrium value, $\overline{\beta^*}$, of the grain boundary sliding is obtained by $\delta(E + V)=0$ or $\sigma_{31} + \sigma_{31}^{A}=0$ on the inclusion-free domain. Therefore, from Eqns (14), (15) and (16),

$$\overline{\beta^*} = \overline{\beta_0^*}/(1 + \pi ar/\lambda^2), \tag{22}$$

where $\overline{\beta_0^*}$ is defined as

$$\overline{\beta_0^*} = (\sigma^A/\mu)(1/A)(a/c). \tag{23}$$

$\overline{\beta_0^*}$ is the equilibrium sliding of the grain boundary without the inclusions. Although $\overline{\beta_0^*}$ depends on c, which is somewhat vaguely defined physically, as well as a, the macroscopic anelastic strain does not depend a and c. This is because the volume fraction assigned for the boundary is proportional to $(1/a^3)(\pi a^2 c)$, and the product of this term with Eqn (23) results in a macroscopic anelastic strain independent of a and c. Therefore, defining Δ_0 as the relaxation strength of an inclusion-free specimen by boundary sliding, we can express the relaxation strength, Δ, of a specimen by sliding on boundaries with inclusions as

$$\Delta = \Delta_0/(1 + \pi ar/\lambda^2) \tag{24}$$

from Eqn (22).

Assuming that grain boundary sliding occurs with an intrinsic viscosity of η, the sliding rate, $\dot{\beta}^*$, is given by

$$\eta\dot{\beta}^* = \sigma_{31}^{A} + \sigma_{31}. \tag{25}$$

From Eqns (16), (22) and (23), this is rewritten as

$$\dot{\beta}^* = (\overline{\beta^*} - \beta^*)/\tau, \tag{26}$$

where τ is the relaxation time of sliding for the boundary with inclusions. It is related to τ_0, which is the relaxation time of the boundary free from inclusions, as

$$\tau = \tau_0/(1 + \pi ar/\lambda^2) \tag{27}$$

and τ_0 is expressed as

$$\tau_0 = (\eta/\mu)(a/c)A. \tag{28}$$

In summary, boundary inclusions reduce the relaxation strength of the sliding and shorten the relaxation time by the same factor, $(1 + \pi ar/\lambda^2)$. Therefore, we expect that when grain boundaries contain slide-blocking inclusions, the peak observed in an internal friction vs temperature curve should occur at a lower temperature and with a smaller height than when the boundaries are free from inclusions.

3.2 UNCONSTRAINED BOUNDARIES (BAMBOO STRUCTURED SAMPLE)

When grain boundaries penetrate through the whole section of a sample (bamboo structure), do not contain inclusions on them and are subjected to an external shear stress, they slide freely with a characteristic rate determined by the boundary viscosity. Since no internal stress is developed, internal friction of a relaxation character is also not expected. However, when inclusions are present on unconstrained boundaries, the situation is changed, because the inclusions oppose free sliding by developing an opposing internal stress. Grain boundary sliding is thus limited to an amount determined by the external stress and the internal stress and the sample shows internal friction of a relaxation character. In the following we will formulate the characteristic quantities for this internal friction.

The equilibrium boundary sliding, $\overline{\beta_B^*}$, is obtained by the condition of $\sigma_{31} + \sigma_{31}^A = 0$. From Eqn (17), we have

$$\overline{\beta_B^*} = (\sigma^A/\mu)(1/A)(\lambda^2/\pi rc). \tag{29}$$

The macroscopic strain, $\overline{\gamma_B^*}$, corresponding to this is calculated to be

$$\overline{\gamma_B^*} = \frac{\sigma^A}{\mu}\,\frac{4(1-\nu)}{(2-\nu)}\,\frac{\lambda^2}{\pi^2 r\ell} \tag{30}$$

where ℓ is the distance between the bamboo "knots". The sliding rate, $\dot{\beta}_B^*$, is proportional to the total stress and is expressed as

$$\dot{\beta}_B^* = (\overline{\beta_B^*} - \beta_B^*)/\tau_B \tag{31}$$

from Eqns (17) and (29). Here τ_B is the relaxation time of a bamboo structured sample and is given by

$$\tau_B = (\eta/\mu)(1/A)(\lambda^2/\pi rc). \tag{32}$$

Grain boundary sliding appears macroscopically as anelastic strain. Further, because of Eqns (29) and (31), it is clear that a bamboo structured sample with inclusions on the boundaries responds to an external stress as a standard linear solid.

4. EXPERIMENT

To examine the validity of the results of the previous sections, we measured the internal friction of Cu polycrystals which had either SiO_2 particles or Fe particles as inclusions on the grain boundary. An inverted torsion pendulum apparatus was used. The employed frequencies were in the range of ~1 Hz. The specimen dimensions were ~0.8 x 0.8 x 100 mm.

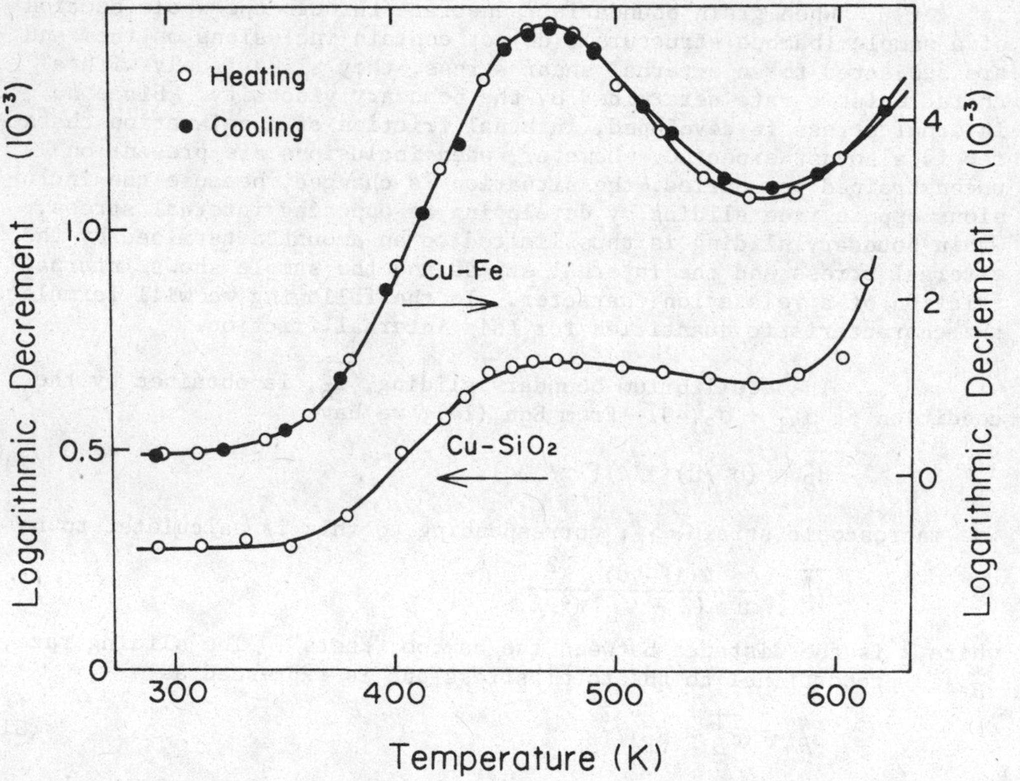

FIG. 3. Internal friction vs temperature curves of polycrystalline Cu with SiO_2 and Fe particles on boundaries.

4.1 NORMAL POLYCRYSTALLINE SAMPLES

In Fig. 3, the logarithmic decrement is plotted against temperature for samples with SiO_2 particles (r=250 nm, λ=1.9 μm, a=0.17 mm, $\omega \cong 8.1\ s^{-1}$) and with Fe particles (r=40 nm, λ=0.9 μm, a= 0.03 mm, $\omega \cong 8.5\ s^{-1}$). Internal friction peaks are seen approximately at 470 K. In a pure Cu sample, a peak occurs around 500 K with a much larger magnitude when the samples with a in the order of 0.1 mm is tested with $\omega \cong 6\ s^{-1}$ (12,13). That is, the height of the peak are smaller and the peak temperature is lower in the inclusion bearing samples. This is in qualitative agreement with the prediction of the preceding analyses, Eqns (24) and (27). In the following, we will quantitatively examine the effect of the inclusions on the boundary sliding.

To test Eqn (24), the measured relaxation strength is plotted against the reciprocal of $(1 + \pi ar/\lambda^2)$ in Fig. 4. We can certainly see the linearity expected from Eqn (24). From the slope of the line we calculate the relaxation strength, Δ_0, of the boundary relaxation of Cu without inclusions; it is 0.14, which agrees with the values reported previously (12,13).

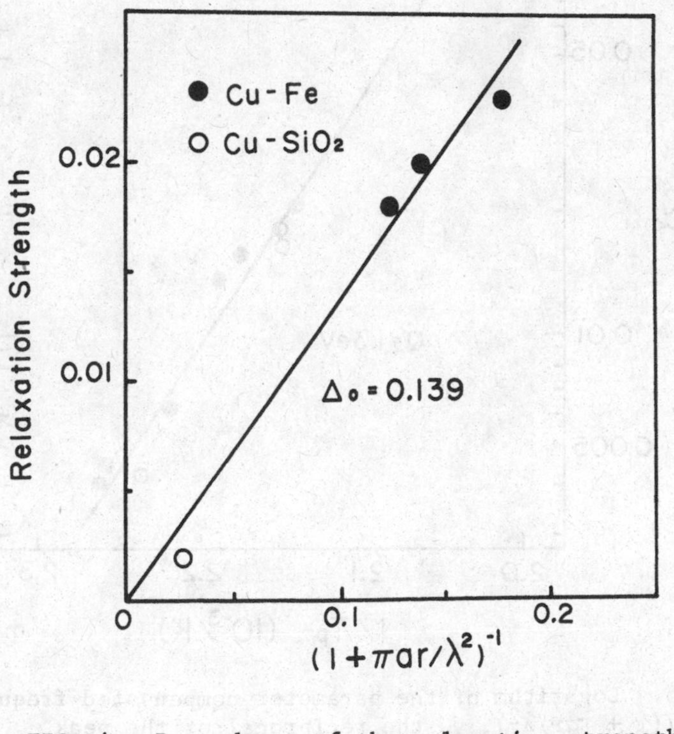

FIG. 4. Dependence of the relaxation strength on $(1 + \pi ar/\lambda^2)^{-1}$.

Next , we will examine Eqn (27). Traditionally, the intrinsic boundary viscosity is described as

$$\eta = \eta_0 T\exp(Q/kT), \qquad (33)$$

where Q is the activation energy and k and T have their usual meanings. A similar form has been theoretically given in the work of Ashby, in which Q is identified as the activation energy for boundary diffusion and η_0 is related to the boundary structure (14). Since the internal friction peak occurs when $\omega\tau=1$ is satisfied (ω is the angular frequency) and the relaxation time τ is given by Eqn (27), we have plotted the logarithm of $\omega_p T_p a/(1 + \pi ar/\lambda^2)$ against the reciprocal of the peak temperature T_p (ω_p is the angular frequency at T_p), as indicated in Fig. 5. The data points are seen to follow the theoretical expectation. The slope of the line gives the activation energy of 1.3 eV, which agrees with the activation energy of the relaxation of the grain

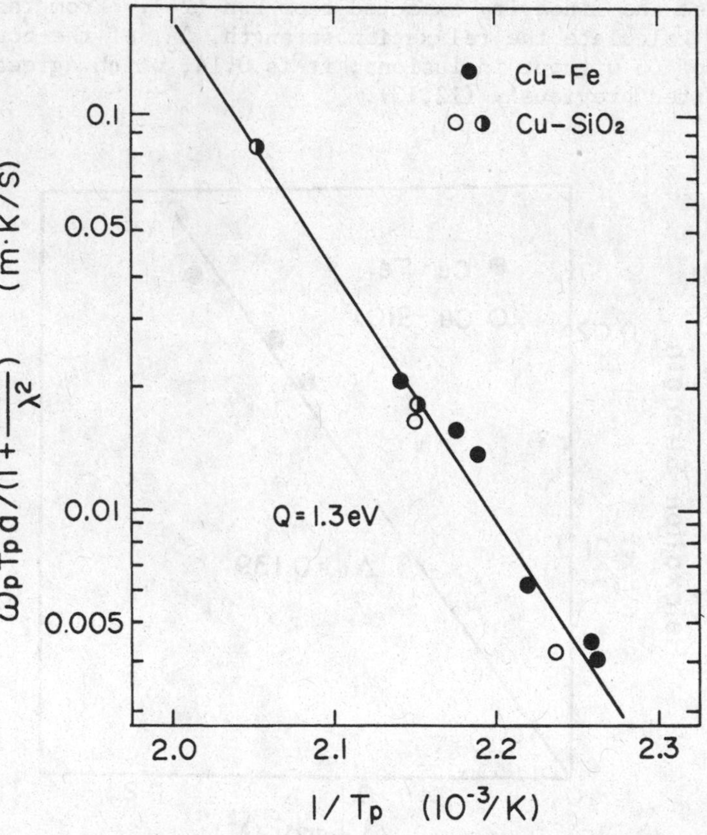

FIG. 5. Logarithm of the parameter compensated frequency, $\omega_p T_p a/(1 + \pi ar/\lambda^2)$, vs the reciprocal of the peak temperature (normal polycrystalline samples).

boundary sliding in pure Cu (12,13). This is natural, because inclusions on the boundary influence (i.e. suppress) sliding through the internal stress but do not change the atomic mechanism responsible for the intrinsic property of boundary sliding. There is another satisfactory point in Fig. 5; from this figure the peak temperature of a pure Cu sample without inclusions is estimated to be 498 K when it has the grain diameter of 0.1 mm and is tested with a frequency of 1 Hz. This is in the range reported for pure Cu (12,13).

4.2 BAMBOO STRUCTURED SAMPLES

As mentioned previously, an interesting point of the sliding suppressing effect of the inclusions is that the internal friction of a relaxation character can exist when a specimen has a bamboo structure and inclusions are present on boundaries. In this section, we will examine this point.

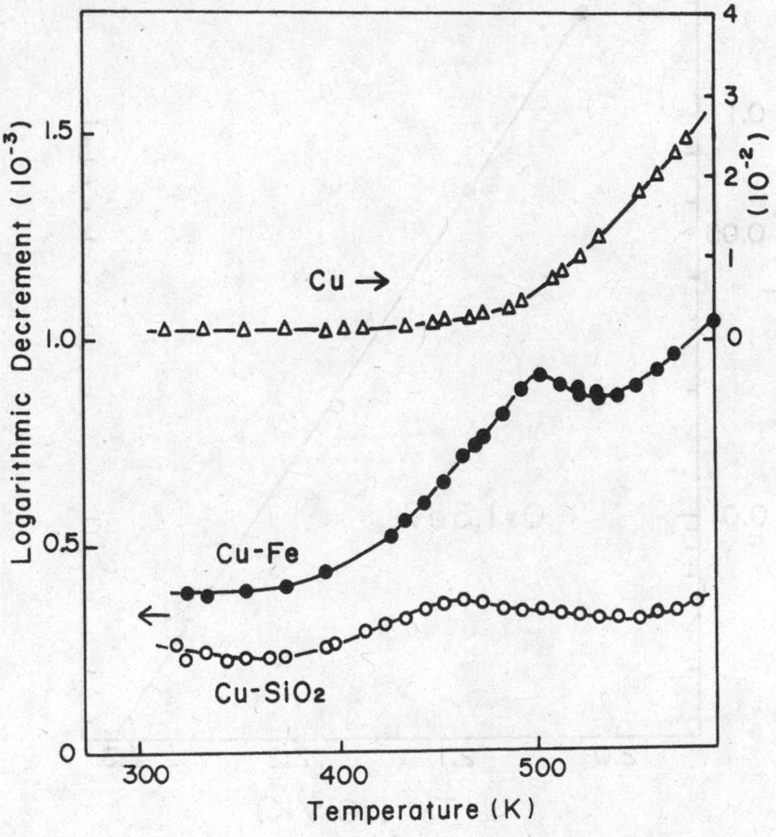

FIG. 6. Internal friction of bamboo structured samples (pure Cu, Cu-Fe and Cu-SiO_2).

Figure 6 shows the logarithmic decrement against temperature for a Cu-Fe (λ=3.9 μm, r=210 nm), a Cu-SiO_2 (λ=1.0 μm, r=330 nm) and a pure Cu sample. By choice of suitable heat treatments to produce coarse grains, all of these samples were given bamboo structures. In pure Cu, the internal friction increases monotonically as the temperature increases. Contrary to this, an internal friction peak is seen at 495 K in the Cu-Fe (ω=7.97 s^{-1}) and at 460 K in the Cu-SiO_2 (ω=8.26 s^{-1}). This observation is in accordance with the theory.

By assuming the temperature dependence of the intrinsic

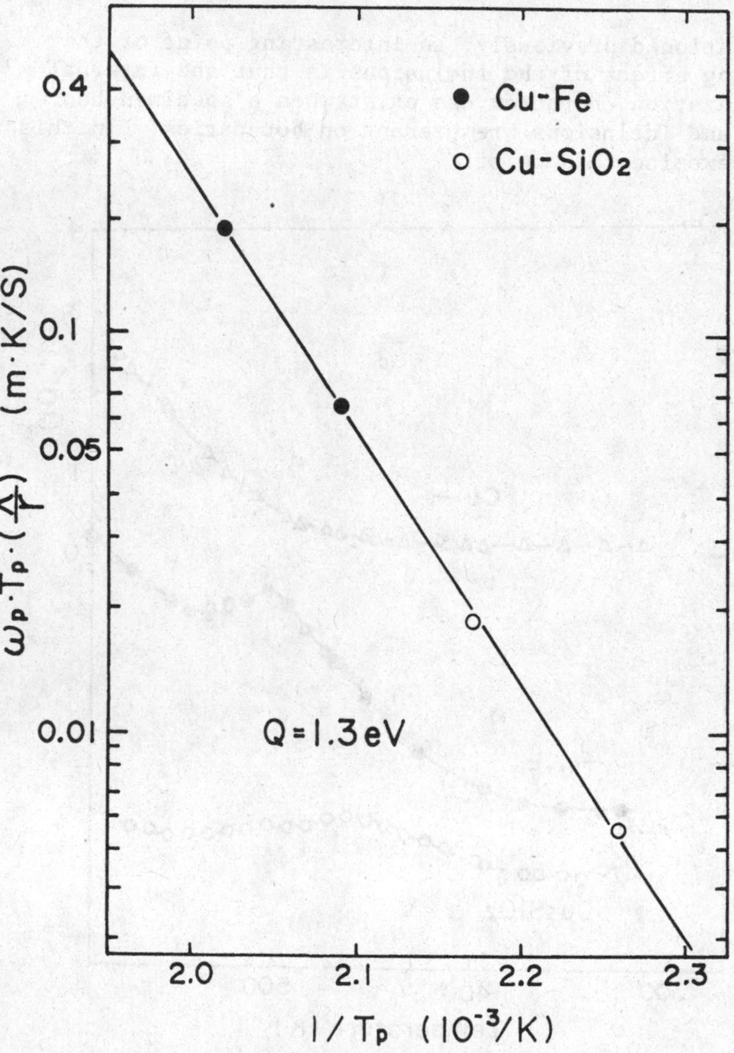

FIG. 7. Logarithm of the parameter compensated frequency, $\omega_p T_p(\lambda^2/r)$, vs the reciprocal of the peak temperature (bamboo structured samples).

boundary viscosity as described in Eqn (33) and considering Eqn (32), the logarithm of $\omega_p T_p(\lambda^2/r)$ was plotted against the reciprocal of the peak temperature T_p. This is shown in Fig. 7. Although the number of the data points is limited, they are on a straight line, which gives 1.3 eV as the activation energy for the boundary sliding. It agrees with the value determined in Fig. 5, as it must.

Because a sufficient number of samples with bamboo structures and inclusions on the boundaries were not tested in the present study, no systematic comparison between the theory and the experiment was performed concerning the relaxation strength. Nevertheless, we would like to report that the measured relaxation strength for the samples indicated in Fig. 6 is within the order of magnitude predicted by Eqn (30).

5. RELATED WORK

The present work is related to some past studies on inclusion bearing materials. As was mentioned in INTRODUCTION, one of them is the study by Raj and Ashby who treated the diffusion-controlled sliding of a boundary which is subjected to an external shear stress and has inclusions on it. When diffusion is allowed around inclusions to accommodate the plastic incompatibility responsible for the internal stress generated by the blocking inclusions, the internal stress is relaxed. A stationary state is achieved by the balance between the accumulation of incompatible strain introduced by the boundary sliding and its loss by diffusion. The resulting expression is the effective boundary viscosity, determined by the diffusion constant and the size and distribution of the inclusions on the boundary. Of course, by incorporating diffusion explicitly into the analysis of the present study, an expression almost identical to that given by Raj and Ashby is obtained (10).

The suppression of grain boundary sliding by inclusions on the boundary is essentially work hardening in two-dimensional plastic deformation in inclusion bearing materials. There is an exactly analogous subject in three-dimensional plastic deformation. As mentioned previously, it is work hardening in dispersion strengthened materials. The stress defined in Eqns (17) and (21) and produced by the blocking inclusions on the boundary is similar to the so-called back stress, image stress or mean stress in the work hardening of a dispersion strengthened material. It is defined as the average internal stress in the matrix, the plastic deformation of which is blocked by inclusions. The nature of this stress is discussed by Brown and Pedersen in this symposium.

We further note the similarity which exists between two-dimensional and three-dimensional deformation if diffusion is allowed. The internal stress responsible for work hardening in a dispersion strengthened alloy is also relaxed by diffusion. It is pipe-diffusion through the core of the dislocations surrounding the inclusions (15~17). And similar to the stationary state of the sliding of the boundary with inclusions achieved by diffusion around the inclusions (9,18), there exists a stationary creep state in a dispersion strengthened alloy, controlled by the dislocation pipe diffusion. This was found experimentally and was also successfully analyzed by the inclusion method with the additional consideration of dislocation pipe diffusion (19,20).

6. CONCLUSION

In the present paper, the suppression of grain boundary sliding by inclusions has been treated quantitatively. When the inclusion-free area of the boundary slides and the sliding cannot penetrate the inclusions, an internal stress is developed. The internal stress state is evaluated using the inclusion method developed by Eshelby. The internal stress suppresses grain boundary sliding and affects the anelastic deformation associated with grain boundary sliding. The theory was tested with internal friction measurements of Cu alloys with SiO_2 or Fe particles on their boundaries. In normal polycrystalline samples, an internal friction peak appears at a lower temperature and with a smaller magnitude than in a sample without inclusions, in qualitative agreement with the theory. The dependence of the relaxation strength and the relaxation time of the boundary sliding on the size and spacing of the inclusions also satisfactorily agrees with that predicted by the theory. Similarly, the theory accounts for the appearance of an internal friction peak of a relaxation character in a bamboo structured sample containing inclusions on its boundaries, as well as the relaxation time of such samples.

ACKNOWLEDGEMENT

I appreciate the cooperation of M. Koda, R. Monzen and N. Shigenaka who enthusiastically participated in the experimental phase of the present study. Also, I express my gratitude to Professor Toshio Mura for his stimulating discussions.

REFERENCES

(1) Eshelby, J. D. (1957). The determination of the elastic field of an ellipsoidal inclusion, and related problems. Proc. Roy. Soc., A241, 376-396.

(2) Eshelby, J. D. (1959). The elastic field outside an ellipsoidal inclusion. Proc. Roy. Soc., A252, 561-569.

(3) Eshelby, J. D. (1961). Elastic inclusions and inhomogeneities. In: Progress in Solid Mechanics, eds. I. N. Sneddon and R. Hill, 2, pp. 89-140. Amsterdam: North-Holland.

(4) Tanaka, K. & Mori, T. (1970). The hardening of crystals by non-deforming particles and fibres. Acta Metall., 18, 931-940.

(5) Brown, L. M. & Stobbs, W. M. (1971). The work-hardening of copper-silica, a model based on internal stress, with no plastic relaxation. Phil. Mag., 23, 1185-1199.
(6) Zener, C. (1941). Theory of the elasticity of polycrystals with viscous grain boundaries. Phys. Rev., 60, 906-908.
(7) Ke, T. S. (1947). Experimental evidence of the viscous behaviour of grain boundaries in metals. Phys. Rev., 71, 533-546.
(8) Nowick, A. S. & Berry, B. S. (1972). Anelastic Relaxation in Crystalline Solids. New York: Academic Press.
(9) Raj, R. & Ashby, M. F. (1971). On grain boundary sliding and diffusional creep. Metall. Trans., 2, 1113-1127.
(10) Mori, T. et al. (1983). Particle blocking in grain boundary sliding and associated internal friction. Acta Metall., 31, 275-283.
(11) Mori, T. & Tanaka, K. (1973). Average stress in matrix and average energy of materials with misfitting inclusions. Acta Metall., 21, 571-574.
(12) Cordea, J. N. & Spretnak, J. W. (1966). Grain boundary relaxation in four high-purity fcc metals. Trans. Met. Soc. AIME, 236, 1685-1691.
(13) Williams, T. M. & Leak, G. M. (1967). High temperature relaxation peaks in copper and aluminum. Acta Metall., 15, 1111-1118.
(14) Ashby, M. F. (1972). Boundary defects, and atomic aspects of boundary sliding and diffusional creep. Surf. Sci., 31, 498-542.
(15) Hazzledine, P. M. & Hirsch, P. B. (1974). A coplanar Orowan loops model for dispersion hardening. Phil. Mag., 30, 1331-1351.
(16) Gould, D. et al. (1974). The Bauschinger effect, work hardening and recovery in dispersion-hardened copper crystals. Phil. Mag., 30, 1353-1377.
(17) Mori, T. & Tokushige, H. (1977). Effect of particle size on low temperature softening of work hardened Cu-SiO_2 crystals. Acta Metall., 25, 635-641.
(18) Raj, R. & Ashby, M. F. (1972). Grain boundary sliding, and the effects of particles on its role. Metall. Trans., 3, 1937-1942.
(19) Mori, T. & Osawa, T. (1979). Softening of work-hardened Al-Si crystals and low temperature creep. Phil. Mag., 40, 445-457.
(20) Okabe, M. et al. (1980). Low temperature recovery creep of a Cu-SiO_2 alloy. Phil. Mag., 41, 615-618.

A CONTINUUM DAMAGE APPROACH TO CREEP CRACK GROWTH

H. Riedel

Max-Planck-Institut fuer Eisenforschung

Max-Planck-Str. 1, 4000 Duesseldorf, FRG

ABSTRACT

The equations of continuum damage mechanics proposed by Kachanov are solved for cracked specimens. In particular, the small-scale damage limit and its range of validity are described using similarity solutions and the finite element method. Crack growth is an inherent part of the solutions. Lifetimes of pre-cracked specimens are estimated by integrating the crack growth laws obtained in the small-scale damage limit. These estimates compare well with complete finite element analyses of Hayhurst et al. thus indicating that the small-scale damage approximation is valid for the greatest part of the lifetime. This means that creep failure of pre-cracked specimens can adequately be described by the appropriate fracture mechanics parameters, K_I, C^*, J or others, depending on the testing conditions.

1. INTRODUCTION

Creep crack growth is the time-dependent extension of a macroscopic crack under more or less constant (rather than cyclic) load at elevated temperatures. The micromechanism of creep crack growth often is grain boundary cavitation ahead of the main crack and coalescence of the cavities with the crack. Corrosive effects also play an important role in many materials [1].

There are two different approaches to deal with creep failure of cracked specimens. The usual way is to describe creep crack growth macroscopically by the appropriate fracture mechanics parameters like K_I, C^*, J or others [2-12]. The problem with this approach is that it relies on the small-scale damage assumption. That means that the effect of damage (in the form of cavities, for example) on the stress fields is neglected. The approach is therefore invalidated once intense damage spreads across the whole ligament rather than being confined to a small process zone. The second approach was developed recently by Hayhurst et al. [13]. They include damage in the constitutive relations between strain rate and stress and solve these continuum damage equations for cracked specimens employing the finite element method. This approach avoids the small-scale damage assumption, but its drawback is that every configuration requires a separate finite element solution, which is computationally difficult. Further, there is a danger of severely overestimating lifetimes by this method in cases where corrosive processes near the crack tip enhance creep crack growth. Local corrosion usually cannot be included in continuum damage equations. On the other hand, the fracture mechanics approach relies on measured crack growth rates which are merely transferred from specimens to structures using the appropriate load parameter. Thus corrosive processes at the crack tip are expected to be automatically taken into account.

Considering the advantages of the fracture mechanics approach it appears worthwhile to examine its range of validity and its limitations resulting from the small-scale damage assumption. The continuum damage model is a suitable tool to study this question.

Another, probably less important, issue to be addressed here is related to micromechanical models of creep crack growth. Creep crack growth by the nucleation, growth and coalescence of cavities has been modelled by several authors (e.g. [12,14-17]). However, the effect of the cavities on the crack-tip stress fields has always been ignored. It will be shown that, under small-scale damage conditions, these models lead to correct results for the crack growth rates (apart from numerical factors) even though the stress distribution in the process zone is completely changed by the evolution of damage. Correspondingly, the simple models have been found previously to describe the dependences of the crack growth rate on C^* (when C^* applies) and on temperature in broad agreement with experimental results [4-6,11,12].

2. THE CONTINUUM DAMAGE EQUATIONS

The analysis of the stress, strain and damage fields around cracks is based on a constitutive model proposed by Kachanov [18] and extended recently by Hayhurst and co-workers [13,19]. A scalar internal variable, the damage parameter, ω, is postulated to exist. It varies from $\omega = 0$ for the virgin material to $\omega = 1$ at failure. Its evolution in time is assumed to be given by the kinetic law:

$$\dot{\omega} = D[\kappa\sigma_I + (1-\kappa)\sigma_e]^{\chi}/[(1+\phi)(1-\omega)^{\phi}] \tag{1}$$

and its effect on the stress-strain rate relation is assumed to be

$$\dot{\varepsilon}_{ij} = \frac{1+\nu}{E}\dot{\sigma}'_{ij} + \frac{1-2\nu}{3E}\delta_{ij}\,\dot{\sigma}_{pp} + \frac{3}{2}B\sigma_e^{n-1}\sigma'_{ij}/(1-\omega)^n. \tag{2}$$

Here, σ_I is the maximum principal tensile stress; further,

$$\sigma_e = (3\,\sigma'_{ij}\,\sigma'_{ij}/2)^{1/2} \tag{3}$$

is the equivalent tensile stress, the prime denotes the deviator, the superposed dot indicates the time derivative and δ_{ij} is the unit tensor. Further, D, κ, χ, ϕ, E (Young's modulus), ν (Poisson's ratio), B (creep coefficient) and n (stress exponent) are material parameters. They can be determined from creep tests using unnotched specimens. The measurement of κ requires tests under at least two different stress states. Table 1 gives the material parameters for an aluminum alloy and for copper as reported in [13].

Note that for simplicity's sake the elastic terms in eqn. (2) are assumed not to be affected by damage. If at failure elastic strains are not negligible compared to creep strains (e. g., possibly, in ceramics) damage must be included in the elastic terms as well.

The damage parameter has no direct physical meaning, although it is somehow, in a not yet well defined manner, related to the cavitated area fraction of grain boundary. An attempt made by Hutchinson [20] to develop continuum damage mechanics equations from principles of cavity growth is confined to a rather special situation. At present the empirical damage mechanics equations (1,2) seem to be preferable.

For a constant-stress uniaxial creep test, eqns. (1) and (2) are readily integrated to give the creep curve shown in Fig. 1. Analytically the result for the creep strain is

$$\varepsilon^{cr} = \varepsilon_f \, [1-(1-t/t_f)^{1-n/(1+\phi)}] \tag{4}$$

with the strain to failure, ε_f, and the time to failure, t_f, defined by

$$\varepsilon_f = B\sigma^n(1+\phi)/[D\sigma^\chi(1+\phi-n)], \tag{5}$$

$$t_f = (D\,\sigma^\chi)^{-1}. \tag{6}$$

The initial (steady-state) strain rate is $\dot{\varepsilon}_{ss} = B\sigma^n$. The Monkman-Grant product

$$\dot{\varepsilon}_{ss}t_f = \varepsilon_f \, (1+\phi-n)/(1+\phi) = B\sigma^{n-\chi}/D \tag{7}$$

is given by the intercept of the dashed lines with $t/t_f=1$ in Fig. 1.

From these relations and from Fig. 1 it is clear that the shape of the creep curve, after normalizing strain and time, is solely determined by the ratio $n/(1+\phi)$. Small values of that ratio mean an extended steady-state, or secondary, stage and only a small tertiary stage. If $n/(1+\phi)$ approaches unity, the tertiary stage becomes

	E	B	n	D	χ	ϕ	κ	$\dot{\varepsilon}_{ss}t_f$
Al	$6\cdot10^4$	$3.2\cdot10^{-21}$	6.9	$5\cdot10^{-18}$	6.48	9.5	0	0.33%/0.59%
Cu	$6.6\cdot10^4$	$3.6\cdot10^{-10}$	2.97	$1.7\cdot10^{-7}$	1.21	3.83	0.7	0.85%/2.9%

TABLE 1. Material parameters for an aluminum alloy at 210°C and for copper at 250°C (after [13]). The Monkman-Grant product (last column) is calculated from eqn. (7) for σ = 50 and 100 MPa (Al) and σ = 30 and 60 MPa (Cu). Units in MPa, sec and absolute strain unless otherwise stated.

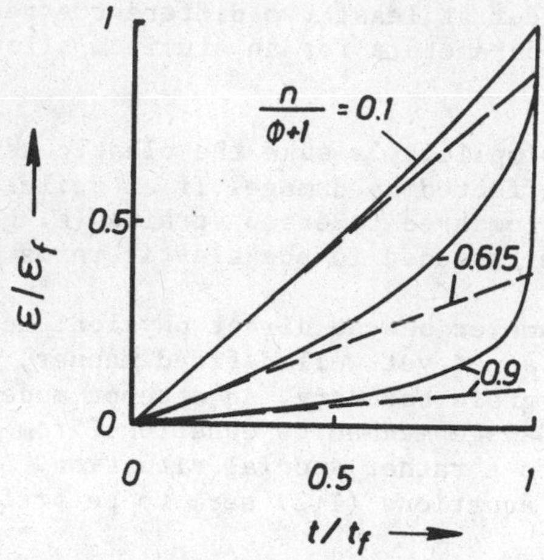

FIG. 1. Creep curves according to eqns. (1) and (2).

predominant. The value $n/(1+\phi)=0.615$ has been chosen in Fig. 1 to represent the case of copper at 250°C (cf. Table 1).

In the following sections, solutions of the continuum damage equations, together with the usual (small strain) equilibrium and compatibility equations, will be derived for two-dimensional (plane strain) crack geometries. It is important to note that crack growth is inherently included in the damage mechanics equations. Whereever $\omega = 1$, the material has failed and no tractions are transmitted. Therefore the contour $\omega = 1$ in the damage field represents the extending crack.

3. THE EVOLUTION OF DAMAGE IN EXTENSIVELY CREEPING SPECIMENS

In this section, elastic strain rates are completely neglected so that the governing equations are eqns. (1) and (2) with the terms containing $1/E$ deleted. A pre-requisite for the neglect of elastic terms is that the specimen spends most of its life in the extensive creep regime. In other words, the lifetime, t_f, must be much greater than the characteristic time, t_1, which characterizes the transition from the initial elastic behavior to extensive creep. Riedel and Rice [7] and Ohji et al. [8] give

$$t_1 = K_I^2(1-\nu^2)/[E(n+1)C^*]. \tag{8}$$

Under given test conditions, K_I and C^* can be calculated from the applied load using formulas of Kumar et al. [21], and hence t_1 can be calculated. Further, in elastic/nonlinear viscous materials, elastic strains are not negligible near growing crack tips [22]. It can be shown, however, that, when damage is included, elastic straining becomes negligible everywhere if $t \gg t_1$.

Inspection of eqns. (1) and (2), with the elastic terms deleted, together with compatibility, shows that the solutions must have the following scaling properties. If P(t) is load per unit specimen thickness, stress and damage fields have the form

$$\sigma_{ij} = (P/W)\,\overline{\sigma}_{ij}(r_i/W,\tau) \tag{9}$$

$$\omega = \omega(r_i/W,\tau) \tag{10}$$

with $$\tau = D\int (P/W)^{\chi}dt \tag{11}$$

or $\tau = D(P/W)^{\chi}t$ for time independent P. Here, W represents the in-plane dimensions of the specimen, r_i represents the in-plane coordinates, and $\overline{\sigma}_{ij}$ is a dimensionless function of its dimensionless arguments. In particular, these scaling properties mean that the lifetimes of cracked, as well as of unnotched, specimens under constant load scale as $t_f \propto (D\sigma_o^{\chi})^{-1}$ independent of the absolute specimen size and of σ_o, where σ_o is the initial applied net section stress.

Within the limiting range of negligible elastic straining, the limiting case of small-scale damage can reasonably be defined as the short-time limit. A well-defined long-time limit does not exist. Rather, when damage spreads across the whole ligament, this leads to failure. Hence the characteristic time for the small-scale damage limit to be valid is a certain, yet to be determined, fraction of the lifetime, t_f.

3.1 SMALL-SCALE DAMAGE IN EXTENSIVELY CREEPING SPECIMENS

The response of a material described by eqn. (2) with the elastic strains deleted is nonlinear viscous immediately upon load application since damage requires time to develop. In a nonlinear viscous material, the C^*-integral is path-independent. Since the nonlinear viscous stress-strain rate relation used here is a power law, the crack-tip field is an HRR-type field (after Hutchinson [23], and Rice and Rosengren [24]):

$$\sigma_{ij} = \left(\frac{C^*}{I_n B r}\right)^{1/(n+1)}\tilde{\sigma}_{ij}(\theta). \tag{12}$$

Here, r and θ are polar coordinates centered at the crack tip with θ = 0 lying directly ahead of the crack. The dimensionless quantities I_n and $\tilde{\sigma}_{ij}(\theta)$ have been given in [23,24] and, more completely, in [25].

In the initial, singular crack-tip stress field, damage develops rapidly according to eqn. (1). For short times compared to the time to failure, substantial damage has only accumulated in a small zone, which will be called the process zone, while further away ω is still negligibly small and the stress field is the practically undisturbed field in a nonlinear viscous body. The evolution of damage in the process zone in the short-time, or small-scale damage, limit can be described by requiring, as a remote boundary condition, that the stress field must asymptotically approach the undisturbed field described by eqn. (12) and ω must approach zero.

If eqn. (12) represents the initial condition at t = 0 and the remote boundary condition, dimensional considerations similar to those used in [7] show that the damage mechanics equations with elasticity terms deleted have the similarity solutions

$$\sigma_{ij} = (Dt)^{-1/\chi}\,\Sigma_{ij}(R,\theta) \qquad \text{and} \qquad \omega = \omega(R,\theta) \tag{13}$$

with

$$R = (Br/C^*)\,(Dt)^{-(n+1)/\chi} \tag{14}$$

where Σ_{ij} and ω are as yet unknown dimensionless functions of their dimensionless arguments. Without using dimensional arguments, it can directly be verified by insertion that the similarity solutions satisfy all equations and boundary conditions.

Due to the similarity of the fields, all contours of constant stress or constant damage expand around the crack tip according to $r \propto t^{(n+1)/\chi}$. In particular, the crack tip which is characterized by $\omega = 1$ moves according to

$$\Delta a = \alpha(n,\chi,\phi,\kappa)\;(C^*/B)\;(Dt)^{(n+1)/\chi}. \tag{15}$$

Time-differentiation and elimination of time using eqn. (15) gives the crack growth rate as a function of the amount of crack growth which has occurred since the beginning of the test, Δa:

$$\dot{a} = ((n+1)/\chi)\;(\alpha C^*/B)^{\chi/(n+1)}\;D\Delta a^{(n+1-\chi)/(n+1)}. \tag{16}$$

The factor $\alpha(n,\chi,\phi,\kappa)$ cannot be determined from similarity arguments alone. It will be determined below in an approximate manner and, for one particular case, employing finite element analysis.

3.2 APPROXIMATE METHODS IN SMALL-SCALE DAMAGE

A very simple approximation for the damage distribution $\omega(R,\theta)$ is obtained by assuming the stress field to be the undisturbed HRR-field (eqn. 12). This is, of course, not exactly correct, since damage accumulation relaxes the crack tip stress concentration. However, if ω is calculated by integrating eqn. (1) for a constant HRR stress field, it turns out that the so-calculated damage field has the similarity properties of the exact small-scale damage solution. In this way, the contour $\omega = 1$ (i.e., the crack) is found to expand in agreement with eqns. (15) and (16) with α being given by

$$\alpha(n,\chi,\phi,\kappa) = [\kappa\tilde{\sigma}_I + (1-\kappa)\tilde{\sigma}_e]^{n+1}/I_n. \tag{17}$$

For plane strain and $n=\phi=\chi=5$, $\kappa=1$ and $\theta=24^\circ$, this gives $\alpha = 29.8$, which will be compared with the finite element result ($\alpha = 37$) in Fig. 2. Intuitively, one would suspect that this approximate treatment gives a lower bound for α and, hence, for the crack growth rate.

An estimate which presumably gives an upper bound is obtained if the HRR-field is attached to the moving crack tip but is otherwise undisturbed. The damage distribution obtained from eqn. (1) has the similarity properties of the exact solution if the crack tip moves in accordance with eqn. (15) with unspecified α. The value of α is now determined by requiring that ω at a material point ahead of the crack tip just reaches unity when the crack tip arrives at that point. This leads to

$$\alpha(n,\chi,\phi,\kappa) = (\pi s/\sin \pi s)^{1/s} \; [\kappa\tilde{\sigma}_I + (1-\kappa)\tilde{\sigma}_e]^{n+1}/I_n \qquad (18)$$

with $s = \chi/(n+1)$. Note that this upper bound estimate differs from the lower bound, eqn. (17), only by the factor $(\pi s/\sin\pi s)^{1/s}$. For $n=\phi=\chi=5$, $\theta=24^\circ$ and $\kappa=1$, eqn. (18) gives $\alpha = 217$, while eqn. (17) gave $\alpha = 29.8$, (cf. Fig. 2).

The approximate analysis above shows that neglecting the effect of damage on the stress distribution does not lead to serious error. Only the factor α is estimated inaccurately, while the dependences on C^*, B, D and Δa are obtained correctly. This observation is important in relation to models for creep crack growth by cavity coalescence which neglect the effect of cavitation on the stress fields [14-17]. It may be concluded that these models give correct results except for errors in the absolute values of the crack growth rates. In fact, the upper-bound result (eqn. 16 with α from eqn. 18) is identical with the result of a crack growth model [12,15] by cavity coalescence under the following conditions: if cavity growth is creep constrained; if in [12,15] the critical strain for cavity coalescence is replaced by the Monkman-Grant product (eqn. 7) and the structural length, x_c, is set equal to zero; and if χ is set equal to n in the continuum damage model.

Arbitrarily, the process zone is defined here as the zone within which the equivalent strain rate is at least doubled by damage compared to undamaged material. Within this zone the stress field can be expected to be relaxed markedly by damage. A rough estimate of the so-defined process zone size is obtained by calculating strain rate and damage from eqns. (1) and (2) using the undisturbed HRR stress field. Then the process zone size, normalized by the amount of creep crack growth that has occurred, is found to be

$$r_p/\Delta a = (1-2^{-(1+\phi)/n})^{-(n+1)/\chi} - 1. \qquad (19)$$

For $n=\phi=\chi=5$, this gives $r_p = 0.99\ \Delta a$, which agrees approximately with the finite element result. For the material parameters of copper, eqn. (19) gives $r_p = 2.6\ \Delta a$. For the aluminum alloy (Table 1), is $r_p = 0.69\ \Delta a$.

3.3 FINITE ELEMENT ANALYSIS OF SMALL-SCALE DAMAGE

The finite element program developed by Ehlers [26,27] for elastic/nonlinear viscous material was extended to include damage according to eqns. (1) and (2). In order to model small-scale damage, the HRR stress field is prescribed at time t=0 on all Gaussian points. The displacement prescribed on a large circle around the crack tip is the HRR displacement field increasing in proportion to time.

The performance of the program is not yet satisfactory in terms of stability of the time-integration scheme. Therefore only a few cases have been analysed so far, and a detailed description of the program will not be given here. Only one of the problems should be mentioned. The stability degraded when ω approached unity at one or several Gaussian points. Therefore ω had to be cut off at some maximum value, ω_{max}, and was subsequently kept constant at that value. Thus the Gaussian points that should actually have failed completely were kept at a very small but finite stiffnes by the program.

The case $n=\chi=\phi=5$, $\kappa=1$ was analysed on a regular grid of 72 eight-noded isoparameteric elements arranged in 6 sectors (for $0<\theta<\pi$) and 12 rings around the crack tip. Figure 2 shows that the numerical program gives the correct behavior required by the similarity solutions. According to eqns. (13) and (14), the fields must expand around the crack tip as $r \propto t^{6/5}$ for the present set of material parameters. In Fig. 2, the symbols represent the distances from the original crack tip at which ω has reached its cut-off value, which was chosen as ω_{max}= 0.39, 0.72 and 0.84 in three different calculations. The limit $\omega_{max} \to 1$

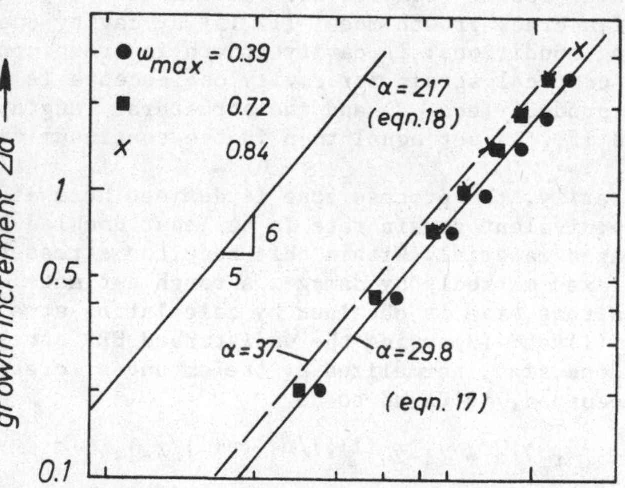

FIG. 2. Amount of crack growth vs. time (arbitrary units). Symbols are finite element results. Dashed line represents extrapolation $\omega_{max} \to 1$; $n = \phi = \chi = 5$, $\kappa = 1$, $\theta = 24^{o}$.

represents the growing crack tip. From the plot, the value of the factor α in eqn. (15) is estimated as $\alpha(5,5,5,1)=37$. Crack growth is predicted to occur at an angle $\theta \simeq 24^o$, which is slightly less than the direction of the maximum tensile stress, $\theta=33^o$, in an HRR-field for n=5. Recall that the present calculation is done for $\kappa=1$, which means that failure is controlled by the maximum principal tensile stress according to eqn. (2). In reality, crack growth along an inclined plane will not necessarily be observed since some parts of the crack front will grow along $\theta=+24^o$ and others along $\theta=-24^o$. Macroscopically, the fracture surface may well be flat in this case.

The stress distribution in the process zone ahead of the crack is shown in Fig. 3, after smoothing the finite element results. The zone where the HRR-stress field is markedly modified by damage is of the order of the process zone size calculated in eqn. (19).

3.4 THE RANGE OF VALIDITY OF THE SMALL-SCALE DAMAGE APPROXIMATION

The small-scale damage approximation becomes increasingly inaccurate as the process zone grows, thus invalidating the HRR-field as a remote boundary condition. In a finite specimen, the HRR-field is valid at distances from the crack tip which satisfy

$$r_p < r < 2.5\ a/M. \qquad (20)$$

Here, the inner range of validity is determined by the process zone, while the outer range scales with the crack length, a, or ligament width, W-a. The factor M has been determined from finite element analyses of nonlinear elastic materials and is found to be of the order

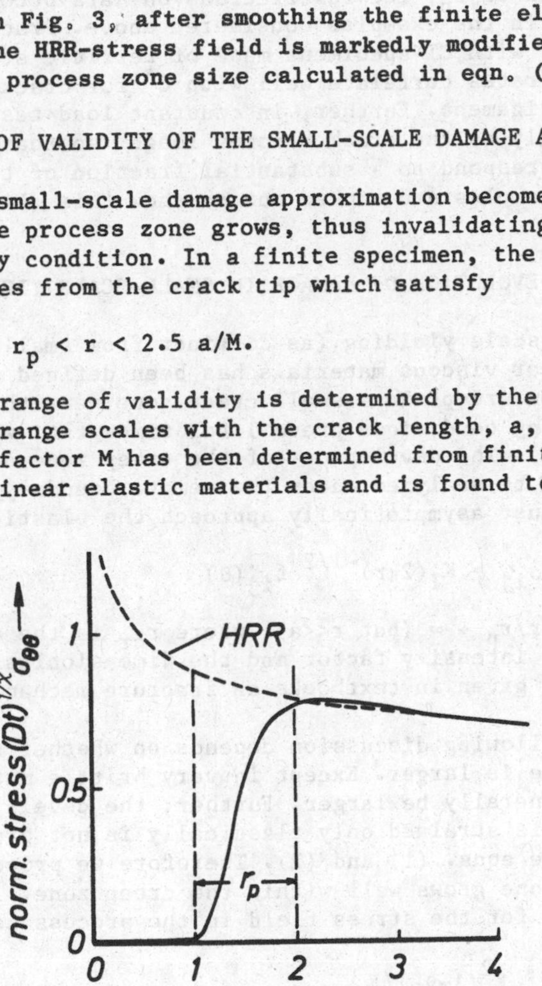

FIG. 3. Stress distribution in process zone ahead of growing crack; $n = \phi = \chi = 5$, $\kappa = 1$, $\theta = 24^o$.

M=25 for bend geometries including the CT-specimen and M=200 for center-cracked plates in plane-strain tension [28,29]. In the latter case, the value of M increases with increasing stress exponent. The value M = 200 is representative for n = 10. According to eqns. (19) and (20), the HRR-field is completely displaced by the process zone when

$$\Delta a/a = 2.5[(1-2^{-(1+\phi)/n})^{-(n+1)/\chi}-1]^{-1}/M. \tag{21}$$

If M = 25, Δa is thus limited to 10% of the crack length or ligament width if $n=\phi=\chi=5$. For copper and aluminum the corresponding limitation is 4% and 14%, respectively. If in eqn. (20) the lower bound is assumed to be $\Delta a+r_p$ instead of r_p, the restrictions on $\Delta a/a$ become 5%, 3% and 6%, respectively, in the examples considered above. Practical experience in this laboratory with CT-specimens made of ferritic steels indicates that crack growth rates correlate well with C^* for crack growth almost across the whole ligament. Further, in constant load tests, crack growth accelerates strongly as the crack becomes longer so that 5 to 10% crack growth usually correspond to a substantial fraction of the lifetime. In Section 5 a few examples are worked out to show this (Fig. 5).

4. THE EVOLUTION OF DAMAGE IN SMALL-SCALE YIELDING

Small-scale yielding (as distinct from small-scale damage) in elastic/nonlinear viscous materials has been defined as predominantly elastic deformation except in a small creep zone near the crack tip [7]. Small-scale yielding conditions prevail for times $t<t_1$ where t_1 was defined in eqn. (8). The development of the creep zone and of the process zone can then be analyzed under the remote boundary condition that the stress field must asymptotically approach the elastic singular field

$$\sigma_{ij} = K_I(2\pi r)^{-1/2} f_{ij}(\theta) \tag{22}$$

for $r/r_{cr} \to \infty$ and $r/r_p \to \infty$ (but $r<<a$), where r_{cr} is the creep zone size. Here, K_I is stress intensity factor and the dimensionless angular function $f_{ij}(\theta)$ is given in textbooks on fracture mechanics [30].

The following discussion depends on whether the creep zone or the process zone is larger. Except in very brittle materials, the creep zone will generally be larger. Further, the development of damage in material which is strained only elastically is not properly modelled by the constitutive eqns. (1) and (2). Therefore we proceed assuming that the process zone grows well within the creep zone. Then the remote boundary condition for the stress field in the process zone is [7]:

$$\sigma_{ij} = \left(\frac{K_I^2(1-\nu^2)/E}{(n+1)I_n Brt} \right)^{1/(n+1)} \tilde{\sigma}_{ij}(\theta). \tag{23}$$

Since elastic straining can be neglected well inside the creep zone, eqns. (1) and (2), with the elastic terms deleted, have similarity solutions having the form of eqn. (13) with

$$R = rEBt(Dt)^{-(n+1)/\chi}/K_I^2. \quad (24)$$

The crack growth increment and the crack growth rate are then found to be

$$\Delta a = \alpha(n,\phi,\chi,\kappa)\ \frac{K_I^2(1-\nu^2)}{(n+1)EBt}\ \left(\frac{n+1}{n+1-\chi}\ Dt\right)^{(n+1)/\chi}, \quad (25)$$

$$\dot{a} = [(n+1)/\chi]\ \left(\alpha\frac{K_I^2(1-\nu^2)}{(n+1)EBt}\right)^{\chi/(n+1)} D\ \Delta a^{(n+1-\chi)/(n+1)}. \quad (26)$$

The dimensionless factor $\alpha(n,\phi,\chi,\kappa)$, in general, is not identical with that used in eqns. (15) and (16). In the following, the two factors are assumed to be equal, which is true within the approximation leading to eqn. (17).

5. LIFETIME ESTIMATES

In this section, the lifetime of pre-cracked specimens is estimated using the crack growth laws which were derived in the preceding sections based on the small-scale damage approximation. The results are compared with finite element results of Hayhurst et al. [13] who made no use of the small-scale damage approximation.

In the examples considered by Hayhurst et al. [13], the lifetime, t_f, is greater than the characteristic time, t_1, but not so much greater that the initial elastic behavior could entirely be neglected. Therefore, an interpolation formula between the small-scale yielding limit and the extensive creep limit, eqns. (26) and (16), is needed. Such an interpolation is obtained if in eqn. (16) C^* is replaced by

$$C^* \rightarrow C^*(1 + t_1/t), \quad (27)$$

which recovers eqn. (16) for long times and eqn. (26) for short times. This form of interpolation accurately reproduces the finite element results for the near-tip stresses in an elastic/nonlinear viscous material [26,27]. Whether it interpolates the crack growth rates equally well is not yet clear.

The next step is to integrate the crack growth law, eqn. (16) with eqn. (27) inserted. To do this it is necessary to know C^* as a function of the crack length. Since the value of C^* is not known to-date for branched cracks, it is assumed here that the growing crack remains macroscopically flat although, on a smaller scale, the crack may grow at an angle θ. The following analysis for copper is done setting $\theta=0$, since the evolution of damage in copper is controlled primarily by the maximum principal tensile stress. This favors forward-directed crack growth. In aluminum, failure is controlled by the von Mises equivalent stress which leads to crack growth at an angle. Therefore, the projection of the crack growth rate along $\theta=60^o$ onto the symmetry plane is taken as the growth rate of a planar crack in aluminum. This may, of course, be a rather crude model of the actual situation. For macroscopically planar cracks, the C^*-integral is given by

$$C^* = a\, B\, \sigma_{net}^{n+1}\, g_1(a/W,n), \tag{28}$$

where σ_{net} is net section stress and g_1 is a dimensionless function of the specimen geometry [10,12,21].

The crack growth law, eqn. (16), can be integrated by separation of the variables. The result is the following implicit relation between crack length and time:

$$F(a/W) = D\sigma_o^{\chi} t\; G(t/t_1)\; (\alpha\ \cos\theta)^{\chi/(n+1)}\ (n+1)/\chi \tag{29}$$

$$\text{with} \quad F(a/W) = \int_{a_o}^{a} \frac{(\sigma_o/\sigma_{net})^{\chi}\, da}{(ag_1)^{\chi/(n+1)}\,(a-a_o)^{(n+1-\chi)/(n+1)}} \tag{30}$$

$$G(t/t_1) = (t_1/t) \int_0^{t/t_1} (1+1/\tau)^{\chi/(n+1)} d\tau\,. \tag{31}$$

Here, σ_o is the net section stress at the beginning of the test, a_o is the initial crack length, and the load is assumed to be constant for simplicity. The factor $\cos\theta$ accounts for the projection of the local crack growth direction, θ, onto the symmetry plane.

The solutions of eqn. (29) will now be given graphically based on numerical evaluation of the integrals. Figure 4 shows the dimensionless function $G(t/t_1)$, which takes account of elasticity effects at short times. If elastic straining can be neglected, this means that

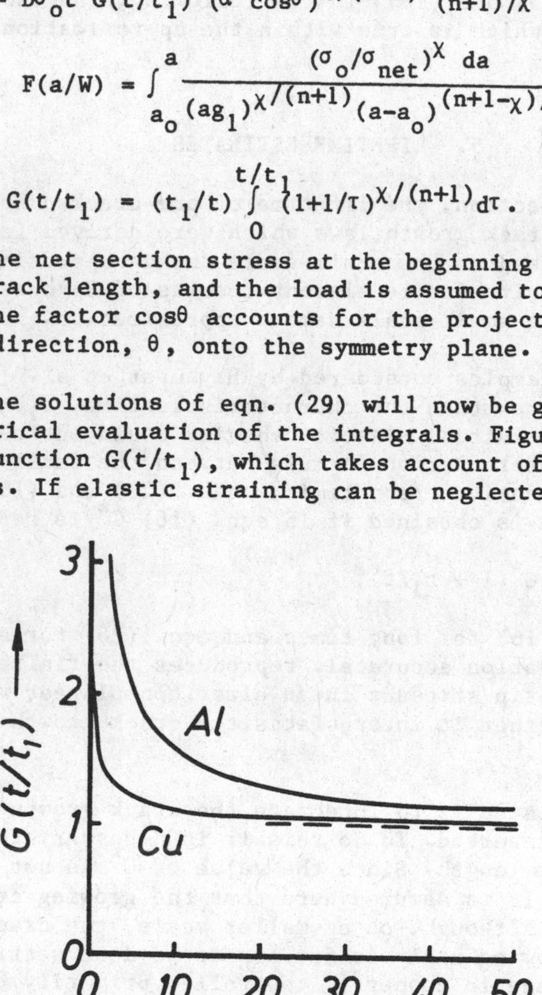

FIG. 4. Elasticity correction function defined in eqn. (31).

G=1. The group $D\sigma_o^{\chi}t$ on the right-hand side of eqn. (29) represents the time in units of the failure time of an unnotched specimen in uniaxial tension. Therefore in Fig. 5, a uniaxially stressed specimen fails at $D\sigma_o^{\chi}t=1$. Failure at longer times means notch strengthening, while failure at shorter times implies notch weakening. The function F(a/W) defined in eqn. (30) depends on the specimen geometry. To compare with the work of Hayhurst et al. [13], double edge cracked panels (DECP) with $a_o/W=2/3$ and center-cracked panels (CCP) with $a_o/W=1/3$ were selected. Both specimen types are considered in plane-strain tension. For the factor α, the upper bound estimate, eqn. (18), was used in preparing Fig. 5. The material parameters for Al and Cu were taken from Table 1.

Having evaluated the integrals F(a/W) and $G(t/t_1)$, the result of eqn. (29) can be plotted in the form a/W vs. $D\sigma_o^{\chi}t$ as in Fig. 5. The dashed lines are obtained if elasticity effects are neglected; that means by setting $G(t/t_1)=1$. The solid lines are computed including elasticity effects. Here the characteristic time t_1 is calculated from eqn. (8) taking $a=a_o$ in the calculation of K_I and C^* as an approximation. For copper, the elasticity correction is small. (In fact, the corrected curve for CCP-specimens was omitted since it deviates by only 2% from the dashed curve). For aluminum, which is less ductile than

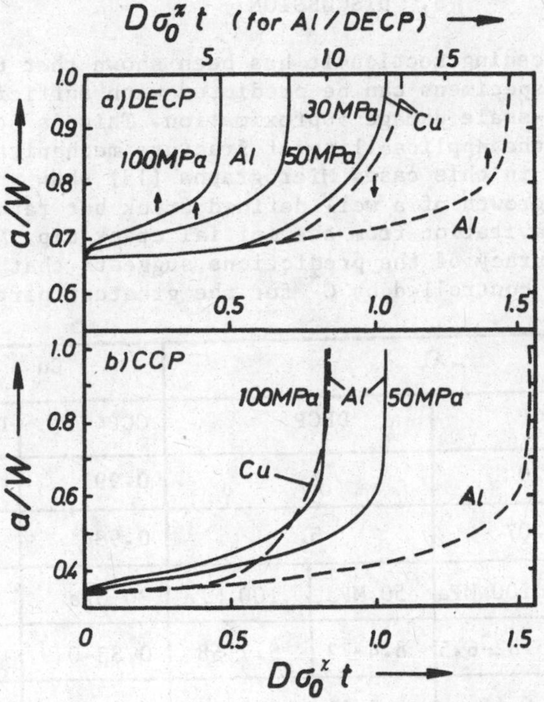

FIG. 5. Normalized crack length vs. normalized time according to eqn. (29) for DECP and CCP-specimens. Failure occurs at a/W = 1. Note separate scale for DECP-specimens of aluminum. Dashed line: without elasticity correction (G = 1).

copper in terms of the Monkman-Grant product (Table 1), elasticity effects play a significant role. They cause rapid crack growth at the beginning of the test and lead to a substantial reduction of the lifetime. It is interesting to note that elasticity effects are so pronounced although the lifetimes are several times greater than t_1. The ratio t_f/t_1 is included in the last row of Table 2.

Also shown in Table 2 is a comparison of the lifetimes calculated above with finite element results [13], which are reported to agree well with observed lifetimes. (The quantity Σ_R tabulated in [13] is related to the present notation by $\Sigma_R^{-\chi} = D\sigma_o^{\chi}t_f$). Since for the material parameters involved, no accurate values of the factor α are available, the upper and lower bounds, eqns. (17) and (18), were used giving lower and upper bounds, respectively, for the lifetimes. The elasticity correction depends on the applied stress, σ_o. In [13], σ_o is not specified. Therefore, values were assumed here which appear to be reasonable. Table 2 shows that the present calculations reproduce the finite element results and the measured lifetimes well as far as copper is concerned. For aluminum, only the lower bound for t_f leads to good agreement with the finite element results.

6. DISCUSSION

In the preceding section it has been shown that the lifetime of pre-cracked copper specimens can be predicted with sufficient accuracy based on the small-scale damage approximation. This is so, although the prerequisites for the applicability of fracture mechanics are not particularly favorable in this case. Micrographs [13] show that failure does not occur by the growth of a well defined crack but rather by the diffuse spreading of cavitation from the initial crack tip. Nevertheless, the relative accuracy of the predictions suggests that the evolution of damage is controlled by C^* for the greatest part of the

	Al				Cu	
	CCP		DECP		CCP	DECP
Exp. [13]	1.0		5		0.99	1.18
FE [13]	1.07		5		0.98	1.17
σ_o =	50 MPa	100 MPa	50 MPa	100 MPa	30 MPa	30 MPa
Eqn. (29)	1.03-6.7	.82-6.5	8.4-73	5.7-68	0.83-0.97	1.04-1.22
t_f/t_1 =	12-75	6-48	4.6-41	2.3-26	41-48	20-24

TABLE 2. Normalized lifetimes, $D\sigma_o^{\chi}t_f$ and ratio t_f/t_1 (bottom line). Experimental and finite element results from [13] (σ_o is not specified). Hyphenated entries represent results for upper and lower bound estimate for α, respectively.

lifetime. Further, the specimen geometries considered, in particular the CCP-geometry, set tight limitations to the applicability of C^* (cf. Section 3.4). However, a description of failure by C^*-controlled crack growth seems still to be possible.

In the case of aluminum, the results are less conclusive. First, the upper and lower bound estimates of the crack growth rates in the small-scale damage limit are not close enough together to allow for accurate predictions. This uncertainty can be removed by a finite element analysis of the type described in Section 3.3. Second, the case of aluminum is a complicated one since elasticity effects have a marked influence on the lifetime. They have been accounted for by an interpolation formula which may be inaccurate in the transition region between K_I-controlled and C^*-controlled crack growth. Finally, crack growth in aluminum occurs at an angle to the symmetry plane, at least locally, and it is not clear whether in a thick specimen the crack remains flat macroscopically. Experimentally, this can be enforced by using side-grooved specimens.

7. CONCLUSIONS

The equations of continuum damage mechanics provide a basis for exploring the range of validity of fracture mechanics parameters (K_I, C^* etc.). These approaches are valid as long as the small-scale damage approximation is sufficiently accurate. Limitations to C^* by the evolution of damage have been described in some detail. Important points in this connection are:

1) The process zone grows in proportion to the crack growth increment, Δa. Therefore C^* is invalidated once Δa reaches a certain fraction of the crack length or ligament width. The numerical value of this fraction depends on material parameters and on specimen geometry. A few estimates have been given in Section 3.4.

2) In terms of time, the range of validity of C^* usually comprises a substantial fraction of the lifetime in a constant load test provided that the material is ductile enough so that $t_f >> t_1$.

3) Lifetimes of pre-cracked copper specimens can be predicted accurately based on C^*-controlled crack growth.

4) The analysis of tests on aluminum specimens was not entirely conclusive with respect to the validity of the small-scale damage approximation. However, it was shown that elasticity effects play a great role in the material considered.

Finally it should be emphasized that the crack growth rate will not only depend on the appropriate load parameter but also on the crack growth increment, Δa, as for example, in eqn. (16). Since n and χ have similar numerical values in many materials, the dependence of $\dot{a}$ on Δa is moderate at larger values of Δa so that it can often be neglected.

REFERENCES

[1] Speidel, M.O. (1981). Influence of Environment on Fracture. In: Advances in Fracture Research, eds. D. Francois et al., vol. 6. pp. 2685-2704, Pergamon Press, Oxford and New York.

[2] Landes, J.D. and Begley, J.A. (1976). A Fracture Mechanics Approach to Creep Crack Growth. In: Mechanics of Crack Growth, ASTM STP 590, pp. 128-148, American Society for Testing and Materials. Philadelphia.

[3] Ohji, K. et al. (1974). Application of J-Integral to Creep Crack Problem. Preprint of Japan. Soc. Mech. Engrs., No. 740-11, 207.

[4] Koterazawa, R. and Mori, T. (1977). Applicability of Fracture Mechanics Parameters to Crack Propagation under Creep Condition. J. Eng. Mater. Technology, 99, 298-305.

[5] Taira, S. et al. (1979). Application of J-Integral to High Temperature Crack Propagation. J. Eng. Mater. Technology, 101, 154-161.

[6] Saxena, A. (1980). Evaluation of C^* for the Characterization of Creep Crack Growth Behavior in 304 Stainless Steel. In: Fracture Mechanics: Twelfth Conference, ASTM STP 700, pp. 131-151, ed. P.C. Paris, American Society for Testing and Materials, Philadelphia.

[7] Riedel, H. and Rice, J.R. (1980). Tensile Cracks in Creeping Solids. In: Fracture Mechanics: Twelfth Conference, ASTM STP 700, pp. 112-130. ed. P.C. Paris, American Society for Testing and Materials.

[8] Ohji, K. et al. (1980). The Stress Field and Modified J-Integral near a Crack Tip under Condition of Confined Creep Deformation. J. Soc. Mater. Sci. Japan, 29, No. 320, 465-471.

[9] Riedel, H. (1981). Creep Deformation at Crack Tips in Elastic-Viscoplastic Solids. J. Mech. Phys. Solids, 29, No. 1, 35-49.

[10] Riedel, H. (1983). The Use and the Limitations of C^* in Creep Crack Growth Testing. In: International Symposium on Fracture Mechanics, eds. Tan Deyan and Chen Daning, pp. 997-1012, Science Press, Beijing.

[11] Riedel, H. and Wagner, W. (1984). Creep Crack Growth in Nimonic 80A and in a 1Cr-1/2Mo Steel. Sixth International Conference on Fracture, New Delhi, Dec 1984, eds. S.R. Valluri et al., Pergamon Press, Oxford and New York.

[12] Riedel, H. (1983). Creep Crack Growth. In: ASM Seminar on High Temperature Deformation and Fracture, ed. R. Raj, to be published by American Society for Metals, Metals Park, Ohio.

[13] D.R. Hayhurst et al. (1983). The Role of Continuum Damage in Creep Crack Growth. University of Leicester, Department of Engineering Report 83-4. To be published.

[14] Raj, R. and Baik, S. (1980). Creep Crack Propagation by Cavitation near Crack Tips. Metal Sci., 14, 385-394.

[15] Riedel, H. (1981). The Extension of a Macroscopic Crack at Elevated Temperature by the Growth and Coalescence of Microcracks. In: Creep in Structures, eds. A.R.S. Ponter and D.R. Hayhurst, pp. 504-519, Springer Verlag, Berlin Heidelberg New York.

[16] Bassani, J.L. (1981). Creep Crack Extension by Grain-Boundary Cavitation. In: Creep and Fracture of Engineering Materials and Structures, eds. B. Wilshire and D.R.J. Owen, pp. 329-344, Pineridge Press, Swansea.

[17] Wilkinson, D.S. and Vitek, V. (1982). The Propagation of Cracks by Cavitation: A General Theory. Acta Metall., 30, 1723-1732.
[18] Kachanov, L.M. (1960). The Theory of Creep. English translation, ed. A.J. Kennedy, Boston Spa, Wetherby.
[19] Hayhurst, D.A. and Leckie, F.A. (1983). Behaviour of Materials at High Temperatures. In: Mechanical Behaviour of Materials, eds. J. Carlsson and N.G. Ohlsson, suppl. vol., pp. 8-24, Pergamon Press, Oxford and New York.
[20] Hutchinson, J.W. (1983). Constitutive Behavior and Crack Tip Fields for Materials Undergoing Creep-Constrained Grain Boundary Cavitation. Acta Metall., 31, 1079-1088.
[21] Kumar, V. et al. (1981). An Engineering Approach for Elastic-Plastic Fracture Analysis. Electric Power Research Institute Report NP-1931, Palo Alto.
[22] Hui, C.Y. and Riedel, H. (1981). The Asymptotic Stress and Strain Field near the Tip of a Growing Crack under Creep Conditions. Int. J. Fracture, 17, 409-425.
[23] Hutchinson, J.W. (1968). Singular Behavior at the End of a Tensile Crack in a Hardening Material. J. Mech. Phys. Solids, 16, 13-31.
[24] Rice, J.R. and Rosengren, G.F. (1968). Plane Strain Deformation near a Crack Tip in a Power-Law Hardening Material. J. Mech. Phys. Solids, 16, 1-12.
[25] Shih, C.F. (1983). Tables of HRR Singular Field Quantities. Brown University, Division of Engineering Report MRL E-147.
[26] Ehlers, R. (1981). Stress Analysis near Stationary Cracks under Creep Conditions by the Finite Element Method. PhD-thesis, Fakultaet fuer Bergbau und Huettenwesen, RWTH Aachen. In German.
[27] Ehlers, R. and Riedel, H. (1980). A Finite Element Analysis of Creep Deformation in a Specimen Containing a Macroscopic Crack. In: Advances in Fracture Research, eds. D. Francois et al., vol. 2, pp. 691-698, Pergamon Press, Oxford and New York.
[28] McMeeking, R.M. and Parks, D.M. (1979). On Criteria for J-Dominance of Crack-Tip Fields in Large-Scale Yielding. In: Elastic-Plastic Fracture, ASTM STP 668, eds. J.D. Landes et al., pp. 175-194, American Society for Testing and Materials, Philadelphia.
[29] Hutchinson, J.W. and Paris, P.C. (1978). Stability Analysis of J-Controlled Crack Growth. In: Elastic-Plastic Fracture, ASTM STP 668, eds. J.D. Landes et al., pp. 37-64, American Society for Testing and Materials, Philadelphia.
[30] Rice, J.R. (1968). Mathematical Analysis in the Mechanics of Fracture. In: Fracture, An Advanced Treatise, ed. H. Liebowitz, vol. 2, pp. 191-311, Academic Press, New York and London.

THERMAL EXPANSION OF POLYCRYSTALLINE AGGREGATES

Z. HASHIN

Dept. of Solid Mechanics, Materials and Strcutures,
Faculty of Engineering,
Tel Aviv University,
Tel Aviv, Israel.

The problem under consideration is the analytical determination of the thermal expansion coefficients of a polycrystalline aggregate (PA) composed of randomly oriented anisotropic crystals, in terms of the thermoelastic properties of the single crystals. The related problem of determination of elastic properties of a PA has been the subject of much work but it appears that the thermal expansion problem of a PA has not been given attention.

In the present work two approaches are developed. The first one is exact but is confined to the case where the constituting crystals have hexagonal, tetragonal or trigonal symmetry - thus, have one preferred axis of anisotropy. In these cases there is an exact relation derived between the effective thermal expansion coefficient α^* and the effective bulk modulus K^* of the PA. If K^* is known, e.g. experimentally, then α^* also becomes known. Another attractive possibility is to exploit upper and lower bounds for K^* derived in the literature to establish bounds for α^* via the exact relation. This procedure gives very close bounds for α^* which are in good agreement with experimental data.

When the crystals are more complicated, thus, orthorhombic, monoclinic, or triclinic, the exact method does not apply. The approach taken in this case is the self consistent approximation to assess α^*.

REFERENCE

Hashin, Z. Thermal expansion of polycrystalline aggregates: I. Exact Analysis, II. Self-consistent approximation. J. Mech. Phys. Solids, 32, 149-157, 159-165 (1984).

THE ROLES OF DISLOCATION MOTION AND VACANCY FLUXES IN CREEP

G W GREENWOOD

Department of Metallurgy, Sheffield University, Sheffield S1 3JD U.K.

ABSTRACT

Deformation by dislocation glide is readily distinguishable from that occurring through directed vacancy diffusion. Distinguishing features arise in fine scale shape changes, in defect creation and rearrangement, in relative displacements of any second phase particles and in the form of internal stress redistribution. Such distinctions have led to useful concepts of regimes of stress and temperature over which a particular mechanism of deformation can have predominating influence and to methods of strengthening to match prevailing conditions.

The transition between one mechanism and another however is known not to be sharp and the paper examines particularly the transition between diffusion creep by directed vacancy flow and power law creep involving dislocation motion. Some significant geometrical features arise from the analysis from which the extent and form of the co-existence of the two mechanisms can be identified.

It is discussed further that mechanistic links do not only occur in transition situations. In the examination of creep phenomena in a broader context, physically distinct processes are seen to be related in ways that can be described by similar mathematical formulation. Experimental observations go some way towards providing meaningful interpretations of these relationships.

1. INTRODUCTION

Dislocation motion during creep results in slip steps on free surfaces and in initial dislocation multiplication within grains leading to the development of a dislocation substructure (1). In the steady state under an applied stress σ, a dislocation network of spacing λ is often formed where $\lambda \sim Gb/\sigma$, where G is the shear modulus and b the Burgers vector (2). In polycrystals grain boundaries may also slide and the extent of this sliding will determine the change in the number of grains observed in the cross section as the area reduces (3). There is relative motion between any second phase particles present but, overall, the form of their distribution remains unaltered.

In contrast, creep by stress-directed vacancy diffusion does not lead to any development of defect substructure (4) and can, in principle, occur in material entirely free from defects except for the vacancies present in thermal equilibrium. In polycrystals, the grain shape change is such that the number of grains in the cross section remains the same and the grain boundary sliding that takes place can be regarded as that necessary to accommodate the diffusion creep process (5). There is no relative motion of any second phase particles except that the plating of matrix atoms (equivalent to vacancy creation at) grain boundaries nearly perpendicular to the applied tensile stress (6) leads to particle free zones adjacent to these boundaries. Conversely, atom removal from (equivalent to vacancy absorption at) longitudinal boundaries leads to an accumulation of second phase particles on them. In this creep process, the strain rate increases linearly with the applied stress, whereas a much stronger stress dependence is typical of the involvement of dislocation movement (7).

All these distinguishing features emphasise the importance of acknowledging the existence of different creep mechanisms and of identifying the conditions under which a particular mechanism may predominate (8). Clearly a process with a linear stress dependence must be favoured at low stress levels over one which depends on a higher power of stress. The present analysis begins with such a low stress situation and shows how a total picture can emerge of preferred vacancy diffusion paths and of equipotential lines of the chemical potential of vacancies to which the internal stress redistribution is directly related. The possibility of observing the pattern of re-distributed stress through surface distortion is commented upon. Associated with this pattern is the identification of the region where the shear stress exceeds a given value and where, on increasing the applied stress, dislocation motion is first likely to be activated. Thus conditions are assessed where both dislocation movement and vacancy fluxes simultaneously contribute to creep.

At higher stresses it is clear that dislocation processes dominate and, in doing so, a stress-dependent dislocation substructure is built up. There is now much evidence that the scale of the dis-location network is of critical importance and it plays a somewhat similar role in dislocation recovery creep to that of grain size at lower stresses where creep proceeds by vacancy diffusion (9). Through this connection, different creep processes can be seen to be more closely related than has generally been recognised. The approach emphasises the main physical features rather than becoming enmeshed in complex analyses. Many details remain to be completed but the main theme suggests that a useful overall view may be taken.

2. VACANCY FLUXES AND STRESS REDISTRIBUTION

The simplest situation to analyse is that in Fig. 1 where numerically equal tensile and compression stresses σ and $-\sigma$ respectively, too small to cause dislocation multiplication, are applied perpendicular to pairs of faces of a cubic grain in the x and y directions (10). The vacancy concentration C is enhanced at the grain boundaries perpendicular to the tensile stress and exceeds the thermal equilibrium vacancy concentration C_v by the value $C - C_v = C_v\,(\sigma\Omega/kT)$ where Ω is the atomic volume, k is Boltzmann's constant and T is the absolute temperature, with $\sigma\Omega \gg kT$. Likewise the vacancy concentration is depressed at the boundaries perpendicular to the compressive stress. This leads to a directed vacancy flux shown by arrows in Fig. 1 and a counterflow of atoms which plate uniformly on the pair of boundaries to elongate the grain in the direction of the tensile stress with contraction of the grain by vacancy absorption along the line of compressive stress and no change (apart from small elastic effects) on the third pair of faces.

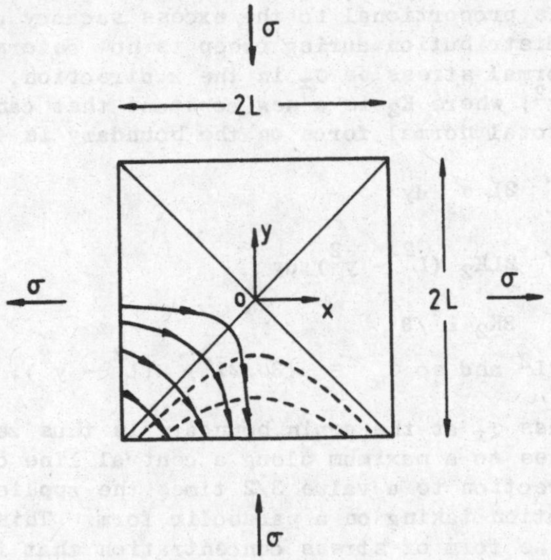

FIG. 1. Numerically equal tensile and compressive stresses σ are applied along the x and y directions respectively, perpendicular to the faces of a cube of side 2L as shown. The arrowed lines illustrate the direction of the vacancy flux and are families of rectangular hyperbolae. These are orthogonal to families of similar hyperbolae shown by dotted lines which represent lines of vacancy equipotential.

These vacancy fluxes lead (11) to a strain rate $\dot{\varepsilon} = D_V\nabla c/L$ in which D_V is the vacancy diffusion coefficient, ∇c the gradient of vacancy concentration and 2L is the length of the grain edge. Now vacancies are only created and destroyed at grain boundaries or free surfaces under these low stress levels and at constant temperature, hence Laplace's equation $\nabla^2 c = 0$ holds. Further, the vacancy concentration everywhere along the pair of unstressed faces perpendicular to the Z axis remains at the value C_V, so we may write $\nabla^2 c = (d^2c/dx^2) + (d^2c/dy^2) = 0$. Since atom plating and removal at the two pairs of faces occurs everywhere at an equal rate, $\delta c/\delta x$ is constant for $x = \pm L$ and $\delta c/\delta y$ constant for $y = \pm L$. It follows additionally that the vacancy concentration is C_V along each of the diagonal planes of the cube described by $x = \pm y$.

With the above boundary conditions, the solution to Laplace's equation is $C - C_V = K_1 (x^2 - y^2)$ where K_1 is a constant. This implies that the vacancy concentration varies along a grain boundary since, for $x = \pm L$, $C - C_V = K_1 (L^2 - y^2)$. The normal stress across the boundary is proportional to the excess vacancy concentration and so the stress redistribution during creep is now determined. Letting this local normal stress be σ_x in the x direction, we may thus write $\sigma_x = K_2 (L^2 - y^2)$ where K_2 is a new constant that can readily be evaluated since the total normal force on the boundary is $4L^2\sigma$.

Thus
$$4L^2\sigma = \int_{-L}^{L} 2L\,\sigma_x\,dy$$
$$= \int_{-L}^{L} 2LK_2\,(L^2 - y^2)\,dy$$
$$= 8K_2\,L^4/3$$

Hence $K_2 = 3\sigma/2L^2$ and so $\sigma_x = (3\sigma/2L^2)\,(L^2 - y^2)$.

The local normal stress σ_x at the grain boundary is thus zero at the grain corners and rises to a maximum along a central line of the boundary in the Z direction to a value 3/2 times the applied stress, with the stress variation taking on a parabolic form. This is in marked contrast to the form of stress concentration that is built up locally where glide processes operate (12). In the y direction the stress is compressive and similarly it is deduced that the local stress σ_y at the pair of faces normal to the y axis is given by

$$\sigma_y = -(3\sigma/2L^2)\,(L^2 - x^2)$$

Although the pair of faces perpendicular to the Z direction is not subjected to stress, these faces are free to distort in response to elastic stresses and do so because of a finite value of Poisson's ratio. Contours of constant excess vacancy concentration which correspond to the lines of vacancy equipotential, can also be related to stress. These equipotential lines take the form of rectangular hyperbolae, shown by the dotted lines in Fig. 1, orthogonal to another set of such hyperbolae shown with arrows which form the lines of vacancy flux. This situation causes a distortion of the pair of faces initially perpendicular to Z, which may be appreciated by considering the strain ε_Z in the Z direction, given by $\varepsilon_Z = -(\sigma_x + \sigma_y)\ \nu/E$ where E is Young's Modulus and ν is Poisson's ratio for the material. It follows that

$$\varepsilon_Z = -(3\sigma\nu/2L^2E)\ (x^2 - y^2) \qquad (1)$$

The form of this distortion is illustrated in Fig. 2 with the pair of square surfaces including A B C D initially lying parallel to the x - y plane now taking a three dimensional shape but with their diagonals $x = \pm y$ remaining in their original positions. Alternate quadrants separated by these diagonals take on convex and concave forms. The lines of vacancy equipotential (which are sections of surfaces intersecting the x - y plane) now emerge as contour lines on the A B C D face as families of rectangular hyperbolae. The shape of lines AB, BC, CD and DA is parabolic corresponding to the normal stress distribution on the faces.

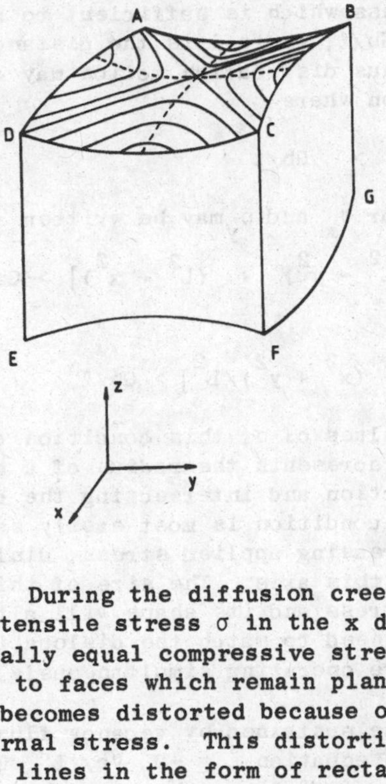

FIG. 2. During the diffusion creep of a cubic crystal with a tensile stress σ in the x direction and a numerically equal compressive stress in the y direction applied to faces which remain plane, the free surface A B C D becomes distorted because of the redistribution of internal stress. This distortion is represented by contour lines in the form of rectangular hyperbolae which correspond to the intersection with the A B C D face of surfaces of constant vacancy concentration lying parallel to the Z direction. Edges in the Z directions also remain straight.

It is interesting to evaluate what may be the order of magnitude of contour depth (or height) with respect to the original surface. This will depend on the parameters constituting Eqn (1) which clearly has a maximum value of ε_Z for $y = L$ and $x = 0$ and a minimum value for $x = L$ and $y = 0$. The greatest differential strain is then given by

$$\Delta \varepsilon_Z = 3 \sigma \nu / E \tag{2}$$

Taking $E \sim 2 \times 10^{11}$ Pa and the stress in a diffusion creep experiment $\sigma \sim 2 \times 10^7$ Pa, with $\nu \sim 0.33$, the maximum differential strain from Eqn (2) gives $\Delta \varepsilon_Z \sim 10^{-4}$. With a large crystal size of $2L \sim 10^{-2}$m, the greatest vertical displacement between points on the top face, e.g. between the mid point of edge DA and the mid point of CD would be $L \Delta \varepsilon_Z \sim (5 \times 10^{-3}) \times 10^{-4} \sim 5 \times 10^{-7}$m. Experimental verification of such a result would appear feasible.

From the point of view of deformation mechanisms, a more important feature which clearly emerges is the identification of a region where $(\sigma_x - \sigma_y)/2$ exceeds some specific value. This can represent the shear stress which is sufficient to activate dislocation motion when it reaches Gb/ℓ, where ℓ is the distance between dislocation pinning points. Thus dislocation motion may contribute to deformation in the region where

$$(\sigma_x - \sigma_y)/2 \quad > \quad Gb/\ell$$

which by substituting for σ_x and σ_y may be written

$$(3\sigma/4L^2) \; [(L^2 - y^2) \; + \; (L^2 - x^2)] > Gb/\ell$$

or

$$(3\sigma/4) \quad [2 - (x^2 + y^2)/L^2] > Gb/\ell \tag{3}$$

For sufficiently high values of σ, this condition can be satisfied when $(x^2 + y^2)^{\frac{1}{2}} < r$ where r represents the radius of a cylindrical region with axis in the Z direction and intersecting the centre of the face A B C D in Fig. 2. The condition is most easily satisfied (Fig. 3) for $x = y = 0$ and so on increasing applied stress, dislocation motion will first be activated near this axis. The size of this region will depend on the applied stress and its shape will alter with increase in its size because of the need to match the dislocation and diffusion creep rates when they are operating simultaneously.

The creep rate sustained by vacancy fluxes is readily determined (13) from the equation $\dot{\varepsilon} = -D_v \nabla c / L$. This can be written in the form $\dot{\varepsilon} = (-D_v/L)(\delta c/\delta x)_{x=L}$ because the rate of arrival of atoms is uniform over the entire grain boundary area. Now it was shown earlier that $C - C_v = K_1 (x^2 - y^2)$ and it is readily deduced that $K_1 = K_2 C_v \; \Omega/kT = 3 \sigma \Omega C_v / 2kTL^2$. Hence $(\delta c/\delta x)_{x=L} = 3 \sigma \Omega C_v / kTL$

and it follows that the creep rate $\dot{\varepsilon} = 3 D_v c_v \sigma \Omega/kTL^2$. But $D_v c_v = D$, the lattice self diffusion coefficient and $2L = d$ the grain or crystal size, so with this stress system and geometry, the Nabarro-Herring equation may be written

$$\dot{\varepsilon} = 12 D \sigma \Omega / kTd^2 \tag{4}$$

Several derivations of Eqn. (4) have been made covering a variety of grain geometries (4). Not all of these have been mathematically correct in the value derived for the numerical coefficient for the particular geometry considered. The value 12 has been arrived at (13) for only a tensile stress acting on a cube and this is in error. The correct value for this case of a tensile stress only is a numerical factor of 8 and departure from a cube geometry leads to substantial variation (14). At first sight it may seem surprising that this value is not one half, i.e. 12/2, of the value where equal tensile and compressive stresses are both present. The reason lies in the different flow patterns for vacancies that are then introduced. In the simple tensile case the fluxes are actually 3-dimensional and are more complex than in the case considered for the derivation of Eqn (4). Only when the crystal dimension along the Z-axis tends to infinity does the coefficient tend to 6 for a tensile stress since the flux lines then revert to a two-dimensional form (Fig. 4).

This approach can also be developed to deal with the situation at lower temperature, or for smaller grain sizes where self diffusion along grain boundaries may make a predominating contribution to the creep rate (15). There is again experimental support (16). Eqn. (4) is then modified by the factor $D_g w/Dd$ where D_g is the grain boundary self diffusion coefficient and w is the grain boundary width. There is also some change in the numerical constant, depending again on geometrical factors.

3. THE GENERAL FORMULATION OF A STEADY STATE CREEP EQUATION

By collection of numerous experimental results, it has been shown (7) that most creep data (for steady state creep) can be incorporated within the general formula

$$\dot{\varepsilon} = A\left[\frac{D G b}{kT}\right]\left[\frac{\sigma - \sigma_0}{G}\right]^n \left[\frac{b}{d}\right]^m \tag{5}$$

where A, n and m are dimensionless constants. For the case of lattice diffusion controlled vacancy creep, the Nabarro-Herring formula (Eqn (4) as above) is clearly of this type when $n = 1$, $m = 2$, $A \sim 10$, σ_0 is negligible and noting that $b^3 = \Omega$ the atomic volume. The Coble modification, when grain boundary diffusion is rate controlling (15), is also clearly incorporated into this general scheme when $n = 1$, $m = 3$, $A \sim 50$ and $b \sim w$ the grain boundary width. Here the value of σ_0 may not be negligible and it has been found experimentally (17) to increase to significant values as the temperature is lowered or the grain size decreased.

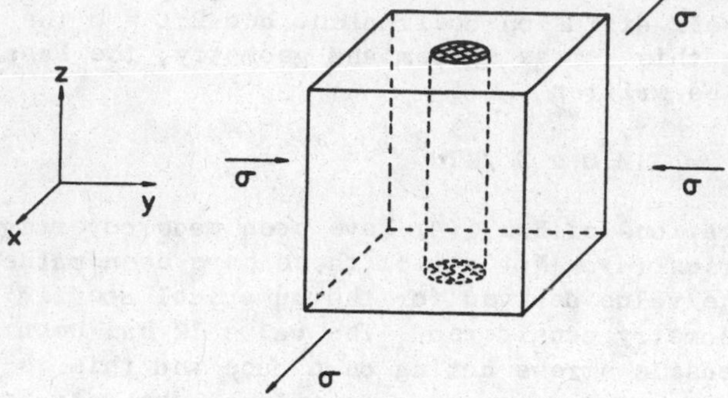

FIG. 3. When the stress level is sufficiently high, dislocation motion is activated within the cylindrical region illustrated with its axis in the centre of the cube perpendicular to the applied stresses. From Eqn. (3) the radius of this cylinder is given by

$$r = (x^2 + y^2)^{\frac{1}{2}} = [2L^2(1 - (2Gb/3\sigma\ell))]^{\frac{1}{2}}$$

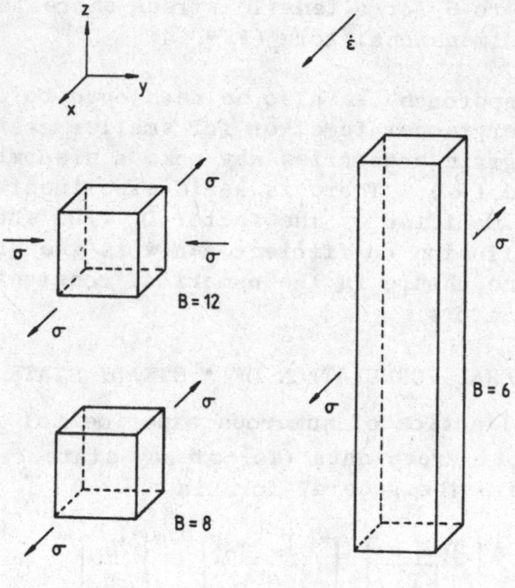

FIG. 4. The value of B in the Nabarro-Herring Eqn. (4) $\dot{\varepsilon} = B\ D\ \sigma\ \Omega/kTd^2$ has the values given above for the particular stress systems and grain shapes illustrated where d is the length of the cube edges and of the small edges of the infinitely long columnar grain. The creep rate $\dot{\varepsilon}$ is measured in the x-direction in which the tensile stress is applied.

Most data used to justify the type of formulation in Eqn (5), however, were collected in the stress and temperature range where dislocation recovery creep was considered to be the operating mechanism. Under these conditions, the effect of variation in grain size is negligible and so $m \sim 0$. In some cases (2) it is found that $n \sim 3$ and $A \sim 1$. Equation (5) may the be reduced to the simplified form

$$\dot{\varepsilon} \sim \left[\frac{D\,G\,b}{kT}\right]\left[\frac{\sigma}{G}\right]^3 \tag{6}$$

In other instances it has been noted that $n \sim 5$ and this has been attributed to the importance of enhanced diffusion along dislocation cores. The effective diffusion coefficient D_e can then be written (8)

$$D_e = D + \beta\,\rho\,b^2\,D_c \tag{7}$$

where ρ is the dislocation density ($\sim 1/\lambda^2$), D_c is the diffusion coefficient along the dislocation core and β is a dimensionless constant. For $\beta\,\rho\,b^2\,D_c >> D$, then $D_e \sim D_c\,\beta\,(b/\lambda)^2$ and, at a stress σ, $\lambda \sim Gb/\sigma$. Thus $D_e \sim D_c\,\beta\,(\sigma/G)^2$ and so Eqn (6) takes the form

$$\dot{\varepsilon} = \beta\left[\frac{D_c\,G\,b}{kT}\right]\left[\frac{\sigma}{G}\right]^5 \tag{8}$$

4. FURTHER LINKS BETWEEN DIFFERENT CREEP PROCESSES

It has recently been shown by Burton and Knowles (9) that the well established relation between dislocation network size and stress links together aspects of diffusion creep and dislocation creep in ways that have not been widely recognised. Such links are best illustrated by considering the ratio of the creep rate to the energy that may be supplied by the applied stress. In the formulation, the creep rate is made dimensionless by dividing it by a frequency factor ν and similarly the energy term involving stress is divided by the thermal energy kT. Then the ratio which is shown to be of some importance takes the form $(\dot{\varepsilon}/\nu)/(\sigma\,b^3/kT)$. If the further step is taken of relating the frequency ν to the diffusion coefficient with $\nu \sim D/b^2$, then this ratio becomes $(\dot{\varepsilon}/D)/(\sigma\,b/kT)$.

The ratio can be approximately evaluated by bringing it into line with Eqn (6). This is done simply by writing

$$\frac{\dot{\varepsilon}/D}{\sigma\,b/kT} \sim \left[\frac{\sigma}{G}\right]^2 \tag{9}$$

so that it is in a form which is now identical with Eqn (6). By invoking the further relation $\sigma/G \sim b/\lambda$, Eqn (9) can be written

$$\frac{\dot{\varepsilon}/D}{\sigma\,b/kT} \sim \left[\frac{b}{\lambda}\right]^2 \tag{10}$$

Comparing Eqn (10) with the Nabarro-Herring Eqn (4), it is readily seen that they become essentially identical when $\lambda \sim d$. On this basis it is apparent that when the stress is too low to form dislocation networks smaller than the grain size, a transition naturally occurs from dislocation recovery creep to diffusion creep. For the former, grain size is relatively unimportant whereas for diffusion creep there becomes a strong grain size dependence with this parameter uninfluenced by the stress level.

When diffusion along dislocation cores takes on a predominating role, then D is replaced by $D_c \beta (\sigma/G)^2$ in Eqn (9) and so, in this case,

$$\frac{\dot{\varepsilon}/D_c}{\sigma b/kT} \sim \beta \left[\frac{\sigma}{G}\right]^2 \left[\frac{\sigma}{G}\right]^2 \tag{11}$$

In this form the equation is identical with Eqn (8).

When temperatures are sufficiently low, or the grain size sufficiently small for grain boundary diffusion creep to predominate, then the above approach can be further extended (9). The effective diffusion coefficient is now given by $D_e = D\,[1 + (w/d)\,(D_g/D)] \sim (w/d)\,D_g$ for $(w/d)\,(D_g/D) >> 1$. Eqn (10) for this case is thus written

$$\frac{\dot{\varepsilon}/D_g}{\sigma\, b/kT} \sim \left[\frac{b}{d}\right]^2 \left[\frac{w}{d}\right] \tag{12}$$

Taking the grain boundary width $w \sim b$, the atom spacing, and with $b^3 = \Omega$, the atomic volume, it is noted that Eqn (11) corresponds to the Coble equation. Thus the important and analogous roles played by grain size in diffusion creep and by dislocation network size in dislocation recovery creep provide a clear link between these mechanisms independently of whether lattice, dislocation core or grain boundary diffusion transport is primarily operating.

The significance of other factors that influence creep, whilst not being considered in this brief general approach, may also be incorporated into an overall interpretation. The much greater influence of stacking fault energy on dislocation recovery creep than on diffusion creep may be reflected in the less well defined dislocation faulting in grain boundaries.

In the first paper to point out and evaluate the contribution to creep that is made by vacancy drift in response to an applied stress, Nabarro (10) also drew attention to the close analogy of this process to fluid flow, as crystal dimensions were reduced to atomic spacings. Using the familiar Nernst-Einstein equation for drift velocity $v = DF/kT$ where F is the driving force equal to the viscous resistance of the medium $6\pi\eta bv$ and putting the viscosity $\eta = \sigma/3\dot{\varepsilon}$ (in tensile instead of the more familiar shear terms), a formula essentially indentical with Eqn (4) was derived with d replaced by b.

It is worth noting that, as Burton has shown, a comparable approach may also be made from the point of view of the kinetic theory of gases. Now the factor ν corresponds to the mean collision frequency $2^{\frac{1}{2}}\pi N b^2 v$ where N is the number of atoms (for a monatomic perfect gas) per unit volume. The gas viscosity $\eta = \sigma/\dot{\varepsilon} = mv/2(2)^{\frac{1}{2}}\pi b^2$ and the velocity is determined from the kinetic energy $mv^2/2 = 3\,kT/2$. Thus, using the ratio previously identified

$$\frac{\dot{\varepsilon}/\nu}{\sigma\, b^3/kT} \sim \left[\frac{2(2)^{\frac{1}{2}}\pi b^2}{mv}\right]\left[\frac{1}{(2)^{\frac{1}{2}}\pi N b^2 v}\right]\left[\frac{kT}{b^3}\right] \sim \frac{1}{Nb^3}$$

Now at high pressures N approaches $1/b^3$ and so that ratio tends to unity, confirming the similarity of behaviour with diffusion creep when grain size is replaced by an atomic dimension.

5. CONCLUSIONS

Creep involving only vacancy movement is clearly distinguishable experimentally from that which arises through the movement of dislocations. In the former case grain boundaries play a key role and the ability of these boundaries to emit and absorb vacancies is a factor of critical importance. In the special case of a cubic grain undergoing diffusion creep with numerically equal tensile and compressive stress applied respectively to two pairs of faces some interesting geometrical factors arise. It is interesting for example to note that this stress system gives a numerical coefficient of 12 in the Nabarro-Herring Eqn (4) whereas if either the tensile or the compressive stress component were removed, this coefficient would be reduced to 8 for a cube, showing the importance of the deviatoric stress component and also of crystal geometry, Fig. (4). These features were implicit in earlier analyses of the problem but do not appear to have been widely recognised. The diffusion creep process is expected to cause small but well defined surface distortion and, more importantly, lead to highest stress levels in the centres of the grains in marked contrast to cases of deformation purely by dislocation glide and shear where stresses rise to highest values at the tips of the sliding interfaces. On increasing the stress level marginally above that which causes only creep by vacancy diffusion, it seems clear that dislocation movement will first arise near grain centres. On further stress increase, this central region of dislocation movement will grow and eventually encompass the grain.

Dislocation motion invariably results in the development of a defect sub structure whose form can be complex. A much improved understanding of such sub structures is required before a detailed interpretation of creep mechanisms becomes possible. Nevertheless in many instances the sub structure can be characterised by the scale of dislocation network which, in turn, is related to the stress. This network appears to play a role in dislocation recovery creep which is analogous to that played by grain size in diffusion creep.

REFERENCES

1. Gittus, J. H., 'Creep, Viscoelasticity and Creep Fracture in Solids' Applied Science Publishers, London, 1975.
2. Takeuchi, S. & Argon, A.S., 1976, J. Mater Sci., 11, 1542.
3. Rachinger, W.A., J. Inst. Metals, 1952-3, 81, 33.
4. Burton, B., 'Diffusional Creep of Polycrystalline Materials' Trans. Tech. Publications, SA, 1977.
5. Raj, R., & Ashby, M.F., Metall. Trans. 1971, 2, 1113.
6. Squires, R. L.et al, J. Nucl. Mater. 1963, 8, 77.
7. Bird, J.E. et al, 1969, 'Quantitative Relationship between Properties and Microstructure,' Israel Univ. Press, p.255 1969.
8. Ashby, M. F., Acta Met., 1972, 20, 887.
9. Burton, B. & Knowles, G., Phil. Mag., A, 1981, 44, 987.
10. Nabarro, F. R. N., Report on Conference on the Strength of Solids The Physical Society, London, 1948, p.75.
11. Gibbs, G. B., Phil. Mag., 1966, 13, 589.
12. Eshelby, J. D. et al, Phil. Mag., 1951, 42, 351.
13. Gibbs, G. B., Mem. Sci., Rev. Met., 1965, 62, 781.
14. Nix, W. D., Metals Forum, 1981, 4, 38.
15. Coble, R. L., J. Appl. Phys., 1963, 34, 1679.
16. Burton, B., & Greenwood, G. W., Met. Sci. J. 1970, 4, 215.
17. Crossland, I. G. & Burton, B., Met. Sci., 1977, 11, 504.

DISLOCATIONS IN SEMICONDUCTORS

P.B. HIRSCH

Department of Metallurgy & Science of Materials, University of Oxford,
Parks Road, Oxford, OX1 3PH, U.K.

ABSTRACT

At relatively low temperatures the motion of dislocations is considered to be controlled by the intrinsic lattice resistance, and to occur by the generation and movement of double kinks. The dislocations are dissociated into partials and structural models and valence force calculations suggest that the cores of the basic partial dislocations may be reconstructed. Kinks may be reconstructed or have dangling bonds; the former will be associated with fairly shallow levels, the latter with deep acceptor and donor levels in the band gap. The existence of acceptor/donor levels at kinks leads to the formation of charged kinks, the concentration of which depends on the position of the Fermi level. The effect of doping on dislocation velocity is attributed to the change of concentration of charged kinks and to the dependence of kink migration energy on the charge state. The application of the microindentation technique to the study of the doping effect on dislocation velocity and on fracture is discussed.

1. INTRODUCTION

Eshelby made seminal contributions to the treatment of the Peierls Force (1), to the motion of kinks (2) on dislocations, and to the problem of charged dislocations in ionic crystals (Eshelby et al (3)). All these studies are relevant to the structure and properties of dislocations in semiconductors. During the last few years there has been a considerable advance in our knowledge and understanding of the structure and properties of dislocations in semiconductors, due to the application of new techniques and theoretical developments. Some of these advances relating to structure and mechanical properties of dislocations in tetrahedrally coordinated semiconductors will be reviewed briefly in this paper.

2. STRUCTURE OF DISLOCATIONS

2.1 DISSOCIATION

Using the "weak beam" method of electron microscopy it has proved possible to demonstrate that glide dislocations in Si, Ge and diamond (4,5,6,7) and in a number of III-V and II-VI compounds, GaAs, GaP, GaSb, InAs, InP, InSb, CdTe (8,9,10,11) lying on (111) planes with $\frac{1}{2}[1\bar{1}0]$ Burgers vectors, are dissociated into two Shockley partials. Glide dislocations in CdS with the hexagonal wurtzite structure have been found to be similarly dissociated (12,13). Several determinations of the intrinsic stacking fault energy, γ_{int} in Si and Ge have been made, and in general the results are consistent with the average values γ_{int} = 54mJ m^{-2} for Si, and γ_{int} = 66 mJ m^{-2} for Ge (for discussion of error limits see Cockayne (14)). For diamond γ_{int} has been determined to be 279$\pm$41 mJ m^{-2} (7). This compares with recent theoretical estimates based on valence potentials, ranging from 992 to 340 mJ m^{-2} (15), and with the results of a first principle calculation of 340 mJ m^{-2} (16).

2.2 GLIDE OR SHUFFLE DISLOCATION?

In the diamond cubic structure the glide dislocation could lie either on the narrowly spaced (111) planes (for example on the Ba plane in fig. 1 which is a projection on to the $(1\bar{1}0)$ plane), or on the widely spaced (111) planes (e.g. plane bB in fig. 1). Since the dislocations are generally dissociated, they will either lie on the narrow planes (glide set) or alternatively take up the associated shuffle set configuration (Hornstra, (17)). A 60° dislocation normal to

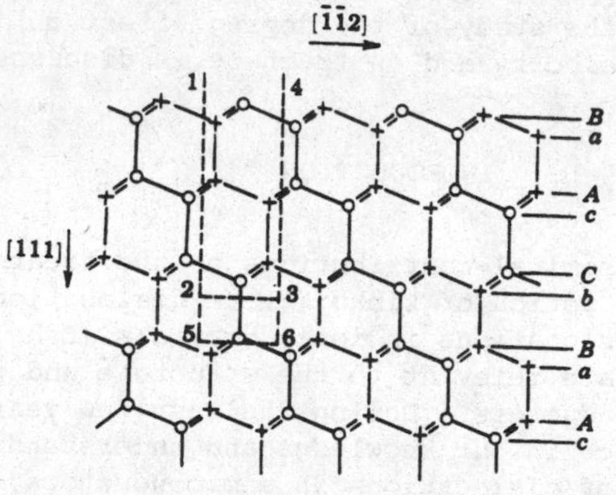

Fig. 1. Diamond cubic lattice projected on to $(1\bar{1}0)$ plane.
o represents atoms in plane of the paper and + atoms in the plane below (from Hirth and Lothe (38)).

fig. 1 dissociates into a 30° and a 90° partial, and in the Hornstra configuration the 30° glide set partial climbs into the neighbouring widely spaced shuffle set plane. High resolution lattice images of well characterised 30° partial dislocations have been shown to be compatible with the simple glide partial structure, but incompatible with the Hornstra associated shuffle configuration derived from the glide partial by vacancy climb (18,19). Although the possibility of the shuffle configuration derived by interstitial climb has not yet been thoroughly tested, the weight of evidence is now in favour of the glide partial configuration.

2.3 CORE RECONSTRUCTION

Shockley's (20) original structure of the core of an undissociated 60° dislocation on the "shuffle set" planes consists of a row of dangling bonds along the core, corresponding to a one dimensional half-filled band (Read (21)). But with the glide partials reconstruction by bonding along the core for 30° partials and across the core for 90° partials is possible (Hirsch (22), Jones (23)). Calculations due to Marklund (24), Jones (23), Altmann et al (25) suggest that for Si reconstruction for 30° partials is likely, but the conclusion for 90° partials is more uncertain. Fig. 2a shows the reconstructed core of a 30° partial running along the length of the diagram, but transferring from one [110] row to the next at a kink; the core is viewed normal to the (111) slip plane, but only one set of atoms

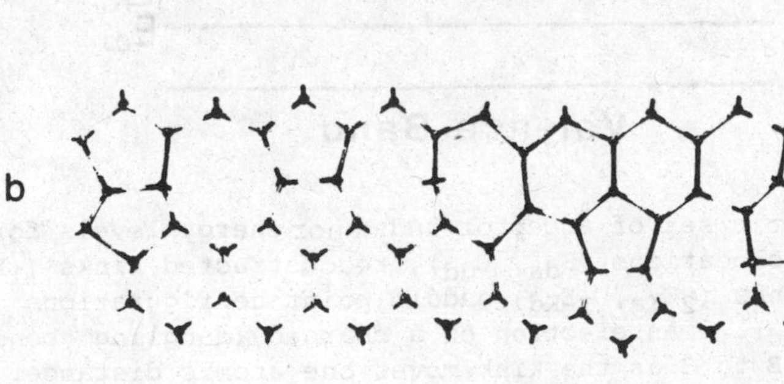

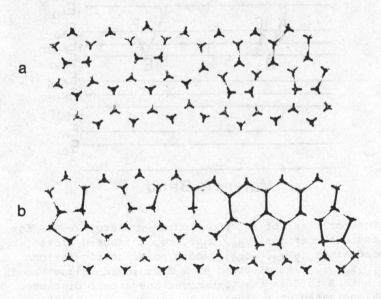

Fig. 2. Reconstructed 30° partials with kinks.
a) reconstructed kink; b) kink with dangling bond.

above and below the slip plane is shown. Reconstruction of this type doubles the periodicity along the core, and any particular atom has the choice of bonding with one of two neighbours in equivalent positions, leading to different relative displacements. Where the bonding changes from one set of neighbours to the other along the core, an antiphase defect (APD) occurs, associated with a dangling bond (Hirsch (26)). This defect has been called a soliton by Heggie and Jones (27). Such an APD/soliton is shown in fig. 2b, where it is situated at a kink. Kinks can be either reconstructed, as in fig. 2a, or associated with dangling bonds, as in fig. 2b (Hirsch (26,28), Jones (29)). An APD can react with a reconstructed kink to form a dangling bond kink (see figs. 2a, b), and conversely APD's can be generated by decomposition of a dangling bond kink.

The effect of reconstruction on electronic energy levels is that the half filled band for the undissociated shuffle dislocation model is replaced by split acceptor and donor levels, the latter filled and the former empty in the neutral state. Fig. 3 shows schematically a set of energy levels which might be expected from reconstructed dislocations and kinks. Acceptor and donor levels for reconstructed dislocations (E_{da}, E_{dd}) and kinks ($_1E_{Ka}$, $_1E_{Kd}$) are likely to be

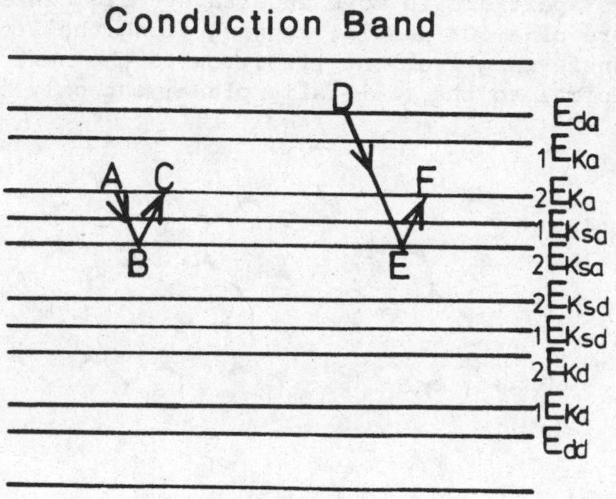

Fig. 3. Schematic set of acceptor and donor energy levels for reconstructed dislocations (E_{da}, E_{dd}), reconstructed kinks ($_1E_{Ka}$, $_1E_{Kd}$) dangling bond kinks ($_2E_{Ka}$, $_2E_{Kd}$), saddle point configurations ($_{1,2}E_{Ksa}$, $_{1,2}E_{Ksd}$). An electron on a charged (dangling bond) kink moves from A to B to C as the kink moves one atomic distance. For a double kink nucleated at a negative charge on a dislocation, an electron moves from D to the saddle point energy level E, to the kink level F.

relatively shallow levels; dangling bond levels associated with dangling bond kinks ($_2E_{Ka}$, $_2E_{Kd}$) are expected to be associated with deeper levels. When a kink migrates along the dislocation line, bonds are broken in the saddle point configuration between neighbouring sites for the kink. The energy levels for the saddle point configurations ($_{1,2}E_{Ksa}$, $_{1,2}E_{Ksd}$) are again expected to be deep levels, and are shown in fig. 3.

There are many results in the literature on energy levels deduced from deep level transient spectroscopy, Hall conductivity data, optical spectroscopy, and other data, but at the present time there is still some uncertainty regarding the identification of various levels. However, for Si there is good evidence for the occurrence of both acceptor and donor levels although the correlation with particular sites on the dislocation is not clearly established.

3. KINK MOBILITY

The activation energy for kink migration is likely to be smaller for dangling bond kinks than for reconstructed kinks, because for the former kink migration occurs by dangling bond exchange compared with the formation of two dangling bonds for the latter. Direct measurements of migration energy have been reported for intrinsic Si (Hirsch et al (30), Louchet (31)). The method consists of finding the length L of dislocation segment below which the dislocation velocity becomes dependent on length of the segment. In fig. 4 consider a segment BE of dislocation segment parallel to the Peierls valley. For small lengths the kinks C, D generated on BE will travel to the ends of the segment before another double kink is generated; under these conditions the dislocation velocity is proportional to the length BE. For long lengths the kinks will be annihilated by other kinks generated on BE before reaching the ends of the segment, and the velocity becomes independent of the segment length. The critical length L is that for which the time taken for a kink to travel to the end is equal to that between successive nucleation events. Thus, if v_K is the kink velocity and v the dislocation velocity, $v_K = \frac{L}{2h}v$, where h is the height of the kink. By in situ electron microscope observations the regime in which the dislocation velocity appears to be

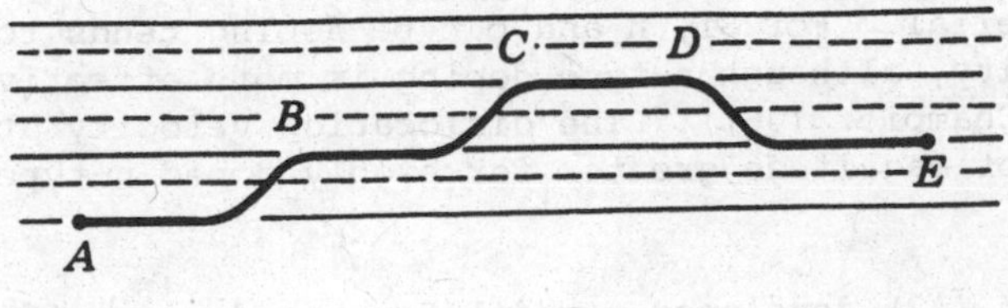

Fig. 4. Movement of dislocations by generation and motion of kinks.

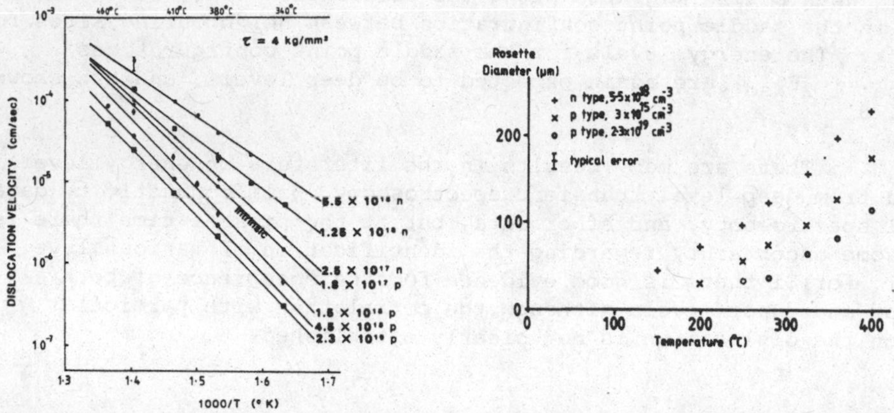

a b

Fig. 5 a. Log dislocation velocity as a function of reciprocal temperature in Ge specimens variously doped with As, Ga (Pirouz, Freeland unpublished); 5b. Rosette size v. temperature (44) for same specimens.

dependent on segment length is identified, and the results suggest a high kink migration energy (∿1.2 ev) for intrinsic Si (30,31), consistent with a reconstructed kink model. No data for doped Si are available. Such high activation energies are confirmed by recent internal friction measurements (32).

4. EFFECT OF DOPING ON DISLOCATION VELOCITY

At relatively low temperatures, in the Peierls controlled regime, the dislocation velocity depends on doping. The effect was first established by Patel and Chaudhuri (33) for Ge, and since then the doping effect has been observed in Ge, Si and in some III-V compounds (34,35,36,37). Fig.5a shows some unpublished results by Pirouz and Freeland on Ge, using a wide range of doping levels, and the single stress pulse dislocation etching technique. These results confirm the earlier data of Patel and Chaudhuri (33) to show that the dislocation velocity is greater for n type and smaller for p type than for intrinsic material. For Si, n and p type doping tends to increase dislocation velocity, although n-type doping is more effective than p type (George and Champier (36)). The dislocation velocity at 340°C is about two orders of magnitude greater for highly doped n-type than for p-type Ge.

Hirsch (22) developed a model for the doping effect within the framework of the drift model discussed by Hirth and Lothe (38), which applied to relatively high temperatures and low stresses, and in which the velocity is proportional to the concentration of kinks (exp $-F_k/KT$, F_k is the kink formation energy), the velocity of the kinks (α exp $-W_m/kT$, W_m is the activation energy for kink migration along the dislocation line assumed to be controlled by the secondary Peierls potential), and the driving force τ^n (τ is the resolved shear stress and $n \sim 1\rightarrow 2$). In the model doping affects the concentration of charged kinks, and the activation energy of migration of charged kinks. The theory follows closely that for the doping effect on diffusion, kinks replacing point defects. If we neglect the concentration of positively charged kinks, $_+c_k$, the concentration of negatively charged kinks, $_-c_k$, relative to that of neutral kinks, $_oc_k$, is given by Fermi-Dirac statistics as

$$_-c_k/_oc_k = \exp\,(E_F - {_iE_{Ka}} - eV)/kT \tag{1}$$

where E_F is the Fermi energy and eV is the electrostatic energy of the charged dislocation, i = 1,2 depending on the type of kink. The concentration of charged kinks therefore depends on doping through E_F. The concentration of neutral kinks $_oc_k$ is constant, at a given temperature, independent of the doping level, provided the interaction between charged and neutral kinks is neglected, and the concentrations of all kinks are small. The velocity of the dislocations controlled by negatively charged kinks (assuming $_+c_k$ to be negligible) is given by

$$v_-/v_o = \exp\,(E_F - {_iE_{Ka}} - eV - \Delta W_m)/kT \tag{2}$$

where $\Delta W_m = W_m' - W_m$ is the difference in migration energy between charged (W_m') and uncharged (W_m) defects. During the motion of the kinks over the saddle point, assuming the Born-Oppenheimer approximation holds, the electron will follow the motion of the ion and move from energy level $_iE_{Ka}$ to $_iE_{Ksa}$ and back to $_iE_{Ka}$ during the activated process (e.g. A to B to C in fig. 3). Thus

$$\Delta W_m = {_iE_{Ksa}} - {_iE_{Ka}} + \Delta W_m' \tag{3}$$

where $\Delta W_m'$ is the contribution to ΔW_m from changes in position of the ions due to charge. Thus, the activation energy term ΔU in (2) becomes

$$\Delta U = (E_F - {_iE_{Ksa}} - eV - \Delta W_m') \tag{4}$$

This expression is similar to that obtained by Jones (29) who assumed that equilibrium with the electron gas is established during the activation at the saddle point. The present treatment avoids this difficulty (Hirsch (39)) since equilibrium is established already in the ground state of the kink. However, as pointed out by Jones (29), the electron energy level important in the doping effect is that corresponding to the saddle point configuration. This energy level is not necessarily the same as that found from DLTS, Hall conductivity or optical data from specimens containing static dislocation distributions.

At higher stresses and lower temperatures, the dislocation velocity is controlled by the nucleation of double kinks (Hirth and Lothe (38)). Hirsch (39), suggested that preferential nucleation occurs at a charged point on the dislocation line; Heggie and Jones ((27), (41)) proposed that preferential nucleation occurs at an APD. Several alternative reaction paths seem possible, but in the simplest case for the former mechanism an electron moves from a dislocation level E_{da}, to a saddle point level ($_2E_{Ksa}$, see fig. 3), corresponding now to the saddle point for nucleation of the double kink, and beyond the saddle point to a charged kink level $_2E_{Ka}$, i.e. corresponding to D to E to F in fig. 3. The result is the nucleation of a double kink, with one charged and the other uncharged. The charged kink is likely to be more mobile, but after equilibrium is reestablished with the electron gas, both kinks may be charged, depending on the position of E_F. The resulting expression for the difference in activation energy of the velocity of dislocations controlled by charged and neutral double kinks is identical to that derived for the drift model, assuming that the velocity is independent of the segment length. Heggie and Jones ((27), (48)) have obtained the same result for their model.

For an extensive set of data for 60° for dislocations in Si (George and Champier (36)) it has been possible to neglect the eV term in (4) and derive a consistent pair of acceptor and donor levels (Hirsch (39), (40)); an acceptor level E_a = 0.67±0.04 ev, with $\partial E_a/\partial T$ = (-1.6±0.5) x 10^{-4} ev K^{-1} and a donor level E_d = 0.28±0.17 ev with $\partial E/\partial T$ = (-2.0±2.1) x 10^{-4} ev K^{-1}, all measured relative to the valence band. For the Ge data in fig.5a it has not been possible to obtain a consistent fit by neglecting eV and this term will probably have to be taken into account.

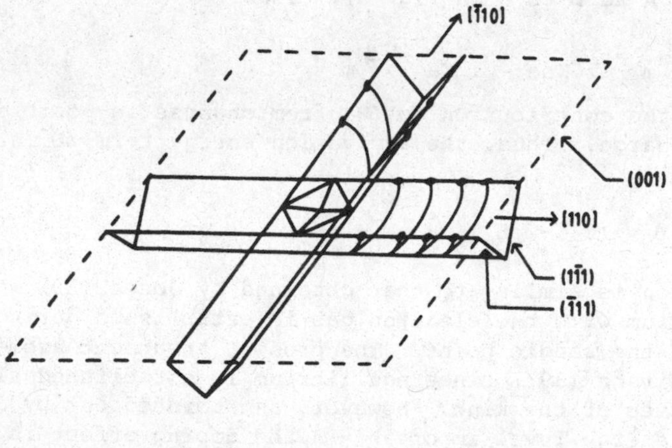

Fig. 6. Form of indentation rosette on a {100} silicon surface (Hu (49)).

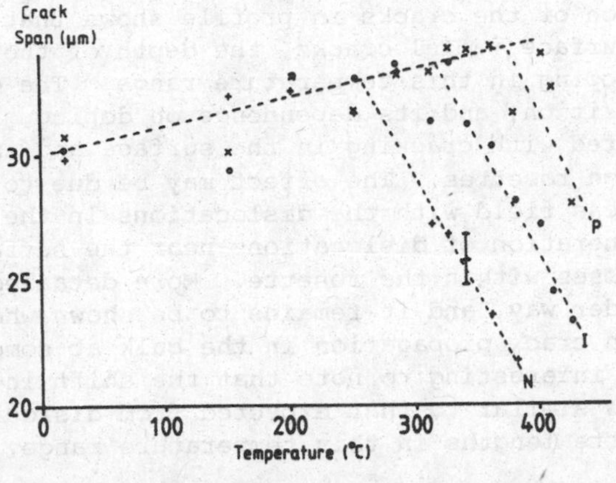

Fig. 7. Radial crack span at 50 g load for three differently doped germanium samples; heavily doped n and p type (N, P) and intrinsic (I) (44).

5. EFFECT OF DOPING ON INDENTATION ROSETTES AND FRACTURE

The effect of doping on dislocation velocity can be followed qualitatively by measuring the extent of dislocation rosettes for a given load as a function of doping. Results for Si have reflected the trends expected from the dependence of dislocation velocity on doping (Roberts et al (43)). More recently results have been obtained for Ge, for indentations on a (100) surface. Fig. 6 shows the geometry of the rosettes, and Fig.5b the rosette diameter as a function of doping and temperature (Roberts et al (44)). The trend with doping is in line with that expected from dislocation velocity measurement (see fig. 5a).

Sharp indentations in brittle materials are often surrounded by cracking. The normal pattern of such cracks has been described by Lawn and Wilshaw (45), those which form on symmetry median planes containing the load axis (median/radial system), and those which form laterally on planes parallel to the surface (lateral system). A characteristic of median/radial cracks is that they generate surface traces emanating radially from the corners of the impression of the indenter. The driving force is the elastic point contact strain field due to the indenter during loading, and the residual strain field due to the mismatch between the plastic zone and the undeformed surrounding material during unloading (Lawn et al (46)). The radial cracks are thought to reach their maximum extent during the unloading part of the cycle. The lengths of the radial cracks (along [110]) were measured as a function of temperature and doping. At low temperatures data from all specimens lie on the same line (fig. 7). With increasing temperature the curves for each doping level diverge from this line, the cracks becoming rapidly shorter. This divergence occurs at 290°C for n-type, 350°C for intrinsic, and 400°C for p-type material. Examination of the cracks in profile shows that, unlike the length of the near surface radial cracks, the depth of the median cracks does not vary with doping in this temperature range. The observed ductile-brittle transition, and its dependence on doping, is therefore specifically associated with cracking in the surface and in particular within the dislocation rosettes. The effect may be due to interaction of the crack tip stress field with the dislocations in the rosettes, to crack blunting by generation of dislocations near the surface, or to the compressive stresses within the rosette. More detailed interpretation is under way, and it remains to be shown whether there is a doping effect on crack propagation in the bulk at somewhat higher temperatures. It is interesting to note that the shift in brittle ductile transition is similar to that expected from dislocation velocity measurements or rosette lengths in this temperature range.

In GaAs the edge dislocations moving in the rosette arms along the $[110]$ and $[\bar{1}10]$ directions in fig. 6 have their extra half planes terminating along rows of Ga and As atoms respectively (α and β dislocations). Such dislocations have different core structures and velocities. Unlike the case for Si and Ge, the rosettes in n-type GaAs at 350°C are asymmetric (Warren et al (47)), the long and short branches reflecting the differences in velocities for α and β dislocations in n-type GaAs (Choi et al (48)). The system of median/radial cracks along the <110> directions is also found to be asymmetric, correlating with the long and short branches of the rosette, as expected from the results for Ge. It is clear that the micro-indentation technique provides a powerful tool for the qualitative assessment of the relative mobility of dislocations of different types, in differently doped crystals, at different temperatures, and it will be particularly interesting to see whether the technique can be used to study the effect of doping on crack propagation.

The existence of quite strong doping effects in GaAs raises the possibility that this effect might also occur in SiC, and might be utilised to change the brittle-ductile transition temperature in this engineering ceramic. Experiments are under way to explore this possibility.

6. CONCLUSIONS

1. Dislocations with Burgers vectors $\frac{1}{2}[1\bar{1}0]$ on (111) planes in tetrahedrally coordinated semiconductors are dissociated into partials, and the weight of evidence suggests that they are lying on the narrowly spaced planes (glide set).

2. Models and calculations suggest that in the elemental semiconductors the 30° and possibly the 90° partials are reconstructed. Reconstructed and dangling bond type kinks can

occur. Reconstructed dislocations are expected to be associated with split bands (relatively shallow acceptor and donor levels). Dangling bond kinks are expected to be associated with deep levels.

3. The activation energy for migration of kinks in intrinsic Si is high and suggests that the kinks are reconstructed.

4. The doping effect on dislocation velocity is thought to be due to the enhanced concentration and mobility of charged kinks.

5. The doping effect on dislocation velocity can be studied by measurements on rosettes around micro-indentations. The different core structures of α and β dislocations in III-V compounds lead to different velocities for these two types of dislocations, and this effect results in asymmetric rosettes on (001) planes.

6. The fracture behaviour around micro-indentations is also dependent on the effect of doping on plasticity.

ACKNOWLEDGEMENTS

My thanks are due to P. Pirouz, S.G. Roberts, and P.D. Warren for providing the recent experimental material for this paper, and for helpful discussions. This work is supported by the B.P. Venture Research Unit.

REFERENCES

(1) Eshelby, J.D. (1949). Phil. Mag. 40, 903.
(2) Eshelby, J.D. (1962). Proc. Roy. Soc. A 266, 222.
(3) Eshelby, J.D. et al. (1958). Phil. Mag. 3, 75.
(4) Ray, I.L.F. & Cockayne, D.J.H. (1971). Proc. Roy. Soc. A 325, 543
(5) Ray, I.L.F. & Cockayne, D.J.H. (1973). J. Micros. 98, 170.
(6) Häussermann, F. & Schaumburg, H. (1973). Phil. Mag. 27, 745.
(7) Pirouz, P. et al. (1983). Proc. Roy. Soc. A 386, 241.
(8) Gai, P. & Howie, A. (1974). Phil. Mag. A 30, 939.
(9) Gómez, A. & Hirsch, P.B. (1978). Phil. Mag. A 38, 733.
(10) Gottschalk et al (1978). Physica Status Solidi a 45, 207.
(11) Hall. E.L. & Vander Sande, J.B. (1978). Phil. Mag. A 37, 137.
(12) Cockayne, D.J.H. et al. (1980). Phil. Mag. A 42, 733.
(13) Pirouz, P. et al (1982). Electron Microscopy and Analysis 1981 (ed. M.J. Goringe), Confer. Series 61, p. 531. (Bristol, London: Institute of Physics).
(14) Cockayne, D.J.H. (1981). Rev. Mater. Sci. 11, 75.
(15) Altmann, S.L. et al. (1983). Phil. Mag. 47, 827.
(16) Persson, A. (1983). Phil. Mag. 47, 835.
(17) Hornstra, J. (1958). J. Phys. Chem. Solids 5, 129.

(18) Anstis, G.R. et al (1981). Microscopy of Semiconductor Materials, Oxford 1981, Confer. Series 60, p. 15. (Bristol, London: Institute of Physics).
(19) Bourret, A. et al (1981). Microscopy of Semiconductor Materials, Oxford 1981, Confer. Series 60, p. 9. (Bristol, London: Institute of Physics).
(20) Shockley, W. (1953). Phys. Rev. 91, 228.
(21) Read, W.T. (1954). Phil. Mag. 45, 1119.
(22) Hirsch, P.B. (1979) J. de Physique, Suppl. C6, 40, C6-117.
(23) Jones, R. (1979). J. de Physique, Suppl. C6, 40, C6-33.
(24) Marklund, D. (1979). Phys. Stat. Sol. (b) 92, 83.
(25) Altmänn, S.L. et al (1983). Int. J. Quantum Chem. 23, 1057.
(26) Hirsch, P.B. (1980). J. Microscopy 118, 3.
(27) Heggie, M. & Jones, R. (1982). J. de Physique, Suppl. C6, 43, C1-45.
(28) Hirsch, P.B. (1981) Proc. Mater. Res. Soc. Symp. on Defects in Semiconductors (ed. J. Narayan and T.Y. Tan), p. 257. (New York: North Holland).
(29) Jones, R. (1980). Phil. Mag. 42B, 213.
(30) Hirsch, P.B. et al (1981). Microscopy of Semiconductor Materials, Oxford 1981, Confer. Series 60, p. 29 (Bristol, London: Institute of Physics).
(31) Louchet, F. (1981). Microscopy of Semiconductor Materials, Oxford 1981, Confer. Series 60, p. 35 (Bristol, London: Institute of Physics).
(32) Möller, H.T. private communication
(33) Patel, J.R. & Chaudhuri, A.R. (1966) Phys. Rev. 143, 601.
(34) Patel, J.R. et al (1976) Phys. Rev. B 13, 3548.
(35) Erofeev, V.N. & Nikitenko, V.I. (1971). Sov. Phys. Solid State 13, 116.
(36) George, A. & Champier, G. (1979). Phys. Stat. Sol. (a) 53, 529.
(37) Erofeev, S.A. & Ossipyan, Yu.A. (1973). Sov. Phys. Solid State 15, 538.
(38) Hirth, J.P. & Lothe, J. (1968). Theory of Dislocations (New York: McGraw Hill).
(39) Hirsch, P.B. (1981). J. de Physique, Suppl. C3, 42, C3-149.
(40) Hirsch, P.B. (1983). Microscopy of Semiconductor Materials, Oxford 1983, Confer. Series 67, p. 1. (Bristol, London: Institute of Physics).
(41) Heggie, M. & Jones, R. (1983). Microscopy of Semiconductor Materials, Oxford 1983, Confer. Series 67, p. 45. (Bristol, London: Institute of Physics).
(42) Hirsch, P.B. (1984). Report of Conference on Plastic Deformation of Ceramic Materials, Penn State 1984, in press.
(43) Roberts, S.G. et al (1983). J. de Physique, Suppl. C4, 44, C4-75.
(44) Roberts, S.G. et al (1984) Poster at Conf. on Dislocations: Core Structure and Physical Properties, Aussois 1984.
(45) Lawn, B.R. & Wilshaw, T.R. (1975). Fracture of Brittle Solids (Cambridge: University Press).

(46) Lawn, R. et al (1980). J. Am. Ceram. Soc. 63, 574.
(47) Warren, P.D. et al (1984). Phil. Mag. in press.
(48) Choi, S.K. et al (1977). Jap. J. Appl. Phys. 16, 737.
(49) Hu, S.M. (1973). J. Appl. Phys. 46, 1470.

DISLOCATIONS IN IONIC CRYSTALS

P. L. PRATT
Department of Metallurgy & Materials Science, Imperial College,
Prince Consort Road, London SW7 2BP, UK.

ABSTRACT

Dislocations in ionic crystals are often electrically charged and in thermal equilibrium they are surrounded by compensating charge clouds. The presence of the charges on dislocations can influence their mobility, but the force required to pull a dislocation away from its charge cloud is too small to be involved in determining the flow stress. The transport properties of alkali halide crystals are strongly influenced by plastic deformation. Near room temperature transient increases and steady-state decreases in conductivity have been observed. At higher temperatures, pipe-diffusion of both cations and anions can occur.

1. INTRODUCTION

In 1933 Stepanov (1) showed that plastic deformation of rock-salt produced a potential difference between the ends of the crystal in the absence of an applied field. He found that the ionic conductivity increased during deformation, attributing this to local melting near to the glide planes. In 1956 Eshelby and I (2) showed that the heating effect of moving dislocations is negligible near the glide planes, except perhaps at very high rates of strain, so that local melting is unlikely to occur; Seitz (3) had earlier accounted for the increased conductivity in terms of the generation of vacancies by moving dislocations. In 1955, Fischbach and Nowick (4) suggested that the potential difference found by Stepanov was due to the motion of dislocations, which became positively charged by the formation of jogs as they moved into the crystal. In 1956 Bassani and Thomson (5) calculated the binding energy, and thus the concentration, of positive ion vacancies at sites on the straight, unjogged dislocation core. From these foundations emerged the theory of charged dislocations in ionic crystals (6,7,8). In this paper a summary of the present state of the theory is followed by consideration of the mobility of dislocations and of the influence of dislocations upon the transport properties of ionic crystals. All of these ideas as well as others were considered in the original paper of Eshelby, Newey, Pratt and Lidiard in 1958 (8).

2. THE THEORY OF CHARGED DISLOCATIONS

In this Section the infinite source/sink model of Eshelby et al (8) is contrasted with the bound vacancy model of Bassani and Thomson (5) in which dislocations do not act as sources or sinks of vacancies. The incorporation of the concept of bound vacancies into the infinite source/sink model by Whitworth (9) has led to the development of a general theory, which has implications for the group II-VI semiconductors. These equilibrium models can only apply at temperatures where all the defects involved are mobile. Particularly at room temperature where many experiments have been carried out, much larger non-equilibrium charges have been observed, up to one vacancy per site along straight edge dislocations.

2.1 EQUILIBRIUM THEORIES

INFINITE SOURCE/SINK MODEL. Eshelby et al (8) extended the model of Frenkel (10) and Lehovec (6) to include dislocations as well as surfaces as a source and sink for vacancies in ionic crystals. If the energies required to form positive and negative ion vacancies are unequal, then in thermal equilibrium dislocations will be electrically charged and surrounded by a Debye-Hückel cloud of vacancies. The presence of divalent impurities modifies the magnitude of the charge on a dislocation, and may even reverse it, provided that the impurity ions are mobile as well as the vacancies.

Following the notation of Whitworth (9), consider a semi-infinite alkali halide crystal with a plane surface, measuring distance x into the crystal from the surface. Let N be the number of cation or anion sites per unit volume, with $n_-(x)$ and $n_+(x)$ the concentrations of cation and anion vacancies per unit volume respectively. If g_- and g_+ are the free energies of formation of cation and anion vacancies, the vacancy concentrations at a depth x are

$$n_-(x) = N \exp\left[-\frac{g_- - e(\phi(x) - \phi(o))}{kT}\right]$$

and

$$n_+(x) = N \exp\left[-\frac{g_+ + e(\phi(x) - \phi(o))}{kT}\right],$$

where $\phi(o)$ and $\phi(x)$ are the potentials at the surface and at depth x. For an intrinsic crystal $n_+(\infty) = n_-(\infty)$. Hence the potential difference between the surface and the interior is

$$\phi(o) - \phi(\infty) = \frac{g_+ - g_-}{2e}.$$

Since $g_+ > g_-$ for alkali halides, the surface is positive relative to the interior.

In the space charge region $\phi(x)$ must satisfy Poisson's equation

$$\nabla^2\phi(x) = -\frac{e(n_+(x) - n_-(x))}{\varepsilon_o\varepsilon_r}$$

Setting $p(x) = e(\phi(x) - \phi(\infty))/kT$, this becomes

$$\nabla^2 p = \kappa^2 \sinh p,$$

where $$\kappa^2 = \frac{2e^2 n_-(\infty)}{\varepsilon_o\varepsilon_r kT}.$$

$p(x)$ falls from a finite value at the surface to zero at a depth given by the Debye-Hückel screening length $\lambda = \kappa^{-1}$. The charge density on the surface is given by

$$q_s = \frac{2\varepsilon_o\varepsilon_r kT}{e\lambda} \sinh \frac{e(\phi(o) - \phi(\infty))}{2kT}.$$

This theory holds for mobile divalent impurity cations with $n_-(\infty)$ as the concentration of free cation vacancies. In equilibrium

$$\phi(o) - \phi(\infty) = -(g_- + kT\ln\alpha)/e,$$

where $\alpha = n_-(\infty)/N$ is the fractional concentration of free cation vacancies inside the crystal. In the extrinsic range of course α can be much greater than the intrinsic value for the temperature concerned and the potential difference $\phi(o) - \phi(\infty)$ can become negative instead of positive. In the absence of detailed calculations, Eshelby et al (8) argued that the free energy of formation at 0 °K, g_-^o probably should lie between 0.6 and 1.0 eV and that the values for surface and dislocation need not be identical.

Whitworth (9) has identified three difficulties that arise when this model is used to estimate the magnitude of the charge on the dislocation:-

(i) As recognised by Eshelby et al (8), the cylindrical space-charge equation cannot be solved analytically; in the limit $|e\phi| << kT$, it can, but this is not a realistic approximation.

(ii) If the dislocation is represented as a uniform line charge, the potential at its core becomes infinite. The core is better treated as a set of point charges.

(iii) The dislocation is not an infinite source or sink of vacancies. Saturation of the available jogs and the configurational entropy of the charged jogs cannot be neglected.

BOUND VACANCY MODEL. Bassani and Thomson (5) earlier had calculated the binding energy of free cation vacancies at sites on the core of an otherwise straight edge dislocation as about 0.4 eV. From this they determined the equilibrium concentration of cation vacancies, allowing for the electrostatic energy required to assemble a line of charged defects, and the minimisation of this by the formation of the Debye-Hückel charge cloud. In this model the dislocation is not allowed to act as a source or sink of vacancies. Whitworth (11) has refined this model into a more precise form, by calculating the change in the electrostatic energy, E_u, for the uniform spacing of M vacancies along a straight edge dislocation, resulting from the addition of one vacancy to the core. The equilibrium concentration f of vacancies at sites on the core becomes

$$\frac{f}{1 - f} = \alpha \exp\left[\frac{1}{kT}\left(g_{-d} - \frac{\partial E_u}{\partial M}\right)\right],$$

where g_{-d} is the free energy of binding at the dislocation. Further refinements of the bound vacancy model lead to a lower concentration than this.

DISLOCATIONS WITH JOGS AND BOUND VACANCIES. Whitworth (12) had realised that the infinite source/sink model, which allows vacancies to be created or destroyed, required fundamental changes if this takes place at jogs. In fact these changes enable the concept of bound vacancies to be included in the infinite source/sink model, although the complications become considerable.

The processes considered by Whitworth include the creation of cation and anion vacancies at half-jogs, thus changing the charge on the jog, the removal of cation and anion vacancies from bound sites on the dislocation to distant sites, the creation of Schottky pairs both in the lattice and at sites on the dislocation and the formation of separated pairs of half jogs of opposite sign. Minimising the free energy of the whole system, assuming all configurations of defects on the core to be equally probable enables an expression for the equilibrium charge per unit length of dislocation to be derived. Part of the charge is carried by jogs and part by the bound vacancies, but no allowance was made for charged impurity ions on the core, either singly or bound to vacancies. There is still very little information about the free energies of these processes and it is difficult to decide how to treat the change in electrostatic energy at the core with more than one kind of defect on it. On this basis it is impossible to draw conclusions about the likely size of the charge on dislocations or how they are made up.

THE ISOELECTRIC TEMPERATURE. Eshelby et al (8) predicted the existence of an isoelectric temperature at which the equilibrium charge on the surface or on the dislocation becomes zero, linking this with the minimum found in the temperature dependence of the yield stress. They introduced the quantity

$$\alpha' = \exp(-g_-/kT) = \exp\left(\frac{s_-}{k} - \frac{h_-}{kT}\right)$$

which is the concentration of cation vacancies to be expected neglecting the presence of impurities and the requirement of bulk neutrality. The difference between this "naive" concentration, α', and the true concentration, α, gives a measure of the charge on the dislocation. α becomes equal to α' at the isoelectric temperature, T_i, and Eshelby et al (8) considered a number of ways in which this could occur. Whitworth (9) refined the discussion by making use of experimental data for the true concentration, α, from Allnatt et al (13) and Huddart and Whitworth (14), together with reasonable assumptions about the components of g_- for α'. In Fig. 1 the full lines for α are derived from ionic conductivity and the extrapolation of these to A is set by $\ln \alpha = s_S/2k$ at $T^{-1} = 0$. The Schottky formation energy $g_S = h_S - Ts_S$, with $s_S =$

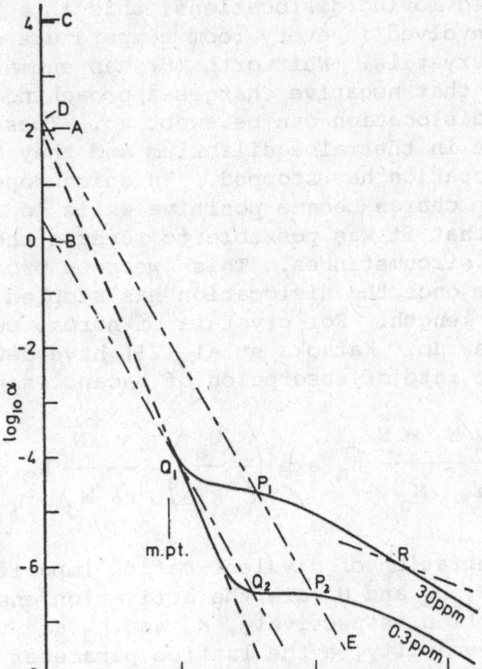

FIG.1. The fractional concentration α of free cation vacancies in NaCl as a function of temperature. After Whitworth (9).

$s_+ + s_-$ and $h_S = h_+ + h_-$. Assuming that both s_+ and s_- are positive, the ln α' line must start between B and C and have a slope $h_- < h_S/2$. For ln α' given by DE, the isoelectric temperatures are P_1 and P_2. If $s_- < s_+$, Q_1 and Q_2 represent two possible isoelectric temperatures in the same crystal. Isoelectric temperatures have been determined experimentally in a number of different ways. Whitworth (9) concludes that they are not consistent with one another, nor are they easy to interpret in view of the high electrical conductivity and diffusion rates at the temperatures of interest. Nevertheless the results he gives in his Fig. 28 (15,16,17,18), and the recent results of Harris (19), appear as almost parallel lines lying not far below the line DE in Fig. 1.

2.2 NON-EQUILIBRIUM THEORIES

At low temperatures the point defects in ionic crystals become less mobile and different ways of acquiring charge can occur. Brown (20) considered the case of NaCl at room temperature where only cation vacancies can move. He found the theory unchanged except in detail, with very nearly the equilibrium charge. He mentions the suggestion by myself of the acquisition of charge by the sweep-up of vacancies by uncharged moving dislocations; this is a valuable and important concept, involved in every room temperature deformation experiment on ionic crystals. Whitworth (9) has summarised the early observations, noting that negative charges approaching one vacancy per two sites along the dislocation can be swept up. These large charges could not possibly be in thermal equilibrium and they decrease after the movement of the dislocation has stopped. In anion doped alkali halide crystals the sweep-up charge became positive while in irradiated NaCl Davidge has shown that it was possible to reverse the sign of the charge under certain circumstances. This sweep-up process must approach a kinetic equilibrium once the dislocation has sampled a vacancy at every site along its length. For crystals of normal purity this amounts to a movement of a few µm. Kataoka et al (21) have developed a simple kinetic model for the rate of absorption of vacancies per unit length

$$\frac{dN}{dt} = \frac{cv}{ba}\,\frac{N_o - N}{N_o}\,K_a \exp\left(-\frac{U_a}{kT}\right) - \frac{v}{ba}\,\frac{N}{N_o}\,K_D \exp\left(-\frac{U_d}{kT}\right)$$

where c is the concentration of divalent cation impurities, N_o is the saturation value of N, U_a and U_d are the activation energies for adsorption and desorption respectively, K_a and K_d are the rate constants, v is the dislocation velocity, a the lattice parameter and b the Burgers vector. When $dN/dt = 0$, the steady-state charge is given by

$$q = q_o/[1 + (K_d/cK_a) \exp (U_a - U_d)/kT].$$

In experiments on crystals of Ca^{++} doped KCl they found good agreement with the form of this expression although the physical reality is obscure. The saturation value of the dislocation charge q_o approached one vacancy per site along the dislocation. This number is twice that found by Huddart and Whitworth (14) perhaps due to the much higher dislocation velocity in the experiments of Kataoka et al. By using the same crystal at different temperatures Kataoka et al found values for the energy difference (U_a - U_d). The lower value at temperatures below 200 °K they suggest is due possibly to sweep-up of isolated vacancies, the higher value to sweep-up from impurity-vacancy pairs.

3. THE MOBILITY OF DISLOCATIONS

The mobility of dislocations in ionic crystals with the NaCl structure depends upon temperature, upon doping with aliovalent ions and upon the slip system on which they are constrained to move. For doped crystals the temperature dependence of the flow stress is complex and under certain circumstances can show a minimum around room temperature. This minimum in the results of Newey led Eshelby et al (8) to suggest that the force between a charged dislocation and its charge cloud could make a contribution to the flow stress. The minimum in the flow stress was associated with the isoelectric temperature at which the dislocations became free from their charge cloud. However, the magnitude of the equilibrium charge on the dislocation is now known to be insufficient to provide a large enough force to account for the yield stress at room temperature even in the most extreme case. Furthermore, at this temperature macroscopic deformation is controlled by the motion of fresh dislocations around which charge clouds will not have time to form (22). These fresh dislocations will acquire charge by sweep-up and their mobility will be determined by their interactions with the lattice and with impurity-vacancy pairs, clusters or precipitates in their path. The minimum in the temperature dependence of the yield stress is better explained by redistribution of the impurity within crystals (23). At high temperatures the Debye-Hückel screening length is reduced and the vacancies become more mobile. Under creep conditions (20), dragging of the dislocation cloud could determine the mobility of dislocations and there is some experimental evidence to support this (24).

3.1 INTERNAL FRICTION

Brown (20) considered the case of a dislocation oscillating within its charge cloud under an applied alternating stress. Provided the displacements are small, as in internal friction experiments, charge cloud damping of the motion should become detectable under certain conditions. Whitworth (9) and more recently Robinson (25) have considered this in some detail. At small displacements in the amplitude independent region (25), the dislocation damping may be due to either phonon drag or charge cloud drag while the restoring force may be due to line tension or to electrostatic interaction with the charge cloud. Robinson (25) has compared the experimental results for the dependence of damping on temperature, on frequency and on dislocation charge with these models. The charge on the dislocations was determined (26) using the ultrasonic composite oscillator over a range of temperatures up to 750 °C. For KCl at room temperature both the amplitude independent damping and the restoring force were due to the charge cloud. At large amplitudes, in the amplitude dependent region, Robinson suggested that the charge cloud radius increased resulting in a rapid increase in the internal friction. At temperatures above the extrinsic isoelectric temperature, corresponding to points P_1 and P_2 in Fig.1, the results suggest that the damping is due to the charge cloud and the restoring force to the line tension. Formation entropies and energies for cation and anion vacancies in KCl are given by Tallon et al (26), with $h_- < h_+$ and $s_- < s_+$. They deduce that no alkali halide should possess an intrinsic isoelectric temperature below its melting-point, thus eliminating the possibility of points like Q_1 in Fig.1.

For crystals of LiF (27) and of AgCl (28), the pinning of dislocations freed in the amplitude dependent region has been studied. In both systems the time dependence of the pinning does not follow that expected for elastic interactions between impurities and dislocations. Rather than $t^{2/3}$, a $t^{1/3}$ dependence is observed and this is interpreted as being due to the presence of a charge cloud.

3.2 THE MOTION OF CHARGED DISLOCATIONS IN AN ELECTRIC FIELD

Eshelby et al (8) suggested that charged dislocations could move under the influence of an applied electric field. Many experiments have been carried out, both with steady fields and with alternating fields. Individual dislocations have been moved a few μm backwards and forwards; small-angle tilt boundaries and large-angle grain boundaries have been moved a few μm at high temperatures. Whitworth (9) has commented on the major problems of interpretation which some of these experiments have produced. Perhaps the clearest demonstration of the effect of an applied field is shown by the experiments of Kataoka et al (21) in crystals of Ca-doped KCl. One surface of the crystal was roughened with emery paper to produce a layer of fresh dislocations some 50 μm deep. Increasing electric fields were applied for times of 1 sec with the crystal immersed in heptane. At a field that depended upon the concentration of Ca^{++}, slip bands appeared from the roughened surface running completely across the crystal. From the measured mechanical yield stress and the critical electric field to cause dislocation yielding, the charges on the dislocations were calculated. Fig.2 shows the variation of this charge with yield stress in doped crystals. The solid circles are the results obtained by this direct method while the open circles were obtained from measuring the charge flow during the compressive deformation of similar pre-bent crystals. Other results for KBr and for NaCl are included on the same figure with no significant difference between the three different materials. In this particular figure the charges appear to have been measured at room temperature and the results are plotted vs. τ_c^2 on the assumption that τ_c^2 is proportional to the concentration c of divalent

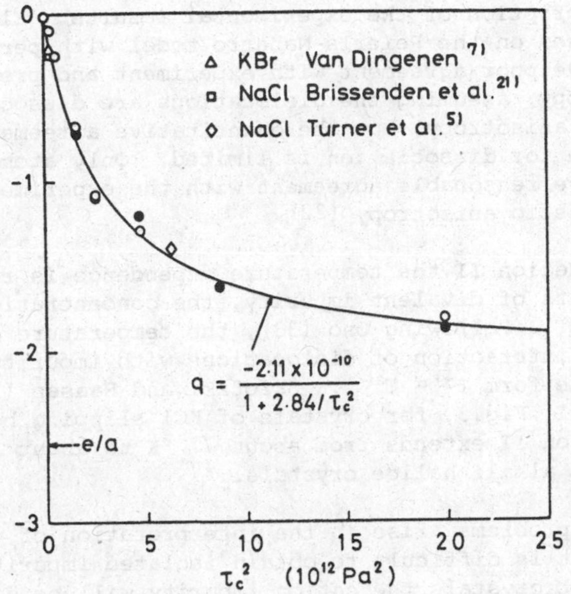

FIG.2. The relationship between dislocation charge and
critical resolved shear stress. After Kataoka et al (21).

impurities. However for the systems shown in Fig.2 at room temperature,
it is τ_c that is proportional to c rather that τ_c^2. Plotted against τ_c,
the charge increases almost linearly with concentration of impurities,
in accordance with the sweep-up model.

3.3 HARDENING MECHANISMS OF IONIC CRYSTALS

        Many attempts have been made to explain the temperature
dependence of the critical resolved shear stress with increasing
temperature. Region I is accepted by most workers as controlled by the
Peierls stress; Region II for a number of years was ascribed to elastic
interactions with impurity-vacancy dipoles and is now associated with
their short range electrostatic interaction, Region III with Snoek-
ordering of dipoles by dislocations and the fall-off in Region IV
variously with faster reorientation of the dipoles, with dragging of the
Snoek atmosphere and with dissociation of the dipoles.

REGION I. Arguments in favour of the Peierls stress being responsible
for the temperature dependence in Region I have been summarised recently
by Skrotzki and Suzuki (29) and Skrotzki and Haasen (30). The strong
temperature dependence for slip on {110} and on {100} and the decrease
in activation volume to below 100 b^3 at low temperature provide good
evidence, with τ_p linearly dependent on $T^{\frac{1}{2}}$ as predicted by the double
kink model. The correlation between τ_p/Z^2 and the sum of the ionic

polarisibilities proposed by Buerger (31) is shown to be only an approximate description of the experimental results. Elastic calculations based on the Peierls-Nabarro model with perfect dislocations give poor agreement with experiment and predict the wrong plastic anisotropy; assuming the dislocations are dissociated gives the correct plastic anisotropy, but the quantitative agreement is still bad and the evidence for dissociation is limited. Only atomistic calculations give reasonable agreement with the experimental results and with the plastic anisotropy (32).

REGION II. In Region II the temperature dependence is reduced and, with small amounts of divalent impurity, the concentration dependence varies as $c^{\frac{1}{2}}$ (30). Following Ono (33), the temperature dependence of the short range interaction of dislocations with impurity-vacancy dipoles takes the form $T^{\frac{1}{2}} \propto T^{2/3}$. Skrotzki and Haasen (30) find a well-defined Region II, Fig.3, for crystals of KCl slipping both on {110} and {100}. Region II extends from about 77 °K to about 250 °K for slip on {110} in most alkali halide crystals.

Two problems arise in the interpretation of these results. Experimentally it is difficult to obtain isolated impurity-vacancy dipoles. In most crystals the cation impurity will be distributed between free dipoles, dimers, aggregates of dipoles, metastable precipitates and stable precipitates of the divalent metal chloride.

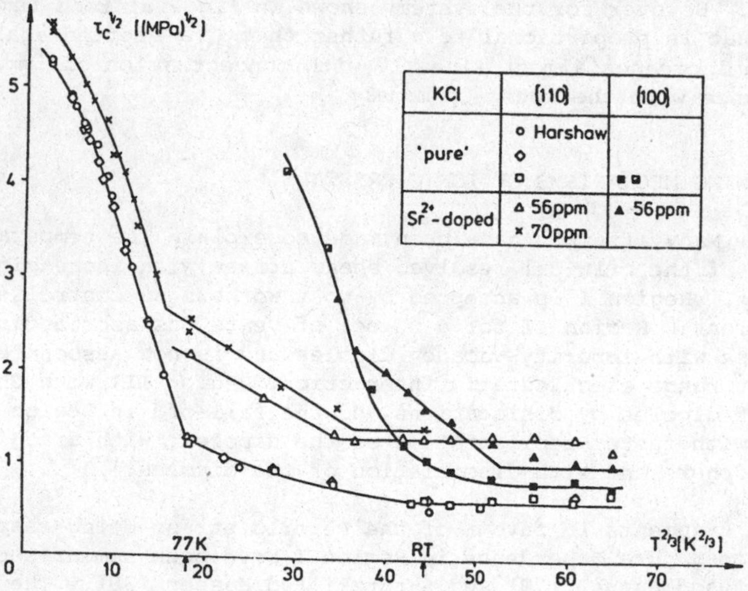

FIG.3. Temperature dependence of the critical resolved shear stress of KCl. After Skrotzki and Haasen (30).

This distribution is determined by the total concentration of the impurity and by the thermal history of the specimen; different systems behave differently. Even for the same system large discrepancies exist between the results of different workers and even more so for the interpretation of the results. Detailed knowledge of the proportions of dimers, trimers, higher aggregates and precipitates and also of their size distribution is required. Increasingly this knowledge is becoming available for aggregates (34,35,36), while precipitates have been observed directly in the electron microscope in a number of systems (37,38). By combining all of these techniques it should be possible to produce reliable measurements of the mechanical properties of doped crystals in a well-characterised condition.

The second problem is the variety of theoretical models with which the experimental results have been compared. Standard theories of solid solution hardening (39) have a concentration dependence as $c^{\frac{1}{2}}$ in agreement with Skrotzki and Haasen (30) whereas Dryden et al (40) found $c^{2/3}$ for a wide range of systems. Fleischer (41) has suggested that the important interaction is with the tetragonal elastic strain field around an impurity-vacancy dipole. Haasen (42) calculated the electrostatic interaction between dipoles and edge dislocations on {110} and {100}, showing that the extrinsic plastic anisotropy can be explained in this way. Including the tetragonal elastic strain field of Fleischer (41), Skrotzki and Haasen (30) suggest that the experimental values can be explained as well.

REGION III. At temperatures above 230-250 °K, the dislocation-induced elastic Snoek ordering of dipoles can account for the linear concentration dependence and for the lack of temperature dependence of yielding, Fig.3. Both Dryden et al (40) and Grau and Fröhlich (36) agree that aggregation of dipoles causes no change in the critical flow stress even up to 500 ppm of divalent cations. This can only be true if the reduced concentration of dipoles is balanced by the formation of a smaller number ($\frac{1}{2}$ for dimers, $^{1}/_{3}$ for trimers) of defects with the appropriately increased maximum interaction force. In fact the aggregation of dipoles does change the initial flow stress in most systems while the formation of precipitates (23,36,43,44) has an even more dramatic effect.

REGION IV. The fall-off in yield stress with temperature in Region IV is much faster than the dissociation of the impurity-vacancy dipoles (45). In addition to dissociation, the increased rate of reordering of the dipoles with increased temperature reduces their retarding influence. Skrotzki and Haasen (30) suggest that at high temperatures and small strain rates the Snoek atmospheres can diffuse sufficiently fast to follow the dislocations.

4. THE INFLUENCE OF DISLOCATIONS ON THE TRANSPORT PROPERTIES

Dislocations are most readily introduced into crystals by plastic deformation and many experiments have attempted to compare the electrical conductivity, or the diffusion of cations or anions, before and after deformation. Care must be taken to distinguish between changes due to the dislocations themselves and changes due to alterations in the density and distribution of point defects and impurity ions in the crystal.

4.1 CHANGES IN IONIC CONDUCTIVITY

TRANSIENT CHANGES. During and immediately after plastic deformation the ionic conductivity can be increased by up to 100 times at temperatures between -170 °C and +70 °C (46,47). This increase can be masked by the charge flow due to sweep-up of defects by moving dislocations first noted by Stepanov (1). Between -170 °C and -10 °C, the only effect that can be observed is the increase during deformation which decays to zero very rapidly. Camagni et al (46) suggest that the increase is due to electrons liberated by the moving dislocations whereas Whitworth (9) has devised an alternative model in which the applied field upsets the balance of charged dislocations moving in opposite directions, so producing a current. At temperatures between -10 °C and +70 °C, conductivity tails appear after the deformation has stopped, decaying in the order of minutes. These are attributed to the recombination of cation and anion vacancies created by moving dislocations. Assuming that the anion vacancies are still immobile, the migration energy for cation vacancies in KCl was found to be 0.45 ± 0.05 eV (46). This low figure was explained by the vacancies moving in highly distorted regions of the lattice, such as near dislocations. At these temperatures, below the isoelectric point, the dislocations will be negatively charged, with divalent cations and anion vacancies in the charge cloud.

Longer lasting transients were found by Gyulai and Hartly (47) as well as by Stepanov (1), with a life-time of minutes to hours. These also are attributed to the recombination of free cation and anion vacancies generated during plastic deformation either by dislocation intersection (3) or by the break-up of existing vacancy complexes (4). Both mechanisms may operate simultaneously and, in addition to impurity-vacancy complexes, divacancy complexes may play a role. Recombination of impurity-vacancy pairs leads to a maximum in the dielectric loss some 30 minutes after deformation, following second order kinetics with an activation energy of 0.7 ± 0.05 eV for NaCl (48). The subsequent diffusion of these impurity-vacancy pairs to the dislocations results in pinning and the recovery of internal friction (49) with a similar activation energy and a time-dependence of $t^{2/3}$. In the first few minutes after deformation, a faster process of pinning was found (49), possibly due to enhanced diffusion near the dislocations as suggested by Camagni et al (46) for their conductivity tail.

STEADY STATE CHANGES. Near room temperature, after the transient changes have decayed, the steady-state conductivity in cation-doped crystals is reduced by up to 50% by small plastic deformations (4,9,50). This suppression anneals out at 200-250 °C. The apparent activation energy for conduction remains unaltered but the number of free cation vacancies has been reduced. Some of these will have been swept up by the edge dislocations acquiring charge as they move; others will have precipitated onto dislocations (4), or onto dislocation debris (9,50) left in their wake throughout the crystal. Taylor (50) showed that no suppression is found for deformations above 200 °C and this supports the role of debris in absorbing free vacancies or pairs. In anion-doped crystals (51) the opposite occurs; the conductivity is increased after deformation, implying sweep-up or precipitation of the free anion vacancies in these crystals.

A further effect of small deformations is found in the high temperature intrinsic region I' (52). Here the slope increases with increasing deformation in the easy-glide region of the stress-strain curve. This enhanced conductivity can be largely annealed out in 24 hr at 700 °C in NaCl. Since it appears only in region I', where anion motion is significant, the effect is interpreted as the movement of anions aided by those dislocations put in by deformation which survive the conductivity run.

4.2 CHANGES IN DIFFUSION DUE TO DISLOCATIONS

More direct evidence exists for enhanced diffusion along dislocations in ionic crystals than that considered in 4.1. Mizuno and Miyamoto (53) showed that the diffusion rates of F-centres in KCl at 450 °C parallel to an array of dislocations were 2-3 times larger than rates perpendicular to the array. The anisotropy of diffusion disappeared at 550 °C. The diffusion of radioactive bromine in KBr was found by Dawson and Barr (54) to be strongly influenced by the presence of dislocations. Curvature in plots of log D against 1/T below about 500 °C were interpreted as a contribution from diffusion along dislocations with an activation energy of 1.49 eV compared with 2.61 eV for intrinsic anion diffusion. A well-defined linear dislocation tail was seen on the concentration profile for a crystal annealed at 460 °C.

Tucker, Laskar and Thomson (55) showed that injected Na^+ ions could diffuse extremely rapidly along dislocations in LiF at temperatures up to 450 °C with an activation energy of 0.31 eV, compared with 0.65 eV for Li^+ ions in region II of the conductivity curve. The minimum density of Na^+ ions along the dislocations was approximately 1% of the lattice sites. They concluded that the core structure of the dislocations is very loose, forming a complex mush of positive ion vacancies. Support for this model came from similar studies by Moment and Gordon (56) on diffusion of injected Na^+ ions in tilt grain boundaries of LiF, with an activation energy of 0.32 eV. The activation energy was constant for all angles of tilt misorientation from 4° - 53°.

McQuhae (57) injected Na^+, K^+ and Cs^+ ions into tilt boundaries of 5° - 40° in pure NaCl, finding activation energies for their motion of 0.32, 0.55 and 0.65 eV respectively, independent of tilt angle. The enhanced diffusion of Na^+ ions injected into NaCl suggests that the injection process itself may be responsible for creating paths of easy diffusion in the grain boundaries and dislocations.

Dislocation pipe diffusion in the absence of injection has been demonstrated by a number of workers (58,59,60), using serial sectioning with a radioisotope. Activation energies for dislocation diffusion of Na^+ ions in pure NaCl were found between 0.4 and 0.6 eV compared with lattice diffusion energies of 1.23 - 1.25 eV in the extrinsic region. The linear dislocation tails found by Ho and Pratt (60), Fig.4, were analysed using the theory of Le Claire and Rabinovitch (61). The two plots in Fig.4 are deliberately at the same temperature with a 4-fold difference in time. The theory for isolated dislocations (61) demands that the slopes of the linear tails should be very nearly the same; in fact the small difference in slope for runs 4 and 5 can be accounted for wholly by the theory. The dislocation densities calculated from the theory agree well with those measured experimentally. The activation energy for dislocation diffusion of Na^+ of 0.57 ± 0.04 eV agrees with the earlier results of Geguzin and Dobrovinskaya (58) and is approximately twice that for Na^+ injected into tilt boundaries. Assuming an upper limit for D', the diffusion coefficient down dislocations, of 10^{-5} cm^2 sec^{-1}, typical of liquid state diffusion, the results of Fig.4 demand an effective dislocation radius of 10^{-5} cms.

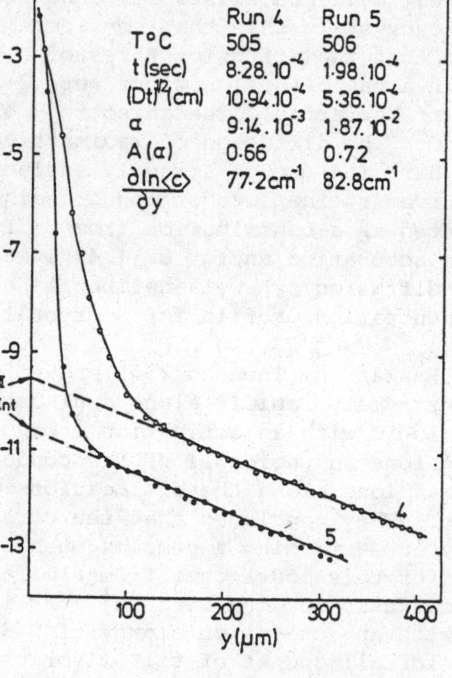

FIG.4. Self diffusion of ^{22}Na in NaCl single crystals.

Ho and Pratt (60) suggested this was compatible with the radius of the charge cloud around the dislocations. However, Whitworth has pointed out that the concentration of vacancies in the charge cloud is far too low to account for the enhancement and an alternative explanation must be sought. The limiting radius of the dislocation might be set at 5 Å, but this would require D' to be as high as 10^{-2} cm² sec^{-1} at 374 °C, implying a jump frequency along the dislocation approaching the phonon frequency. If the dislocation core truly is "a soft mushy region of considerable complexity" perhaps the lattice jump model for diffusion must be abandoned for diffusion down dislocation cores in ionic crystals.

4.3 ATOMISTIC CALCULATIONS

Rabier and Puls (62) have calculated the activation energy for diffusion of anions along the core of edge dislocations in MgO. Assuming a vacancy diffusion mechanism, they found that the energy for pipe diffusion is some 75% of that for the bulk, mainly due to the decrease in the vacancy formation energy at some positions in the core. Their atomistic calculations failed to take into account the presence of impurities or of jogs, in view of the difficulties involved. Since jogs are actually the sites at which vacancies are created and destroyed it seems likely that the energy for pipe diffusion down jogged dislocations might even lower. Narayan and Washburn (63) found a ratio of 0.55 between pipe and bulk diffusion for the anions in MgO, very close to the ratio of 0.57 of Barr and Dawson (54) for Br^- in KBr. For cations in NaCl, Ho and Pratt (60) found a ratio of 0.46.

Similar atomistic calculations for tilt boundaries (64) predict the existence of open channels rather than dislocation cores, down which rapid diffusion might take place rather easily. McQuhae's experiments (57) however show that the ease of diffusion in the channels is dependent on the size of the moving ion. It would be interesting to discover whether Na^+ ions injected into isolated dislocations in NaCl move with an activation energy of 0.32 eV or 0.6 eV.

5. CONCLUSIONS

1) The maximum charge that an edge dislocation in an ionic crystal can acquire by sweep-up at room temperature approaches e per cation site along the dislocation in doped crystals. The equilibrium charge appears to be much less than this, although theory is not yet able to say by how much.

2) The yield stress of ionic crystals is determined by the Peierls mechanism at low temperatures, by short-range electrostatic and elastic interactions with impurity-vacancy dipoles and clusters at intermediate temperatures and by long range Snoek-ordering of the dipoles at room temperature. The form of the aliovalent impurities, as dipoles, clusters, or precipitates, determines the magnitude of the yield stress.

3) The transient changes found in ionic conductivity after plastic deformation at room temperature are due to recombination of point defects. The mobile cation vacancies can diffuse down dislocations or through the lattice in the process of recombining.

4) The steady-state conductivity is strongly influenced by the charges acquired by dislocations during deformation.

5) There is good experimental evidence for enhanced anion and cation diffusion down dislocations in ionic crystals. Atomistic calculations are beginning to model the core structure of dislocations more realistically.

ACKNOWLEDGEMENTS

I am grateful to Dr R W Whitworth for his valuable comments on a number of the topics related to charged dislocations in this paper. In particular I should like to acknowledge the fruitful discussions both with Dr Whitworth and with Dr A D Le Claire on dislocation pipe diffusion in ionic crystals.

REFERENCES

(1) Stepanov, A. V. (1933). Phys. Z. Sowj., 4, 609-627.
(2) Eshelby, J. D. & Pratt, P. L. (1956). Acta Metall., 4, 560-561.
(3) Seitz, F. (1952). Adv. Phys., 1, 43-90.
(4) Fischbach, D. B. & Nowick, A. S. (1955). Phys. Rev., 99, 1333-1334.
(5) Bassani, F. & Thomson, R. (1956). Phys. Rev., 102, 1264-1275.
(6) Lehovec, K. (1953). J. Chem. Phys., 21, 1123-1128.
(7) Pratt, P. L. (1958). In: Vacancies and other point defects in metals and alloys. London: Institute of Metals, pp 99-130.
(8) Eshelby, J. D. et al. (1958). Phil. Mag., 3, 75-89.
(9) Whitworth, R. W. (1975). Adv. in Phys., 24, No.2, 203-304.
(10) Frenkel, J. (1946). In: Kinetic Theory of Liquids. Oxford: Clarendon Press.
(11) Whitworth, R. W. (1972). Phys. Stat. Sol.B, 54, 537-550.
(12) Whitworth, R. W. (1968). Phil. Mag., 17, 1207-1221.
(13) Allnatt, A. R. et al. (1971). J. Phys. C, 4, 1778-1788.
(14) Huddart, A. & Whitworth, R. W. (1973). Phil. Mag., 27, 107-119.
(15) Davidge, R. W. (1963). Phys. Stat. Sol., 3, 1851-1856.
(16) Spencer, O. S. & Plint, C. A. (1969). J. Appl. Phys., 40, 168-172.
(17) Kliewer, K. L. & Koehler, J. S. (1967). Phys. Rev., 157, 685-692.
(18) Strumane, R. & De Batist, R. (1964). Phys. Stat. Sol., 6, 817-821.
(19) Harris, L. B. (1983). Radiation Effects, 75, 211-216.
(20) Brown, L. M. (1961). Phys. Stat. Sol., 1, 585-599.
(21) Kataoka, T. et al. (1983). Radiation Effects, 75, 227-234.
(22) Pratt, P. L. (1964). Proc. Br. Ceram. Soc., 1, 177-182.
(23) Pratt, P. L. et al. (1964). Discuss. Faraday Soc., 38, 211-217.
(24) Menezes, R. A. & Nix, W. D. (1973). Phil. Mag., 27, 1201-1209.
(25) Robinson, W. H. (1983). Radiation Effects, 74, 339-346.

(26) Tallon, J. L. et al. (1983). Radiation Effects, 74, 299-305.
(27) Carpenter, S. H. (1968). Acta Metall., 16, 73-79.
(28) Kim, J. S. et al. (1974). J. Phys. Chem. Solids, 35, 741-751.
(29) Skrotzki, W. & Suzuki, T. (1983). Radiation Effects, 74, 315-322.
(30) Skrotzki, W. & Haasen, P. (1981). J. de Physique, C-3, 119-148.
(31) Buerger, M. J. (1930). Am. Mineralogist, 15, 174.
(32) Granzer, F. et al. (1968). Phys. Stat. Sol., 30, 587.
(33) Ono, K. (1968). J. Appl. Phys., 39, 1803.
(34) Unger, S. & Perlman, M. M. (1977). Phys. Rev. B, 15, 4105.
(35) Guerrero, A. L. et al. (1978). Phys. Stat. Sol.A, 49, 353-362.
(36) Grau, P. & Fröhlich, F. (1979). Phys. Stat. Sol.A, 55, 479-487.
(37) Yacaman, M. J. et al. (1977). Phys. Stat. Sol.A, 39, K85.
(38) Guerrero, A. L. et al. (1981). Phil. Mag.A, 43, 6, 1359-1376.
(39) Labusch, R. (1972). Acta Metall., 20, 917.
(40) Dryden, J. S. et al. (1965). Phil. Mag., 12, 379.
(41) Fleischer, R. L. (1962). Acta Metall., 10, 835.
(42) Haasen, P. (1973). J. de Physique, C-9, 205.
(43) Harrison, R. P. & Newey, C. W. A. (1968). Phil. Mag., 17, 525-533.
(44) Khan, A. R. (1967). PhD Thesis, London University.
(45) Pratt, P. L. et al. (1963). Appl. Phys. Letters, 3, 83.
(46) Camagni, P. et al. (1960). J. Phys. Chem. Solids, 17, 165-168.
(47) Gyulai, Z. & Hartly, D. (1928). Z. Phys., 51, 378-387.
(48) Shine, M. C. & MacCrone, R. K. (1968). Phys. Rev., 176, 1076-1088.
(49) Phillips, D. C. & Pratt, P. L. (1970). Phil. Mag., 21, 217-243.
(50) Taylor, A. (1958). PhD Thesis, Birmingham University.
(51) Fröhlich, F. et al. (1971). Phys. Stat. Sol.A, 6, 165-172.
(52) Pratt, P. L. (1973). J. de Physique, C-9, 213.
(53) Mizuno, M. & Miyamoto, S. (1962). Phys. Rev., 125, 833-836.
(54) Dawson, D. K. & Barr, L. W. (1967). Proc. Br. Ceram. Soc., 9, 171-179.
(55) Tucker, R. et al. (1963). J. Appl. Phys., 34, 445-452.
(56) Moment, R. L. & Gordon, R. B. (1964). J. Appl. Phys., 35, 2489-2492.
(57) McQuhae, K. G. (1969). PhD Thesis, London University.
(58) Geguzin, Y. E. & Dobrovinskaya, E. R. (1965). Soviet Phys. Solid St., 7, 2826-2828.
(59) Schlederer, L. J. (1974). PhD Thesis, University of New South Wales, Australia.
(60) Ho, Y. K. & Pratt, P. L. (1983). Radiation Effects, 75, 183-192.
(61) Le Claire, A. D. & Rabinovitch, A. (1984). In: Diffusion in Crystalline Solids, eds G. E. Murch & A. Nowick, Academic Press, in press.
(62) Rabier, J. & Puls, M. P. (1983). Radiation Effects, 75, 193-201.
(63) Narayan, J. & Washburn, J. (1979). Acta Metall., 21, 533.
(64) Duffy, D. M. & Tasker, P. W. (1983). Phil. Mag. A, 47, 817.

IMAGE STRESS AND THE BAUSCHINGER EFFECT IN DISPERSION-HARDENED ALLOYS

L. M. Brown

Cavendish Laboratory, Madingley Road,

Cambridge CB3 OHE

ABSTRACT

Eshelby's model of an elastic inclusion can be applied direct to understand the pattern of interphase stresses which builds up in a dispersion-hardened alloy as a result of plastic deformation. The resulting picture of the work-hardening of such materials relates typical dislocation concepts, such as pile-ups and forest hardening, to the concepts of continuum plasticity widely used in composite theory. Some simple examples are given to illustrate the ideas, and some experimental evidence is presented to indicate to what extent the concepts may be useful in practice.

1. INTRODUCTION

This short review is concerned with two of Eshelby's most famous papers: one in which he, with F. C. Frank and F. R. N. Nabarro, introduced the concept of the dislocation pile-up (1), and the other in which he introduced the notion of internal stresses resulting from a general transformed inclusion (2). The former paper has had a profound influence in theories of deformation produced largely by physical metallurgists or physicists: one can instance the Petch theory of yielding in polycrystals (3) and Seeger's theory of work-hardening (4). The latter has had very considerable impact on theories of phase transitions, fibre-reinforcement and other applications too numerous to cite in detail. This paper attempts to synthesize the two points of view.

2. PILE-UPS AND CONTINUUM MODELS

The viewpoint of the continuum theory of plasticity is that plastic deformation is homogeneous, and that slip-lines do not exist. This basic fallacy has not prevented continuum plasticity from playing a useful role in science and engineering. On the other hand, the theory and observation of dislocations is potentially capable of supplanting and including the continuum theory of plasticity, by showing directly how the microstructure of a material determines its strength properties. Yet, at the time of writing, dislocation theories for plastic behaviour do not command general acceptance, because there is no agreed basis for such theories. This is in contrast with continuum mechanics, which has become a branch of applied mathematics in which all writers share assumptions about the basis of the subject: definitions of stress and strain, flow criteria, techniques for solving problems of non-uniform flow, etc. It happens that the Bauschinger effect in two-phase alloys is an ideal subject in which to display the fundamental relationship between the continuum and dislocation models. To see this, consider the simplest model of a two-phase alloy: an 'elastic element' welded to a 'plastic element' and both deformed in shear: Fig. 1. The interface between the elements cannot slip. The geometry of this thought-experiment is chosen to be as simple as possible. The continuum model yields a straightforward answer for the stress-strain behaviour. At low stresses, both elastic and plastic elements deform elastically with shear modulus μ, until the stress exceeds the flow stress of the plastic element, σ_o. Thereafter, the plastic element can contribute no resistance to deformation, but the elastic element continues to provide elastic resistance, so the slope of the stress-strain curve changes to $2f\mu$, where f is the volume fraction of elastic element. If the composite is unloaded, continuum mechanics again provides a simple prediction for its behaviour: the stress σ_o acts as a friction stress, so when the applied stress has been reduced by $2\sigma_o$ reverse flow begins along the line CD in Fig. 1b.

Dislocation mechanics produces rather different answers to the problem. Let us suppose (fig. 1a) that slip occurs on slip-planes SS' with spacing h. Let us also suppose that the plastic element or 'matrix' possesses a friction stress σ_o, so that the flow begins when the stress exceeds σ_o. At this point, the dislocations form pile-ups in their slip planes. The elastic 'element' in this model becomes a 'barrier' to dislocations. One answer to the problem is to approximate the dislocation interactions by 'independent pile-ups' - each slip plane SS' contains two non-interacting pile-ups of length ℓ. Under an effective stress $\sigma-\sigma_o$, each pile-up contains n dislocations, where n is given by $(\sigma-\sigma_o)\ell/\mu b$. The plastic displacement of the slip plane is $3nb/4$, because the dislocations repel each other from the interface (5). Thus the equation of stress-strain curve AB is given by

total strain = elastic strain + plastic strain

$$\varepsilon = \frac{\sigma}{2\mu} + (\sigma - \sigma_o)\frac{\ell}{\mu b} \times \frac{3b}{4h}(1-f)$$

$$\text{or} \qquad \sigma = \sigma_o \times \frac{1}{1+2h/3\ell(1-f)} + 2\mu\varepsilon \times \frac{2h/3\ell(1-f)}{1+2h/3\ell(1-f)} \qquad (2.1)$$

This model predicts work-hardening which depends mainly on the arrangement of pile-ups, but does not depend sensibly upon any parameter directly related to the amount of elastic component in the composite. This model is wrong in principle, as can be seen from the following argument: Whatever its faults, the continuum model provides us with a rigorous lower bound for the slope of the stress-strain

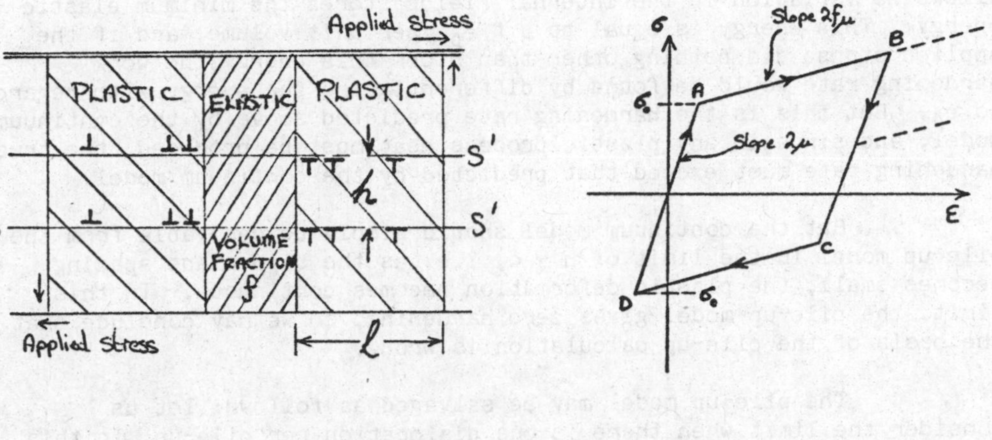

(a) (b)

FIG. 1. The simplest model of a two-phase alloy deformed in pure shear.

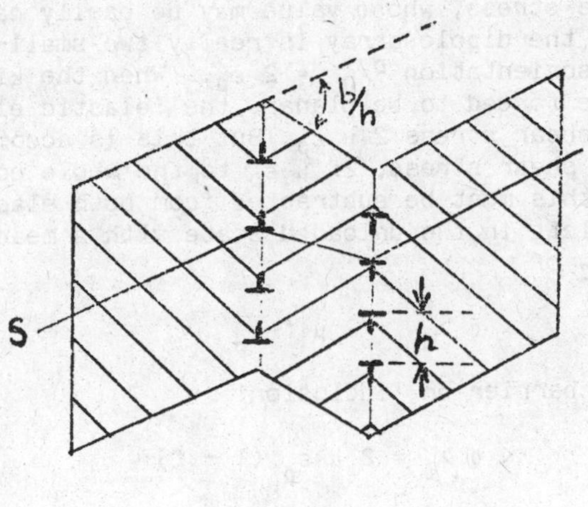

FIG. 2. The shape of the two-phase alloy after shear and unloading.

curve. The elastic energy in this strained body with uniform applied stresses is a minimum when the internal stress and strain fields are uniform (in other words, when you stretch a rubber band, the elastic strain is uniform along the band). Thus the continuum model, which allows no variation in the internal fields stores the minimum elastic energy. This energy is equal to $\mu f \varepsilon_\rho^2$ per unit volume, and if the applied stress did nothing other than store this energy, the work hardening rate would be found by differentiating the energy with regard to ε_ρ. But this is the hardening rate predicted above by the continuum model, and since in any plastic process heat must be produced, the true hardening rate must exceed that predicted by the continuum model.

But the continuum model should simply be derivable from the pile-up model in the limit of $h \to o$, i.e. as the slip-plane spacing becomes small, the plastic deformation becomes continuous. In this limit, the pile-up model gives zero hardening, so we may conclude that the basis of the pile-up calculation is wrong.

The pile-up model may be salvaged as follows: let us consider the limit when there is one dislocation per pile-up. In this limit, the dislocation array takes the form of 'shear dipoles', one dipole on each slip plane. The stress-field due to such a dipole has a 'local' component, whose mean value is zero; the local component opposes slip on the planes SS' outside the elastic barrier, but aids it inside the barrier. In addition to these local stresses, the dipoles produce a long-range stress, whose value may be easily calculated with reference to fig.2: the dipole array is really two small-angle grain boundaries, with misorientation $b/h = 2\,\varepsilon_\rho$. When the kinked surfaces of the composite are forced to be planar, the 'elastic element' experiences a pure shear stress $2\,\mu\,\varepsilon_\rho$. But this is accomplished by applying an average shear stress, $2f\,\mu\,\varepsilon_\rho$ to the whole composite; in the unloaded state this must be subtracted from both elastic barrier and matrix. We are left in the unloaded state with a mean stress in the matrix <u>opposing flow</u> of

$$\langle \sigma \rangle_M = 2\,\mu\,f\,\varepsilon_\rho \qquad (2.2)$$

and a stress in the barrier or 'inclusion'

$$\langle \sigma \rangle_I = 2\,\mu\,\varepsilon_\rho\,(1 - f) \qquad (2.3)$$

<u>aiding flow</u>. Thus when the elastic field of the dislocations is found in the <u>finite</u> body, an internal stress in the matrix is obtained which agrees with the continuum model. Note that the average internal stress, $\langle\sigma\rangle$,

$$\langle \sigma \rangle = f\langle \sigma_I \rangle + (1 - f)\,\langle \sigma \rangle_M\,, \qquad (2.4)$$

is equal to zero. The important point is that the theory of dislocation pile-ups, developed for infinite bodies, is a misleading basis for the plastic behaviour of composites.

Now it is clear that a pile-up model with n dislocations per pile-up instead of one will still produce long-range stresses in the finite body; indeed, the angle of the small-angle boundary is independent of the number of dislocations in the pile-up, so we can see that the average internal stress in the matrix probably does not depend very strongly on the number of dislocations in the pile-ups.

The three-dimensional version of this theory, based on Eshelby's work (2), has received attention from a number of authors (6-10). The results for small volume fraction can be stated succinctly as follows:

In the continuum model, closely approached when there is only one dislocation per pile-up, the mean stress in the matrix is

$$\langle \sigma \rangle_M = 2D\, \gamma\, \mu\, f\, \varepsilon_\rho^* \tag{2.5}$$

In this equation γ is a dimensionless 'accommodation' coefficient; for spherical particles it is for all practical purposes equal to $^1/_2$. Values for ribbons, plates and fibres have been calculated. The quantity D depends on the elastic constants of the inclusion and matrix; for spherical particles of modulus μ^*,

$$D = \frac{\mu^*}{\mu^* - \gamma(\mu^* - \mu)} \tag{2.6}$$

The quantity ε_ρ^* measures the 'incompatibility' between the particle and the matrix. If the particle is spherical, of radius r_o, then the number of Orowan loops around the particle is

$$n_o = \frac{4 r_o\, \varepsilon_\rho^*}{b} \tag{2.7}$$

If the loops have not cross-slipped or climbed or penetrated the particle, or otherwise disappeared, then every dislocation which has passed the particle has left one Orowan loop behind, and $\varepsilon_\rho^* = \varepsilon_\rho$, the imposed plastic strain. More often, however, plastic relaxation will have occurred, and ε_ρ^* will be less than ε_ρ. The spacing of slip steps on the surface of a crystal is h, and the average number of loops per slip plane n_s is given by

$$n_s = 2\varepsilon_\rho\, h/b \tag{2.8}$$

In the unrelaxed state, n_s is equal to the number of loops in each pile-up of Orowan loops around the particle; in fig. 3(a) n_s is unity, appropriate to the continuum model, but in fig. 3(b) n_s is 2. But if plastic relaxation occurs, the number of loops in each pile-up is not simply related either to the spacing of slip steps or to ε_ρ^*; some of the Orowan loops will have vanished and others will remain, so that on average there may be n_{os} of them in each pile-up. This quantity, n_{os}, is another independent parameter describing the state of the Orowan loops; the first parameter is ε_ρ^*, which describes the net

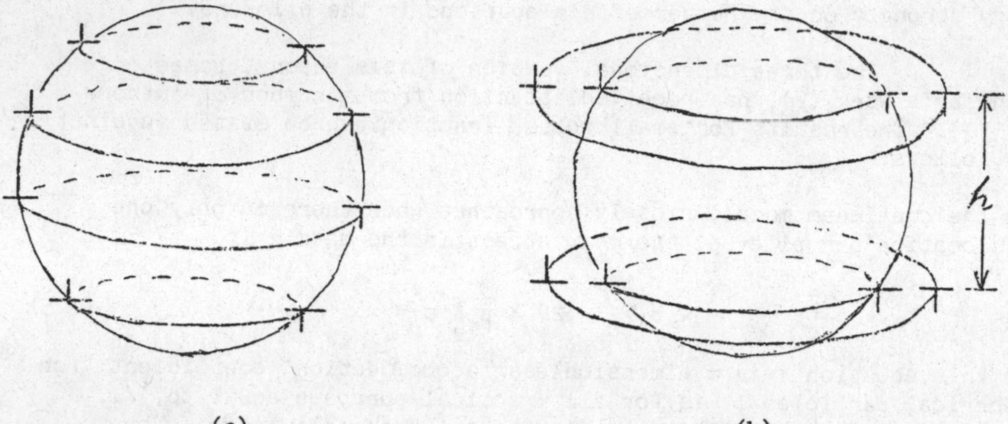

FIG. 3. Orowan loops around a precipitate

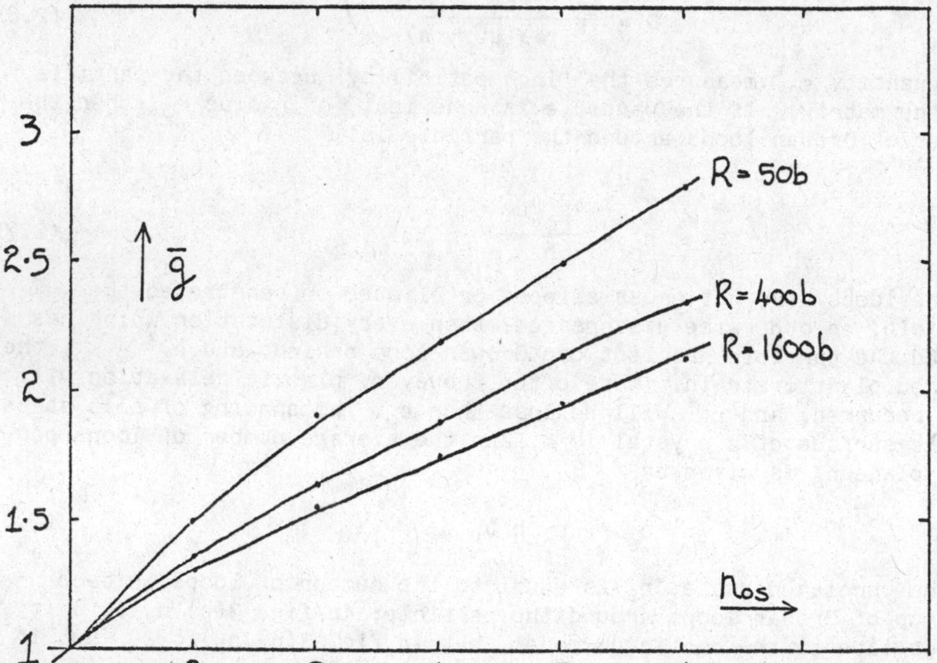

FIG. 4. Correction to the continuum model according to Hazzledine & Hirsch (11)

incompatibility between particle and matrix (the angle of the small-angle boundary, in our simple model or the geometrically necessary dislocations, in Ashby's model), and the second parameter is n_{os}, the number of Orowan loops in each pile-up.

The mean stress in the matrix can be written as

$$\langle \sigma \rangle_M = 2\, D(n_{os}, r_o)\, \bar{g}\, (n_{os}, r_o)\, \gamma\, \mu\, f\, \varepsilon_\rho^* \qquad (2.9)$$

This equation contains factors which correct the continuum model to allow for more than one loop per pile-up. Hazzledine & Hirsch (11) have calculated the factor $g(n_{os}, r_o)$ making the assumption of elastic homogeneity (i.e. D = 1) and one pile-up per particle. They have also assumed that the mean stress in the matrix is effectively controlled by the total area of Orowan loops. The results of their calculation are shown in fig. 4, where it can be seen that the correction to the continuum model can amount to a factor of nearly three for small particles with six loops per pile-up.

The assumptions made by Hazzledine & Hirsch nearly all tend to overestimate the correction factor. If the 'matrix' is allowed to extend to the particle as it should, stress inside some of the loops should count in the average, and this reduces the stress; if pile-ups interact on the same particle, they tend to close each other up and reduce the stress; also in a real crystal, the pile-ups are unlikely to exist on a single plane and hence will be smaller.

Thus it seems likely that the data in fig. 4 provide an upper limit to the true correction factor. Hazzledine & Hirsch have not calculated $D(n_{os}, r_o)$, but give a simple argument to show that it should lie between unity and the fully-corrected continuum value.

To sum up: the deformation of two phase materials produces a characteristic set of internal stresses. The stress in the 'matrix' or plastic component on average impedes forward flow, but aids reverse flow. The stress is calculated by an elastic model which, while firmly based in dislocation theory, provides values for the stress which are close to what one would calculate from the continuum model. In addition to the volume fraction, shape, and the elastic constants of the second phase, the mean internal stress is controlled by two parameters of the dislocation distribution: ε_ρ^*, the unrelaxed strain or incompatibility between matrix and particles, and n_{os}, the number of Orowan loops in each slip-plane around the particle.

3. THE SLIP LINE SPACING

Recently it has become clear that it is possible to estimate the slip line spacing in dispersion-hardened alloys, starting from the continuum model (12). Let us suppose we know the dependence of the flow stress on the strain and on the slip-line spacing through the relations

$$\sigma_{os} = \sigma(\varepsilon_\rho, n_{os})$$
$$\varepsilon_\rho^* = n_{os}\ b/h \qquad (3.1)$$

In order to write these equations, some model for the plastic relaxation must be assumed: in other words, we must know ε_ρ^* as a function of ε_ρ. We expect that when $n_{os} = 1$ the flow stress is given very closely by the continuum model, and that for small values of n_{os} it does not depart very markedly from it. Now if we also assume that the flow stress is associated with a characteristic slip-plane spacing, viz.

$$\sigma_{int} = \frac{\mu b}{4\pi h} \qquad (3.2)$$

we can solve for the slip-plane spacing and for the work-hardening.

The argument leading to these equations is given by Pedersen & Brown (12); it is closely related to the principle first put forward by Hirsch & Humphreys (13), although the latter authors were not concerned with the mean stress in the matrix. The point of view we adopt is to accept that dislocation storage is determined by the pattern of flow, local flows being part of a plastic relaxation process; also, if relaxation is incomplete, dislocations may be stored as 'Orowan loops', which cause long range internal stress. The flow pattern is independent of the slip-line spacing,although the slip line spacing affects the stress levels as described earlier. But long-range slip on planes of spacing h requires the break-up of dipoles of spacing h, so the overall flow stress is determined by

$$\sigma = \sigma_{os} = \sigma_{int} \qquad (3.3)$$

Because we do not expect the overall stress levels to depend strongly on the number of dislocations in a pile-up, we may approximate the solution of equations 3.3 by putting $n_{os} = 1$ and taking the slip-line spacing as

$$h^{-1} = \frac{4\pi}{\mu b}\ \sigma\ (\varepsilon_\rho, 1) \qquad (3.4)$$

This approximation overestimates the spacing by a factor which is difficult to assess, but which in the limit of large obstacles will be quite small, perhaps 50%. Figure 5 shows schematically the solution of equations 3.1 and 3.2.

Thus one arrives at a simple approximate theory which enables one to estimate the work-hardening, the internal stress distribution, and the slip-line spacing. The main empirical ingredient in the theory is the pattern of flow, which seems to be a fundamentally unpredictable aspect in crystals as it is in turbulent fluids.

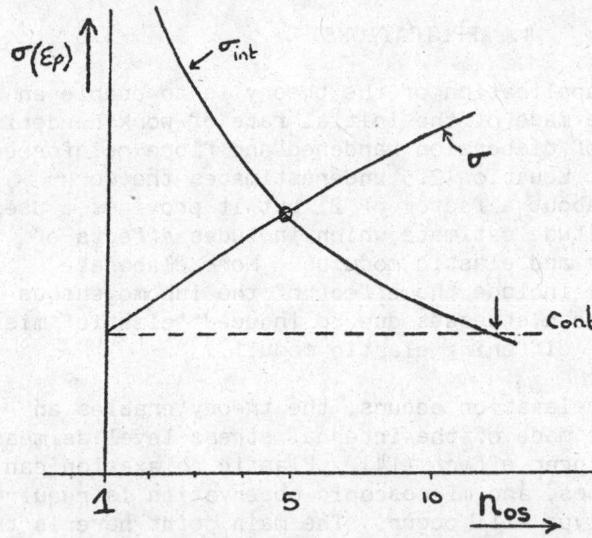

FIG. 5. The slip line spacing is determined by equating eqn. 3.1 with eqn. 3.2. The continuum limit is shown. Curves are only schematic.

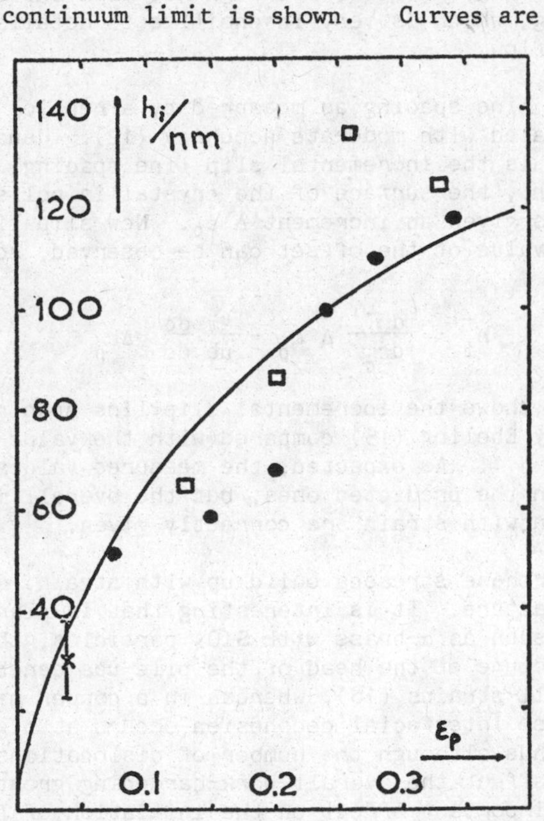

FIG. 6. Predicted and observed incremental slip-line spacing. Squares and circles from Ebeling (15); cross from Brown & Stobbs (7). Line from constitutive equation quoted in ref.(14).

4. APPLICATIONS

1. The simplest application of the theory is to enable an estimate to be made of the initial rate of work-hardening in a wide range of dispersion-hardened and fibre-reinforced materials (9). Equation 2.5 underestimates the work-hardening by about a factor of 2, but it provides a useful order-of-magnitude estimate which includes effects of particle shape and elastic modulus. More elaborate estimates (10) include the effect of the inhomogeneous stresses and mean stresses due to induced 'elastic' misfit as a result of differing elastic moduli.

2. When plastic relaxation occurs, the theory enables an estimate to be made of the internal stress level as measured by the Bauschinger effect (14). Plastic relaxation can be of several types, and microscopic observation is required to decide which type will occur. The main point here is that the internal dislocation structure can be correlated with the Bauschinger effect, although not with the overall work-hardening, which is very insensitive to details of the plastic flow.

3. The slip line spacing as measured by a replica technique can be estimated with moderate accuracy (12). Usually what is measured is the incremental slip line spacing, h_i. To measure h_i, the surface of the crystal is polished, and the strain is given an increment $\Delta\ \varepsilon_\rho$. New slip lines with the current value of the offset can be observed, so one measures

$$h_i^{-1} = \frac{dh^{-1}}{d\varepsilon_\rho}\,\Delta\ \varepsilon_\rho = \frac{4\pi}{\mu b}\,\frac{d\sigma}{d\varepsilon}\ \Delta\varepsilon_\rho \qquad (4.1)$$

Figure 6 shows the incremental slip-line spacing as measured mainly by Ebeling (15) compared with the value predicted by equation 4.1. As expected, the measured values are somewhat less than the predicted ones, but the overall magnitude and variation with strain are correctly given.

4. The interphase stresses build up with strain, and eventually cause fracture. It is interesting that in planar slip alloys, such as α-brass with SiO_2 particles, the microfracture at the head of the pile-ups can be observed at very small strains (16), whereas in a copper matrix a kind of diffuse interfacial decohesion occurs at higher strains (17). Thus although the number of dislocations in a pile-up may not affect the overall work-hardening greatly, it can have an important effect on the initiation of fracture.

5. The internal stresses depend on the shape of the particles, and can often cause 'plastic anisotropy', that is, the

specimen may deform to produce a shape different from that produced by the same stress applied to an isotropic body (9). This effect is of considerable importance in the development of texture in polycrystals. The presence of plate-like undeformable particles has a profound influence on the development of rolling texture, which may be understood by detailed application of the theory (18,19).

REFERENCES

(1) Eshelby, J. D., Frank, F. C. & Nabarro, F. R. N. (1951). Phil. Mag. 42, 351-364.
(2) Eshelby, J. D. (1957). Proc. Roy. Soc. A241, 376-396.
(3) See, for example, Armstrong, R. W. (1983) in Yield, Flow and Fracture of Polycrystals, T. N. Baker, Ed., Applied Science Publishers, Barking, pp 1-31.
(4) Seeger, A. (1958). Hdb. der Phys. VII, 2, Springer, Berlin.
(5) Cottrell, A. H. (1953). Dislocations and Plastic Flow in Crystals, Oxford University Press, pp 104-107
(6) Tanaka, K. & Mori, T. (1970). Acta Met. 18, 931-941.
(7) Brown, L. M. & Stobbs, W. M. (1971). Phil. Mag. 23, 1185-1199.
(8) Mori, T. and Tanaka, K. (1973). Acta Met. 21, 571-574
(9) Brown, L. M. & Clarke, D. R. (1975). Acta Met. 23, 821-830.
(10) Pedersen, O.B. (1983). Acta Met. 31, 1795-1808.
(11) Hazzledine, P. M. & Hirsch, P. B. (1974). Phil. Mag. 30, 1331-1351.
(12) Pedersen, O. B. & Brown, L. M. (1983). 'Deformation of Multi-Phase and Particle-Containing Materials', ed. J. B. Bilde-Sorensen et al., Riso National Laboratory, Roskilde, Denmark, p 83.
(13) Hirsch, P. B. and Humphreys, F. J. (1970). Proc. Roy. Soc. A318, 45-72.
(14) For a brief, but incomplete, review of the many references, see Brown, L. M. (1979) in 'Strength of Metals and Alloys' Proc. of ICSMA-5, Aachen (Pergamon Press, Oxford) p 1551. An especially elegant demonstration is one by G. D. Moan and J. D. Embury (Acta Met.(1979) 27, 903).
(15) Ebeling, R. (1970). Z. fur Metallkunde 61, 42-52.
(16) Humphreys, F. J. & Stewart, A. T. (1972). Surface Science 31, 389.
(17) Brown, L. M. & Embury, J. D. (1973) in 'Microstructure and Design of Alloys', ICSMA-3, Cambridge (Inst. of Metals) p 164.
(18) Bate, P., Roberts, W. T. & Wilson, D. V. (1981). Acta Met. 29, 1797-1814.
(19) Bate, P., Roberts, W. T. & Wilson, D. V. (1982). Acta Met. 30, 725-737.

GEOMETRY AND BEHAVIOUR OF PRISMATIC DISLOCATION LOOPS AND STACKING FAULT TETRAHEDRA

B.L. EYRE

Department of Metallurgy and Materials Science,
The University, P.O. Box 147, Liverpool L69 3BX

ABSTRACT

Prismatic dislocation loops and stacking fault tetrahedra (SFT) form in crystals as a result of a number of processes, but most commonly by the aggregation of point defects generated, for example, by quenching, irradiation or deformation. Whatever the formation mechanism, such defects have a major influence on evolving defect structures and a range of mechanical and physical properties. The principal experimental technique used to study loops and SFT has been transmission electron microscopy (TEM). This paper will review the progress that has been made in understanding their geometries and behaviour.

1. INTRODUCTION

Since the formation of prismatic loops was first proposed by Nabarro (1) and Frank (2), considerable progress has been made in understanding their geometries and properties. The early experimental work was carried out on FCC metals and transmission electron microscopy (TEM) enabled the first direct observations to be made of prismatic vacancy loops in quenched Al (3). Since then experimental research into loops and SFT in a wide range of metals and alloys has almost entirely been based on TEM and the progress made, particularly in understanding their geometries, is due to the application of diffraction contrast theory to the analysis of observed images.

As well as being of fundamental importance, an understanding of the geometries and properties of prismatic loops and SFT has direct practical relevance to the problem of irradiation damage in reactor materials. More specifically, the nucleation, growth and interactions of such defects have a major influence on how damage structures evolve during irradiation and this, in turn, governs the changes in a wide range of mechanical and physical properties. Thus, much of the research on loops and SFT has been carried out on irradiated metals, although important contributions have come from studies of quenched metals, particularly in the earlier work. The main objective of this paper is to summarise the more recent results and understanding that has come out of this research and to highlight outstanding areas of uncertainty. No attempt is made to comprehensively review the subject and only selected references are given. More complete reviews of the experimental work have been published elsewhere (e.g. 4-7).

2. NUCLEATION AND GROWTH OF DISLOCATION LOOPS

The formation and growth of dislocation loops requires two conditions to be satisfied, namely, the formation of a critical sized nucleus and an excess flux of point defects of the appropriate type to the loops. The driving force to form vacancy loops and SFT is low compared to interstitial loops, and whether or not they form depends critically on a number of factors, particularly vacancy super-saturation and stacking fault energy (SFE). Thus, because of the combination of comparatively low melting temperatures and SFE's, quenched-in vacancy super-saturations are sufficient to nucleate vacancy loops and SFT in many FCC metals. In contrast, planar vacancy defects have not been found in quenched BCC metals, due to the much higher SFE's and generally higher melting temperatures. Moreover, collapse of vacancy rich cascade centres to form vacancy loops in irradiated metals generally occurs more readily in the heavier FCC metals, e.g. Cu, Ag and Au, than in BCC metals. Much less work has been done on hexagonal metals, but vacancy loops have been observed in the lower melting point metals following quenching, e.g. Mg. Moreover, cascade collapse to vacancy loops has been demonstrated in heavy-ion irradiated Co, Ti and Ru, and large vacancy loops have also been observed in neutron and electron irradiated Ti and Zr. Thus, all of the evidence is consistent with CPH metals having significantly lower SFE's than BCC metals.

The driving force for interstitials to aggregate to form loops in all irradiated metals is clearly much higher than for vacancy loops. Although it has been established that under most conditions nucleation occurs heterogeneously, the evidence is consistent with the minimum stable nucleus being a di-interstitial.

Whether or not interstitial and vacancy loops continue to grow during irradiation is of fundamental importance to damage structure evolution. It is established that dislocations interact more strongly with interstitial point defects and this results in a biased flow of interstitials to both interstitial and vacancy loops during irradiation. Moreover, the reduction in line tension provides an additional strong driving force for the shrinkage of small vacancy loops by thermal vacancy emission during irradiation at high temperatures. Consequently, interstitial loops are intrinsically more stable than vacancy loops and, even if the latter form, they are not expected to grow during irradiation (5). The majority of the experimental evidence is consistent with this overall picture, but there are important exceptions. For example, large perfect $^{a}/2\langle 111\rangle$ loops have been observed in TZM (a molybdenum alloy containing Ti and Zr) following neutron irradiation in the temperature range 923-1123K. A second more striking example is the continuous growth of vacancy loops during neutron and electron irradiation of Ti and Zr. These two exceptions will be discussed further in sections 2.2 and 2.3 respectively.

2.1 FCC METALS

Point defect clusters have been studied most extensively in FCC metals and the major factors influencing their geometries are relatively well understood. Since the results have been extensively reviewed at earlier conferences and elsewhere, we will concentrate on summarising the more recent developments here. The main area of interest has been the formation of planar vacancy defects in displacement cascades, and particularly the role played by solute content and surface image forces. A comparison of these results with the earlier work on quenched FCC metals emphasises the fundamentally important role played by initial point defect distributions in governing nucleation mechanisms for planar vacancy clusters.

Considering first, very briefly, the formation of interstitial loops, results from in situ electron irradiation experiments in an HVEM on a number of FCC metals are consistent with a di-interstitial being the minimum stable nucleus and with nucleation occurring heterogeneously on impurity atoms, except in very high purity metals irradiated at low temperatures (<100K), when homogeneous nucleation can become the dominant mechanism (8). In the absence of stresses or intersections with surfaces and other defects, faulted interstitial loops grow to large sizes, even in a moderately high SFE metal such as Ni. Thus, it appears that the unfaulting reaction involving the nucleation of two Shockley partials is the limiting step. Nevertheless, unfaulting can occur, even in the lowest SFE metals such as Au and Cu, if a driving force exists which triggers off the unfaulting reaction.

It has been established that the formation of vacancy loops during irradiation of FCC metals requires the generation of locally high vacancy concentrations at displacement cascade centres. Displacement cascade generation is the dominant process during neutron irradiation and a widely used method of simulating this process is to irradiate with heavy-ions in the relevant energy range, i.e. 10-100KeV. The incident ion range is typically 10-20nm and the adjacent surface plays an important role in acting as a sink, particularly for the mobile interstitial defects, and as a source of image forces induced by the collapsed defects. Much of the detailed work on defect geometries, number and sizes has been carried out on Cu and Cu alloys (e.g. 9,10,11), but experiments have also been carried out on other FCC metals, including Ag Au, Ni, Al and Cu_3Au. Calculations of defect self energies predict that a SFT is the preferred geometry for small vacancy clusters in all of these metals. However, it is observed that the vacancy rich centres collapse initially to Frank loops. In Cu and Cu alloys these loops undergo secondary dissociation towards SFT by the so-called Silcox-

Hirsch mechanism, involving the glide of Shockley partials across intersecting {111} planes and interacting at the corners to form $\frac{1}{6}$<110> stair rod dislocations. It is clear that surface image forces provide the driving force for the secondary dissociation and this is supported by the observation that this process occurs to a much smaller extent in bulk neutron irradiated Cu (12). As expected, the degree of secondary dissociation increases with decreasing SFE. This is demonstrated by a comparison of Ag (SFE = 16mJm^{-2})(13) with Cu (SFE = 45mJm^{-2}) and, more systematically, by comparing defect geometries in a series of Cu alloys (11).

An important exception to the above behaviour is shown by Al for which there is no clear evidence of cascade collapse to visible vacancy loops during self ion or neutron irradiation. This is probably due to the diffuseness of the cascades resulting in a lower vacancy concentration and a higher rate of recombination with interstitials. Thus, the conditions for forming visible loops are not achieved. In support of this conclusion, small vacancy defects are observed in Al irradiated with a much heavier ion, i.e. 150KeVHg+, which generated considerably more compact cascades (16).

Brief mention should be made of the results from quenched metals since in a number of cases they are consistent with SFT nucleating directly from the more uniformly distributed vacancy population. The most systematic studies have been carried out on Au (14,15) and here there is convincing evidence that the SFT grow from a hexa-vacancy cluster. In general nucleation is heterogeneous, but it is also necessary to have a balanced population of mobile mono-, di- and tri-vacancies in order to form hexa-vacancy clusters. Any factor influencing the balance between the three mobile vacancy complexes, and particularly the tri-vacancy concentration, affects the relative probability of forming Frank loops rather than SFT. Thus, depending on ageing temperature and solute content, it is possible to obtain any proportions of SFT to Frank loops in the vacancy cluster population. The main point we wish to emphasise here is the contrast with the situation in displacement cascades, where presumably the required balance between the mobile vacancy complexes to form SFT is not achieved. It should be noted that whereas SFT have also been observed in quenched Cu, Ag and Ni, they have never been observed in Al. Vacancies cluster readily in quenched Al and Al alloys, but the initial cluster geometries are always Frank loops or voids, depending on conditions (4).

2.2 BCC METALS

Considerable progress has recently been made in understanding the nucleation and geometry of dislocation loops in BCC metals. A majority of the evidence has come from irradiation experiments and particularly from heavy ion irradiation, in which vacancy loops form as a result of cascade collapse. Two outstanding features of the results are the sensitivity of the collapse process to surface image forces and the formation of perfect a<100> type loops in α-iron.

The closest packed planes in BCC metals are {110}, and point defects are expected to aggregate on these planes to form faulted $^a/2$<110> type loops which can unfault by shearing to $\frac{1}{2}$<111> or <100> Burgers vectors

$$\tfrac{1}{2}\langle 110\rangle + \tfrac{1}{2}\langle 001\rangle \rightarrow \tfrac{1}{2}\langle 111\rangle \qquad 1(a)$$

$$\tfrac{1}{2}\langle 110\rangle + \tfrac{1}{2}\langle 1\bar{1}0\rangle \rightarrow \tfrac{1}{2}\langle 100\rangle \qquad 1(b)$$

Vacancy loop geometries have been studied most extensively in heavy ion irradiated Mo and W (e.g. 17,18). The results are in good agreement and consistent with the loops forming according to reaction 1(a), i.e. initial collapse onto {110} to form faulted loops with b = $^a/2$<110> which unfault at an early stage to $^a/2$<111>. There is evidence that a small fraction of the loops in W have b = a<100> and a significantly higher fraction of these loops have been identified in Mo, but only after irradiation with molecular Sb_3^+ ions.

Surface image forces have a major influence on both the unfaulting reaction and glide of perfect loops to the surface. The interaction energy between a loop and its image is given by:

$$E^{(i)}_{int} = A^2 \, b_i \, \sigma_{ij} \, n_j \qquad (2)$$

where A is the loop area, b_i and n_j its Burgers vector and loop plane normal respectively, and σ_{ij} is the induced stress field. Assuming that the infinitesimal loop approximation is valid and using the equations derived by Eshelby for the associated displacement and stress fields (19), values for E_{int} can be computed for different combinations of b and n (20). Thus, it is more favourable for a faulted loop to shear to the perfect $^a/2$<111> b that has the largest component parallel to the surface normal. The loops are also subject to an image shear stress $\tau^{(i)}$, given by:

$$\tau^{(i)} = (2\pi rb)^{-1} \; \frac{dE^{(i)}_{int}}{dZ} \left[b_z\right] \qquad (3)$$

Clearly if $\tau^{(i)}$ exceeds the loop glide stress, loops can glide to the surface, resulting in significant loop loss which must be taken into account in any quantitative analysis of loop numbers (20). The experimental results from Mo and W are entirely consistent with the theoretical predictions.

Results from bulk neutron irradiated Mo are consistent with those from ion irradiated Mo, regarding both vacancy loop formation by cascade collapse and loop geometries. Moreover, irradiation at elevated temperatures results in the vacancy loop population disappearing, and this aspect has been discussed elsewhere. However, as already meantioned, an unexpected result is the observation of large vacancy loops in TZM-Mo neutron irradiated at 923-1123K (21). The circumstantial evidence is consistent with oversized Zr solute atoms diffusing to the dilatational sides of the loops, making both interstitial adsorption and vacancy emission more difficult. Thus, the vacancy loops are stabilised and able to grow during irradiation in spite of the interstitial bias.

Less work has been done on the geometries of interstitial loops in irradiated refractory BCC metals, but the results are consistent with the observations on vacancy loops. Thus, all of the visible loops are perfect $^{a}/2\langle 111\rangle$ type, but the analysis of loop normals is again consistent with them nucleating on {110} planes. Although loop nucleation occurs heterogeneously under a wide range of conditions, an important feature distinguishing BCC metals from FCC metals is that, depending on irradiation conditions, interstitial loops tend to aggregate into rafts. We will return to this aspect in section 3.

Iron and steels differ from other BCC metals in two important respects. First, cascades generated by self ions or primary recoils do not collapse to form visible vacancy loops (22,23). In contrast, the more compact cascades generated by much heavier ions, i.e. Xe and W, do collapse to vacancy loops and this result highlights very clearly the critical role played by vacancy supersaturation at the cascade centres in the collapse process. The second major difference is that both vacancy and interstitial loops can have $b = a\langle 100\rangle$ as well as the expected $^{a}/2\langle 111\rangle$. More specifically, in room temperature W or Xe irradiated iron, vacancy loop populations consist of approximately equal proportions of $a\langle 100\rangle$ and $^{a}/2\langle 111\rangle$ perfect loops (20). Both families of loops are nucleated on {110} and the unfaulting reactions and loop loss by glide are again influenced by surface image interactions. Irradiation of α-iron and ferritic steels with heavy ions, electrons and neutrons at elevated temperatures results in the formation of large $a\langle 100\rangle$ perfect interstitial loops (4), although some $^{a}/2\langle 111\rangle$ type loops are also formed in room temperature neutron irradiated Fe (23). It is interesting that in this latter case small vacancy loops were also observed, but only in the immediate vicinity of dislocations, indicating that the conditions for vacancy loop nucleation can be created by the biased flow of interstitials to dislocations.

The observed loop geometries in BCC metals raise important questions regarding both the nucleation and unfaulting reactions, particularly the shearing to $a\langle 100\rangle$ loops in iron. Clearly the aggregation of self-interstitials to form {110} platelets can only result in the initial formation of an $^{a}/2\langle 110\rangle$ loop, although this shears to a perfect loop at an early stage. However, there is some uncertainty about whether a {110} vacancy disc will collapse fully to a $\frac{1}{2}\langle 110\rangle$ loop. Results from computer simulations indicate that vacancy discs containing up to 35-40 vacancies (r = 6-7nm) may only undergo partial collapse to a loop (24). The barrier to collapse comes from the very high SFE. For example, this is calculated to be 1.7Jm^{-2} and 2.8Jm^{-2} respectively for Fe and Mo. Thus, even if collapse were to occur, these values are consistent with unfaulting occurring before the loops reach a visible size. More specifically, the critical loop radius for unfaulting to $^{a}/2\langle 111\rangle$, r_c, is related to SFE by:

$$\text{SFE} = \frac{\mu a^2}{8\pi r_c(1-\nu)}\left[\ln\frac{4r_c}{\rho} - 1\right] \qquad (4)$$

where μ is the shear modulus, a is the lattice parameter, ν is Poissions ratio and ρ is the core radius. Substituting values for α-Fe gives a value for r_c ∿0.5nm, which is well below the minimum visible size of 1nm.

The crucial question remains as to why ½<110> loops or discs ever unfault to the high energy a<100> perfect loop configuration. A notable feature distinguishing α-Fe from other BCC metals is its elastic anisotropy. The anisotropy factor $A = 2C_{44}/C_{11}-C_{12}$ is 2.4 for Fe, and the only other BCC metal to have a value >1 is Ta (A = 1.6). Masters (25) suggested that elastic anisotropy might be responsible for a<100> loops having a lower self energy than a/2<111> loops in Fe, but this is not supported by more recent calculations (24). It is necessary, therefore, to examine the unfaulting process, and a simple approach is to relate the resistance to shear to a/2<111> and a<100> to the shear moduli along <100> and <110> in a {110} plane respectively. This ratio is given by A and an Fe crystal is therefore considerably softer in shear along <110> than along <100>, Thus, if this were the sole criterion, a<100> loops should be preferred over a/2<111> loops. On this basis the only other BCC metal in which the shear to a<100> loops should be favoured is Ta, and there are no experimental results for loop geometries in this metal.

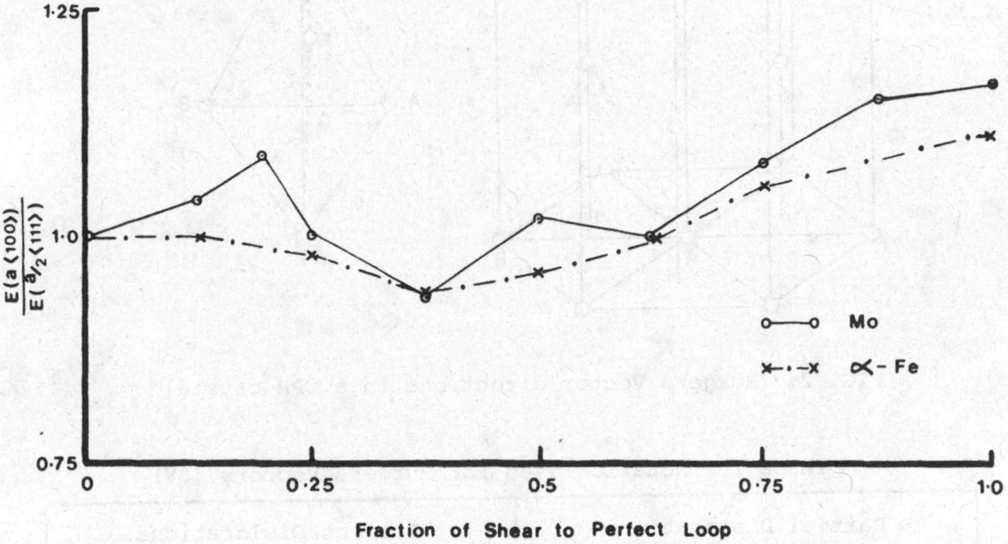

FIG. 1. Plot of loop energy ratio versus shear to perfect loops in Fe and Mo

A more detailed analysis of the influence of elastic anisotropy has been carried out by calculating the change in loop energy as it unfaults (24). Results obtained for a small vacancy loop containing 37 vacancies in Fe and Mo are shown in Fig. 1., in which the ratio of loop energy as it unfaults to a<100> and a/2<111> respectively is plotted against the amount of shear expressed as a fraction of the total shear required. It can be seen that during the early stages the shear to <100> is favoured in Fe, but not in Mo, which is consistent with the experimental results. Although at later stages the <110> shear becomes increasingly unfavourable in both metals, it seems unlikely that once the reaction has started its direction will change at some intermediate stage.

2.3. CPH METALS

Although the observed loop geometries in CPH metals do not fall into such a clear pattern as those in FCC and BCC metals, there is an underlying consistency in the results. Major factors that have been shown to be important are the $^{c}/a$ ratio and stress. Solute atoms also appear to be important, but their role is less well understood.

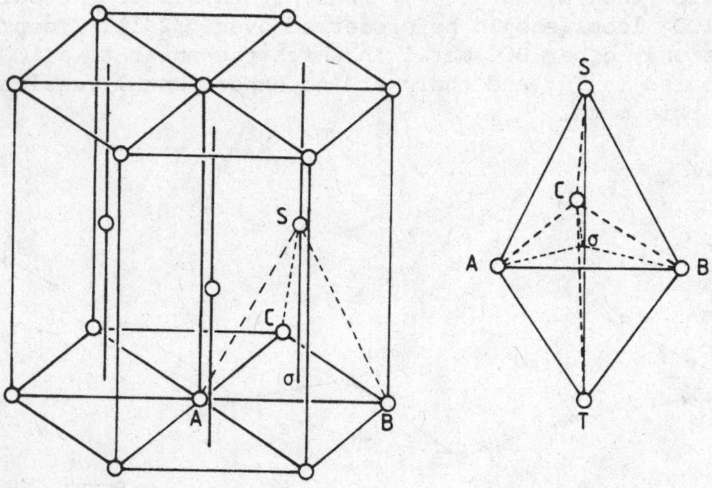

FIG. 2. Burgers Vector directions in a CPH crystal

TABLE 1. Notation used for Burgers Vectors (BV)

Partial Dislocations			Perfect Dislocations		
b	Vector in Fig. 2	notation	b	Vector in Fig. 2	notation
$\frac{1}{2}[0001]$	Sσ or Tσ	$\frac{c}{2}$	$[0001]$	ST	c
$\frac{1}{6}<20\bar{2}3>$	AS	$(\frac{c}{2} + p)$	$\frac{1}{3}<\overline{11}23>$	AB + ST	(c + a)
$\frac{1}{2}<10\bar{1}0>$	$\frac{3}{2}$Aσ	p^{1}	$\frac{1}{3}<11\bar{2}0>$	AB	a

The directions of the possible BV in CPH crystals are shown schematically in Fig. 2. and the notation that will be used is presented in Table 1. The closest packed planes depend on the $^c/a$ ratio and values for a number of metals are listed in Table 2. For $^c/a>\sqrt{3}$, the closest packed planes are (0001) whereas for $^c/a<\sqrt{3}$ they are the $\{10\bar{1}0\}$ prism planes.

TABLE 2. $^c/a$ Ratios for CPH metals

$^c/a<\sqrt{3}$							$^c/a>\sqrt{3}$	
Be	Ru	Ti	Zr	Re	Co	Mg	Zn	Cd
1.568	1.582	1.587	1.593	1.615	1.623	1.623	1.856	1.886

The only analyses of small loops in CPH metals, which provide a direct insight into loop nucleation mechanisms, have been carried out on Co, Ti and Ru following R.T. heavy ion irradiation (26-28). In all cases the loops are vacancy, having formed as a result of cascade collapse. The dominant nucleation planes in the three metals are the closest packed $\{10\bar{1}0\}$ prism planes to give faulted p^1 type loops. In Co and Ti these loops unfault at an early stage, so that the visible loop population is predominantly perfect with a type Burger's vectors:

$$\tfrac{1}{2}\langle 10\bar{1}0\rangle + \tfrac{1}{6}\langle \bar{1}2\bar{1}0\rangle \rightarrow \tfrac{1}{3}\langle 11\bar{2}0\rangle \qquad (5)$$

The behaviour of Ru differs from that of Co and Ti in two respects (28). First, a majority of the loops are faulted $\frac{1}{2}\langle 10\bar{1}0\rangle$ type, which is consistent with a lower SFE. A comparison of the size spectra of the faulted and perfect loops indicates that the critical radius for unfaulting is ~1.2nm, and substituting this into Eqn (4) gives a value for the SFE of ~$1\,Jm^{-2}$. Second, although the faulted and perfect prism loops remain in the majority, faulted basal loops with $^c/2$ and $(^c/2 + p)$ type Burger's vectors are formed in foils with the surface normal, N, perpendicular to [0001]. The reason for these loops forming in Ru and not in Co and Ti, even when N is perpendicular to [0001], is not clear, particularly in view of its lower $^c/a$ ratio. One factor could be the lower SFE, but it is more likely that stresses, perhaps generated by a surface oxide, are playing the decisive role. The stabilisation of basal vacancy loops normal to the foil surface would require the stresses to be compressive.

A considerable amount of work has been carried out on the geometries of large loops in Zr and Ti irradiated at elevated temperatures (e.g. 29-32). There is a particular interest in Zr and its alloys because of their use for fuel cladding and pressure tubes in water reactors. All of the loops in Ti and a majority in Zr are perfect prism loops with a type Burger's vectors. The loops are mostly non-edge with habit planes close to $\{10\bar{1}0\}$ which is consistent with $^c/a<\sqrt{3}$ and with the results from Co, Ti and Ru. An interesting feature that is not properly understood is the greater ellipticity of the vacancy loops compared with the interstitial loops. A minority population of basal loops have also been identified in neutron and electron irradiated Zr. In neutron irradiated specimens they are only observed following irradiation at temperatures $\gtrsim$723K, while in HVEM electron irradiated foils they are observed at temperatures $\gtrsim$650K. In both cases ($^c/2$ + p) faulted and (c + a) perfect loops have been identified and there is evidence that the latter are formed by the combination of two ($^c/2$ + p) type loops, e.g.:

$$\frac{a}{6}[\bar{2}0\bar{2}3] + \frac{a}{6}[0\bar{2}\bar{2}\bar{3}] \rightarrow \frac{a}{3}[\bar{1}\bar{1}2\bar{3}] \qquad (6)$$

Examples of $^c/2$ type vacancy loops have also been found in some electron irradiated specimens.

The HVEM electron irradiation results from both Zr and Ti are particularly interesting in showing the sensitivity of loop geometries to internal stresses, generated in this case by surface oxide films (30). In both metals there is a tendency for the interstitial loops to have b perpendicular to N, while vacancy loops tend to have b parallel to N, thereby relieving the oxide induced tensile stresses. In the specific case of Zr the nature of the basal loops also depends on foil orientation. For example, when N = <1$\bar{2}$11>, perfect interstitial loops with (c + a) type Burger's vectors normal to N are formed, while for N = [0001], faulted $^c/2$ vacancy loops are formed.

The influence of stress may also account for the simultaneous growth of vacancy and interstitial loops in Ti and Zr during irradiation. This observation is in contrast to the behaviour of most other metals, including CPH metals, and is not consistent with a preferred drift of interstitials to all loops. Models have been proposed to explain the observed vacancy loop growth based on preferred solute segregation to the loops and the weaker attraction loops have for interstitials when they are small, but they do not account for all aspects of the results. An alternative explanation is that anisotropic growth induced stresses modify the normally weak induced interactions between point defects and dislocations (19) so as to allow the growth of vacancy loops on certain planes. More specifically, irradiation growth will generate internal stresses in a polycrystalline material. Because of the crystallographic nature of growth (contraction along the c-axis and expansion along the a-axis), the stresses will be bi-axial compression normal to the prism planes and uni-axial tension normal to the basal planes. The resulting stress induced interactions modify the total interaction between dislocation loops and point defects (33):

$$E^{O}_{\alpha} = \left[\frac{(1 + \nu)\mu b \Delta V_{\alpha}}{3\pi(1 - \nu)} + \frac{\mu b \varepsilon V_{\alpha} A^{(\beta)}_{\alpha}}{\pi}\right] \frac{\sin\theta}{r} \qquad (7)$$

where E^{O}_{α} is the total interaction energy, ΔV_{α} is the point defect relaxation volume (subscript α refers to defect type), V_{α} is their formation volume and $A^{(\beta)}_{\alpha}$ is a function of the difference in elastic constants between matrix and defect and of the orientation b with respect to the stress axis, denoted by superscript (β). The first term in the brackets of Eqn (7) is that due to the first order size effect and the second is the so-called SIPA term (33). For vacancies the size effect interaction is negligible, but for the compressive stress the SIPA term is positive. Conversely, although the size effect interaction is dominant for interstitials, it will be partially offset by a negative SIPA term when the stresses are compressive. Due to elastic anisotropy and the random grain shapes and orientations, we can expect the compression stresses normal to the prism planes to be asymmetrically distributed about the c axis. Thus, the necessary imbalance in the interactions between the different sets of prism loops and point defects to allow simultaneous growth of vacancy and interstitial loops can be set up. In order to test the validity of this model there is a need for quantitative analyses of vacancy and interstitial loop Burgers vector distributions.

Turning briefly to the low melting point CPH metals, results from Mg, with $^c/a<\sqrt{3}$, show that loop nucleation can occur on basal planes as well as the closer packed prism planes. For example, prism and basal loops have been observed in quenched Mg, with the tendency for $^c/2$ and $(^c/2 + p)$ basal loops increasing with decreasing purity. In contrast to Zr and Ti, only interstitial loops have been identified in HVEM electron irradiated Mg (34,35). Although a majority of the loops are perfect a type, some basal loops with $^c/2$ and $(^c/2 + p)$ type BV are formed and it again appears that oxide generated stresses have a crucial influence. Thus, the loop geometries tend to be such as to relieve the resulting tensile stresses and, for example, only a type loops are observed when N ∿ [0001], whereas some basal loops are found in foils with N ∿ $[1\bar{2}10]$.

Cd and Zn are the only CPH metals to have $^c/a>\sqrt{3}$ and, in contrast to Mg, the observed loop geometries are always consistent with their nucleating on the closest packed basal planes. The most detailed studies have been carried out on HVEM electron irradiated specimens (36) and, again, only interstitial loops are found. The loops nucleate initially as $^c/2$, but in many cases the fault energy is reduced by the formation of double loops or by shearing to $(^c/2 + p)$ types, which have lower SFE's.

3. DISLOCATION - LOOP INTERACTIONS

3.1 LOOP-LOOP INTERACTIONS

An important aspect of damage structure development in a number of bulk irradiated metals is the aggregation of dislocation loops into so-called 'rafts'. The rafts are generally aligned along close packed directions and consist of planar arrays of interstitial loops having identical Burgers vectors. Although loop rafting was first noted in Al, it is not generally observed in FCC metals. Rafts consisting of glissile $^a/2$<111> loops occur in Mo neutron irradiated at temperatures in the range 473-673K, but their size and spacing is sensitive to both purity and temperature (21). Rafts lying on prism planes and consisting of $^a/2$<11$\bar{2}$0> interstitial loops have also been observed in neutron irradiated Zr (32).

It is generally agreed that rafting occurs as a result of elastic interactions between loops, but the details of the process are not properly understood. A complete description of rafting must include factors affecting loop nucleation as well as their subsequent glide and climb. More specifically, if loop nucleation is random, particularly in terms of Burgers vectors, it is difficult to understand the very effective partitioning of loops into rafts having identical BV. It is more reasonable to assume that nucleation results in colonies of loops having identical BV that can then aggregate by glide and climb. Qualitatively we envisage that the Burgers vectors of the first loops to nucleate are randomly distributed amongst the possible variants. However, provided the nucleation rate is not too high, these initial loops can grow and increasingly influence the nucleation of other loops in their vicinity. This can be treated using a simple binary loop interaction model based on the infinitesimal loop approximation (19). Thus, the change in work done in forming a second loop with Burger's vector $b_i^{(2)}$ and loop normal $n_i^{(2)}$ in the vicinity of a primary loop is given by the interaction energy:

$$E_{int} = -A^{(2)} \sigma_{ij}^{(1)} b_i^{(2)} n_j^{(2)} \tag{8}$$

where $A^{(2)}$ is the area of the second loop and $\sigma_{ij}^{(1)}$ is the stress field associated with the primary loop. By defining a primary loop of given $b^{(1)}$ and $n^{(1)}$ at the origin, the regions within which the second loops having identical BV are preferred can be defined. If the nucleation process involves a faulted loop which then shears to a perfect loop, the calculation involves two steps: (a) determination of the regions within which the possible faulted loop geometries are preferred; and (b) determination of the preferred unfaulting shear for loops in any given region. The third step is then to determine the range over which such loops will aggregate into rafts by a combination of glide and self climb. This involves resolving the force exerted on a second loop along the glide and self climb directions; the relevant equations have been presented elsewhere (38).

The above model of rafting exposes a number of factors which will have a major influence on rafting. First, a high rate of nucleation on a fine scale will inhibit the development of colonies of loops with identical BV. Second, any factor influencing loop glide will clearly affect the range over which loops can aggregate. Third, although some self climb is necessary to permit the necessary three dimensional movement of the loops required for aggregation, excessive migration normal to b will lead to loop coalescence in the rafts. Clearly SFE is an important parameter, since if the loops remain faulted they are not able to aggregate by glide, and this accounts for the absence of rafting in most FCC metals. Impurity content is also important because of the effect it has on loop nucleation and on loop glide. Thus, a high impurity level is expected to suppress rafting and this is supported by results from neutron irradiated Mo (21). Irradiation temperature is particularly important because of its effect on the scale and rate of loop nucleation, loop glide stress and self climb rate. Thus, rafting is not expected at low irradiation temperatures because of the fine scale of loop nucleation and high loop glide stress. Conversely, if the irradiation temperature is too high the loops in rafts will coalesce to form large loops that can then climb into a network. Again, the results from neutron irradiated Mo are consistent with this picture in exhibiting rafting over a fairly narrow temperature range of 473-673K. Lastly, dose rate and damage generation processes during irradiation have an important influence on loop nucleation. An important aspect here is the formation of vacancy loops by cascade collapse. Loop nucleation is dominated by the highly non-equilibrium conditions existing in the cascades and any effects due to loop-loop interactions will be negligible. This probably accounts for the absence of evidence for rafting of vacancy loops and it highlights the importance of the interactions in promoting the formation of loop colonies with identical BV as a necessary precursor to rafting.

The link between rafting and loop coarsening has an important effect on loop growth rates and dislocation network development during both elevated temperature irradiation and post irradiation annealing. It is a major factor distinguishing the behaviour of low SFE metals (FCC) and high SFE metals (BCC) and between interstitial and vacancy loop behaviour in the latter case. In the absence of rafting, loop coarsening occurs by a bulk diffusion controlled Ostwald ripening process and the mean loop radius increases linearly with time. In contrast, loop coarsening by glide and self climb occurs very rapidly in the initial stages with the mean loop growth rate following a $t^{1/6}$ law (38) and, because self climb is the rate limiting step, it occurs down to lower temperatures. Thus, interstitial dislocation networks are both coarser and develop much more rapidly during irradiation at elevated temperatures. This basic difference in loop growth mechanisms in, for example, many FCC metals and BCC metals, has a major influence on the evolution of damage structure as a whole, including the nucleation and growth of voids.

3.2 DISLOCATION SEGMENT-LOOP INTERACTIONS

A further manifestation of the influence elastic interactions have on dislocation loop geometries and distributions is the observation of enhanced loop concentrations in the vicinity of dislocation segments in Cu, Fe and Mo following comparatively low dose nuetron irradiation (12,23,39). In Cu and Mo these enhanced concentrations, referred to as high damage regions (HDR's), are super-imposed on more uniformly distributed defect populations, referred to as low damage regions (LDR's). The loops in the LDR's are predominantly vacancy in nature and result from cascade collapse, which accounts for their absence in Fe. The loops in the HDR's are perfect even in Cu and are, in general, interstitial in nature, the exception being Cu in which 10% of the loops are vacancy. There is some evidence for local groups of loops having identical BV to form, but overall, loops having all possible variants of BV (including a<100> in Fe) have been found. A systematic analysis of HDR loop distributions in Mo has shown that whereas they are symmetrically distributed around screw dislocations, they form predominantly on the dilatation side of edge dislocations, leaving a denuded zone on the compressive side. An example of this is shown in Fig. 3a and the corresponding results from a stereo-microscopy analysis are presented in Fig. 3b, which demonstrates the wedge shape distributions of the loops on the dilatation side of the dislocation.

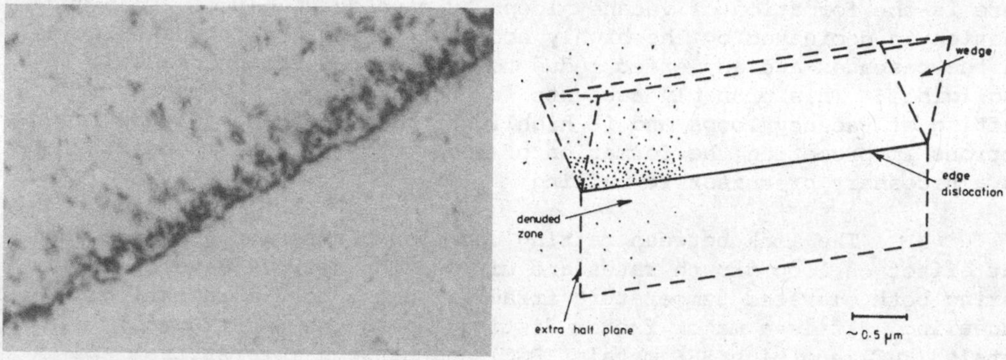

FIG. 3. Dislocation loop distribution adjacent to an edge dislocation in neutron irradiated Mo (39)

The enhanced nucleation and growth of loops in HDR's can be attributed to the combined effect of a number of factors, all of which have their origins in the elastic interactions between dislocations and other defects. First, the first order size effect interaction will cause a drift of interstitials and misfitting impurities to dislocations resulting in locally enhanced heterogeneous nucleation of interstitial loops. The wedge of interstitial loops on the dilatational sides of edge dislocation, illustrated in Fig. 3, is consistent with this process. Second, a positive interaction energy will reduce the work done in nucleating loops of given b and n. Third, such an interaction will also result in the loops being subject to an attractive force which may lead to their gliding towards dislocations if they are perfect. An important consequence of these interactions is that both pre-existing dislocations and dislocation networks that develop during irradiation will have an important influence on how the damage structure evolves.

4. CONCLUSIONS

We have attempted to outline the current understanding of loop geometries and behaviour in metals and to identify the main areas of uncertainty. The major conclusions can be summarised as follows:

(a) Although an increasingly complex picture emerges on going from FCC to BCC to CPH metals, many of the factors governing dislocation loop properties are reasonably well understood.

(b) As well as intrinsic properties such as crystallography, elastic constants and SFE, stress, both internally and externally generated,

plays a major role in determining loop geometries and distribution.
(c) Solute atoms are also very important in their effect on both nucleation and stability of loops, and there is a need for more systematic research on this aspect.

REFERENCES

(1) Nabarro, F. R. N. (1948). Conf. Rept., Strength of Solids, Phys. Soc., p.38.
(2) Frank, F. C. (1949). Proc. Phys. Soc., London, A62, 202.
(3) Hirsch, P. B. et al. (1959). Phil. Mag., 3, 897.
(4) Eyre, B. L. (1973). J. Phys. F., 3, 422.
(5) Eyre, B. L. et al. (1977). Vacancies '76: The Metals Society, London, p.63.
(6) Wilkens, M. (1970). Modern Diffraction and Imaging Techniques: North-Holland, Amsterdam, p.233.
(7) Wilkens, M. (1970). Vacancies and Interstitials in Metals: Ibid, p.485.
(8) Kiritani, M. (1976). Fundamental Aspects of Radn. Damage in Metals: O.R.N.L., U.S.A., p.695.
(9) Wilson, M. and Hirsch, P. B. (1972). Phil. Mag., 25, 983.
(10) Haussermann, F. (1972). Phil. Mag., 25, 537.
(11) Stathopoulos, A. Y. (1981). Phil. Mag., A44, 309.
(12) Muncie, J. W. (1979). D. Phil. Thesis: University of Sussex.
(13) Jenkins, M. L. (1974). Phil. Mag., 29, 813.
(14) Quader, M. A. and Dodd, R. A. (1968). Phil. Mag., 17, 575.
(15) Hussein, A. A. and Dodd, R. A. (1971). Phil. Mag., 24, 1441.
(16) Norris, D. I. R. (1969). Phil. Mag., 19, 527.
(17) Haussermann, F. (1972). Phil. Mag., 25, 561.
(18) English, C. A. et al. (1980). J. Phys. F., 10, 1065.
(19) Eshelby, J. D. (1957). Proc. Roy. Soc. A241, 376.
(20) Jaeger, W. and Wilkens, M. (1975). phys. stat. sol. (a) 32, 89.
(21) Bentley, J. (1974). Ph.D. Thesis: University of Birmingham.
(22) Jenkins, M. L. et al. (1978). Phil. Mag. A38, 97.
(23) Robinson, I. M. (1982). D. Phil. Thesis: University of Oxford.
(24) Matthai, C. and Bacon, D. J. (1984). J. Nucl. Matls., in press.
(25) Masters, B. C. (1965). Phil. Mag., 11, 881.
(26) Foll, H. and Wilkens, M. (1977). phys. stat. sol. (a) 39, 561.
(27) Woo, O. T. et al. (1983). EMAG '83: Inst. of Phys. Conf. Series, p.179.
(28) Phythian, W. et al. (1983). EMAG '83: Inst. of Phys. Conf. Series, p.335.
(29) Jostons, A. et al. (1980). Phil. Mag., 41, 903.
(30) White, J. et al. (1983). Proc. 7th Intnl. Conf. on HVEM: Berkeley.
(31) Northwood, D. O. et al. (1979). J. Nucl. Matls., 79, 379.
(32) Jostons, A. et al. (1979). Radn. on Matls. Paper in ASTM STP 683, p.46.
(33) Heald, P. T. and Bullough, R. (1976). Vacancies '76: The Met. Soc., p.134.

(34) Hossein, M. K. and Brown, L. M. (1977). Acta Met., 25, 257.
(35) Salisbury, I. G. et al. (1980). Proc. 6th Intnl. Conf. on HVEM, p.232.
(36) Whitehead, M. E. et al. (1978). Acta Met., 26, 983.
(37) Foreman, A. J. E. and Eshelby, J. D. (1962). Harwell Rept., AERE-R4170.
(38) Eyre, B. L. and Maher, D. M. (1971). Phil. Mag., 24, 767.
(39) Riley, B. (1983). Ph.D. Thesis: University of Birmingham.

DIRECT OBSERVATIONS OF ANISOTROPIC AND SURFACE EFFECTS ON DISLOCATIONS

P.M. HAZZLEDINE

Department of Metallurgy and Science of Materials,
Parks Road, OXFORD OX1 3PH.

ABSTRACT

The effects of elastic anisotropy on dislocations observed in the electron microscope are described, in particular the instabilities introduced by negative line tensions.

The most frequently observed surface effect is Lothe's force which causes dislocations to rotate to lower their line energy. Observations, using X-ray topography, electron microscopy and etch hillocks, of the consequences of this force to the growth of crystals, the velocity of dislocations, the structures of grain and sub-grain boundaries and to cross slip of dislocations are presented.

1. INTRODUCTION

The scope of this paper could include any dislocation behaviour which is different from that expected in an infinite isotropic solid. However, two of the most striking effects, one the soft skin effect of Fourie (1,2) and the other the dependence of the Peierls stress in anisotropic bcc metals on non-glide components of the stress tensor (3,4) are not measured by direct observation of individual dislocations. Both effects are measured by variations in the flow stress of single crystals. The soft skin effect results from the fact that dislocations within one slip line length of the free surface escape to the surface during deformation and hence do not interact with other dislocations strongly. The result is that when a work-hardened crystal is sliced into subcrystals the flow stress of those subcrystals from near the surface of the original crystal is lower than that of the subcrystals from the bulk. The effect of non-glide stresses on the Peierls stress in bcc metals results from the fact that non-glide stresses may modify the core structure of the ⟨111⟩ screw dislocation sufficiently to alter the Peierls stress by orders of magnitude. The effect may be detected by measuring the orientation dependence and absolute magnitude of the flow stress of single crystals of different bcc metals.

In a sense all direct observations of dislocations involve surface and anisotropic effects. The metallographic techniques used in the observation of dislocations: etch pitting, X-ray topography, electron microscopy all reveal dislocations either at a surface or close to the two surfaces of a thin foil. Unless the dislocations have been carefully pinned before observation they are at least perturbed by the image stresses of the surface and at most drawn out of the foils and rendered unobservable. Within the specimens the dislocations lie in an elastically anisotropic solid; the displacements, strains and stresses differ from those in a hypothetical isotropic solid (5). The presence of surfaces and anisotropy alter the dislocation-dislocation interactions and hence the dislocation configurations observed. They also alter the interaction between a dislocation and a probe such as an electron beam. For example in two-beam electron microscopy a dislocation is invisible if its strain field leaves the reflecting plane flat. In this way dislocation Burgers vectors may be determined by the $\mathbf{g}.\mathbf{b}$ and $\mathbf{g}.\mathbf{b} \times \mathbf{u}$ rules. In anisotropic solids, not only are the detailed forms of the dislocation contrast modified, but also dislocations leave low index planes flat in only rather restricted circumstances (6). Consequently it is frequently difficult to determine the Burgers vectors of dislocations in strongly anisotropic solids. Likewise the proximity of a surface modifies the strain field of a dislocation (7) and invalidates the $\mathbf{g}.\mathbf{b}$. and $\mathbf{g}.\mathbf{b} \times \mathbf{u}$ rules. The effects of a surface on dislocation contrast are generally smaller than anisotropy effects but can be important (8). A further complication is that the forces exerted by a surface on a dislocations often prevent the dislocations from being straight and therefore make its core structure difficult to observe by lattice resolution (end-on) electron microscopy.

Eshelby (54) wrote "many dislocation calculations are concerned with problems relating to specific configurations of one or more dislocation lines. There is, of course, indefinite scope for the solution of such problems, but if they are not to serve simply as an excuse for a display of mathematical weight-lifting they must be chosen carefully, either to cover a wide range of special cases, or to refer to specific situations of interest (e.g. some configuration observed in the electron microscope), or they should be concerned with some simple situation chosen so as to bring out a physical point". In this paper the early work on the effects of surfaces and elastic anisotropy is very briefly reviewed and some recent observations, all of which can be understood with the minimum of theory, are presented.

2. ANISOTROPIC EFFECTS

2.1 MEASUREMENTS OF STACKING FAULT ENERGIES

The most accurate way to measure stacking fault and antiphase domain boundary (APB) energies is to image dissociated dislocations, usually by weak beam electron microscopy (9), and to measure the separation between the partial dislocations. The value of the interfacial energy is inversely proportional to the separation of the dislocations but the proportionality constant depends on the character of the dislocations and on the elastic constants. Many measurements of this kind have been made (see e.g. 10,11,12). One of the earliest, and the one which shows the effect of anisotropy most clearly, is the measurement of the two APB energies in DO_3 ordered Fe 26% Al of Crawford et al (13). In this alloy dislocations dissociate into four partials all with the same character. The two outer partials are joined to the two inner partials by an APB of specific energy γ_1 and the two inner partials are joined by an APB of specific energy γ_2 (Fig. 1). In the isotropic approximation the equilibrium equations for two of the partials are

$$\alpha\left(\frac{1}{R_1} + \frac{1}{R} + \frac{1}{R + R_1}\right) = \gamma_1$$

$$\alpha\left(\frac{1}{R} + \frac{1}{R - R_1} - \frac{1}{R_1}\right) = \gamma_2 - \gamma_1$$

where $\alpha = \frac{Gb^2}{2\pi}\left[\cos^2\theta + \frac{\sin^2\theta}{(1-\nu)}\right]$, G is the shear modulus, ν = Poissons ratio, b = Burgers vector and θ is the character. If anisotropic elasticity is used, α becomes $(C_{44}\,H_{ij}\,b_i\,b_j)/2\pi$ where H_{ij} is a matrix which depends on the line direction and degree of anisotropy. Taking γ_1 to be 77 mJm^{-2} and γ_2 to be 85 mJm^{-2} the values of R_1 and R as functions of θ in the isotropic and anisotropic approximations are compared with the measurements in Fig. 1. Despite the scatter in the

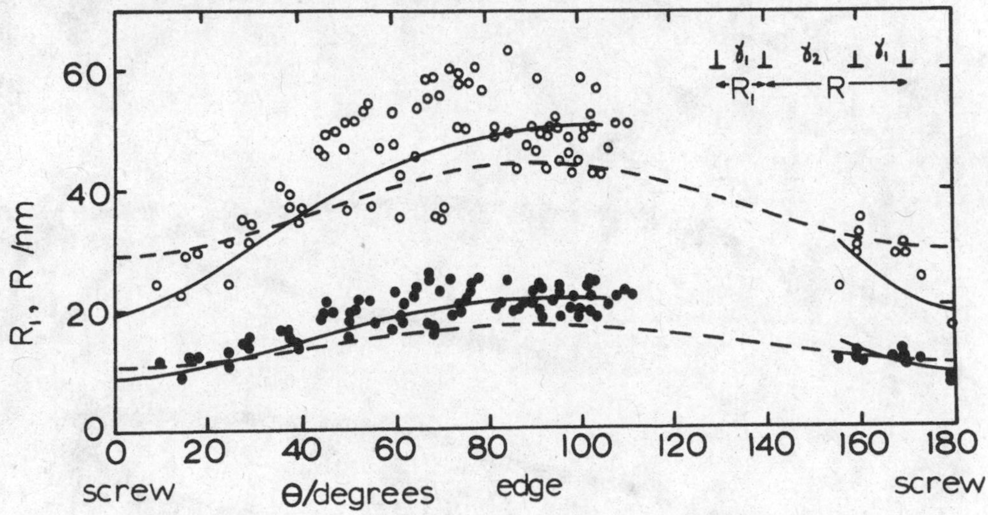

Fig. 1 Stacking fault widths versus dislocation character in Fe 26% Al. Full line, anisotropy; dashed line isotropy (ν = 1/3).

measurements (caused partly by surface effects) the fit with the anisotropic theory (full line) is distinctly better than with the isotropic theory (dashed line) both with regard to the ratio R_1/R and with regard to the variation of R with θ. In addition, there is a distinct gap in the data between $\theta = 110°$ and $\theta = 155°$ which agrees with the anisotropic but not with the isotropic calculation. The dislocation with $\underline{b} = \frac{a}{4}[111]$ is unstable in this character range and is consequently absent from the micrographs.

2.2 DISLOCATION LOOP SHAPES AND INSTABILITIES

The approximate shape of a dislocation loop in equilibrium with an applied stress may be found from the Wulff construction if the energy of the dislocation $E(\theta)$ can be calculated as a function of θ. The direct Wulff construction is based on a polar plot of $E(\theta)$. A point on the polar plot has coordinates $E\cos\theta$, $E\sin\theta$ and this point corresponds to a point $E\cos\theta - (\partial E/\partial\theta)\sin\theta$, $E\sin\theta + (\partial E/\partial\theta)\cos\theta$ on the curve which represents the shape of the dislocation loop (14). At any point the curvature of the dislocation shape is inversely proportional to the line tension. In isotropic elasticity the line tension varies smoothly from screw to edge character and consequently the shape of the dislocation loop is also smooth and roughly elliptical. In anisotropic elasticity, however, the line tension, $E + \partial^2E/\partial\theta^2$ may vary rapidly with character and may become negative. In the latter case a dislocation instability results and sharp bends develop in dislocation loops.

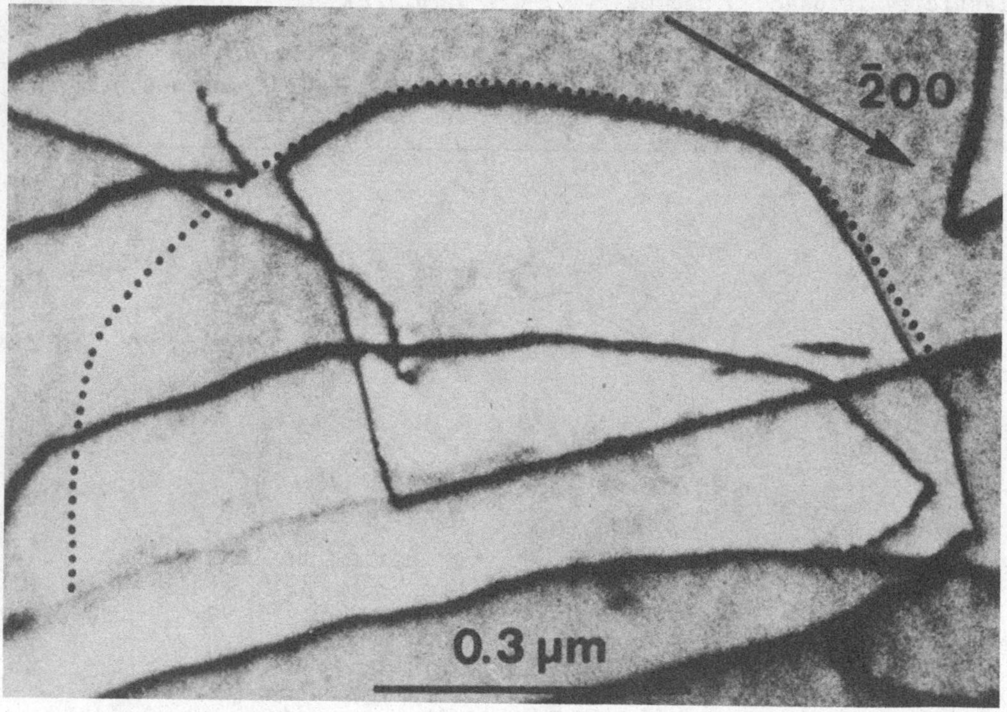

Fig. 2. Predicted (dotted) and actual loop shape for Lomer dislocation in Fe_3Ni.

Fig. 2 shows an example of a dislocation loop in face-centred cubic disordered Fe_3Ni from the work of Korner and Karnthaler (15). The shape of the loop, which is a Lomer dislocation spreading on a {100} plane, predicted from anisotropic elasticity is drawn on the micrograph. The dislocation has a low line tension, and consequently high curvature, near characters 35° and 90° but is nowhere unstable. However, in Cu 15% Al the equivalent dislocation is unstable near 36° and 90° character (16,17) and sharp bends are observed near these characters (18).

Head (19) made the first systematic analysis of instabilities in dislocation lines. In particular he showed that in β brass $(1\bar{1}0)[001]$ and $(1\bar{1}0)[111]$ dislocations contain unstable ranges of characters but that $(1\bar{1}0)[110]$ dislocations do not. A very clear example of unstable behaviour in β brass from the work of Saka (20) is shown in Fig. 3a. [010] dislocations are only stable over a narrow character range in this material (19) and these ranges are indicated in Fig. 3a. Fig. 3b shows a Wulff plot W and the predicted diamond shape S of a complete glide loop. The shape of the slightly curved segments matches the micrograph exactly except for one feature which is that screw dislocation segments are observed although they are not predicted to be stable. A conceivable explanation is given in Fig. 3b - that the screw dislocation (which would have a four-fold symmetric core) has a lower energy than expected from elasticity theory; this is represented by an inward spike P on the Wulff plot at $\theta = 0$. The necessary length of the spike (30% of the screw energy) is so large, however, that a more likely explanation must be sought. If the mobility of the screw dislocations is much lower than that of the stable dislocations it is possible that screw dislocations would be laid down as the diamond-shaped loop expands even though they would not be present at equilibrium.

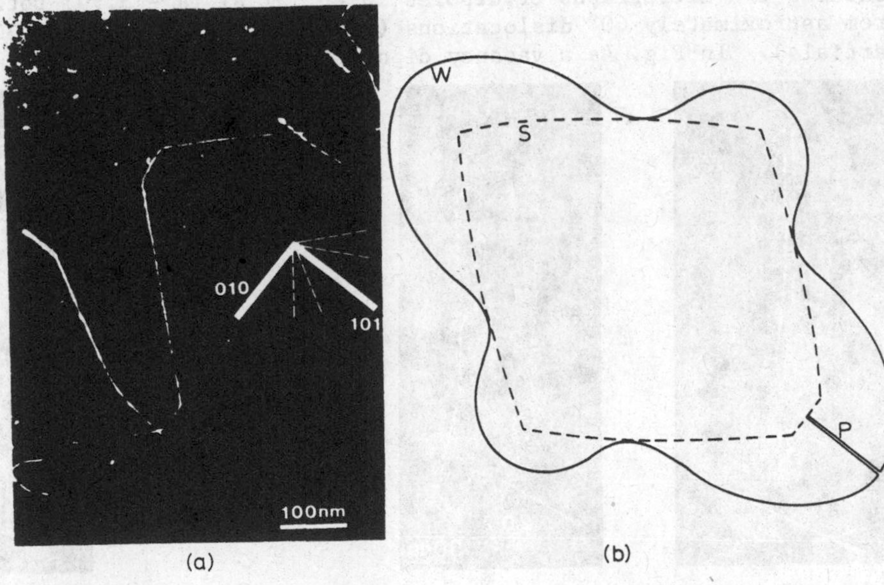

Fig. 3 (a) $(\bar{1}01)[010]$ dislocation in β brass, (b) Wulff plot W and predicted shape S with artificially lowered screw energy (P) at one end.

In many face-centred cubic metals with quite low Zener anisotropy ratios e.g. Cu (A = 3.21) the Shockley partial with $b = \frac{a}{6}[\bar{1}\bar{1}2]$ is elastically unstable near edge orientation. A 60° glide dislocation in these metals dissociates into a 30° partial which is stable and an edge partial which is not. The presence of the 30° partial (with a high line tension) and the stacking fault stabilize the edge partial to a certain extent but it still forms a zig zag to avoid pure edge character (16,21). The effect is demonstrated in Cu 15% Al by the right hand partials in Fig. 4d.

2.3 DIPOLE CONFIGURATIONS

Dislocation dipoles in the mildly anisotropic Ni (A = 2.44) were studied by Forwood and Humble (22) who used bright field electron microscopy and the image-matching technique to extract the details of the configurations of partial dislocations. Although the individual partials were not directly imaged these authors were able to show that their positions were very different from those predicted by isotropic elasticity and they were even able to measure, for the first time reliably, the stacking fault energy of Ni (73-100 mJm^{-2}). When the weak beam technique became available all four partials in a dipole could be imaged and the positions of the dislocations deduced. Wintner and Karnthaler (23), continuing the Australian work (22,24) demonstrated one important difference between isotropic and anisotropic predictions: that vacancy and interstitial dipoles in f.c.c. metals do not behave in the same way. An example is shown in Fig. 4 which contains two micrographs of dipoles in Cu 15% Al (A = 3.90) both formed from approximately 60° dislocations (dissociated into 90° and 30° partials). In Fig. 4a a vacancy dipole forms with its stacking faults

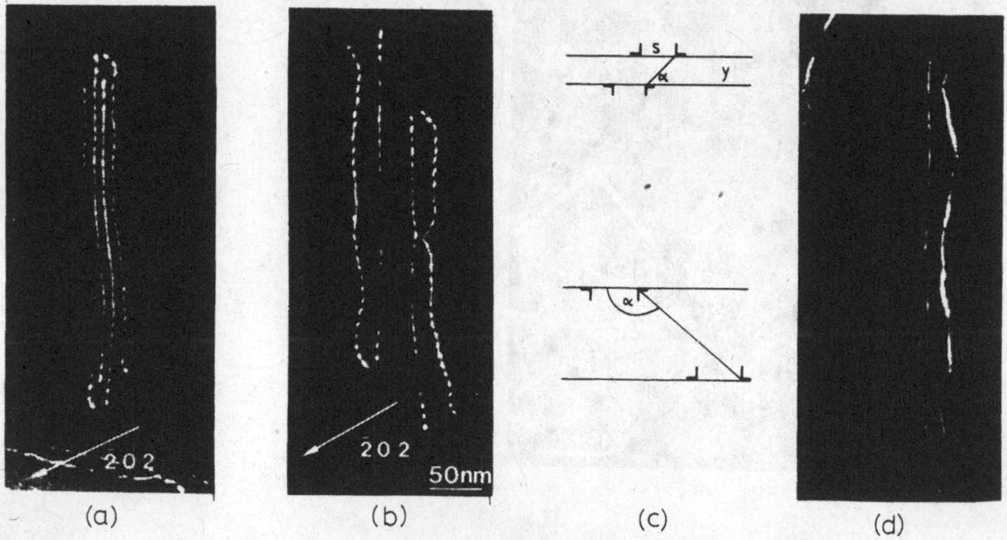

Fig. 4 Cu 15% Al (a) vacancy dipole, (b) interstitial dipole, (c) anisotropic dipole configurations, (d) unstable edge partial on right, stable 30° partial on left.

overlapping and in Fig. 4b an interstitial dipole forms in a much wider configuration. In both cases the experimentally determined positions of the partials are in good agreement with the anisotropic (Fig. 4c) but not with the isotropic calculations.

3. EFFECTS OF SURFACES ON DISLOCATIONS

3.1 SPECIAL GEOMETRY

Early work on the effect of a surface on dislocations concentrated on dislocations either perpendicular or parallel to either a single surface or to both surfaces of a plate. Eshelby and Stroh (25) worked out the stress field of a screw dislocation lying perpendicular to the surface of a thin plate; they showed in particular that the stress is localized essentially to a cylinder of radius equal to the plate thickness and that it falls exponentially with distance r rather than as r^{-1} as it would in the bulk. The strain field of the screw dislocation can be used (26) to image screw dislocations when viewed end-on in the electron microscope even though such dislocations would be invisible in all reflexions if they had the bulk strain field.

If the screw dislocation lies along the axis of a slender cylinder, in order to annul the surface stresses, the crystal develops a twist, the Eshelby twist (27). The twist about the axis of the cylinder has been observed in the electron microscope (28) in AlN whiskers and its magnitude measured by electron diffraction.

When a screw disloction lies parallel to a thin plate the surface tractions may be annulled by an infinite array of image screw dislocations (29). If the original screw dislocation is close to one surface of the plate, it and the nearest image dislocation effectively form a dipole and so again the stresses of the screw dislocation fall off rapidly with distance. In the case of an edge an image dislocation is not sufficient to annul the surface tractions and another stress field (with no singularities in the solid) must be superimposed (30). The problem of an edge dislocation in a thin plate has been solved (31) when the dislocation lies on the midplane of the plate. In this case the forces between two like dislocations are not only short ranged but even change from repulsion to attraction at some separations.

The reduction of the interaction forces between dislocations near a surface is most evident in the behaviour of partials in a dissociated dislocation near a surface (32,33). Siems et al (33) observed that the partials of dissociated dislocations in thin films of SnS_2 become closer wherever the dislocation passes a surface step i.e. wherever the plate becomes thinner and that the dislocation eventually constricts when it comes near to the surface. Another indication of the same effect is that when alloy foils containing extended dislocation nodes are heated in the electron microscopes the nodes generally shrink irreversibly (34). The most probable explanation is that nodes are formed at their equilibrium size in the bulk materials and are pinned by the solute atoms during foil preparation. Upon heating, the dislocations are unlocked from the solute atoms and reach the new smaller equilibrium size appropriate to the thin foil.

When dynamic experiments are carried out in the electron microscope to observe the development of dislocation structures there are two reasons why the behaviour will not be similar to that in the bulk. First, the dislocations feel a force drawing them to the nearest surface and this may reduce the mobile dislocation density; second, two dislocations barely interact until they are closer together than the foil thickness. These effects, which have been observed but rarely quantified are both reduced by making the observations in the thick foils of a high voltage electron microscope (35).

3.2 LOTHE'S FORCE

The end of any terminating dislocation of character α and with line energy $E(\alpha)$ requires forces E along the dislocation to prevent it from shrinking and $\partial E/\partial\alpha$ at right angles to the dislocation to prevent it from rotating to a character with a lower energy (Fig. 5a). At the surface, therefore, the dislocation will reach equilibrium at that value of θ for which the horizontal components of these forces, $-E\cos\theta$ and $\sin\theta\ \partial E/\partial\alpha$ are equal. Since $\partial E/\partial\alpha = -\partial E/\partial\theta$, the dislocation is at equilibrium when $-E\cot\theta + \partial E/\partial\theta$ is zero. In agreement with this Lothe (36), Fig. 5b, showed that an element $d\lambda$ of dislocation, λ from the surface, would experience an image force of $(-E\cot\theta + \partial E/\partial\theta)\ d\lambda/\lambda$. If the dislocation is confined to a slip plane these results remain true, regardless of the orientation of the surface plane, so long as the angles α and θ are measured in the slip plane. Lohne (37), using X-ray topography, has measured the angles at which dislocations in Al approach the surface and has found good agreement with the result that $E\cot\theta = \partial E/\partial\theta$. Despite the presence of an oxide film the experiments and the simple theory should, and do, agree (38).

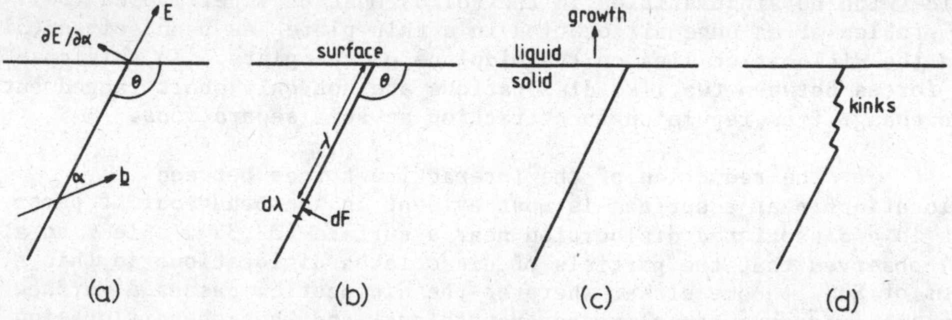

Fig. 5 (a) Forces on a terminating dislocation. (b) Lothe force, (c) dislocations extending at an angle to the growth direction, (d) generation of kinks at a surface.

The calculation for the free surface applies equally well to the interface between a solid growing crystal and either the melt or solution from which it is growing. Klapper (39), also using X-ray topography, has observed that, when a crystal (e.g. of benzil) grows, its dislocations do not elongate along the growth direction but at an angle θ to it such that $\theta = \tan^{-1}(E/(\partial E/\partial \theta))$, as expected (Fig. 5c). In this way the increase in dislocation energy per unit growth is minimised.

Louchet (40), in his in situ studies of the deformation of Si in the 3MeV electron microscope made an observation of a different kind which also illustrates the presence of the Lothe force on a dislocation at the surface. He measured the velocity of a 60° dislocation as it moved through the foil and found that the velocity suddenly doubled when one part of the dislocation broke through to the surface of the foil. In the geometry of his experiment the segment of dislocation at the surface experienced a Lothe force in the direction which would promote the generation of kinks at the surface. The dislocation moved faster because it was no longer necessary to create double kinks in the middle of the dislocation line (Fig. 5d).

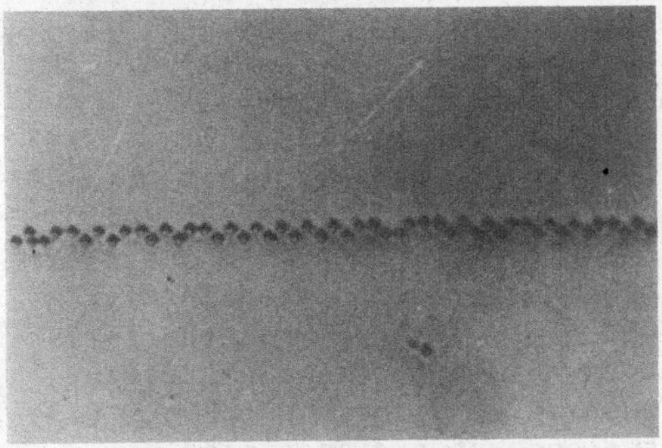

(a)

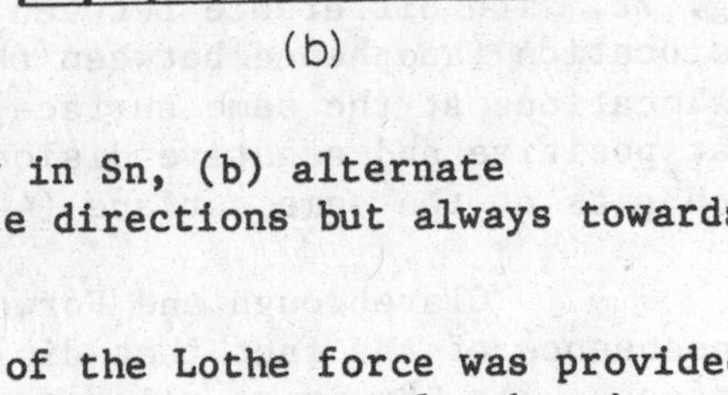

(b)

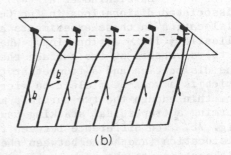

Fig. 6 (a) Double subgrain boundary in Sn, (b) alternate dislocations bend in opposite directions but always towards their Burgers vector.

Another clear demonstration of the Lothe force was provided by Honda (41) who used an etch hillock technique to reveal subgrain boundaries in deformed and recovered Sn (Fig. 6). He observed that some subgrain boundaries, consisting of $\frac{1}{2}\langle 111 \rangle$ dislocations which all had the same edge component of Burgers vector but which had alternating positive and negative screw components, formed double walls at the surface even though the wall was planar beneath the surface. Since a screw dislocation has a lower energy than an edge dislocation, the dislocations bend at the surface towards their Burgers vector and alternate dislocations bend in opposite directions (Fig. 6(b)) giving the impression, on the surface, of a double wall.

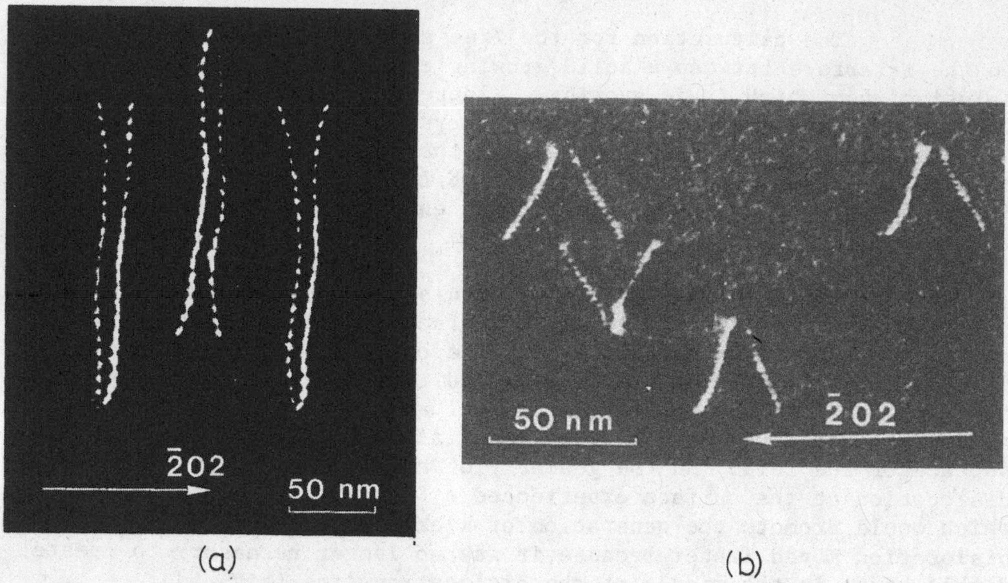

Fig. 7 Cu 10% Al (a) Dislocations close at one surface and open at the other, (b) trapezoidal stacking faults in a thin foil caused by rotation of the partials.

Hazzledine et al (42) found a similar behaviour in dissociated dislocations in fcc CuAl alloys. The two kinds of dislocation in these experiments are the two Shockley partials of a glissile ½⟨110⟩ dislocation. When the partials bend towards their Burgers vectors at the surface they form a constriction at one end of the dislocation and a trumpet at the other end, illustrated in Fig. 7 which is a weak-beam electron micrograph of Cu 10%Al. When the foils are thin enough the two partials may both rotate but remain straight, forming a trepezoidal stacking fault instead of a rectangular one (43) Fig. 7b. The difference between the two ends of a dissociated dislocation (and hence between the ends of positive and negative dislocations at the same surface) may possibly explain the observation that positive and negative dislocations produce distinctly different etch pits at the same surface (44).

Clarebrough and Forwood (45) observed an interesting consequence of the fact that dislocations may dissociate widely at a surface. Fig. 8a is an electron micrograph of a grain boundary in Cu which migrated in the microscope in order to become more nearly perpendicular to the foil, Fig. 8b. Near the lower surface the migrating boundary left behind a pair of stacking faults (extrinsic on (111) and intrinsic on ($\bar{1}$11)) at an obtuse angle to one another and joined by a low energy 1/6⟨110⟩ stair rod SR. The fault pair was formed by two different 1/2⟨110⟩ dislocations DD in the grain boundary dissociating at the surface, one on (111) and the other on ($\bar{1}$11). One Shockley from each dislocation combined to form the stair rod SR while the other two, Sh, moved with the grain boundary. These and similar defects are commonly observed and their creation impedes the migration of grain boundaries (46).

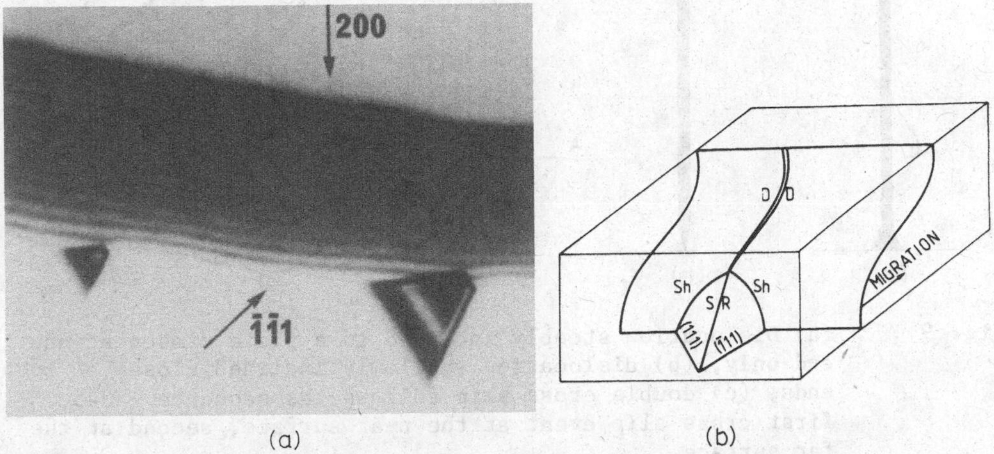

Fig. 8 Obtuse stacking fault bend left behind a migrating grain boundary in Cu

3.3 CROSS SLIP OF DISLOCATIONS AT A SURFACE

When a dissociated screw dislocation in a fcc material comes to a surface two simple arguments may be advanced to describe the behaviour of the partials:

(a) Each partial is acted on principally by the force from its own image, the Lothe force. In this case the partials swing towards their Burgers vectors, one end of the dislocation closes and the other end opens, Fig. 9a.

(b) If the dislocation is nearly parallel to the surface, each partial plus its image constitutes one dipole; the two dipoles repel one another weakly and the stacking fault causes both ends of the screw dislocation to constrict, Fig. 9b.

When the full stress field of the dislocations and all surface effects are taken into account it is found (47) that result (a) is correct almost always. Both ends of a screw dislocation only constrict if the slip plane is inclined at less than ~10° to the surface plane.

A series of experiments by George and Champier (48) involving X-ray topographic observations of dislocations in Si and a similar series by Moeller and Haasen (49) on Ge confirm the conclusion that screws constrict at only one end by observing that cross slip is nucleated on specific free surfaces of the plate specimens. One of their results is illustrated in Fig. 9c. A dissociated screw dislocation moving to the right on the primary plane ABCD cross slips onto the plane BCEF. The cross slip is observed to start at the closed end of the original screw dislocation i.e. the near surface DCFH.

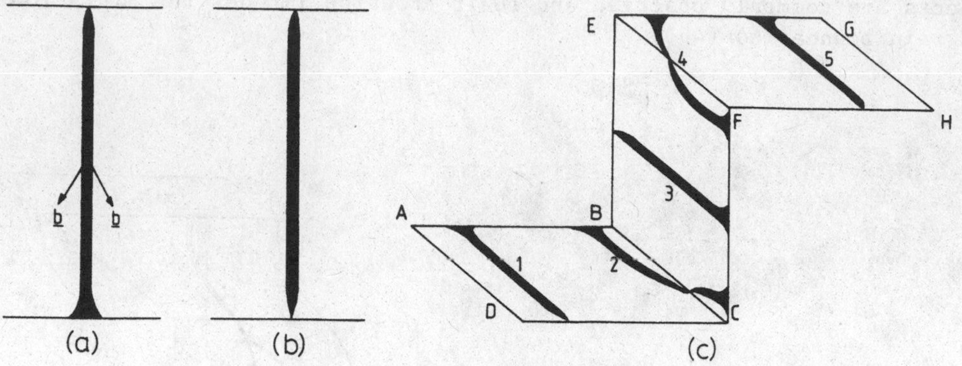

Fig. 9 (a) Dislocation steeply inclined to a plate closes at one end only, (b) dislocation shallowly inclined closes at both ends, (c) double cross slip follows the sequence 12345, first cross slip event at the near surface, second at the far surface.

The screw on the cross slip plane now constricts on the far surface ABEG and when it cross slips back onto the primary plane the process starts on the far surface. These, and other observations like them, are in full agreement with the idea that cross slip only occurs on the surface at which the Lothe force tends to close the partials. A further suprising observation (48,50) is that cross slip can even be nucleated at the surface at the closed end of a 60° dislocation.

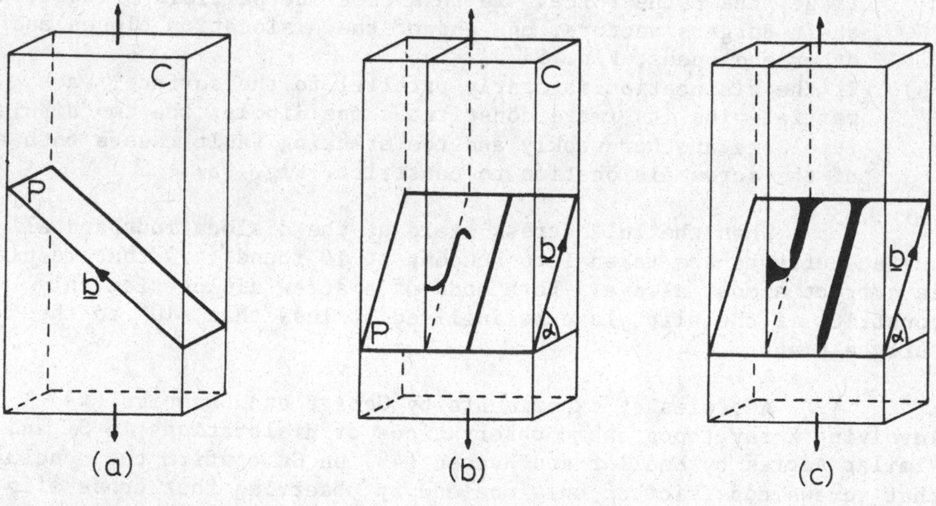

Fig. 10 Violation of Schmid law. Slip plane (a) is favoured over (b) and (c) because screws in (b) and (c) can cross slip to the surface, (b) high stacking fault energy case, (c) low stacking fault energy case.

Cross slip of screw dislocations at a surface can explain the observed violations of Schmid's law immediately after yield. Lohne (51) observed that, in the early stages of deformation of Al, the slip system to operate is not necessarily the one with the largest Schmid factor but one in which the Burgers vector is nearly parallel to the surface of the crystal. Young and Sherrill (52) made the same observation in Cu and Vesely (53) noted the same effect in Mo deformed at room temperature. Lohne's explanation, backed by X-ray topographic experiments, is illustrated schematically in Fig. 10 (47). In this figure a tensile specimen with large front and rear faces has two rival primary slip planes (a) and (b) each with a cross slip plane at right angles to it. Even if the Schmid factor in (b) is larger than in (a), system (a) is favoured because a screw dislocation in (b) can cross slip and reach the surface with an energy saving, Fig. 10(b). The cross slip pins the primary dislocation and prevents multiplication. Thus slip systems with $\underline{b}$ nearly parallel to the nearest surface are favoured. Cross slip only starts at the surface where the screw partials constrict and as soon as it starts it will continue because the image of the screw dislocation exerts a Lothe force per unit length of $Gb^2\tan\alpha/4\pi\lambda$ on the cross slip plane.

4. CONCLUSIONS

When electron micrographs of a nearly isotropic metal like tungsten are compared with micrographs from a very anisotropic metal like β brass there is no immediately obvious difference between them. In detail, though, the dislocation images are quite different and even the Burgers vectors are difficult to determine in β brass without extensive image matching. The anisotropy of β brass affects also the dislocation interactions and shapes; in particular smoothly rounded glide loops are replaced by sharply angled ones displaying only very limited character ranges.

Near a surface dislocations in both isotropic and anisotropic metals experience forces which are not present in the bulk. In particular Lothe's force causes any dislocation to rotate to take up a character with a lower line energy. This rotation can affect such diverse properties and behaviours as the form of dislocations in a growing crystal, the velocity of a dislocation at a surface, cross slip at the surface (and consequently violations of Schmid's law) and the structure of grain and sub-grain boundaries.

It is hoped that the effect which is potentially the greatest of them all, the interaction between non-glide stresses and screw dislocations in anisotropic b.c.c. metals will become directly observable in the near future.

ACKNOWLEDGEMENTS

I am very grateful to Drs. L.M. Clarebrough, K. Honda, H.P. Karnthaler and H. Saka for providing me with the micrographs for this paper.

REFERENCES

(1) Fourie, J.T., (1967). Canad. J. Phys. 45, 777.
(2) Fourie, J.T., (1970). Phil. Mag. 21, 977.
(3) Duesbery, M.S., (1983). Acta Metall. 31, 429.
(4) Basinski, Z.S. et al. (1981). Acta Metall. 29, 801.
(5) Steeds, J.W. Anisotropic Elasticity Theory of Dislocations. O.U.P., Oxford.
(6) Head, A.K. et al. (1973). Computed Electron Micrographs and defect identification. North Holland, N.Y.
(7) Shaibani, S.J. & Hazzledine, P.M., (1981). Phil. Mag. 44A, 657.
(8) Ohr, S.M., (1977). 35th Ann. Proc. Electron Microscopy Soc. Amer., 52. Boston, Mass. G.W. Bailey, editor.
(9) Cockayne, D.J.H. et al. (1969). Phil. Mag. 20, 1265.
(10) Saka, H. et al. (1978). Phil. Mag. A37, 273.
(11) Saka, H. et al. (1983). Phil. Mag. A47, 859.
(12) Ray, I.L.F. & Cockayne, D.J.H., (1971). Proc. Roy. Soc. A325, 543.
(13) Crawford, R.C. et al. (1973). Phil. Mag. 27, 1.
(14) Hazzledine, P.M. et al. (1984). Phys. Stat. Sol. (a) 81, 473.
(15) Korner, A. & Karnthaler, H.P., (1980). Phil. Mag. A42, 753.
(16) Holmes, S.M. et al. (1979). Phil. Mag. A39, 277.
(17) Korner, A. et al. (1979). Phys. Stat. Sol. 51(a), 613.
(18) Korner, A., & Karnthaler, H.P., (1983). Phys. Stat. Sol. 75(a), 525.
(19) Head, A.K., (1967). Phys. Stat. Sol. 19, 185.
(20) Saka, H., (1984). Phil. Mag. A49, 327.
(21) Clarebrough, L.M. & Head, A.K., (1969). Phys. Stat. Sol. 33, 431.
(22) Forwood, C.T. & Humble, P., (1970). Austral. J. Phys. 23, 697.
(23) Wintner, E. & Karnthaler, H.P., (1977). Phil. Mag. 36, 1317.
(24) Morton, A.J. & Forwood, C.T., (1973). Crystal Lattice Defects 4, 165.
(25) Eshelby, J.D. & Stroh, A.N., (1951). Phil. Mag. 42, 1401.
(26) Tunstall, W.J. et al, (1964). Phil. Mag. 9, 99.
(27) Eshelby, J.D., (1953). J. Appl. Phys. 24, 176.
(28) Drum, C.M., (1965). J. Appl. Phys. 36, 816.
(29) Leibfried, G. & Dietze, H.D., (1949). Z. fur Physik 126, 790.
(30) Head, A.K., (1953). Proc. Phys. Soc. 56B, 793.
(31) Nabarro, F.R.N. & Kostlan, E.J., (1978). J. Appl. Phys. 49, 5445.
(32) Spence, G.B., (1962). J. Appl. Phys. 33, 729.
(33) Siems, R. et al, (1962). Phys. Stat. Sol. 2, 636.
(34) Gallagher, P.C.J., (1970). Met. Trans. 1, 2429.
(35) Louchet, F. et al, (1979). Phil. Mag. A39, 433.

(36) Lothe, J., (1969). Fundamental Aspects of Dislocations Theory, N.B.S. Special Publication 317, 11. Simmons, J.A. et al., editors.
(37) Lohne, O., (1973). Phys. Stat. Sol. (a)18, 473.
(38) Lohne, O. & Lothe, J., (1977). Scripta Metall. 11, 23.
(39) Klapper, H., (1972). Phys. Stat. Sol. (a)14, 99.
(40) Louchet, F., (1981). Phil. Mag. A43, 1289.
(41) Honda, K., (1979). Jap. J. Appl. Phys. 18, 215.
(42) Hazzledine, P.M. et al, (1975). Phil. Mag. 32, 81.
(43) Korner, A. et al, (1977). Phys. Stat. Sol. (b)81, 191.
(44) Hockey, B.J. & Mitchell, J.W., (1972). Phil. Mag. 26, 409.
(45) Clarebrough, C.M. & Forwood, C.T., (1978). 9th Int. Cong. on Electron Microscopy, ed. J.M. Sturges, vol. III, 380. Toronto: Microscopical Society of Canada.
(46) Clarebrough, L.M. & Forwood, C.T., (1980). Phys. Stat. Sol. (a) 60, 409.
(47) Hazzledine, P.M & Shaibani, S.J., (1982). 6th International Conference on the Strength of Metals and Alloys, ed. R.C. Grifkins, Vol. 1, 45. Oxford: Pergamon.
(48) George, A. & Champier, G., (1980). Scripta Metall. 14, 399.
(49) Moeller, H-J. & Haasen, P., (1976). Phys. Stat. Sol. (a)33, K59.
(50) Wintner, E. et al, (1976). 4th Int. Conf. on Strength of Metals and Alloys, Vol. 2, 927. Nancy.
(51) Lohne, O., (1974). Phys. Stat. Sol. (a)25, 709.
(52) Young, F.W. & Sherrill, F.A., (1972). J. App. Phys. 43, 2949.
(53) Vesely, D., (1968). Phy. Stat. Sol. 29, 675.
(54) Eshelby, J.D., (1969). Fundamental Aspects of Dislocation Theory, N.B.S. Special Publication 317, 1. Simmons, J.A. et al, editors.

DISLOCATIONS IN ANISOTROPIC ELASTIC MEDIA

D.J. BACON

Department of Metallurgy and Materials Science,
The University, P.O. Box 147, Liverpool L69 3BX

ABSTRACT

Brown first showed how the complexity of the general solution for the elastic field of a dislocation in an anisotropic continuum can be reduced to a geometrical relation between the field of a curvilinear dislocation and that of an infinite straight line. Field data for the latter are reasonably easy to compute, and the geometrical formula has now been used to treat anisotropy in many problems. The theoretical framework is summarized here, and recent papers on strain energy and self-force are reviewed. Some effects of anisotropy are discussed, and effective isotropic elastic constants which can give good approximations to anisotropic behaviour for common b.c.c. and f.c.c. metals are tabulated.

1. INTRODUCTION

Crystalline solids are elastically anisotropic to various degrees, and the analysis for the elastic field of a dislocation in such media should, in principle, treat this effect. Although the analysis of Volterra (1) for the distortion in multiply-connected bodies is general in this respect, closed-form analytical expressions for dislocation fields cannot be obtained in general for anisotropic media. Thus, most applications of continuum mechanics to dislocation theory have employed the assumption of isotropic behaviour. The errors involved in this approximation are frequently difficult to judge, however, and there have been many attempts to obtain the more accurate anisotropic results. The first successful treatment was that of Eshelby et al. (2), who considered the infinite straight dislocation in a medium of general anisotropy and derived expressions for the elastic field in terms of the roots of a sixth-order polynomial involving the elastic stiffness constants. With the exception of certain high-symmetry cases, a solution has to be found numerically, for the roots of a sextic cannot be found by algebraic methods. Later, Foreman (3) extended the analysis to obtain the strain energy per unit length of line. Surprisingly, it was found that in three f.c.c. metals, a perfect dislocation with Burgers vector $\underset{\sim}{b} = \frac{1}{2}\langle 110\rangle$ on the $\{110\}$ planes has lower energy than one of the same character on {111}. This is contrary to known behaviour, and a clear indication that core effects can be much more important than elastic anisotropy. Nevertheless, the method of Eshelby et al. was used widely in the following decade to examine dislocation energy and the interaction of one dislocation with another and with other defects in a wide range of crystals.

It should be noted that in a few situations, the anisotropic theory gives results which differ qualitatively from their isotropic counterparts. A striking example was afforded by Head (4), who observed zigzag line shapes on {110} and {211} planes for dislocations with $\underset{\sim}{b} = \frac{1}{2}\langle 111\rangle$ in β-brass. These were shown to arise from instability resulting from negative line tension, in good agreement with anisotropic elasticity. In general, however, anisotropic effects found by the method of (2) are of secondary importance and the errors produced by use of the isotropic theory are usually tolerable, particularly when the assumption of a straight-line shape can of itself be a gross approximation. Fortunately, substantial advances have been made in the continuum theory of dislocations over the past few years, and it is now possible to meet the latter point by applying anisotropic elasticity to polygonal and curved dislocations. The purpose of this paper is to review the theoretical background to these advances and assess the importance of anisotropy. Several comprehensive reviews of this subject have been published recently (5-8), and so no attempt will be made here to either derive the many equations or discuss all the papers relevant to this topic. Also, only infinite, homogeneous media will be considered. It is necessary, therefore, to apologize in advance to the many authors whose important work has been omitted by such a selective view.

2. THEORETICAL DEVELOPMENTS

2.1 INTRODUCTION

Consider a dislocation L of Burgers vector $\underset{\sim}{b}$ in an homogeneous continuum with elastic stiffness constants C_{kpim}. (Subscripts here take values 1,2,3 corresponding to three rectangular Cartesian axes.) The stress σ_{nr} and the distortion, i.e. the gradient of displacement $\underset{\sim}{u}$, at a point with position vector $\underset{\sim}{x}$ is given in its simplest form as a line integral over L by Mura's formula (5,6,8):

$$\sigma_{nr}(\underset{\sim}{x}) = C_{nrjs}\,\partial u_j/\partial x_s = -\,b_k C_{nrjs} C_{kpim}\,\varepsilon_{qps} \oint_L \partial G_{ij}(\underset{\sim}{x}-\underset{\sim}{x}')/\partial x_m\, dx'_q \qquad (1)$$

where ε_{qps} = +1 or -1 if qps is an even or odd permutation of 1,2,3 respectively and is zero otherwise. Throughout this paper, repetition of subscripts implies summation over 1,2,3, i.e. seven summations are implied in Eqn (1). The Green's function G_{ij} represents the displacement response of the medium to a point force acting within it, and is the solution to a second-order partial differential equation. It can be reduced to either a single integral or (equivalently) a sum involving the eigenvectors of a six-dimensional matrix eigenvalue equation (5,7). An exact analytical solution can be obtained only for isotropic or transversely isotropic media, otherwise G_{ij} has to be obtained numerically. The derivatives of G_{ij} can be expressed in a similar fashion.

This emphasizes the fact that the evaluation of the elastic field of a dislocation is an anisotropic medium is in principle a complicated task. Furthermore, the expression for the dislocation strain energy, which is given by integration of the stress over a surface S bounded by L, is more involved again. The reduction of the formulae to more manageable forms stems, in large part, from the analysis of Brown (9).

2.2 BROWN'S FORMULA

The geometry relevant to Brown's formula is shown in Fig.1.

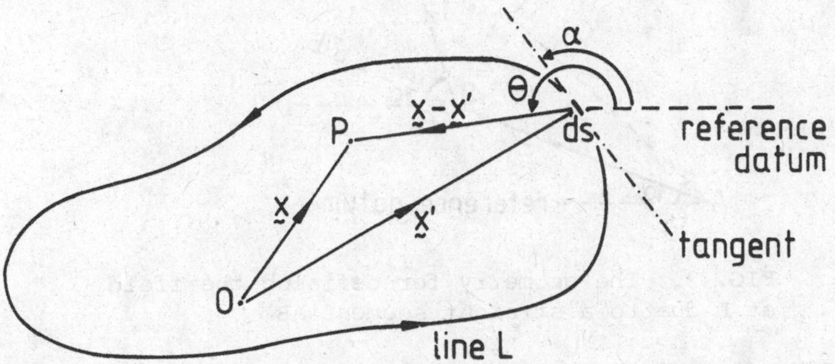

FIG. 1. The geometry used in the text for the elastic field at P due to dislocation line L

The dislocation is a plane curvilinear line L and $\underset{\sim}{x}'$ is the position vector from a fixed origin to element ds on L. Starting with an alternative form of Eqn (1), Brown showed that the stress at point P in the plane of the line is

$$\sigma_{ij}(\underset{\sim}{x}) = \frac{1}{2}\oint_L \{\Sigma_{ij}(\theta)+\Sigma''_{ij}(\theta)\} \frac{\sin(\theta-\alpha)}{|\underset{\sim}{x}-\underset{\sim}{x}'|^2} ds, \quad (2)$$

where the angles α and θ are measured from a reference datum as shown in the Figure. The function $\Sigma_{ij}(\theta)$ is the angular part of the stress field of an infinite straight line of direction $\underset{\sim}{x}-\underset{\sim}{x}'$ and the same Burgers vector as L, i.e. the in-plane stress a distance d from an infinite straight line of orientation θ is $\Sigma_{ij}(\theta)/d$; and $\Sigma''_{ij}(\theta)$ is

$d^2\Sigma_{ij}(\theta)/d\theta^2$. An exactly equivalent formula holds for the distortion. By comparison with Eqn (1), it is seen that Brown's formula incorporates the Burgers vector and elastic properties into the angular factors $\Sigma(\theta)$ and $\Sigma''(\theta)$. Although these require numerical evaluation in general, the geometrical form of Eqn (2) has a particular advantage. That is, it may be partially integrated to find simple formulae for the fields of finite straight segments, and these may then be used by superposition to construct the fields of complicated polygonal dislocations. Various forms are possible, but possibly the most useful for the field of a straight segment is

$$\sigma_{ij} = \frac{1}{2d}\left[-\cos(\theta-\alpha)\Sigma_{ij}(\theta)+\sin(\theta-\alpha)\Sigma'_{ij}(\theta)\right]_{\theta_1}^{\theta_2}, \tag{3}$$

where the geometry is shown in Fig. 2.

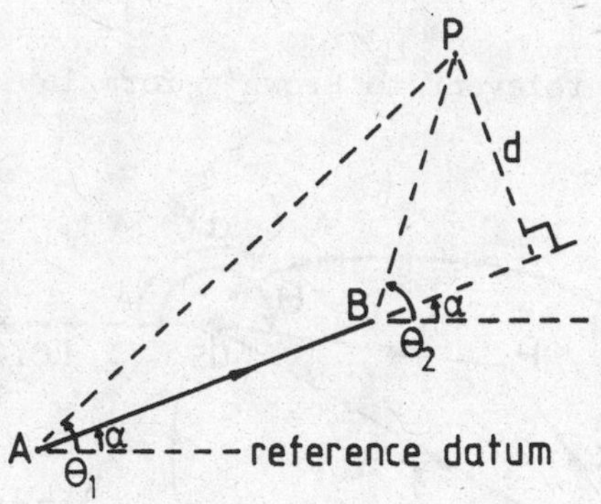

FIG. 2. The geometry for defining the field at P due to a straight segment AB

Indendom and Orlov (10) derived a more general form of Eqn (2) which is applicable to non-planar dislocations. It may be shown (5), however, that Eqn (3) is also valid for a segment in a three-dimensional configuration, provided the geometrical variables are defined in the plane of AB and P. Thus, Brown's formula leads to simple geometrical forms for the elastic fields of curved and polygonal lines which only require knowledge of the field factors for infinite straight lines having orientation angles θ (Fig. 1) or θ_1 and θ_2 (Fig. 2).

2.3 FIELD FACTORS FOR INFINITE STRAIGHT LINES

The field factors can only be obtained in analytical form for isotropic media, and numerical evaluation is required for anisotropy. There are, in fact, two convenient but different ways of computing them: one involves numerical integration, the other solution of a sextic polynomial. Both formalisms are equivalent, and a full description is given in (5), where alternative expressions due to Willis (11) are also discussed.

It has proved important in applications of Brown's formula to note that the field factors are periodic functions of θ with period π. They can therefore be written as Fourier series, e.g.

$$\Sigma_{ij}(\theta) = a_o + \sum_{r=1}^{\infty} (a_r \cos 2r\theta + b_r \sin 2r\theta). \quad (4)$$

With θ measured from a low-index crystallographic direction, Σ_{ij} is often an even function with $b_r = 0$. The Fourier coefficients can be determined by evaluating the factor for a range of angles using the methods referred to above. Once this preliminary computation has been carried out, the factor can be handled in any subsequent studies as an analytic function of known accuracy. This has great advantages in applications involving the geometrical formulae of section 2.2. Coefficients for the stresses resolved in the plane of the dislocation and its Burgers vector have been evaluated for many metals (5), and in most cases the factors are given to an accuracy of better than 0.5% by series terminated at $r \leq 5$. Isotropic solids require only one or two coefficients.

It will be seen in the next section that the (prelogarithmic) energy factor for an infinite straight line in the orientation θ is an important function. It is related to stress resolved in the direction of $\underset{\sim}{b}$ by

$$E = \frac{1}{2} b_i N_j \Sigma_{ij}, \quad (5)$$

where $\underset{\sim}{N}$ is a unit vector normal to the line, and it can also be written as a Fourier expansion

$$E(\theta) = A_o + \sum_{r=1}^{\infty} (A_r \cos 2r\theta + B_r \sin 2r\theta). \quad (6)$$

For an isotropic medium with the geometry shown in Fig. 3, all the coefficients are zero except

$$A_o = \frac{\mu b^2}{8\pi(1-\nu)} (2-\nu\cos^2\beta), \quad A_1 = \frac{\mu b^2}{8\pi(1-\nu)} (-\nu\cos^2\beta), \quad (7)$$

where μ and ν are the shear modulus and Poisson's ratio respectively. Anisotropic coefficients for the common metals for cases with β = 0 are contained in (5), and the effects of dissociation and core energy on these coefficients have been analysed by Prinz et al. (12). Coefficients for some geometries with β ≠ 0 in the b.c.c. and f.c.c. metals have recently been computed (13).

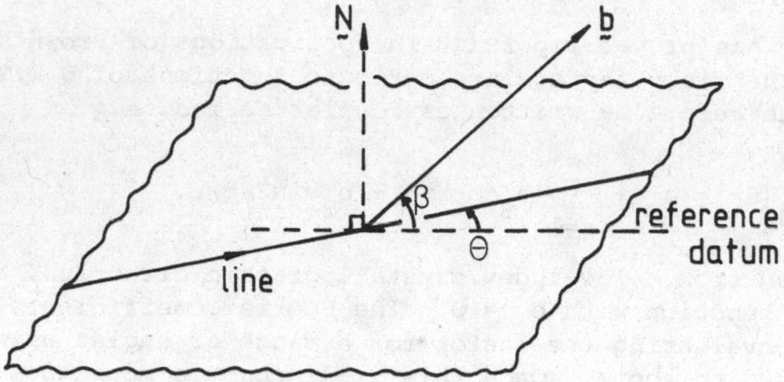

FIG. 3. The geometry used for the field factor of an infinite straight line lying in a plane with normal $\underset{\sim}{N}$. The datum, $\underset{\sim}{b}$ and $\underset{\sim}{N}$ are coplanar.

3. ENERGY AND SELF-FORCE

3.1 THE NATURE OF THE PROBLEM

In addition to values of stress and strain, it is frequently necessary to know either the strain energy produced by a dislocation or the virtual force acting on it. The former quantity might be needed for the theoretical treatment of structures produced by radiation damage, for example, and the latter is usually an essential ingredient in the analysis of dislocation behaviour under external and internal stresses. The strain energy ξ is given formally by the surface integral

$$\xi = \tfrac{1}{2}\iint_S \sigma^s_{ij} u_i \, dS_j , \tag{8}$$

where S is a surface bounded by L (and on which $\underset{\sim}{u} = \underset{\sim}{b}$) and σ^s_{ij} is the stress produced by L. The force per unit length acting perpendicular to the line in a plane with unit normal vector $\underset{\sim}{N}$ is given by the Peach-Koehler relation

$$f = \sigma_{ij} b_i N_j , \tag{9}$$

where σ_{ij} is evaluated at point P on L where f is required. It is convenient to split σ_{ij} into the part σ^s_{ij} due to L itself - this gives a self-force f^s - and another σ^a_{ij} due to all other internal or external sources. The dislocation will be in equilibrium when the condition

$$f^s + f^a = (\sigma^s_{ij} + \sigma^a_{ij}) b_i N_j = 0 \tag{10}$$

is satisfied at all points on the line.

Unfortunately, evaluation of both ξ and f^s is hampered by the divergence of σ^s_{ij} in linear elasticity when a field point approaches the line. For ξ, singular behaviour is conventionally avoided by terminating S a distance r_o from L and adding to it a tubular surface S_o of radius r_o centred on L. The total energy is then the linear elastic contribution ξ plus the contribution of the core material within S_o. The self-force f^s is related to the strain energy, for if L experiences an arbitrary virtual displacement such that an infinitesimal element ds of L is displaced perpendicularly to itself by δr, the change in ξ is (14)

$$\delta\xi = -\oint_L f^s \delta r ds. \tag{11}$$

In order that equilibrium as defined by Eqn (10) be properly treated, it is therefore necessary to find a prescription for a non-singular σ^s_{ij} such that the resulting self-force f^s is consistent with Eqn (11). This problem has only recently been resolved.

3.2 TREATMENTS OF THE SELF-FORCE

A definition of σ^s_{ij} for use in Eqn (10) was first given by Brown (15), who proposed that for a planar dislocation L, σ^s_{ij} at a point P on the line is the average of the appropriate resolved stress evaluated at points in the plane perpendicular distances $\pm r_o$ from P:

$$\sigma^s_{ij} = \frac{1}{2}\{\sigma_{ij}(+r_o) + \sigma_{ij}(-r_o)\} \tag{12}$$

This stress diverges as $\ln(\rho/r_o)$, where ρ is the radius of curvature of the line at P, and r_o is the core cut-off parameter of order b. Brown (15) compared f^s obtained from Eqns (9) and (12) with that derived from Eqn (11) for circular loops in isotropic media, and noted that although the expressions agree in the $\ln(\rho/r_o)$ term, they do not agree in the non-singular terms. Subsequently, Gavazza and Barnett (16) reanalysed the self-force as a variational problem, and showed that

$$f^s = \frac{1}{\rho}\left[E-\{E+E''\}\ell n(\frac{8\rho}{r_o})-\{F+F''\}\right] -J \tag{13}$$

where E is the energy factor for an infinite straight dislocation tangent to the line at P (Eqn (5)). Factor F is the energy per unit length arising from the tractions on the core surface S_o, i.e. the tube of radius r_o, and may be approximated by the value for an infinite straight line. Of the four terms on the right-hand side of Eqn (13), the first three depend on the local radius of curvature ρ of the line at P, whereas J, which may be evaluated from Eqn (2), depends on the distant line shape: it is of secondary importance except when another part of the line approaches close to P. The three local terms are seen to be an arc-lengthening term, a line tension arising from the orientation-dependence of E and a core-traction-induced tension respectively: the second is the term deduced by Brown (Eqn (12)) and is usually dominant. Eqn (13) is consistent with Eqn (11), but suffers, as do all linear-elastic dislocation energy calculations, with the uncertainty surrounding the value of r_o.

Recently, Kirchner (17) and Scattergood (18) have independently reconsidered dislocation self-energy and self-force in anisotropic elasticity, with the object of developing a procedure which is computationally-convenient and internally consistent. Although ξ can be computed from Eqn (8) by the analyses reviewed in section 2, the method is demanding in computer time and more complicated than for isotropic media. Further simplification of the anisotropic form is therefore very desirable. By consideration of the interaction of one segment ds of L with another segment ds' (Fig. 4) Kirchner (17) used Brown's formula (Eqn (2)) to obtain

$$\xi = \xi_1 + \xi_2 + \xi_3, \tag{14}$$

where
$$\xi_1 = \oint_L E(\alpha)\{\ln[\frac{8\rho(\alpha)}{r_o}]-2\}ds \tag{15}$$

$$\xi_2 = \oint_L \int_0^\pi \{E(\alpha+\phi)-E(\alpha)\} \frac{\cos^2\phi}{\sin\phi} d\phi ds \tag{16}$$

$$\xi_3 = -\frac{1}{4}\oint\oint_L \frac{E(\alpha+\phi)\sin^2(\phi-\psi)}{r\sin\phi\sin\psi} ds ds'. \tag{17}$$

Since E can be expressed as a Fourier series (section 2.3), ξ_3 is the only real double integral: it makes only a small contribution ($\lesssim$10%) and is actually zero for a circle (ϕ=ψ). Kirchner's analysis omits the contribution to the energy from the core surface tractions on S_o, and ξ, which corresponds to S being planar inside L, is therefore not complete. This could be rectified by the addition of F as in (16), but Kirchner argues that the contribution of the tractions on S_o and the non-linear effects within the core are best incorporated by adding to ξ a single term $\xi_c = \oint_L E_c(\alpha)ds$, where $E_c(\alpha)$ is the core energy per unit length

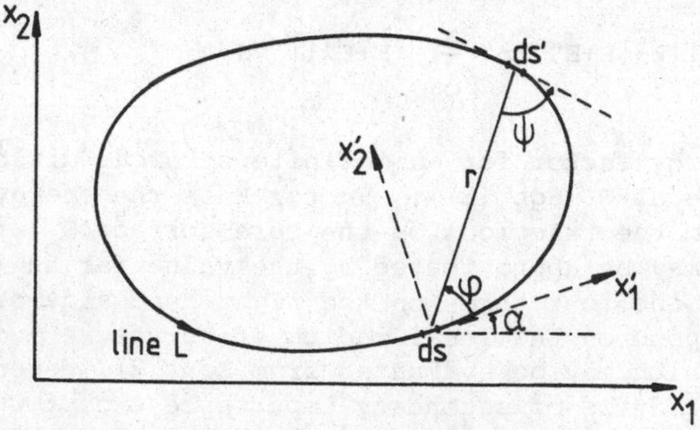

FIG. 4. The geometry of two elements ds and ds' on line L.

for an infinite straight line of orientation α, chosen such that $\xi+\xi_c$ is independent of r_o. Although Prinz et al. (12) have shown how this can be done for several f.c.c. and h.c.p. metals using the Peierls model, it is not clear how such a procedure can be developed generally. By treating self-force f^s as the variational derivative of ξ, Kirchner obtains terms consistent with Eqns (14-17): viz $f^s = f^s_1 + f^s_2 + f^s_3 + f^s_c$, where only f^s_3 contains a line integral. In comparison with Eqn (13), Kirchner's result treats the core-traction contribution to the linear elastic energy differently, but all the remaining elastic terms are given explicitly. They are all easy to handle, and a particular advantage of his treatment is that a line integral involving the geometry of interaction with distant parts of the line only occurs in ξ_3 and f^s_3: it makes a minor contribution in most cases.

Scattergood (18) has derived alternative expressions by use of a different cut-off procedure for avoiding the stress singularity. In earlier modelling of dislocation behaviour in isotropic materials (19,20), it was found that an equilibrium line shape can be readily computed by evaluating σ^s_{ij} at a finite number of points on a trial, non-equilibrium line and relaxing the shape iteratively until Eqn (10) is satisfied. This is achieved very efficiently by treating the line close to a given point P as the arc of a circle, for which Eqn (12) can be expressed analytically for isotropy, and approximating more distant parts by straight segments. In the case of anisotropy, Scattergood and Bacon (21) were unable to obtain a closed-form expression for σ^s_{ij} on a circular arc from Eqn (12), and used instead the assumption that parts of the line closer to P than a distance ε make no contribution to the stress at P. With this cut-off procedure, which is illustrated in Fig. 5, the circular arc self-stress was reduced to non-integral form (21). Although not based on the variation of ξ, this method gave results for isotropy which were very close to those of Eqn (12) with $\varepsilon = r_o/2$, and it has been employed in several papers (21-24). Scattergood's (18) self-consistent analysis for ξ and f^s uses the same cut-off procedure, and provides a rigorous basis for these earlier studies. He reduced ξ from Eqns (2) and (8) to a double line integral, which, using the geometry of Fig. 4, may be written

$$\xi = \frac{1}{2}\oint_L \oint_{L_\varepsilon} \{E(\alpha+\phi) + E''(\alpha+\phi)\} \ell n\left(\frac{r}{\varepsilon}\right)\frac{1}{r} \sin\phi \sin\psi \, ds' ds, \qquad (18)$$

where L_ε means the integral omits the part of L within a distance ε of point $x'_1 = x'_2 = 0$. When E is written as a Fourier series, Eqn (18) is amenable to further reduction. For the case of a circular dislocation loop of radius R in an anisotropic medium, Scattergood has derived the remarkably simple result

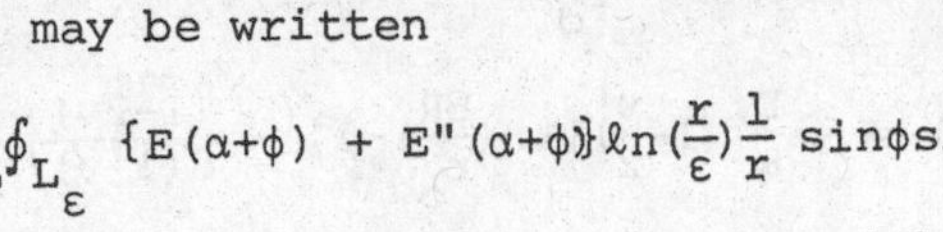

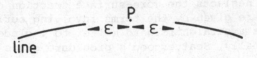

FIG. 5. The cut-off used in (21) to avoid the core singularity

$$\xi = 2\pi RA_o \{\ell n(\frac{4R}{\varepsilon})-1\}, \tag{19}$$

where A_o is defined in Eqn (6). As explained in section 2.3, values of A_o have already been computed for many metals. For polygonal dislocations in isotropic media, it has been found desirable to obtain analytical expressions for the self-energy and interaction energy of straight segments (8), thereby avoiding considerable labour in calculations for particular configurations. Corresponding partial integration of Eqns (14) and (18) has not yet been made for anisotropy.

3.3 DISCUSSION

Since it appears that there are several ways of calculating f^s, each of which is consistent with Eqn (11) when a suitable meaning of ξ is defined, it would seem appropriate to consider the differences between them. They are, perhaps, best demonstrated by considering the circular loop in an isotropic continuum, a situation for which analytical results can be obtained. First, it should be remarked that Eqn (8) gives a unique result for ξ, independent of the choice of S, provided the traction contribution on S_o is included. The energy for pure edge and shear loops respectively is (25)

$$\xi(\text{edge}) = CR\left[\{\ell n(\frac{8R}{r_o})-2\} + \frac{(3-2\nu)}{4(1-\nu)}\right] \tag{20a}$$

$$\xi(\text{shear}) = CR\left[(1-\frac{\nu}{2})\ \{\ell n(\frac{8R}{r_o})-2\} - \frac{(1-2\nu)}{8(1-\nu)}\right] \tag{20b}$$

where $C = \mu b^2/2(1-\nu)$. The term involving curly brackets in the contribution from the integral over a planar surface inside the loop and the final term arises from the integral over S_o. Gavazza and Barnett's analysis is entirely consistent with this, i.e. with Eqn (11), giving a self force for each loop of

$$f^s(\text{edge}) = -\frac{D}{R}\left[\{\ell n(\frac{8R}{r_o})-1\} + \frac{(3-2\nu)}{4(1-\nu)}\right] \tag{21a}$$

$$\bar{f}^s(\text{shear}) = -\frac{D}{R}\left[(1-\frac{\nu}{2})\{\ell n(\frac{8R}{r_o})-1\} - \frac{(1-2\nu)}{8(1-\nu)}\right] \tag{21b}$$

where $D = C/2\pi$ and the mean value of f^s has been taken for the shear loop. Brown's prescription (Eqn (12)) is not consistent, giving

$$f^s(\text{edge}) = -\frac{D}{R}\ \ell n(\frac{8R}{r_o});\ \bar{f}^s(\text{shear}) = -\frac{D}{R}(1-\frac{\nu}{2})\ \ell n(\frac{8R}{r_o}). \tag{22}$$

Kirchner's analysis neglects the core-surface traction contribution to ξ, and his results are given by the terms involving curly brackets in Eqns (20-21). It is a straightforward matter to include the missing terms, however. Finally, Scattergood's procedure gives different results: from Eqns (7) and (19), they are

$$\xi(\text{edge}) = CR\{\ell n(\frac{4R}{\varepsilon})-1\};\ \xi(\text{shear}) = CR(1-\frac{\nu}{2})\{\ell n(\frac{4R}{\varepsilon})-1\} \quad (23)$$

$$f^s(\text{edge}) = -\frac{D}{R}\ell n(\frac{4R}{\varepsilon}) \quad ;\ \bar{f}^s(\text{shear}) = -\frac{D}{R}(1-\frac{\nu}{2})\ell n(\frac{4R}{\varepsilon}). \quad (24)$$

It is clear that with $\varepsilon = r_o/2$, the difference between the sets of ξ and f^s values is small. Even for loops as small as $R = 10r_o$, the largest difference in f^s (for the shear orientation) is only 20%, and this reduces to 15% for $R = 100r_o$. This reasonable agreement between the various forms is not surprising, because in all analyses the largest contribution arising from line curvature is

$$f^s \simeq -\frac{1}{\rho}(E + E'')\ell n(\frac{8\rho}{r_o}), \quad (25)$$

i.e. it has the orientation dependence expected from the conventional line-tension approximation. Bearing in mind the uncertainty associated with choosing a value for r_o in dislocation theory, the formulae for ξ and f^s discussed here are probably equally valid. It is only for very strong curvature ($\rho<10r_o$) that the different analyses for f^s lead to significant changes in shape. Even then, it is not clear which prescription is to be preferred, for this is precisely the situation in which the validity of linear elasticity is called into question. For these reasons, the earlier studies of dislocation behaviour under internal and external stress (15,19-24) are considered to be sound within this approximation. The incorporation of core-energy effects (12,17) would be required to improve the earlier results to a significant extent.

4. DISCUSSION

Although the effect of anisotropy can be of first order when, for example, the factor E+E" is negative, in most situations anisotropy merely provides a small perturbation on the isotropic result, provided the isotropic moduli are chosen in an appropriate way. It should not be assumed that the development of anisotropic theory is unimportant, however, for conclusions regarding anisotropy require that it be tested first in calculations for realistic dislocation configurations. The developments reviewed in sections 2 and 3 have allowed this to be done in a number of areas, and, furthermore, they sometimes enable anisotropic calculations to be performed with only slightly greater effort than that required using isotropic formulae. The subject was reviewed in some detail in (5): here, a few topics are briefly touched upon.

Whereas a single crystal may have considerable anisotropy in its elastic moduli, the dislocation field factors seldom exhibit the same degree of anisotropy because they depend on the elastic properties in more than one direction. Consider, for example, α-iron and molybdenum: both are moderately anisotropic b.c.c. metals, but the anisotropy ratio $A=2C_{44}/(C_{11}-C_{12})$ is 2.35 in iron and 0.77 in molybdenum. The stifness constants measuring resistance to shear on the (110) plane in the [001] and [1$\bar{1}$0] directions are C_{44} and $\frac{1}{2}(C_{11}-C_{12})$ respectively, so that their ratio is A. It might be supposed, therefore, that the ratio of the energy factors E/b^2 for straight screw dislocations lying along [001] and [1$\bar{1}$0] would reflect this considerable difference between the metals. In fact, the ratio is 1.55 and 0.84 for iron and molybdenum respectively (13), demonstrating that the energy of the infinite dislocation is less anisotropic than the crystal shear stiffness. To explore the influence of anisotropy further, consider the variation of $E(\theta)$ with θ for an infinite straight line in the (110) plane, with θ measured from the [001] datum (Fig.6(a)). Plots of $[E(\theta)/E(0)-1]$ vs θ for various $\underset{\sim}{b}$'s in iron, molybdenum and tungsten, which is almost isotropic, are shown in Fig.6(b). There are considerable differences between the metals, the most striking being that for $\underset{\sim}{b} = \frac{1}{2}[110]$, the geometry for which the line is pure edge in character for all θ. The extent to which this behaviour can be approximated by isotropic elasticity will be returned to shortly.

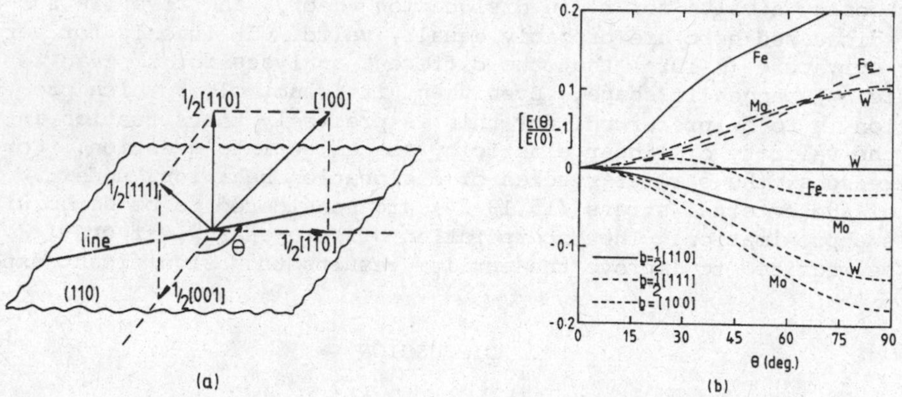

FIG. 6. (a) The geometry of the Burgers vectors and dislocation line used for the energy factors for iron, molybdenum and tungsten plotted in (b).

Interest in the data of Fig. 6 is not purely academic, for the properties of dislocation loops on the {110} planes of the b.c.c. metals has important consequences for the evolution of radiation-damage structure (B.L. Eyre: these Proceedings). It is probable that interstitial and vacancy clusters nucleate as faulted loops on {110} with $\underset{\sim}{b}$ = $\frac{1}{2}$<110> and unfault by one of the two shear reactions

$$\tfrac{1}{2}\langle 110\rangle + \tfrac{1}{2}\langle 001\rangle \rightarrow \tfrac{1}{2}\langle 111\rangle;\ \tfrac{1}{2}\langle 110\rangle + \tfrac{1}{2}\langle 1\bar{1}0\rangle \rightarrow \langle 100\rangle . \qquad (26)$$

The first reaction leads to the shorter b^2, but surprisingly a high proportion of perfect loops in iron have $\underset{\sim}{b}$ = <100>. As discussed above, the ratio of the shear stiffnesses in the <001> and <110> directions on {110} is A, and it is tempting to speculate that the difference in unfaulting behaviour between the metals is related to this ratio. However, it has already been noted that the anisotropy of infinite, straight dislocation properties is generally less than that of the crystal, and this is even more true of loops. Scattergood's relation for the energy of circular loops (Eqn (19)) has been used (13) to examine the ratio of ξ for the two possible shear loops involved in reactions (26), i.e.

$A_o(\underset{\sim}{b}=\frac{1}{2}[1\bar{1}0])/A_o(\underset{\sim}{b}=\frac{1}{2}[001])$: it is 1.66 for iron and 2.12 for molybdenum, compared with 2 for isotropy. Thus, the preference for $\frac{1}{2}\langle 001\rangle$ shear is less strong in iron, but not to the extent suggested by the anisotropy ratio. The ratio of ξ for the two perfect loops that can result is similarly less anisotropic, being 0.83 for iron and 0.75 for molybdenum compared with 0.75 for isotropy. It would seem, therefore, that elastic anisotropy alone cannot account for the observed loop crystallography, although its effects are far from negligible.

Scattergood and Bacon (21-24) used anisotropic elasticity to model various dislocation processes and found that the line shape can exhibit a marked dependence on the anisotropy. Also, the stress at which a dislocation breaks through obstacles by, say, the Orowan mechanism, varies considerably from element to element. To examine the extent to which these effects can be modelled in the isotropic approximation, it was noted that for an isotropic continuum, the constants μ and ν are uniquely determined if a field factor for an infinite straight line is known for two line orientations (except in degenerate circumstances). For example, for the line of Fig. 3, Eqns (6) and (7) give

$$E(\pi/2) = \mu b^2/4\pi(1-\nu),\ E(0) = (1-\nu\cos^2\beta)E(\pi/2), \tag{27}$$

so that ν is given by $E(0)/E(\pi/2)$ and μ by either relation. Scattergood and Bacon only considered slip geometries (β=0), but for all the crystals examined, isotropic results with μ and ν defined in this way were generally satisfactory approximations to their anisotropic counterparts. Although the line tension can be very anisotropic and give marked deviation from isotropy in the local curvature (Eqn (25)), E is less anisotropic, and so in terms of stress and overall line shape the isotropic approximation is usually good to within a few percent. A similar result holds for geometries with $\beta>0$.

Table 1 compiles μ and ν calculated within this framework for the common cubic metals. They have been obtained from the field factors in (5) and (13) for the stress resolved on the plane in the direction of $\underset{\sim}{b}$. They also give the corresponding energy factor by virtue of Eqn (5). The high ν values for the f.c.c. [110](111) system result from a strong θ-dependence of E: the line tension is not negative. When employed in isotropic formulae, the effective μ and ν data enable the anisotropic stress field to be estimated to within a few percent for most of the metals. The isotropic approximation is less good for the starred columns, however, for the isotropic stress is a constant independent of θ, whereas the anisotropic factor can vary considerably, e.g. curves for $\underset{\sim}{b}=\frac{1}{2}[110]$ in Fig. 6(b). Nevertheless, the biggest absolute error, which is that for iron, is only 12%. The possible effects of dissociation and core energy on some of the data for f.c.c. metals in Table 1 have been discussed elsewhere (12,26). Note also that for stresses resolved in other directions, different values of μ and ν are required to fit the anisotropic data (5), and some stresses which are non-zero in anisotropic crystals are identically zero in the isotropic approximation.

TABLE 1. The effective shear modulus μ (in GNm^{-2}) and Poisson's ratio ν for dislocations in some common cubic metals. Each column heading indicates the direction of $\underset{\sim}{b}$ and the plane of the dislocation. Starred entries are values of $\mu/(1-\nu)$ obtained from the mean of E(0) and E(π/2).

f.c.c.	A		$[1\bar{1}0](111)$	$[110](111)$	$[11\bar{2}](111)$	$[111](111)$
Ag	3.01	μ =	26.6	4.35	27.8	48.7*
		ν =	0.449	0.918	0.434	
Al	1.21	μ =	25.9	22.6	25.9	39.7*
		ν =	0.360	0.447	0.359	
Au	2.90	μ =	24.7	6.70	25.9	48.9*
		ν =	0.498	0.872	0.484	
Cu	3.21	μ =	42.1	10.1	44.2	69.0*
		ν =	0.431	0.864	0.413	
Ni	2.51	μ =	78.6	64.4	80.7	126.3*
		ν =	0.363	0.526	0.351	

b.c.c.	A		$[\bar{1}11](110)$	$[111](110)$	$[100](110)$	$[110](110)$	$[\bar{1}11](211)$
Cr	0.69	μ =	128.4	120.8	114.6	150.4*	128.4
		ν =	0.102	0.187	0.260		0.103
Fe	2.35	μ =	62.5	80.5	98.1	115.9*	62.5
		ν =	0.473	0.341	0.060		0.485
Mo	0.77	μ =	130.1	122.7	116.2	175.6*	130.1
		ν =	0.248	0.291	0.358		0.249
Nb	0.51	μ =	44.3	37.4	34.6	63.0*	44.3
		ν =	0.270	0.391	0.474		0.278
Ta	1.56	μ =	61.2	68.3	74.9	104.9*	61.2
		ν =	0.428	0.365	0.254		0.432
V	0.78	μ =	50.1	47.5	45.8	74.5*	50.1
		ν =	0.313	0.355	0.397		0.314

To summarize, it is often possible to model dislocations in moderately anisotropic crystals by use of an isotropic continuum, but the effective elastic constants vary from plane to plane (Table 1), so that an overall average (such as the Voigt approximation (8)) is inappropriate. However, caution must be exercised in using these constants when the features under investigation are possibly sensitive to the degree of anisotropy. For example, the minimum in energy versus orientation for prismatic loops in f.c.c. metals was found to be influenced quite strongly by anisotropy (27). It is usually necessary, therefore, to check on the validity of the isotropic approximation by first using anisotropic elasticity. To this end, the advances reviewed here are most significant, for they show how the energy and fields of dislocation configurations can be obtained from formulae involving the field factors for infinite straight lines. These factors can be computed readily from first principles or, better still, from a previous compilation of the Fourier coefficients. Many phenomena have been studied already, and there is no doubt that the theoretical basis exists for much wider application in the future.

REFERENCES

(1) Volterra, V. (1907). Annls. Scient. Ec. Norm. Sup. Paris, 24, 401.
(2) Eshelby, J.D. et al. (1953). Acta Metall., 1, 251.
(3) Foreman, A.J.E. (1955). Acta Metall., 3, 322.
(4) Head, A.K. (1967). Phys. Stat. Sol. (b), 19, 185.
(5) Bacon, D.J. et al. (1978). Prog. in Mater. Sci., 23, 51.
(6) Mura, T. (1982). Micromechanics of Defects in Solids. The Hague: Martinus Nijhoff.
(7) Barnett, D.M. (1980). In: Dislocation Modelling of Physical Systems, eds. M.F. Ashby et al., pp. 224-237. Oxford: Pergamon Press.
(8) Hirth, J.P. and Lothe, J. (1982). Theory of Dislocations. New York: John Wiley.
(9) Brown, L.M. (1967). Philos . Mag., 15, 363.
(10) Indenbom, V.L. and Orlov, S.S., (1967). Sov. Phys. JEPT, 6, 274.
(11) Willis, J.R. (1970). Philos. Mag., 21, 931.
(12) Prinz, F. et al. (1983). Philos. Mag. A, 47, 441.
(13) Matthai, C.C. and Bacon, D.J. (1984). Submitted for publication.
(14) Eshelby, J.D. (1956). Sol. St. Phys., 3, 79.
(15) Brown, L.M. (1964). Philos. Mag., 10, 441.
(16) Gavazza, S.D. and Barnett, D.M. (1976). J. Mech. Phys. Solids, 24, 171.
(17) Kirchner, H.O.K. (1981). Philos. Mag. A, 43, 1393.
(18) Scattergood, R.O. (1980). Acta Metall., 28, 1703.
(19) Foreman, A.J.E. (1967). Philos. Mag., 15, 1011.
(20) Bacon, D.J. et al. (1973). Philos. Mag. 28, 1241.
(21) Scattergood, R.O. and Bacon, D.J. (1975). Philos. Mag., 31, 179.
(22) Scattergood, R.O. and Bacon, D.J. (1974). Phys. Stat. Sol. (a), 25, 395.
(23) Scattergood, R.O. and Bacon, D.J. (1976). Acta Metall., 24, 705.
(24) Scattergood, R.O. and Bacon, D.J. (1982). Acta Metall., 30, 1665.
(25) Bacon, D.J. and Crocker, A.G. (1965). Philos. Mag., 12, 195.
(26) Bacon, D.J. (1978). Philos. Mag. A, 38, 333.
(27) Bacon, D.J. et al. (1970). Philos. Mag., 22, 31.

A DISLOCATION METHOD FOR SOLVING 3-D CRACK AND INCLUSION PROBLEMS IN LINEAR ELASTIC SOLIDS

G.K. WONG
Civil Engineering Department, Stanford University, Stanford, CA 94305

D.M. Barnett
Materials Science Department, Stanford University, Stanford, CA 94305

ABSTRACT

Both crack and transformed inclusion problems (in the sense of Eshelby) may be formulated rather easily in terms of integral equations whose kernels are the appropriate elastic Green's functions. For cracks and inclusions of arbitrary shape solutions must be obtained using numerical methods which ultimately require some regularization of these highly singular kernels. In the present work we devise a scheme for treating both classes of problems which avoids the need for regularization. Essentially we recognize at the outset that the crack or the inclusion surface is merely a general Somigliana dislocation surface. We discretize the surface by subdivision into a finite number of polygonal Volterra dislocation cells. Brown's Theorem is used to convert the original integral equation into a system of linear algebraic equations relating surface tractions to surface displacement discontinuities at selected points within the cell network. The method is restricted to linear elastic solids of infinite extent, but it may be applied to media of arbitrary anisotropy containing cracks or inclusions of arbitrary shape. The problem of a transformed inclusion with a "slipping interface" may be easily formulated and treated.

1. INTRODUCTION

1.1 PROLOGUE

It seems not inappropriate for one of us (D.M.B.) to first offer a few reflections about the man whose passing has brought us together at this conference. I can add but little to the eloquent remarks of those who were more intimately acquainted with Professor Eshelby than I, but I would like to at least express my appreciation for the positive influences which he and his work had on my own meager career. I was fortunate to have met with him on three occasions; if memory serves me correctly, I had the good sense to do more listening than talking in his presence. I like to think we would have gotten along well had the opportunity for more extended interaction existed.

I believe I have read all of his published work, and I have always been impressed with his ability to determine "what had to be" with an economy of analysis and a total avoidance of an encyclopaedic style. He is gone now, and I think it unlikely that we will see his equal in my lifetime. What is even sadder is the apparently diminishing number of scientists who can or will read his works in the original to keep alive the methodology he left us. Perhaps these conference proceedings will help draw attention to the beautiful legacy of Jock Eshelby.

1.2 THE TRANSFORMED INCLUSION PROBLEM

Consider a homogeneous anisotropic linear elastic solid of infinite extent. Denote as the "inclusion" some closed surface S enclosing a volume V. Now imagine that in the absence of the constraints imposed by the surrounding matrix the inclusion undergoes a uniform stress-free strain e_{ij}^{T} in the sense of Eshelby (1). We wish to find the elastic fields everywhere and the energy of deformation when the inclusion attempts to undergo the same transformation in the presence of a constraining matrix. There are two different, but physically interesting, situations which may be envisioned, namely, (a) the inclusion is coherent with the matrix after the transformation (this is the classical problem treated so beautifully by Eshelby), or (b) the inclusion-matrix interface is incapable of supporting shear tractions; we shall refer to the latter case as that of an incoherent inclusion and merely remark that this case might be realized in solids at temperatures high enough to permit significant diffusion or boundary sliding to occur. In either case Eshelby's work convincingly demonstrates that the transformed inclusion problem is perfectly equivalent to that of a Somigliana dislocation surface S on which is distributed an appropriate variable Burgers vector $\vec{b}(\vec{x})$. A formal representation of the displacement field solution is

$$u_m(\vec{x}') = \int_S C_{ijk\ell} G_{km,\ell}(\vec{x}-\vec{x}')\, b_j(\vec{x})\, n_i(\vec{x})\, dS(\vec{x}) \tag{1}$$

where $\vec{n}$ is the unit outer normal on S and $G_{km}(\vec{x}-\vec{x})$ is a component of the infinite medium elastic Green's tensor (as usual $,\ell$ means $\partial/\partial x_\ell$). If $\vec{b}$ is known at all points on S, $\vec{u}$ is everywhere determined. Implicit in (1) are the facts that the traction vector is continuous across S, all elastic fields vanish at infinity, and $\vec{u}$ is continuous except at all points on S where it suffers a jump discontinuity $\vec{b}$ (having taken $\vec{n}$ as the outer normal, the jump is $\vec{u}$ (matrix) $-$ $\vec{u}$ (inclusion)). Obviously displacement gradients can be found by differentiating (1) with respect to primed coordinates; stresses are related to the gradients by Hooke's law. In what follows we shall always be integrating over unprimed coordinate and thus will dispense with explicitly indicating arguments of the integrands. Relative to the undeformed state the potential energy change of the solid containing the transformed inclusion is simply

$$\Delta E = \frac{1}{2} \int_S \sigma_{ij} n_i b_j dS \ . \tag{2}$$

For a coherent inclusion $\vec{b}(\vec{x})$ is known at the outset and is given by

$$b_i(\vec{x}) = -e^T_{ij} x_j \ ; \qquad \vec{x} \text{ on } S \tag{3}$$

and the problem of complete determination of the elastic fields is already in quadrature form. The incoherent inclusion presents an additional complication. When the matrix-inclusion interface supports no shear tractions, only the normal component of Burgers vector may be specified at the outset, namely

$$b_i(\vec{x}) n_i = -e^T_{ij} x_j n_i \ ; \qquad \vec{x} \text{ on } S \ . \tag{4}$$

The tangential component of $\vec{b}$ at any point on S is no longer predetermined, since additional slip at the interface occurs to allow the shear tractions to relax to zero. The remaining unknown components of $\vec{b}(\vec{x})$ must be found by ensuring that the shear tractions on S indeed vanish, i.e., that

$$\sigma_{kj} n_j (\delta_{ik} - n_i n_k) = 0 \quad \text{on } S . \tag{5}$$

This requirement is equivalent to

$$C_{\alpha\beta mp} n_\beta (\delta_{\alpha s} - n_\alpha n_s) \int_S C_{ijk\ell} G_{km,\ell p} b_j n_i \, dS = 0 ;$$

$$\vec{x}', \vec{x} \quad \text{on } S . \tag{6}$$

For the incoherent inclusion one must solve (6) subject to the constraint (4) to fully determine $\vec{b}(\vec{x})$.

1.3 CRACK PROBLEMS

The problem of a freely-slipping crack in an anisotropic linear elastic solid subjected to uniform remote loading can be formulated in a similar fashion. Bilby and Eshelby (2) have already given an excellent account of the representation of cracks in terms of Somigliana dislocations. Here we need only recognize that if equation (1) is used for an open surface S, it gives the displacement field of a Somigliana dislocation loop (the boundary of S) with a prescribed displacement discontinuity $\Delta u_i = b_i(\vec{x})$ on S. Let us regard S as open and add to the fields generated by (1) the fields due to a uniform stress state σ^A_{ij} in the infinite solid. The sum of these two states yields the elastic fields associated with a traction-free crack (the open surface S is the crack in the undeformed solid) in a solid subjected to uniform far-field loading σ^A_{ij} provided that we can arrange for S to be traction-free. This is possible if $\vec{b}(\vec{x})$ is chosen to satisfy the integral equations

$$\sigma^A_{\alpha\beta} n_\alpha = n_\alpha C_{\alpha\beta mp} \int_S C_{ijk\ell} G_{km,\ell p} b_j n_i dS \ ;$$

$$\vec{x}', \vec{x} \quad \text{on} \quad S \ . \tag{7}$$

Given the remote loading σ^A_{ij}, (7) represents three integral equations to be solved for the unknown displacement discontinuity associated with the crack. We remark that since S is an open surface the choice of $\vec{n}$ on S is not unique. Our convention is such that (regarding S as a two-sided surface, say S^+ and S^-) choosing $\vec{n} = \vec{n}^+$ defines the displacement discontinuity $\Delta\vec{u}$ as $\vec{u}^+ - \vec{u}^- = \vec{b}$. The problem of a crack of arbitrary shape in a solid loaded remotely by σ^A_{ij} and by equal and opposite tractions on S^+ and S^- given by

$$T^{\pm}_i = \sigma^*_{ij} n^{\pm}_j \tag{8}$$

is obtained by solving (7) with $\sigma^A_{\alpha\beta} + \sigma^*_{\alpha\beta}$. For a traction free crack the total potential energy change accompanying the introduction of the crack into the solid stressed uniformly by σ^A_{ij} is merely

$$\Delta E = -\frac{1}{2} \int_S \sigma^A_{ij} b_j n_i dS \ . \tag{9}$$

2. DISLOCATION FORMULATION OF CRACK AND INCLUSION PROBLEMS

2.1 DISCRETIZATION USING VOLTERRA DISLOCATION CELLS

If one actually attempts to directly solve the integral equations (6) or (7), one immediately confronts kernels which involve second derivatives of the elastic Green's functions; these kernels are known to vary as $|\vec{x}-\vec{x}'|^{-3}$ (Barnett (3)). In either set of integral equations both $\vec{x}$ and $\vec{x}'$ refer to points on S, so that these kernels are highly singular. Weaver (4) and Sládek and Sládek (5) have attempted solving (7) for cracks with curved boundaries, but their handling of the singular kernels is less than convincing. For anisotropic problems the angular dependence of the Green's tensor must be determined numerically, so that the complete structure of the singular kernels is anything but transparent. The solution technique we shall propose allows us to ultimately dispense with the Green's function kernels and avoids the need for either regularization or singularity extraction.

Any numerical treatment of (6) or (7) will involve a discretization scheme to reduce the integral equations to a system of linear algebraic equations. The simplest discretization scheme involves first replacing the smooth surface S by a polyhedral surface whose faces are plane polygons (in the cases to be studied we shall use triangles and quadrilaterals). Next, one may, in the simplest case, take $\vec{b}(\vec{x})$ to be constant over any polygonal face ($\vec{b}$ may, of course, be different from face to face). The boundary of each plane polygonal face is then a polygonal Volterra (constant Burgers vector) dislocation loop, and we have merely approximated the original Somigliana dislocation surface by a polyhedral collection of polygonal Volterra loops. Figure 1 depicts such a discretization for a portion of a plane surface S (only a normal displacement discontinuity is indicated for clarity).

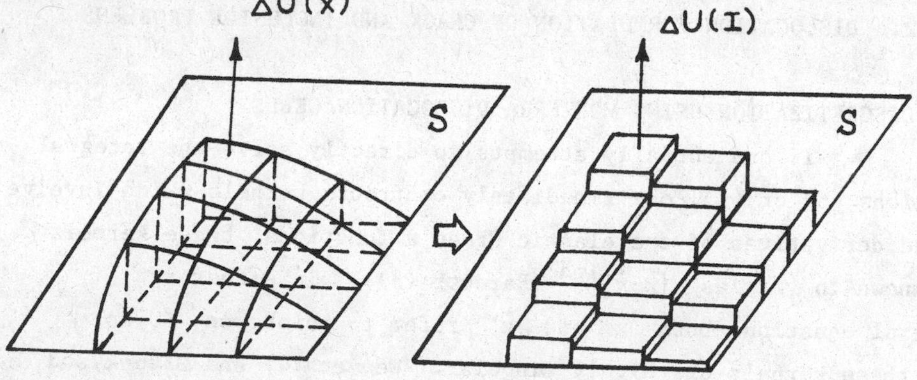

Figure 1: Discretization of a Somigliana surface

Having discretized in the fashion described, if our polyhedral surface has N faces, we elect to satisfy the appropriate integral equation(s) at N "collocation" points (one collocation point located in the interior of each plane polygonal Volterra cell). This is the customary method for reducing the original integral equation to a system of linear algebraic equations.

Now it is well-known that the displacement gradient and the stress fields of a Volterra dislocation loop are non-singular at field points which do not lie on the loop! Hence we immediately see that using collocation points within the plane Volterra polygonal loops provides the possibility of circumventing the problem of singular kernels.

2.2 THE TRACTION AT A POINT ON A POLYHEDRAL DISLOCATION SURFACE

Let us imagine that we have discretized either a crack or inclusion problem as described in the last section, i.e., we are now dealing with a polyhedral dislocation surface whose N faces are plane polygons. The boundary of any single polygon is a polygonal Volterra dislocation loop. Let us choose each of the N collocation points to lie at the centroids of the polygonal faces. For a linear elastic problem the stress field at the centroid of the Ith loop is merely

$$\sigma_{ij}(I) = \sum_{J=1}^{N} \sigma_{ij}(I,J) \; ; \quad I = 1, 2, \ldots, N$$

where $\sigma_{ij}(I,J)$ is the stress field of the Jth polygonal loop evaluated at the centroid of the Ith polygonal face. Furthermore, we know that for a Volterra dislocation loop the loop stress field is a linear function of the Burgers vector of the loop, so that

$$\sigma_{ij}(I,J) = V_{ijk}(I,J)\, b_k(J) \tag{10}$$

where $\vec{b}(J)$ is the Burgers vector of the Jth loop. We shall later show that V_{ijk} can be determined without recourse to elastic

Green's functions; for the present it is sufficient to point out that $V_{ijk}(I,J)$ will not be singular even when $I = J$ since all collocation points reside within the respective loop interiors.

Finally, the traction vector at the centroid of the Ith polygonal face can be written as

$$t_i(I) = \sum_{J=1}^{N} a_{ik}(I,J)\, b_k(J) \tag{11}$$

where

$$a_{ik}(I,J) = V_{ijk}(I,J)\, n_j(I) \tag{12}$$

and $\vec{n}(I)$ is the unit normal to the Ith polygonal face. The $a_{ik}(I,J)$ play the role of a stiffness matrix since they provide the linear connection between traction at a point on the polyhedral surface and the Burgers vectors associated with the polyhedral faces.

Without going into the details of exactly how $a_{ik}(I,J)$ is easily computed (this will be done in a later section), let us now proceed to show how the relation (11) allows us to solve the crack and inclusion problems posed in the Introduction.

2.3 THE COHERENT INCLUSION PROBLEM

The solution to this problem is already in quadrature form as noted in the Introduction; nevertheless, our dislocation formulation of this problem leads to a particularly simple method for evaluating the energy of deformation of the transformed solid. For the coherent inclusion, the Burgers vector $\vec{b}(\vec{x})$ is already known and is given by (3); in order to use (2) to find the associated energy we must know the surface tractions on S. Within our discretized approximation to S, the surface tractions are given by (11) with

$$b_k(J) = -e^T_{km} x_m(J) \tag{13}$$

where the $x_m(J)$ are cartesian coordinates of the centroid of the Jth face. In practice one would evaluate the tractions by matrix multiplication as the boundary integral equation researchers do it, namely

$$\begin{bmatrix} \underline{t}(1) \\ \underline{t}(2) \\ \vdots \\ \underline{t}(I) \\ \vdots \\ \underline{t}(N) \end{bmatrix} = \begin{bmatrix} \underline{a}(1,1) & \underline{a}(1,2) & \cdots & \underline{a}(1,I) & \cdots & \underline{a}(1,N) \\ \underline{a}(2,1) & \underline{a}(2,2) & \cdots & \underline{a}(2,I) & \cdots & \underline{a}(2,N) \\ \vdots & & & & & \\ \underline{a}(I,1) & \underline{a}(I,2) & \cdots & \underline{a}(I,I) & \cdots & \underline{a}(I,N) \\ \vdots & \vdots & & \vdots & & \\ \underline{a}(N,1) & \underline{a}(N,2) & \cdots & \underline{a}(N,I) & \cdots & \underline{a}(N,N) \end{bmatrix} \begin{bmatrix} \underline{b}(1) \\ \underline{b}(2) \\ \vdots \\ \underline{b}(I) \\ \vdots \\ \underline{b}(N) \end{bmatrix}$$

where each $\underline{a}$ is a 3 × 3 block matrix and each $\underline{t}$ and $\underline{b}$ is a column vector with 3 components, i.e.,

$$[\underline{t}(I)] = \begin{bmatrix} t_1(I) \\ t_2(I) \\ t_3(I) \end{bmatrix} .$$

Thus, in matrix form,

$$[T] = [A][B] \tag{14}$$

where $[A]$ is now a $3N$ by $3N$ matrix. The associated energy is then given by

$$\Delta E = \frac{1}{2} \sum_{I=1}^{N} t_i(I)\, b_i(I)\, \Delta S(I) \tag{15}$$

where $\Delta S(I)$ is the area of the Ith polyhedral face.

2.4 THE TRACTION-FREE CRACK

As discussed in the Introduction, the traction-free crack solution is generated as the sum of a uniform stress state and that due to a Somigliana dislocation surface; one requires that the sum of the applied tractions and the dislocation tractions vanish at all points on the crack surface. If we discretize as described earlier so that the crack surface is now an open polyhedral surface with N polygonal faces, the condition that the total traction vanish at the N collocation points is

$$t_i^A(I) + t_i(I) = 0 \tag{16}$$

where $t_i(I)$ is given by (11) and

$$t_i^A = \sigma_{ij}^A n_j(I) \ . \tag{17}$$

Using the matrix notation discussed in the last section, (16) is equivalent to

$$[T^A] = -[A][B] \ . \tag{18}$$

In this problem $[T^A]$ and $[A]$ are known, so that the unknown column vector $[B]$ of crack face displacement discontinuities must be obtained by solving (18) by, say, Gauss elimination. Once $[B]$ is known then, according to the discretized form of (9), the potential energy change due to the introduction of the crack into the uniformly stressed solid is

$$\Delta E = -\frac{1}{2} \sum_{I=1}^{N} t_i^A(I)\, b_i(I)\, \Delta S(I) \ , \tag{19}$$

where $\Delta S(I)$ is the area of the Ith polygonal face.

2.5 THE INCOHERENT INCLUSION

The relation between traction and Burgers vector at each of the N collocation points of a discretized incoherent inclusion is still given by (11). Unlike the coherent inclusion problem or the crack problem, this problem is of "mixed" character in that the normal traction and the tangential displacement discontinuity on S are unknown. The normal displacement discontinuity is known, so that at each of the N collocation points

$$n_k(J)\ b_k(J) = -\ e^T_{sm}\ n_s(J)\ x_m(J)$$

$$= Q(J)\ . \tag{20}$$

Since the shear traction at each of the collocation points vanishes, (11) may be recast in the form

$$T(I)\ n_i(I) = \sum_{J=1}^{N} a_{ik}(I,J)\ b_k(J) \tag{21}$$

where $T(I)$ is the (unknown) normal component of traction on the Ith polyhedral face. Equations (20) and (21) represent a system of 4N linear equations in 4N unknowns (N $T(I)$'s and 3N $b_k(J)$'s) which can be represented in block matrix form as

$$\left[\begin{array}{c|c} A & -N \\ \hline N^T & 0 \end{array}\right] \left[\begin{array}{c} B \\ \hline T \end{array}\right] = \left[\begin{array}{c} \tilde{0} \\ \hline Q \end{array}\right] . \tag{22}$$

The reduction scheme used to obtain (22) is similar to that used to obtain (14). [A] is a 3N × 3N matrix, [0] is the N × N null matrix, $[\tilde{0}]$ is the 3N × 1 null column matrix, [B] is a 3N × 1 column matrix, and [T] and [Q] are N × 1 column matrices. [N] is a 3N × N matrix and $[N]^T$ is its transpose. In explicit form [N] is given by

$$
[N] = \begin{bmatrix}
n_1(1) & 0 & \cdots & 0 \\
n_2(1) & 0 & \cdots & 0 \\
n_3(1) & 0 & \cdots & 0 \\
0 & n_1(2) & \cdots & 0 \\
0 & n_2(2) & \cdots & 0 \\
0 & n_3(2) & \cdots & 0 \\
\cdot & \cdot & & \cdot \\
\cdot & \cdot & \cdots & \cdot \\
\cdot & \cdot & & \cdot \\
0 & 0 & & n_1(N) \\
0 & 0 & & n_2(N) \\
0 & 0 & \cdots & n_3(N)
\end{bmatrix}
$$

Equation (22) may be solved for [B] and [T] by elimination. Finally, the potential energy change of the transformed solid is

$$
\Delta E = \frac{1}{2} \sum_{I=1}^{N} T(I)\ Q(I)\ \Delta S(I)\ . \tag{23}
$$

3. DETERMINATION OF THE STIFFNESS MATRIX

3.1 DISLOCATION ANGULAR STRESS FACTORS

By now it should be clear that the missing ingredient required to actually implement the proposed solution procedure is the specification for computing $V_{ijk}(I,J)$ and thus $a_{ik}(I,J)$. There exists a very compact representation of the stress field of a straight dislocation segment of finite length (and thus the field of a polygonal dislocation loop) which does not involve elastic Green's functions per se. This representation is due to a beautiful theorem proved by Brown (6); the interested reader is also referred to Asaro and Barnett (7), Bacon, et al. (8), and Hirth and Lothe (9) for extensions of the original formula derived by Brown. Space limitations preclude any discussion of Brown's formula here; we shall simply state the result for the field of a dislocation segment.

Before doing so, we must briefly digress to introduce the notion of dislocation angular stress factors.

Consider an infinitely long straight Volterra dislocation line L of Burgers vector $\vec{b}$ in an infinite anisotropic linear elastic medium. From the work of Stroh (10) it is known that the stress field of such a dislocation at any point P in space can be written as

$$\sigma_{ij}(P) = \frac{\Sigma_{ij}}{r} \tag{24}$$

where r is the normal distance from P to the dislocation line. Σ_{ij} is linear in $\vec{b}$ and otherwise depends only on the elastic constants and the direction of L in the plane defined by P and L. Bacon, et al. (8) have given formulae for computing Σ_{ij} in media of arbitrary anisotropy. Σ_{ij} is called the dislocation angular stress factor. For our purposes we need only realize that

$$\Sigma_{ij} = f_{ijk}b_k \tag{25}$$

and that Σ_{ij} or f_{ijk} may be computed more easily than can Green's function derivatives, i.e., the angular stress factors arise from 2-D elastostatic solutions.

Figure 2 depicts a straight dislocation segment AB of Burgers vector $\vec{b}$.

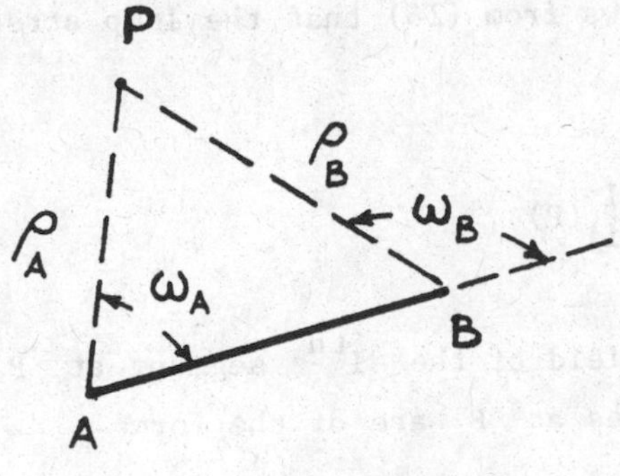

Figure 2: Geometry associated with the formula for the stress field of a dislocation segment.

If one applies Brown's theorem to the segment AB one finds that the stress field of the segment at a point P is given by

$$\sigma_{ij}(P) = \frac{1}{2}\left\{ \frac{\Sigma_{ij}(\omega_B)\cot\omega_B - \Sigma'_{ij}(\omega_B)}{\rho_B} - \frac{\Sigma_{ij}(\omega_A)\cot\omega_A - \Sigma'_{ij}(\omega_A)}{\rho_A} \right\} . \qquad (26)$$

As indicated in Figure 2, ρ_A and ρ_B are the distances from P to A and B, respectively. $\Sigma_{ij}(\omega_A)$ is the angular stress factor in the plane defined by P and AB of an infinitely long straight dislocation passing through P and A and having the same Burgers vector $\vec{b}$ as the segment AB. A similar definition applies for $\Sigma_{ij}(\omega_B)$. $\Sigma'_{ij}(\omega_A)$ means $d\Sigma_{ij}/d\omega$ evaluated at ω_A in the plane of P and AB. Bacon, et al. (8) have shown how $\Sigma'_{ij}(\omega)$ may be computed. Implicit in the use of (26) is the fact that it will ultimately be used for segments joined end to end to form a closed polygonal loop. In the event that P does not lie on the segment AB but is collinear with AB, one must set $\sigma_{ij}(P) = 0$ (see Bacon, et al. (8)).

3.2 THE STRESS FIELD OF A POLYGONAL LOOP

If we now join M dislocation segments (with the same Burgers vector) end to end to form a closed polygonal Volterra dislocation loop, it follows from (26) that the loop stress field at P is

$$\sigma_{ij} = \sum_{I=1}^{M} \sigma_{ij}^{I}(P) \qquad (27)$$

where $\sigma_{ij}^{I}(P)$ is the field of the Ith segment at P. By virtue of (31) the loop stresses at P are of the form

$$\sigma_{ij} = V_{ijk}b_k \qquad (28)$$

which is essentially equation (10). We now have the necessary ingredients to solve the class of problems discussed earlier. It is clear that V_{ijk} is a sum over loop vertices which involves only geometry and angular stress factors (and their first derivatives) for infinitely long dislocations passing through P and the loop vertices. Nowhere do elastic Green's functions enter, and the V_{ijk} are not singular as long as P does not lie on the loop. It should be pointed out that Asaro and Barnett (7) have shown that if P lies in the plane of a plane polygonal loop (and is not collinear with any portion of the loop), the angular stress factor derivatives in (26) cancel in the sum over vertices; this result implies that for planar crack problems the angular derivatives of Σ_{ij} need not be computed to determine V_{ijk} or $a_{ik}(I,J)$!

4. NUMERICAL RESULTS

4.1 A CIRCULAR CRACK PERTURBING REMOTE APPLIED TENSION

Space limitations preclude presenting the complete set of results of a collection of problems studied by one of us (G.K.W.) for a doctoral dissertation (11). We have chosen to present partial results for an illustrative group of problems to indicate the effectiveness of the dislocation technique proposed earlier; the results of our more complete study will be given elsewhere.

Consider a traction-free circular crack of radius "a" in an isotropic solid subjected to remote uniform tension normal to the crack surface; this is a classical problem whose analytical solution is available and well-known (12). The discretized crack surface is shown in Figure 3 for the case in which the crack front is approximated by an inscribed regular polygon of 18 sides ($\Delta\theta = 20°$) and all radial lines are divided into 5 equal segments.

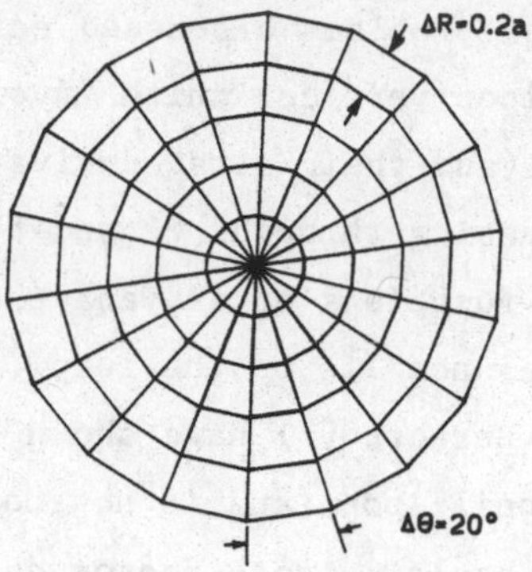

Figure 3: A typical dislocation mesh for a circular crack

Equation (18) may be solved for [B], the matrix of displacement discontinuities associated with the Volterra cells. Symmetry dictates that only a normal displacement discontinuity exists (this is borne out in our numerical study) and that this discontinuity depends only on radial position. Figure 4 depicts the (normalized) normal displacement discontinuity as a function of radius computed for $\Delta\theta = 10°$ using a coarse radial mesh (NR, the number of radial segments, equal to 5) and for a finer mesh (NR = 20). The exact solution is given by the solid curve in Figure 4, and one concludes that the finer mesh produces good agreement between the numerical and exact computations. One may compute the crack stress intensity factor by any of several methods. Using the technique proposed by Chan, et al. (13) we find that for the finer mesh ($\Delta\theta = 10°$, NR = 20) our computed stress intensity factor differs from the exact value ($K_I = 2\sigma\sqrt{a/\pi}$, where σ is the remote applied tension) by 7.2%.

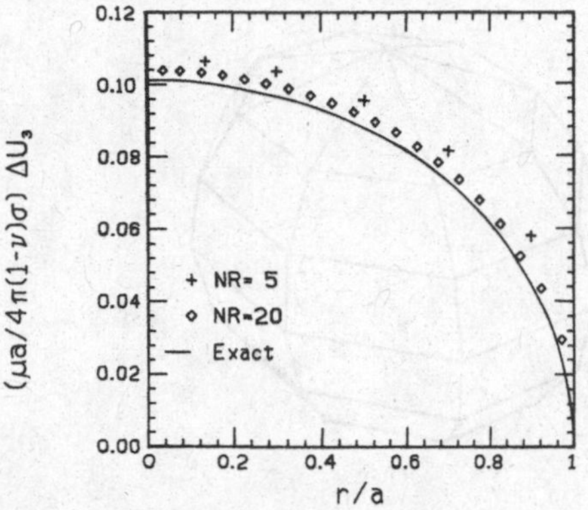

Figure 4: Normal displacement discontinuity vs. radius for the traction-free circular crack ($\Delta\theta = 10°$).

4.2 TRANSFORMED COHERENT AND INCOHERENT INCLUSIONS

If one considers a coherent inclusion of any shape which undergoes a purely dilatational transformation strain $e_{ij}^T = \varepsilon\delta_{ij}$ in an infinite isotropic linear elastic medium, the results of Eshelby (1) show that the strain energy change accompanying the transformation is given by

$$\Delta E_\varepsilon = 2\mu\varepsilon^2 V_0 \frac{1+\nu}{1-\nu} \tag{29}$$

where μ is the shear modulus, ν is Poisson's ratio, and V_0 is the (undeformed) inclusion volume. No solutions exist in the literature for the strain energy change associated with transformed incoherent inclusions of non-spherical shape (for a purely dilational e_{ij}^T there is no difference between the coherent and the incoherent transformed spherical inclusion).

Consider first a coherent transformed spherical inclusion with $e_{ij}^T = \varepsilon\delta_{ij}$. Figure 5 shows a discretized inclusion using 8 azimuthal divisions ($N\theta = 8$) and 6 polar divisions ($N\phi = 6$).

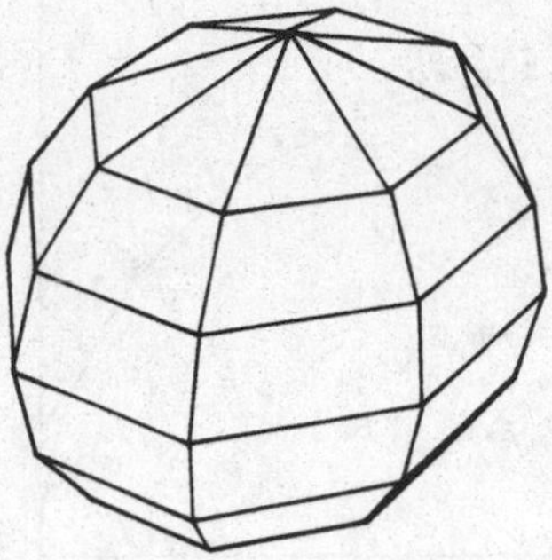

Figure 5: Polyhedral discretization of a spherical inclusion

Using (14) and (15) to compute ΔE we find for the discretized sphere with $N\phi = 24$:

Nθ	$\Delta E/\Delta E_\varepsilon$	% Error
16	0.928	7.2
32	0.975	2.5

The computed energy appears to be converging toward the exact value from below (presumably because the polyhedron used was inscribed within the sphere).

The cubic inclusion is interesting to study since the use of planar square dislocation cells on each face can discretize the surface geometry exactly. Using NE square cells per cube face leads to a total mesh of $6 \times NE$ surface cells. The following table lists results for both coherent and incoherent cubic inclusions for different choices of NE:

Inclusion Type	NE	$\Delta E/\Delta E_\varepsilon$	% Error
Coherent	16	0.929	7.1
Coherent	64	0.972	2.8
Incoherent	16	0.666	?
Incoherent	64	0.732	?
Incoherent	144	0.762	?

For the coherent cuboid the finer mesh produces better (and quite good) estimates of the energy of transformation. No error can be assigned to the incoherent cuboid computations since an exact value of ΔE is unavailable. If the above computation is correct, the energy of transformation of the incoherent cuboid is about 24% less than that of the coherent cuboid (one intuitively expects a lower energy to be attached to the inclusion with a "slipping" interface).

5. DISCUSSION

We would hope that ability of the proposed technique for treating certain classes of crack and inclusion problems has been demonstrated. The technique is closely related to a boundary integral approach; the utility of the method appears to lie in the fact that it circumvents the need for dealing with Green's functions and singular kernels. It is rather obvious how one may construct the elastic fields everywhere once all components of traction and displacement discontinuities on S have been obtained. We have not attempted to here discuss, among other points, optimal choices of collocation points or convergence. For many metallurgical applications a knowledge of the energy changes associated with cracks or transformed inclusions (particularly incoherent inclusions) may be sufficient. The technique is adaptable to transformed inhomogeneities (different elastic constants for matrix and inclusion); this will be the subject of a future study.

REFERENCES

(1) Eshelby, J.D. (1961). Elastic inclusions and inhomogeneities. In: Progress in Solid Mechanics, eds. I.N. Sneddon & R. Hill, pp. 89-140. Amsterdam: North-Holland.

(2) Bilby, B.A. & Eshelby, J.E. (1968). Dislocations and the theory of fracture. In: Fracture, an Advanced Treastise, ed., H. Liebowitz, vol. 1, pp. 99-182. New York: Academic Press.

(3) Barnett, D.M. (1972). The precise evaluation of derivatives of the anisotropic elastic Green's functions. Physica Status Solidi (b), 49, 741-748.

(4) Weaver, J. (1977). Three dimensional crack analysis. Int'l. J. Solids and Structures, 13, no. 4, 321-330.

(5) Sládek, V. & Sládek, J. (1983). Three dimensional curved cracks in an elastic body. Int'l. J. Solids and Structures, 19, no. 5, 425-436.

(6) Brown, L.M. (1967). A proof of Lothe's theorem. Phil. Mag., 15, no. 134, 363-370.

(7) Asaro, R.J. & Barnett, D.M. (1976). Applications of the geometrical theorems for dislocations in anisotropic elastic media. In: Computer Simulation for Materials Applications, vol. 2, eds. R.J. Arsenault, et al., pp. 313-324. AIME: Nuclear Metallurgy Series.

(8) Bacon, D.J., Barnett, D.M., & Scattergood, R.O. (1978). Anisotropic continuum theory of lattice defects. In: Progress in Materials Science, ed. B. Chalmers, et al., 23, nos. 2-4, pp. 51-262. Oxford: Pergamon Press.

(9) Hirth, J.P. & Lothe, J. (1982). Theory of Dislocations, 2nd Edition, pp. 140-146. New York: J. Wiley & Sons.

(10) Stroh, A.N. (1962). Steady state problems in anisotropic elasticity. J. Math. and Physics, 41, 77-103.

(11) Wong, G.K. Three dimensional crack and inclusion problems. Ph.D. thesis, Civil Engineering Department, Stanford University (to be submitted, June, 1984).

(12) Sneddon, I.N. (1951). Fourier Transforms, p. 490. New York: McGraw-Hill.

(13) Chan, S.K., Tuba, T.J., & Wilson, W.K. (1970). On the finite element method in linear fracture mechanics. Engineering Fracture Mechanics, 2, 1-17.

EDITORS' NOTE

In the discussion following the presentation of this paper, a number of participants expressed reservations about the quality of the results that could be expected from the computational scheme that was employed. These centred upon the fact that the basic constant Burgers vector elements inevitably generate such rapidly varying stresses (becoming singular at element boundaries) that associating the actual boundary traction over an element with that generated at a single point in the element was not necessarily accurate. The authors had no theoretical justification for their procedure but defended it by reference to the numerical evidence reported in their paper.

INTERACTIONS OF POINT DEFECTS AND DISLOCATIONS

R. BULLOUGH

Theoretical Physics Division, AERE, Harwell, OX11 ORA, UK.

J.R. WILLIS

School of Mathematics, University of Bath, Bath, BA2 7AY, UK.

ABSTRACT

Point defects and their interactions can be studied by modelling the defects as misfitting inhomogeneities embedded in an elastic continuum. The essential features of this representation are outlined and general expressions concerned with interactions with applied stress fields given. The explicit long-range interaction of a point defect with a dislocation is derived as a special case. A new calculation for the short-range interaction of a point defect with a dislocation is presented in which the dislocation singularity is distributed throughout a cylindrical region of inhomogeneity; the modification of the long-range interaction due to the presence of an external stress is also calculated. The paper is concluded with a brief discussion of the physical relevance of the results obtained.

1. INTRODUCTION

There is extensive evidence that point-defects present in a super saturated solid solution will segregate to sinks such as dislocations (1). Such segregation is the fundamental cause of important macroscopic phenomena such as strain ageing (2,3,4), void swelling (5,6) and irradiation creep (7). Strain ageing is due to the segregation of solute atoms, such as carbon or nitrogen in steels, to dislocations and the two irradiation phenomena arise because of the loss of intrinsic point-defects (vacancies or self-interstitials) at the dislocations. From various studies of the kinetics of such segregation, beginning with the early work of Cottrell & Bilby (2), it is apparent that the long-range elastic interaction between the stress-field of the dislocation and the migrating point-defects plays an important role in defining the effectiveness of the dislocations as a sink for a particular point-defect type. To obtain the form and magnitude of this elastic interaction it is convenient to represent the point-defect as a small misfitting inclusion with elastic constants that differ from those of the host material and then follow Eshelby (8,9) by replacing it by an equivalent inclusion together with an associated transformation strain. Early results of such calculations together with a comprehensive discussion of the ensuing kinetics are given in a review by Bullough & Newman (1) and a later more general expression for the interaction energy is given by Bullough & Willis (7).

In these calculations the stress-field of the dislocation is presumed to be unaffected by the presence of segregated point-defects. If the point-defects are solute atoms then they can segregate around the dislocation as an equilibrium Maxwellian atmosphere or they can form a precipitate along the dislocation core. Direct visual evidence for solute atmospheres surrounding dislocations in iron-carbon alloys using a combined FIM/Atom Probe has recently been obtained by Chang et al (10) and precipitation of solute on dislocations has been frequently observed, for example in suitably doped silicon by infra-red transmission techniques (1,11,12).

In irradiated materials it is known that minor impurities are transported by the intrinsic point defect flux to sinks such as dislocations, grain boundaries, voids or the free surface (13); in the analysis of the evolution of such irradiated microstructures allowance is often made for solute segregation to the voids by supposing that they are all surrounded by a thin shell of material with a modulus differing from that of the host (14); the presence of this shell then modifies the elastic interaction between the point-defects and the void. Such an analysis would be more consistent if allowance were also made for local modulus changes due to solute segregation at the dislocations. In addition, in unirradiated materials, the atmosphere formation or precipitate growth on dislocations will also be influenced by the local modulus changes as the segregation proceeds. For these reasons we believe that kinetic calculations of solute segregation could be improved if we had an interaction energy that included the effect of local modulus changes at the dislocations. A model to include these effects at the dislocations and certain other explicit results are described in the present paper. Thus in Section 2 we present a formulation of the general inclusion problem that extends the work of Eshelby (8) to a finite anisotropic body with arbitrary surface loading and in an arbitrary state of inhomogeneous transformation. The change in energy which occurs upon the transformation from a homogeneous state is then developed in Section 3 and used to obtain, as a special case, the previous expression given by Bullough & Willis (7) together with some equivalent variants. Two applications are then given in Section 4:

(i) The interaction energy when the core is modified due to segregation, in which the presence of solute atoms, as either an atmosphere or a continuous precipitate, is represented by introducing a cylindrical misfitting inclusion, with elastic moduli that differ from the host moduli, lying adjacent to the actual dislocation core.

(ii) Explicit results, using the original expression of Bullough & Willis (7), for the interaction energy between a dislocation and intrinsic point-defects when the body is subjected to external stress. These results, which incidentally explain the fundamental origin of irradiation creep (7,15), are included to demonstrate the interesting fact that an external stress can induce additional first order long-range energy terms even though the energy is quadratic in the total applied field.

Finally, in Section 5, we draw attention to some of the physical limitations inherent in the present 'continuum' analysis.

2. INCLUSION PROBLEMS

The prototype inclusion problem, as considered by Eshelby (8), may be stated as follows. A region V_I of an infinite homogeneous elastic body, with tensor of elastic moduli $C_{ijk\ell}$, tends to undergo a spontaneous change of shape, which is described by a strain tensor e^T_{ij}, which is uniform over V_I. Thus, if the region V_I were cut out of the body, it would display the strain e^T_{ij} when free of stress, relative to its state prior to the transformation. In situ, however, the region V_I (called the inclusion) is constrained by the surrounding material, and the problem is to find the stresses and strains generated throughout the body. Eshelby solved this problem for the case that V_I is an ellipsoid, and exploited properties of the solution to generate a solution for the perturbation of an otherwise uniform applied stress, caused by the presence of an ellipsoidal inhomogeneity.

Here we consider a generalization of these problems. An elastic body with tensor of moduli $C_{ijk\ell}$ occupies a finite region V. It is subjected to fairly general conditions (described later) on its boundary ∂V and, perhaps, to body forces f_i per unit volume, which would generate displacement, stress and strain fields u^A_i, σ^A_{ij}, e^A_{ij} respectively. Now, additionally, the body undergoes a spontaneous transformation in which elastic moduli change to $C^*_{ijk\ell}$, and each element of volume undergoes a change of shape described by the strain tensor e^*_{ij}. Neither $C^*_{ijk\ell}$ nor e^*_{ij} need be uniform in general. We call e^*_{ij} the misfit strain. The additional strain (that is, the strain measured relative to the configuration after the application of σ^A_{ij} but prior to the transformation) is denoted e^C_{ij}. The total strain, measured relative to the configuration prior both to application of σ^A_{ij} and transformation, is then

$$e_{ij} = e^A_{ij} + e^C_{ij} \tag{1}$$

but the elastic strain, measured relative to the unstressed state of an element, is $e_{ij} - e^*_{ij}$. If the total stress is denoted σ_{ij}, we have $\sigma_{ij} = C^*_{ijk\ell}(e_{ij} - e^*_{ij})$. Essentially following Eshelby (8), we consider a body which has undergone no modulus change, but has undergone a transformation strain e^T_{ij} which produces the same total stress and strain fields σ_{ij}, e_{ij}. Thus, using a notation in which suffixes are suppressed, e^T_{ij} is defined so that

$$\sigma = C^*(e^A + e^C - e^*) = C(e^A + e^C - e^T) . \tag{2}$$

It is convenient to define also

$$p^* = C^*e^* , \quad p^T = Ce^T , \tag{3}$$

so that application of the stress $-p^*$ would eliminate the misfit e^* in the transformed body, and $-p^T$ would do the same in the body which suffers e^T without change of moduli. The equilibrium equations now require that

$$\mathrm{div}(Ce^C) - \mathrm{div}\, p^T = 0 , \tag{4}$$

so that e^C is the strain that would be generated in the original body, if it were subjected to homogeneous boundary conditions and to the body force $-\mathrm{div}\, p^T$. If Green's function G for this body is known, it follows that

$$e^C = \Gamma p^T , \tag{5}$$

where Γ is an integral operator whose kernel has components

$$\Gamma_{ijk\ell}(x,x') = \left. \frac{\partial^2 G_{ik}}{\partial x_j \partial x'_\ell} \right|_{(ij)(k\ell)} , \tag{6}$$

where (ij) denotes symmetrization on these suffixes. A derivation appears in Willis (16), for example.

It follows from Eqns (2), (3) and (5) that p^T satisfies the integral equation

$$(C^* - C)^{-1} p^T + \Gamma p^T = (C^* - C)^{-1} p^* - e^A , \tag{7}$$

wherever $C^* \neq C$. We assume for simplicity that $(C^* - C)$ is non-singular: if it is singular, a technical modification is required. With the term p^* absent, this equation has been used frequently in the study of composite media: Willis (16,17) gives references. If Eqn (7) can be solved for p^T, then e^C follows from Eqn (5) and the entire problem is solved.

Before proceeding to particular examples, it is worth noting that, if the body is subjected to traction boundary conditions, the stress $\sigma^C = C(e^C - e^T)$ is associated with zero tractions and body forces and so has mean value zero. Correspondingly, the mean value of the strain e^C is the same as that of e^T so that e^T yields the average shape change of the whole body: this is observable directly, from the boundary displacements.

We now record results for some special cases.

Let $C^* = C$ and $e^* = 0$ except over an ellipsoid V_I, where C^* and e^* are uniform, let the body V be infinite and let the "applied stress" σ^A be uniform. Then, even if C is anisotropic, it can be shown that p^T is uniform over V_I and, within V_I,

$$e^C = Pp^T , \tag{8}$$

where P is a constant tensor. Correspondingly, to satisfy Eqn (7),

$$p^T = \left[(C^* - C)^{-1} + P\right]^{-1} \left[(C^* - C)^{-1} p^* - e^A\right]. \tag{9}$$

The (anisotropic) problem has a long history, starting with work of Kneer (18). References are given by Willis (16,17).

When C is isotropic, and so specified by bulk modulus κ and shear modulus μ, it follows from Love (19) that Green's function G has components

$$G_{ik}(x) = \frac{1}{4\pi\mu}\frac{\delta_{ik}}{r} - \frac{3\kappa + \mu}{8\pi\mu(3\kappa + 4\mu)} r,_{ik} \tag{10}$$

where $r = |x|$ and the suffix $,i$ denotes $\partial/\partial x_i$. Correspondingly, Eqn (8) gives

$$e_{ij} = -\frac{1}{2\mu}\left(\phi,_{j\ell}\, p^T_{i\ell} + \phi,_{i\ell}\, p^T_{j\ell}\right) + \frac{3\kappa + \mu}{2\mu(3\kappa + 4\mu)}\psi,_{ijk\ell}\, p^T_{k\ell} \tag{11}$$

where

$$\phi(x) = \frac{1}{4\pi}\int_{V_I} \frac{dx'}{|x - x'|} , \quad \psi(x) = \frac{1}{4\pi}\int_{V_I} |x - x'|\,dx' \tag{12}$$

Eshelby (8) gave the representation (11) and evaluated ϕ and ψ for ellipsoidal V_I. In the special case where V_I is a sphere, radius a, centred at the origin,

$$\phi = \frac{a^2}{2} - \frac{r^2}{6} , \quad \psi = \frac{a^4}{4} + \frac{r^2a^2}{6} - \frac{r^4}{60} \quad \text{when } r < a \tag{13}$$

and

$$\phi = \frac{a^3}{3r} , \quad \psi = \frac{a^3r}{3} + \frac{a^5}{15r} \quad \text{when } r > a . \tag{14}$$

The tensor P can be deduced for a spherical inclusion using Eqn (11). It is isotropic and can be characterised by a "bulk modulus" κ_P and "shear modulus" μ_P :

$$P_{ijk\ell} = \kappa_P\, \delta_{ij}\, \delta_{k\ell} + \mu_P(\delta_{ik}\, \delta_{j\ell} + \delta_{jk}\, \delta_{i\ell} - 2/3\delta_{ij}\, \delta_{k\ell}), \tag{15}$$

where

$$(3\kappa_P, 2\mu_P) = \left(\frac{1}{3\kappa + 4\mu} , \frac{3(\kappa + 2\mu)}{5\mu(3\kappa + 4\mu)}\right). \tag{16}$$

An equivalent result was given by Eshelby (8). He introduced parameters α, β defined so that $\alpha = 9\kappa\kappa_P$, $\beta = 4\mu\mu_P$. Eqn (9) now gives

$$p^{T}_{kk} = \left(\frac{3\kappa + 4\mu}{3\kappa^* + 4\mu}\right)\left[p^*_{kk} - 3(\kappa^* - \kappa)\ e^{A}_{kk}\right] ,$$

$$'p^{T}_{ij} = \frac{5\mu(3\kappa + 4\mu)}{6\mu^*(\kappa + 2\mu) + \mu(9\kappa + 8\mu)}\left['p^*_{ij} - 2(\mu^* - \mu)\ 'e^{A}_{ij}\right] \qquad (17)$$

where $'p_{ij} = p_{ij} - (1/3)\delta_{ij}p_{kk}$. This result was given, in terms of Eshelby's α and β by Bullough & Willis (7).

In the particular case $e^A = 0$, the field e^C is a property of the inclusion alone and it is convenient to call it e^o. Thus, from Eqns (8) and (17),

$$e^{o}_{kk} = \frac{3\kappa^*}{3\kappa^* + 4\mu}\ e^*_{kk}\ , \quad 'e^{o}_{ij} = \frac{6\mu^*(\kappa + 2\mu)}{6\mu^*(\kappa + 2\mu) + \mu(9\kappa + 8\mu)}\ 'e^*_{ij} \qquad (18)$$

having also used Eqn $(3)_1$. Bullough & Willis (7) described e^o as the "actual strain within the inclusion", but this is open to interpretation as the elastic strain, measured relative to the unstressed state, or $e^o - e^*$. Therefore, we now prefer to describe e^o as the "hole strain", because it describes precisely the strain of the hole into which the inclusion fits.

3. ENERGY RELATIONS

We revert to the general problem introduced in Section 2 and take, as boundary conditions,

$$\sigma_{ij} n_j = T_i - B_{ij}(u_j - U_j) , \qquad (19)$$

with $B_{ij} = B_{ji}$. Strictly, one of T_i or $B_{ij}U_j$ is redundant, but this form allows easy specialization to either traction or displacement conditions. The "machine" which may be considered to produce Eqn (19) experiences the reaction force $-\sigma_{ij}n_j$, so its potential energy is

$$\int_{\partial V} \left\{\tfrac{1}{2}(u_i - U_i)B_{ij}(u_j - U_j) - T_i(u_i - U_i)\right\} dS$$

$$= -\tfrac{1}{2}\int_{\partial V} (\sigma_{ij}n_j + T_i)(u_i - U_i)\, dS . \qquad (20)$$

Correspondingly, the total energy of the body and the loading system is, in symbolic notation,

$$E_{tot} = \int_V \left[\tfrac{1}{2}\sigma(e - e^*) - fu\right] dV - \tfrac{1}{2}\int_{\partial V} (\sigma.n + T)(u - U)\, dS . \qquad (21)$$

Now expand these integrands, using Eqns (1), (2) and the additional definition

$$\sigma^C = \sigma - \sigma^A = C(e^C - e^T) . \tag{22}$$

This gives

$$E_{tot} = \int_V \left[\tfrac{1}{2}\sigma^A e^A - fu^A + \sigma^A e^C - fu^C + \tfrac{1}{2}\sigma^C e^C - \tfrac{1}{2}(\sigma^A e^T + \sigma e^*) \right] dV - \int_V \left[\tfrac{1}{2}(\sigma^A.n + T)(u^A - U) + (\sigma^A.n) u^C + \tfrac{1}{2}(\sigma^C.n) u^C \right] dS, \tag{23}$$

since $\sigma^A = Ce^A$ and σ^C satisfies the boundary condition $\sigma^C.n = -Bu^C$. The result simplifies, upon use of Gauss' theorem to transform the surface terms that involve u^C. The result is

$$E_{tot} - E^A{}_{tot} = -\tfrac{1}{2}\int_V (\sigma^A e^T + \sigma e^*) \, dV , \tag{24}$$

where $E^A{}_{tot}$ is the energy associated with the untransformed body, obtained from Eqn (21) with e^* deleted and the other fields given the superscript A.

Expression (24) was derived previously, but under less general conditions, by Bullough & Willis (7). It is useful because e^T and e^* are different from zero only over the inclusions. In the particular case of a homogeneous inclusion, so that $C^* = C$ and $e^T = e^*$, it reduces to a formula given by Eshelby (8). An alternative form of Eqn (24) can be found by using Eqn (2) to eliminate e^T in favour of e^C. This gives

$$E_{tot} - E^A{}_{tot} = \tfrac{1}{2}\int_V \left[e^A (C^* - C)(e^A + e^C - e^*) - (\sigma^A + \sigma) e^* \right] dV. \tag{25}$$

The integrand is again zero except over the inclusions. For the case of a pure inhomogeneity, so that $C^* \neq C$ but $e^* = 0$, Eqn (25) reduces to a result derived by Eshelby (9).

Final variants of Eqn (24) are obtained by use of the relation

$$\sigma = (S - S^*)^{-1} (e^* - e^T) , \tag{26}$$

which follows from (2). Here, S, S^* are the compliance tensors C^{-1}, C^{*-1} respectively. Thus,

$$E_{tot} - E^A{}_{tot} = -\tfrac{1}{2}\int_V \left[\sigma^A e^T + e^*(S - S^*)^{-1}(e^* - e^T) \right] dV \tag{27}$$

or, alternatively, in terms of p^T, p^*,

$$E_{tot} - E^A{}_{tot} = -\tfrac{1}{2}\int_V \left[e^A p^T + p^*(C^* - C)^{-1}(p^* - p^T) - p^* e^* \right] dV . \tag{28}$$

These results are not available elsewhere. They are closely connected to the Hashin-Shtrikman variational principle, which is associated with the integral equation (7). Willis (20,16,17) discusses the principle from this standpoint.

In the particular case of a spherical inclusion, with e^* uniform, perturbing a field e^A which can be adequately approximated as uniform over the inclusion, Eqns (17) and (27) give

$$E_{int} = -V_I \left\{ \mu A\, 'e^A_{ij}\, 'e^A_{ij} + \tfrac{1}{2}\kappa\, C (e^A_{kk})^2 + 2\mu B\, 'e^A_{ij}\, 'e^*_{ij} + \kappa D\, e^A_{kk}\, e^*_{\ell\ell} \right\}, \qquad (29)$$

where

$$A = \frac{\mu - \mu^*}{\mu - \beta(\mu - \mu^*)}, \quad B = \frac{\mu^*}{\mu - \beta(\mu - \mu^*)}, \quad C = \frac{\kappa - \kappa^*}{\kappa - \alpha(\kappa - \kappa^*)}$$

and $D = \dfrac{\kappa^*}{\kappa - \alpha(\kappa - \kappa^*)}$. E_{int} represents the interaction energy, or that part of E_{tot} which is sensitive to the position of the inclusion, and V_I represents the volume of the inclusion. The constants α and β are dimensionless; in terms of Poisson's ratio ν for the matrix, they take the form $\alpha = (1+\nu)/3(1-\nu)$, $\beta = 2(4-5\nu)/15(1-\nu)$. Notation in Eqn (29) has been chosen so as to conform with that of Bullough & Willis (7), who gave the result in the special case of a dilatational misfit strain e^* (so that $'e^*_{ij} = 0$). It is perhaps worth remarking that the inclusion is still referred to as spherical, even when $'e^*_{ij} \neq 0$. In this general case, the misfitting inclusion is slightly ellipsoidal once it has undergone e^*, but the hole into which it is made to fit is spherical, at least when the matrix is stress-free.

4. DISLOCATION-POINT DEFECT INTERACTIONS

Here we apply the general formulae of the preceding sections to study the energy of interaction between a point defect, modelled as a spherical inclusion, and a dislocation, with allowance for two effects not fully discussed elsewhere.

4.1 EFFECT OF CORE MODIFICATION DUE TO SEGREGATION

The basic "$1/\rho$" field of a dislocation should strictly always be modified to allow for conditions at the dislocation core; for example, the core might be modelled as a hollow tube, whose surface is traction-free. Such modifications are usually ignored, because they are significant only within a very small region surrounding the core. If, however, appreciable segregation of impurities has occurred, it might be appropriate to model their effect by placing, next to the dislocation, a cylindrical region of material with elastic moduli C^*_D and misfit strain e^*_D (the subscript D referring to the dislocation), as shown in Fig. 1. The significance of the effect will depend upon the magnitudes of $C^*_D - C$ and e^*_D, and upon the radius R of the cylinder. The "hollow tube" model is recovered if $C^*_D = 0$ and $e^*_D = 0$; but then the radius R would be small, say equal to b, the Burgers vector of the dislocation.

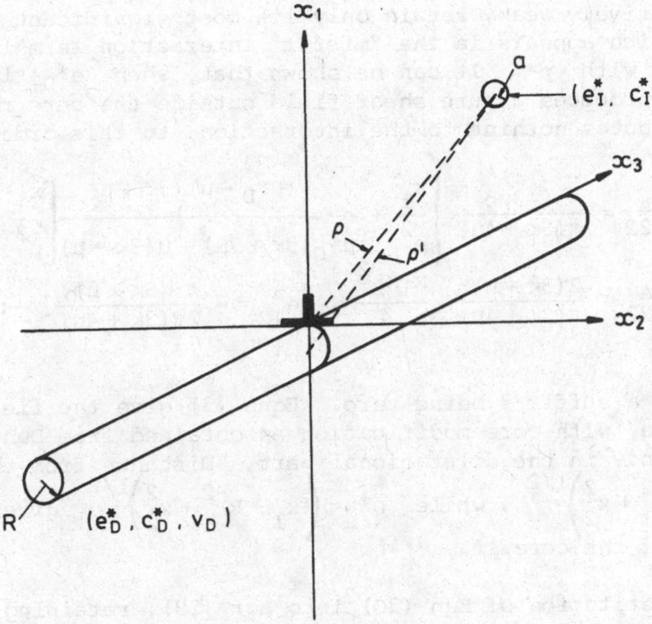

FIG 1. Illustration of an edge dislocation with its associated cylindrical inclusion; the position of the spherical inclusion is also indicated, together with the coordinate systems employed

The point defect is modelled as a spherical inclusion, of small radius a, with elastic moduli C^*_I and misfit e^*_I (the suffix I referring to the inclusion).

The elastic field of a dislocation adjacent to an inhomogeneous cylindrical core has been found by Dundurs & Mura (21) and the field due to e^*_D can be found by the two-dimensional analogue of the reasoning at the end of Section 2. The basic field e^A is taken as the sum of these so that, in the notation of earlier sections, the tensor C is now non-uniform, since it takes the value C^*_D over the core region. The total energy of the system is given by Eqn (28), with the integral extending over the volume of the inclusion and $C^* = C^*_I$. The field p^T is difficult to calculate exactly because, in Eqn (7), the operator Γ is derived from Green's function for a medium made inhomogeneous by the presence of the core. Since the inclusion is taken small, however, a first approximation to p^T can be obtained by ignoring "multiple scattering" effects between the inclusion and the core. In this approximation, Γ becomes the simple "infinite medium" operator, p^T is given by Eqn (17) and, correspondingly, the interaction energy is given by Eqn (29).

We specialize further by taking e^*_I, e^*_D to be pure dilatations and, assuming the perturbation of the dislocation field by the core to be relatively weak, retain only its most significant contribution to Eqn (29), which appears in the "misfit" interaction term involving the product of e^A with p^*. It can be shown that, when e^*_D is a pure dilatation, it produces a pure shear field outside the core region, and so e^*_D contributes nothing to the interaction, to this order. Thus, we take

$$e^A_{11} + e^A_{22} = \frac{-3\mu b}{\pi(3\kappa+4\mu)}\left\{\frac{x_1}{\rho^2} + \frac{(\mu^*_D-\mu)(3\kappa+\mu)}{\mu^*_D(3\kappa+7\mu)+\mu(3\kappa+\mu)}\left[\frac{x_1}{\rho^2} - \frac{x_1+R}{\rho'^2}\right]\right\},$$

$$e^A_{11} - e^A_{22} = \frac{2(3\kappa+\mu)b}{\pi(3\kappa+4\mu)}\,\frac{x_1x_2^2}{\rho^4}, \quad e^A_{12} = -\frac{(3\kappa+\mu)b}{2\pi(3\kappa+4\mu)}\,\frac{x_2(x_1^2-x_2^2)}{\rho^4}, \tag{30}$$

components with a suffix 3 being zero. Eqns (3) give the field of an edge dislocation, with core modification as obtained from Dundurs & Mura (21) retained only in the dilatational part. Distance from the dislocation is $\rho = \left(x_1^2+x_2^2\right)^{1/2}$, while $\rho' = \left((x_1+R)^2+x_2^2\right)^{1/2}$ gives distance from the axis of the core.

Substitution of Eqn (30) into Eqn (29), retaining the effect of core modification only in the "misfit" term, now gives

$$\begin{aligned} E_{int} \sim \mu V_I \Bigg\{ & \left[\frac{x_1}{\rho^2} + \frac{(\mu^*_D-\mu)(3\kappa+\mu)}{\mu^*_D(3\kappa+7\mu)+\mu(3\kappa+\mu)}\left(\frac{x_1}{\rho^2} - \frac{x_1+R}{\rho'^2}\right)\right]\frac{b\,e^o_{kk}}{\pi} \\ & + \frac{9\mu(\kappa^*_I-\kappa)}{2\pi^2(3\kappa^*_I+4\mu)(3\kappa+4\mu)}\,\frac{b^2x_1^2}{\rho^4} \\ & + \frac{5\mu^2(\mu^*_I-\mu)b^2}{2\pi^2[6\mu^*_I(\kappa+2\mu)+\mu(9\kappa+8\mu)](3\kappa+4\mu)}\left[\frac{3x_1^2}{\rho^4} + \left(\frac{3\kappa+\mu}{\mu}\right)^2\frac{x_2^2}{\rho^4}\right]\Bigg\} \end{aligned} \tag{31}$$

the "hole strain" e^o_{kk} being related to e^*_I by Eqn (18). The expression (31) is correct to order $1/\rho^2$ and so has precise validity when the inclusion is far from the dislocation. At closer separations, however, it has only the status of an approximation.

4.2 INFLUENCE OF EXTERNAL STRESS

For simplicity we now omit the presence of the cylindrical inclusion adjacent to the dislocation and consider the variation of the interaction energy between the edge dislocation and a small misfitting inclusion as we vary the direction of an external applied tensile stress τ. The three orthogonal directions of applied stress are indicated in Fig. 2 where the position of the inclusion is conveniently identified by the cylindrical coordinates (ρ,θ). The applied strain at the inclusion now consists of the sum of the dislocation strain field (42) and the homogeneous strains generated by the external stress in each of the three orientations.

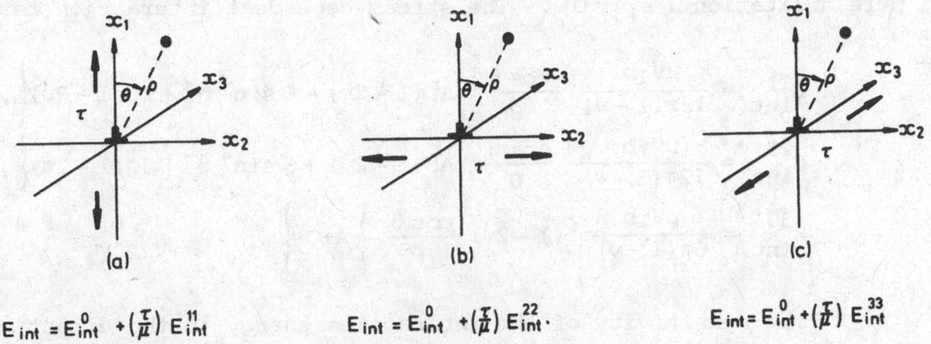

FIG. 2 The three orthogonal orientations of externally applied tensile stress, in relation to the edge dislocation

The interaction energy, which follows directly from Eqn (29), can be conveniently written as;

$$E_{int} = E^{o}_{int} + \left(\frac{\tau}{\mu}\right) E^{ii}_{int} \tag{32}$$

where E^{o}_{int} is the interaction energy in the absence of the external stress τ and E^{ii}_{int} is the coefficient of (τ/μ) in the total interaction energy, when the uniaxial tension is parallel to the x_i axis. We find:

$$\begin{aligned} E^{o}_{int} = &- \frac{V_I b^2}{8\pi^2}\left(\frac{1-2\nu}{1-\nu}\right)^2 \frac{\cos^2\theta}{\rho^2}\left[2\mu A + \kappa C\right] \\ &- \frac{\mu V_I b}{6\pi(1-\nu)\rho}\Big\{B\Big[(1-2\nu)(2e^*_{33} - e^*_{11} - e^*_{22})\cos\theta \\ &+ 6(e^*_{22} - e^*_{11})\cos\theta\sin^2\theta + 6e^*_{12}\sin\theta\,(2\sin^2\theta - 1)\Big] \\ &- 2(1+\nu)\ e^{v}\cos\theta\Big\}, \end{aligned} \tag{33}$$

where ν is Poisson's ratio and

$$e^{v} = D(e^*_{kk}) \tag{34}$$

is the volume strain associated with the isolated inclusion and corresponds to the observable relaxation volume strain associated with the point-defect in a finite crystal; the angle θ is defined in Fig. 2. The first term in Eqn (33), of $O(\rho^{-2})$, is the inhomogeneity interaction; the last term, proportional to e^V, is the usual first-order size effect interaction and the remaining terms arise from the tetragonality of the misfit strain e^*_{ij}; all these last terms vanish if the misfit strain is a pure dilatation $('e_{ij}=0)$. The stress dependent interaction terms are:

$$\left.\begin{aligned} E^{11}_{int} &= \frac{\mu V_I b}{12\pi(1-\nu)} \frac{\cos\theta}{\rho} \left\{A[(1-2\nu) - 6\sin^2\theta] + 2C(1-2\nu)\right\}, \\ E^{22}_{int} &= \frac{\mu V_I b}{12\pi(1-\nu)} \frac{\cos\theta}{\rho} \left\{A[(1-2\nu) + 6\sin^2\theta] + 2C(1-2\nu)\right\}, \\ E^{33}_{int} &= \frac{\mu V_I b}{6\pi(1-\nu)} (1-2\nu) \frac{\cos\theta}{\rho} \left\{-A+C\right\}. \end{aligned}\right\} \quad (35)$$

The sensitivity of the interaction energy to the direction of the applied tensile stress τ has important implications for the response of irradiated materials to such stress. To see that this is so we consider the inclusion representation of the dumbell interstitial. It is known from measurements of the defect modulus in irradiated materials (22,7) that such interstitials have a characteristic soft shear mode and therefore it is appropriate to represent the dumbell interstitial by a small inclusion that is soft in shear $(\mu^* \to 0)$ but highly incompressible $(\kappa^* \to \infty)$ with a positive relaxation volume strain $(e^V \gtrsim 1)$. With this, albeit rather extreme, representation,

$$\left.\begin{aligned} A &\to \frac{15(1-\nu)}{7-5\nu}, \\ B &\to 0, \\ C &\to -\frac{3(1-\nu)}{1+\nu} \end{aligned}\right\} \quad (36)$$

and (33) and (35) become

$$E^{0}_{int} = \frac{\mu V_I b^2}{8\pi^2\rho^2} \frac{(1-2\nu)(25\nu-8)}{(7-5\nu)(1-\nu)} + \frac{\mu V_I b}{3\pi} \frac{(1+\nu)}{(1-\nu)} e^V \frac{\cos\theta}{\rho}, \quad (37)$$

and

$$\left.\begin{aligned} E^{11}_{int} &= -\frac{3\mu V_I b}{2\pi} \frac{\cos\theta}{\rho} \frac{(4-3\nu+5\nu^2)}{(7-5\nu)(1+\nu)}, \\ E^{22}_{int} &= \frac{3\mu V_I b}{2\pi} \frac{\cos\theta}{\rho} \frac{(1+8\nu-5\nu^2)}{(7-5\nu)(1+\nu)}, \\ E^{33}_{int} &= -\frac{6\mu V_I b}{\pi} \frac{\cos\theta}{\rho} \frac{(1-2\nu)}{(7-5\nu)(1+\nu)} \end{aligned}\right\} \quad (38)$$

respectively, where, for simplicity, we have replaced $\cos^2\theta$ and $\sin^2\theta$ in Eqns (33) and (35) by their mean value of ½.

In this continuum model all these energies must be terminated at some core 'cut-off' radius, say $\rho = b$. It is clear from Eqn (37) that, in E^o_{int} for the interstitial, the size effect term always greatly exceeds the inhomogeneity term (except for an extremely small angle around $\theta = \pm 90°$) and therefore the latter term can be safely ignored. From Eqn (38) we see that only E^{22}_{int} is positive and thus only when the uniaxial tension τ is parallel to the Burgers vector of the edge dislocation (Fig. 2(b)) does the presence of such an external stress *increase* the total interaction energy. It follows that more interstitials will migrate to dislocations disposed relative to τ as in Fig. 2(b) than to dislocations disposed as in Figs. 2(a) or 2(c). Such an excess flux of interstitials to such oriented edge dislocations will cause the body to extend preferentially in the direction of the external tension; this is the fundamental source of the so called Stress Induced Preferred Absorption (SIPA) mechanism of irradiation creep (7).

In contrast to the dumbell interstitial a vacancy is probably best represented in the continuum model by a small inclusion that is soft in compression $(\kappa^* \to 0)$, with only a small defect modulus in shear $(\mu^* \simeq \mu)$ and with a negative relaxation volume strain that is smaller in magnitude than that of the interstitial $(e^v \sim -0.5)$. With this representation

$$A \to 0$$

$$B \to 1,$$

$$C \to \frac{3(1-\nu)}{2(1-2\nu)}$$

and Eqns (33) and (35) then yield for the vacancy

$$E^o_{int} = -\frac{\mu V_I b^2}{16\pi^2\rho^2} + \frac{\mu V_I b}{3\pi}\left(\frac{1+\nu}{1-\nu}\right)\frac{e^v\cos\theta}{\rho} \tag{39}$$

and

$$E^{11}_{int} = E^{22}_{int} = E^{33}_{int} = \frac{\mu V_I b}{4\pi}\frac{\cos\theta}{\rho}\ . \tag{40}$$

For the vacancy it is clear from Eqn (39) that the inhomogeneity term is only negligible compared to the size-effect term for $\rho \gtrsim 2b$. However, the external stress yields a contribution that is independent of the orientation of the edge dislocation and therefore we expect little contribution to a SIPA mechanism of irradiation creep from asymmetries in the vacancy fluxes to the dislocations present in the medium.

5. DISCUSSION

In a typical irradiation creep experiment the ratio of external stress to shear modulus for steel (τ/μ) as indicated in Eqn (32) is ~ (1/500) (15) and therefore from Eqns (37) and (38) the presence of such a stress only modifies the stress-free interstitial-dislocation interaction energy $E^o{}_{int}$ by a similar amount. Nevertheless, even though this modification is small, its asymmetry is sufficient to provide a quantitive explanation of the observed and technologically important irradiation creep at low irradiation doses. At higher irradiation doses, when the onset of swelling occurs we frequently observe an apparently correlated increase in the irradiation creep. A possible source of this increase may lie in the modification of $E^o{}_{int}$ due to solute segregation at the dislocation sinks as derived in Section 4.1. We intend to explore these synergistic effects and thereby simultaneously exploit both the solute segregation results in Sections 4.1 and the stress effects discussed for the intrinsic point-defects in Section 4.2.

The continuum model as used here to obtain the interaction energies between atomic sized point-defects and dislocations is inadequate in several particulars: it takes no account of the dispersive nature of the host material and the microscopic symmetry of the point-defect is not present. We therefore would only expect this model to yield reliable interaction energies when the point-defect dislocation separation is reasonably large. Nevertheless the model, for intersititials and vacancies at any rate, does yield interaction energies that are in substantial agreement with the results of atomic simulation studies even when the point-defect is close to the core of the dislocation. The atomic simulation of these interactions in iron by Bullough & Perrin (23) indicated that very close to the core of a dislocation $(\rho \sim b)$ the size-effect is dominant for the interstitial whereas the inhomogeneity interaction is dominant for the vacancy in precise agreement with the expections from the present model suggested by Eqns (37) and (39) respectively. We therefore consider that the present model could also yield qualitatively useful insight concerning the effects of segregation of solute on the interactions with other solute or intrinsic point-defects. In any case it is difficult to see how such phenomena could be studied with a discrete atomic model without requiring an impossibly large atomic ensemble.

REFERENCES

(1) Bullough, R. & Newman, R.C. (1970) Rep. on Prog. in Physics, 33, 101.
(2) Cottrell, A.H. & Bilby, B.A. (1949) Proc. Phys. Soc., 62, 49.
(3) Ham, F.S. (1959) J. Appl. Phys., 30, 915.
(4) Bullough, R. & Newman, R.C. (1962) Proc. Roy. Soc., A266, 198.
(5) Brailsford, A.D. & Bullough, R. (1972) J. Nucl. Mater., 44, 121.
(6) Brailsford, A.D. & Bullough, R. (1981) Phil. Trans. Roy. Soc., A302, 87.
(7) Bullough, R. & Willis, J.R. (1975) Phil. Mag., 31, 855.
(8) Eshelby, J.D. (1957) Proc. Roy. Soc., A241, 376.
(9) Eshelby, J.D. (1951) Phil. Trans. Roy. Soc., A244, 87.
(10) Chang, L. et al. In: Proc. of the 30th Internat. Field Emission Symposium (Aug. 1-5, Philadelphia, Pa., USA).
(11) Bullough, R. & Newman, R.C. (1963) Prog. in Semiconductors, 7, 100.
(12) Newman, R.C. (1982) Rep. on Prog. in Physics, 45, 1163.
(13) Marwick, A.D. (1981) Nucl. Instr. and Methods, 182/183, 827.
(14) Mansur, L.K. & Wolfer, W.G. (1978) J. Nucl. Mater., 69/70, 825.
(15) Bullough, R. et al. (1981) In: Dislocation Modelling of Physical Systems, eds. M.F. Ashby et al., 116-141. Oxford: Pergamon Press.
(16) Willis, J.R. (1981) Advances in Applied Mechanics, 21, 1.
(17) Willis, J.R. (1982) In: Mechanics of Solids, eds. H.G. Hopkins and M.J. Sewell, 653-686. Oxford: Pergamon Press.
(18) Kneer, G. (1965) Physica Status Solidi 9, 825.
(19) Love, A.E.H. (1944) A Treatise on the Mathematical Theory of Elasticity. New York: Dover Publications.
(20) Willis, J.R. (1977) J. Mech. Phys. Solids, 25, 185.
(21) Dundurs, J. & Mura, T. (1964) J. Mech. Phys. Solids 12, 177.
(22) Holder, J. et al. (1974) Phys. Rev., B 10, 363.
(23) Bullough, R. & Perrin, R.C. (1968) In: Dislocation Dynamics, eds. G.T. Hahn and A. Rosenfield, 174-190. New York: McGraw-Hill.

MATERIAL FORCES AND CONFIGURATIONAL FORCES IN THE INTERACTION OF ELASTIC SINGULARITIES

F.R.N. NABARRO

Department of Physics, University of the Witwatersrand,
Johannesburg, South Africa

ABSTRACT

It is generally held that the forces acting on and between elastic defects are "configurational forces" analogous to, but entirely different in kind from, the Newtonian mechanical forces which may act on the material in which the elastic defects are present. However, the rate at which the process of Cottrell locking of a dislocation by solute atoms occurs is successfully calculated by assuming that a mechanical force acts on the diffusing solute atoms which is equal and opposite to the configurational force which these atoms exert on the dislocation. Such a mechanical force cannot appear in linear elasticity. It is shown that it does appear when second-order elastic theory is applied to a simple model system. The force is accompanied by a quadrupole of forces which produce the backflow of the matrix which is displaced as the solute atom moves.

The details of the calculation will be published elsewhere. This brief account emphasizes the contributions of J.D. Eshelby to the development of the ideas involved.

1. INTRODUCTION

The idea of "the force on an elastic singularity" was first introduced by J.D. Eshelby in a paper of that title (1). Eshelby calculated the energy of an elastic body containing the defect and some other elastic field. Cross terms appear in that energy. "The negative gradient of the total energy with respect to the position of an imperfection may conveniently be called the force on it. This force, in a sense fictitious, is introduced to give a picturesque description of the energy changes, and must not be confused with the ordinary surface or body forces acting on the material." The term "configurational force" has been used to describe forces of this kind, which act on configurations such as dislocations. Eshelby maintained this distinction rigorously. When he calculated the force between two parallel disclinations in a nematic liquid crystal and found that "the supposedly configurational force on a disclination in a nematic is in fact a real force exerted on the core of the dislocation by the surrounding medium", he was very disturbed, and he circulated the draft of his paper (2) to many colleagues before publishing it.

The distinction between mechanical and configurational forces is not merely one of terminology. Consider an edge dislocation moving through a solid solution. As the dislocation passes a misfitting solute atom, the solute "plucks" the dislocation, vibrations are set up in the dislocation line and are ultimately dissipated. The result is a viscous drag on the moving dislocation (3). One may consider the converse effect. As the dislocation passes a solute atom, the dislocation exerts on it the force which would after a long time lead to Cottrell locking. The solute atom is elastically displaced for a short time, and as it is displaced and then returns to its original position when the dislocation has passed, it radiates elastic waves which are ultimately dissipated. This is an independent mechanism of viscous drag on a moving dislocation. I calculated this drag and found it to be much smaller than that produced by the plucking of the dislocation. I sent my calculation to Eshelby for comment, and he replied that it seemed to him mostly correct and probably original. However, he said that the true answer was precisely zero. No real mechanical force acted on the solute atom as the dislocation passed it, and the atom was not displaced (except that it was convected in the displacement field of the dislocation). I cited Cottrell locking as evidence for a real mechanical force, and the matter was still unsettled when Eshelby died. What had become clear was that such a mechanical force could not exist in homogeneous linear elasticity, because both the dislocation and the misfitting solute atom can be represented by prescribed displacements on the surfaces of cuts in the medium.

Formally, such a force is expected to appear in quadratic elasticity, in which the displacements and not only the energy density contain cross terms in two applied elastic fields. The solutions of problems in quadratic elasticity usually contain the higher-order elastic constants, whereas the force between a dislocation and a misfitting solute atom depends only on the ordinary linear elastic constants. However, as may be seen in the book of Green and Adkins (4), the results of quadratic elasticity differ from those of linear elasticity by two kinds of terms. The one kind involves the higher-order elastic constants. The other kind depends only on the geometry of finite deformations, and the components of displacement and stress appearing in them involve only the linear elastic constants. This happens, for example, in problems of plane strain. In fact, one of the few "interaction" problems treated in quadratic elasticity is that of the change in the twist of a crystal whisker containing an axial screw dislocation when the whisker is stretched. Eshelby (5) had shown that the result given by elementary arguments was wrong both in sign and in magnitude, and that the correct answer involves only the linear elastic constants.

2. THE MODEL

It is convenient to consider an edge dislocation interacting with a line of misfitting solute atoms lying parallel to its own line, and in the conventional "extra half plane". To reduce the problem to one of plane strain, we replace the line of atoms by a dilated cylinder (Fig. 1). The Burgers vector of the dislocation is b, the relaxed radius of the cavity into which the cylinder fits is also b and $\delta A = \alpha b^2$ is its (constrained) increase in area. The cylinder axis is distant h from the dislocation. We take the small expansion parameter ε of the theory of Green and Adkins to grow from 0 to b/h, so that the strengths of the dislocation and of the cylinder of dilatation attain the values stated when $\varepsilon = b/h$. To simplify the calculation, we take the material of the matrix to be incompressible.

3. THE CALCULATION

The calculation follows the prescription of Green and Adkins. The elastic field is examined in the neighbourhood of the boundary of the dilated cylinder. The complementary functions which appear in the theory are determined uniquely to the lowest non-vanishing order in b/h by the requirements that the displacement and a derivative of the stress function are continuous across the boundary between the dilated cylinder and the matrix. The elastic field first calculated has terms independent of h, whereas the configurational force per unit length conventionally calculated is, with the usual notation,

$$F = b\mu\delta A/\pi h^2. \tag{1}$$

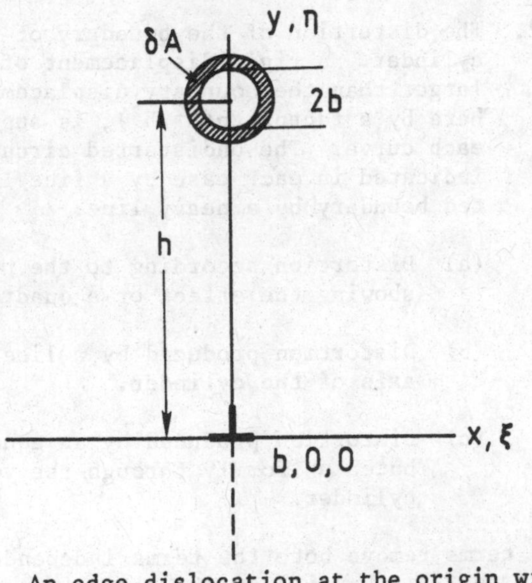

Fig. 1. An edge dislocation at the origin with Burgers vector (b,0,0) and a line of dilatation with a constrained increase in area δA over an original area πb^2. The dislocation cut is taken along the negative y axis. The "small parameter" is $\varepsilon = b/h$.

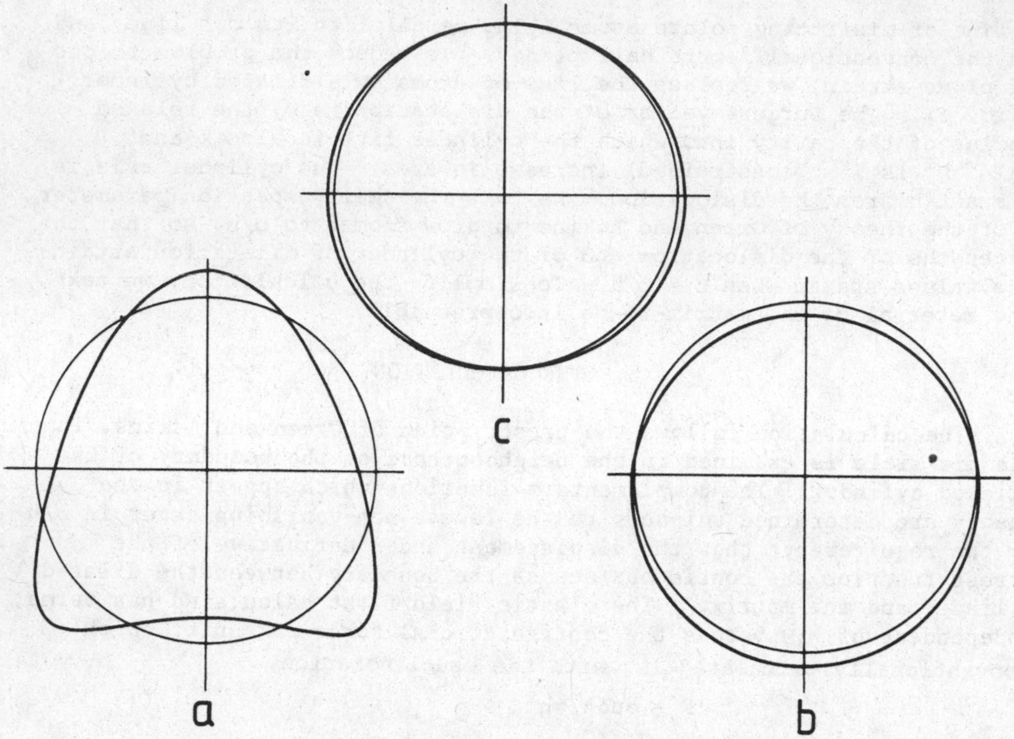

Fig. 2. The distortion of the boundary of the dilated cylinder. A rigid displacement of the cylinder, larger than the boundary displacements illustrated here by a factor $\ell n(h^2/b^2)$, is suppressed in each curve. The undistorted circular boundary is indicated in each case by a fine line, the distorted boundary by a heavy line.

(a) Distortion according to the present calculation, showing the effect of a quadrupole of forces.

(b) Distortion produced by a line force along the axis of the cylinder.

(c) Distortion produced by an equal force distributed uniformly through the volume of the cylinder.

The complementary terms remove both the terms independent of h and those proportional to 1/h. Details of the calculation will be published elsewhere.

4. THE RESULT

The leading term in the second-order displacement of the material on the boundary of the dilated cylinder, the dislocation line being held fixed, is a radial displacement away from the dislocation line of magnitude

$$V = \frac{b\delta A}{8\pi^2 h^2} \ln \frac{h^2}{b^2} \, . \qquad (2)$$

This coincides with the leading term in the displacement which would be produced on the boundary of the dilated cylinder by a mechanical force F given by Eqn. (1), acting either along the axis of the cylinder or uniformly distributed over its area. This result confirms the view that there is a real mechanical force between the dislocation and the dilated cylinder.

Either the force acting along the axis or the distributed force produces in addition to the displacement given by Eqn. (2) a displacement of order $b\delta A/8\pi^2h^2$ on the boundary, with components which are independent of the polar angle θ around the axis of the cylinder, or vary as sin 2θ or cos 2θ. The second-order elastic displacement has components which are similar but larger. They correspond to a quadrupole of forces, the central force assisting the force of Eqn. (1), while the opposing forces produce a backflow in which the material of the matrix is forced into the region vacated by the dilated cylinder. This corresponds to the physical process of the diffusion of a misfitting atom; the solute atom does not move independently through the matrix, but changes place with atoms of the matrix. The displacements corresponding to the second-order calculation, and to localized or spread forces in the linear theory, are shown in Fig. 2.

REFERENCES

1. Eshelby, J.D. (1951). The force on an elastic singularity. Phil. Trans. R.Soc. Lond. A244, 87-112.

2. Eshelby, J.D. (1980). The force on a dislocation in a liquid crystal. Phil.Mag. A 42 359-367.

3. Schwarz, R.B. (1980). Dislocation stress-velocity dependence in alloys. Phys.Rev. B 21 5617-27.

4. Green, A.E., and Adkins, J.E. (1970). Large Elastic Deformations (2nd ed.). Oxford; Clarendon Press.

5. Eshelby, J.D. (1958). Discussion (pp. 130-132) to Gomer, R., Some observations on field emission from mercury whiskers (pp. 126-130). In: Growth and Perfection of Crystals, ed. R.H. Doremus, B.W. Roberts and D. Turnbull, pp. 130-132. New York: John Wiley.

MATERIAL BALANCE LAWS IN ONE-DIMENSIONAL STRENGTH-OF-MATERIALS THEORIES

G. HERRMANN

Division of Applied Mechanics
Department of Mechanical Engineering
Stanford University
Stanford, California 94305

One of the important tasks of fracture mechanics is the calculation of crack extension forces (i.e., energy release rates) which via Irwin's relation determine the stress intensity factors. Quite often extensive analytical and/or numerical work may be involved in accomplishing this task.

It is the aim of the present contribution to show that some problems of determining crack extension forces may be formulated on the level of strength-of-materials theories. The basis for this application is formed by material balance (or conservation) laws which can be established for strength-of-materials theories. Since we shall be concerned with one-dimensional continua only, rotational invariance in material space is not applicable.

A suitable application of the virtual work theorem in (one-dimensional) material space, expressing translational invariance, leads to the expression

$$B' + b = 0$$

For a bar in tension

$$B = W + V - Nu'$$

$$b = n'u - (EA)'u'^2/2$$

for a bar in torsion

$$B = W + V - T\theta'$$

$$b = t'\theta - (GJ)'\theta'^2/2$$

and for a bar in bending

$$B = W + V - M\psi' - Qw'$$

$$b = q'w - (EI)'\psi'^2/2$$

Here W is the strain energy (per unit of length), V is the potential of applied forces, N is the axial force, u the axial displacement, n the shear load, t the tortional load, q the normal load (per unit of length), T is the twisting moment, M the bending moment, Q the transverse shear force, θ the angle of twist, ψ the angle of rotation, and w the transverse displacement. EA, GJ and EI are the extensional, torsional and bending stiffnesses, respectively. A prime designates differentiation with respect to the axial coordinate x.

The invariance of a self-similar transformation in material space leads to the relationship

$$(Bx - H)' = 0$$

where $H = -Nu/2$ for extension, $H = -T\theta/2$ for torsion and $H = -M\psi/2 - 3Qw/2$ for bending.

To obtain the stress intensity factor, one may postulate first (say for bending), that the energy release rate K^2/E is $-2\,[B] - M^2\,[C]$ where [ ] indicates a jump, which leads to

$$K = \frac{6M(\xi)}{eh^{3/2}}\, g\left(\frac{a}{h}\right) \quad , \quad g\left(\frac{a}{h}\right) = \sqrt{\frac{1}{3}\left(\frac{I}{I^*} - 1\right)} \quad , \qquad \text{(A)}$$

for a beam of rectangular cross-section eh. Here I is the nominal moment of inertia and I* is the same quantity at the cracked cross-section. This simple formula can be applied to beams with symmetrical central and edge cracks. The results are compared graphically (in Fig. 1) with results obtained in ref. (1). It is seen that the curves are very close to each other.

Much of the work reported here has been carried out in collaboration with R. Kienzler, visiting scholar in the Division of Applied Mechanics, Stanford University.

Reference:

(1) Benthem, J. P. & Koiter, W. T. (1973). Asymptotic approximations to crack problems. In: Mechanics of Fracture I, Sih, G. C. (ed.), Noordhoff Inter. Publ. Leyden, pp. 131-178.

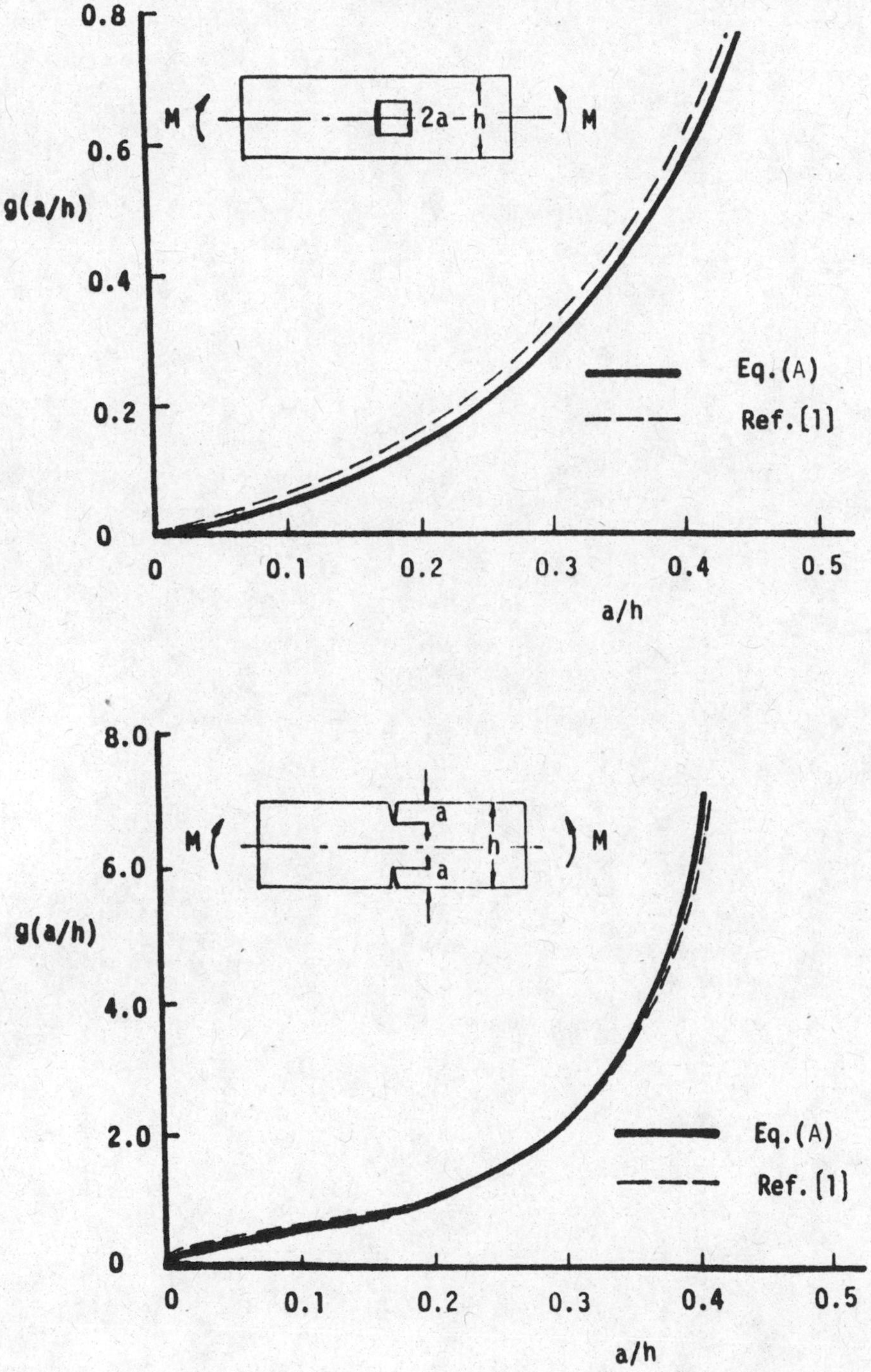

FIG. 1. Stress-intensity factor coefficient g versus nondimensional crack length a/h for the cracked beams shown in the inset

ON THE ELASTIC T-TERM

G. E. CARDEW, M. R. GOLDTHORPE, I. C. HOWARD and A. P. KFOURI

Department of Mechanical Engineering, University of Sheffield, U.K.

ABSTRACT

A theorem due to Eshelby is proved and its use in calculating values for the elastic T-term is demonstrated by the development of two analytical results and a range of numerical values for common specimen geometries.

1. INTRODUCTION

Conventional elastic fracture mechanics attempts to characterize the state of the crack tip material by the value of the applied stress intensity factor K_I. It has been known for some time that the levels of load in elastic plastic material at which this is accurate may be rather low. Furthermore, the work of Larsson and Carlsson (1) and of Rice (2) has shown that the inclusion of the elastic T-term, the second term in Williams' (3) eigen-function expansion of the elastic field at the crack tip, should allow the state of the crack-tip material to be characterized well at much higher levels of load. Edmunds and Willis (4) have subsequently put these ideas on a consistent mathematical foundation, and several workers have demonstrated deviations from conventional LEFM control due to "biaxial" effects (for example (5-7)) whilst others (1,8) have calculated T for various geometries of specimen.

However, there is, to our knowledge, no method in the literature of calculating T that has the simplicity and accuracy of calculating stress intensity factors via the J-integral. This paper proves a previously unpublished theorem due to Eshelby which gives T in terms of the values of path independent integrals. The theorem is easy to use in conventional finite element programs, and numerical results derived with its aid are presented and discussed.

In what follows use will be made of the properties of path independent integrals, and our notation is laid out here. A linear elastic body containing a crack is loaded by forces remote from the crack tip, these forces being denoted by the symbol F. The crack may, or may not, be loaded by the presence of a point force f at its tip, this force being reacted at places remote from the tip.

The static part

$$P_{\ell j} = W\delta_{\ell j} - u_{i,\ell}\sigma_{ij} \tag{1}$$

of the elastic energy-momentum tensor, when suitably integrated around any singularity in the elastic field, gives the force on it (9), the components of that force being

$$F_\ell = \int_\Sigma P_{\ell j} dS_j \, . \tag{2}$$

Here, W is the strain energy density, $\frac{1}{2}\sigma_{ij}e_{ij}$, and σ_{ij}, e_{ij} and u_i are the components of stress, strain and displacement. In Eqn (2) Σ is a surfaoe surrounding the singularity and $dS_j = n_j dS$ where $\underline{n}$ is the outward pointing normal to Σ.

If we now confine our attention to plane deformation, symmetric loading of the crack tip and the component of Eqn (2) in the 1-direction, the line of the crack, we obtain

$$J(F,f) = \int_c P_{1j} ds_j \, , \tag{3}$$

modifying slightly the notation of Rice (10) to indicate the dependence of the J-integral on the independent modes of loading. The path c is a line in the plane of the body and $ds_j = n_j ds$ where ds is arc length along c. An integral that will prove useful later is the M-integral (11)

$$M = \int_c x_\ell P_{\ell j} ds_j \, , \tag{4}$$

originally discovered by Günther (12).

This work was born during coffee-time discussions between the present writers, J. D. Eshelby and B. A. Bilby. It arose from Eshelby's recognition of the potential usefulness of the T-term in extending conventional elastic fracture mechanics, and from his desire to encapsulate the discussion of it in the simple and elegant way that is so evident in the rest of his work. The conception of the theorem given in §2 is his alone, as is the spirit of its proof. To him also is due the idea of the results given in §3. The computations reported in the rest of the paper were performed after his death.

2. A THEOREM DUE TO ESHELBY

Suppose that the external loading produces fields σ_{ij}, e_{ij} and u_i in the body and the application of the point force f at the tip produces fields σ_{ij}', e_{ij}' and u_i'. The simplest form of the theorem arises when the force at the tip is reacted by tractions on the boundary equivalent to those due to the application of the force at the tip in an infinite body. Then the field of stress due to the crack tip force is purely radial (13) with the radial stress

$$\sigma_{rr} = -\frac{f\cos\theta}{\pi r} \tag{5}$$

in terms of polar coordinates r and θ at the crack tip.

The substitution of these fields into Eqns (1) and (2) gives

$$J(F,f) = \int_c \{\tfrac{1}{2}(\sigma_{ik} + \sigma'_{ik})(e_{ik} + e'_{ik})\delta_{ij}$$

$$- (\sigma_{ij} + \sigma'_{ij})(u_{i,l} + u'_{i,l})\}\, ds_j \quad , \tag{6}$$

which may be rewritten as

$$J(F,f) = J(F) + J(f) + J_x \tag{7}$$

where

$$J_x = \int_c \{\tfrac{1}{2}(\sigma_{ik}e'_{ik} + \sigma'_{ik}e_{ik})\delta_{ij} - \sigma_{ij}u'_{i,l} - \sigma'_{ij}u_{i,l}\}ds_j \tag{8}$$

is the integral associated with the "cross-terms", and J(F) and J(f) are the values of the J-integral for each of the applied loadings alone. We note that

$$J(f) = 0 \tag{9}$$

for the method of reaction we have chosen because the r^{-1} stress field is too singular to contribute anything other than zero to J.

Now

$$\sigma_{ik}e'_{ik} = \sigma'_{ik}e_{ik} \tag{10}$$

because of the relations between stress and strain in elastic materials, so that

$$J_x = \int_c (\sigma'_{ik}e_{ik}\delta_{ij} - \sigma_{ij}u'_{i,l} - \sigma'_{ij}u_{i,l})ds_j \ . \tag{11}$$

Near the tip of the crack

$$\sigma_{ij} = Kr^{-\frac{1}{2}}g_{ij}^{(-\frac{1}{2})}(\theta) + Tg_{ij}^{(0)}(\theta) + O(r^{\frac{1}{2}}) \ , \tag{12}$$

$$\sigma'_{ij} = fr^{-1}g_{ij}^{(-1)}(\theta) \ , \tag{13}$$

and the only terms that contribute to J_x when the contour c is shrunk enough is the cross-term between T and f. This may be evaluated by putting

$$\sigma_{ij} = T\delta_{1i}\delta_{1j} , \qquad e_{11} = \frac{T}{E'} ,$$
$$e_{22} = -\frac{\nu(1+\nu)T}{E} , \qquad e_{12} = 0 \tag{14}$$

where $E' = E/(1-\nu^2)$. Then

$$J_x = \int_c \left[\left(\sigma'_{11} \frac{T}{E'} - \sigma'_{22} \frac{\nu(1+\nu)T}{E} \right) n_1 - Tu'_{1,1}n_1 - \sigma'_{ij}u_{i,1}n_j \right] ds . \tag{15}$$

But

$$u'_{1,1} = \frac{\sigma'_{11}}{E'} - \frac{\nu(1+\nu)\sigma'_{22}}{E} , \tag{16}$$

so that the first three terms in the integrand of Eqn (15) cancel out, leaving

$$J_x = -\int_c \sigma'_{ij}n_ju_{i,1}ds = -\frac{T}{E'}\int_c \sigma'_{1j}n_jds . \tag{17}$$

The force associated with integrating the traction around the crack tip balances the applied force f so that

$$J_x = \frac{Tf}{E'} . \tag{18}$$

The theorem is, therefore,

$$\frac{Tf}{E'} = J(F,f) - J(F) . \tag{19}$$

If the crack-tip force is reacted differently, the J-integrals associated with the conjoint application of the applied load and the crack-tip force will contain a contribution from the image field of the tip-force in the surface of the body. So will that due to the application of the tip-force alone. We denote these new values by $J_1(F,f)$ and $J_1(f)$. The field, Eqn (13), of the force near the tip becomes

$$\sigma'_{ij} = fr^{-1}g_{ij}^{(-1)}(\theta) + K_fr^{-\frac{1}{2}}g_{ij}^{(-\frac{1}{2})}(\theta) + T_fg_{ij}^{(0)}(\theta) + O(r^{\frac{1}{2}}) \tag{20}$$

involving now the stress intensity factor K_f and T-term T_f arising from the image of f in the surface of the body. There is now a contribution involving K_f^2 and fT_f to $J_1(f)$, and a cross-term between K and K_f in the two fields. This latter contribution is $2KK_f/E'$ and the theorem becomes

$$J_1(F,f) = J(F) + J_1(f) + \frac{Tf}{E'} + \frac{2KK_f}{E'} . \tag{21}$$

3. SOME SIMPLE RESULTS

3.1 POINT LOADING OF A CRACKED HALF-SPACE

The easiest proof of this result follows by using the M integral in the way indicated by Eshelby (14). Freund (15) shows that the contribution to M from the arc Γ (Fig. 2) at the vertex of a wedge of angle 2α is

$$M(\Gamma) = -\frac{1}{E'}\left\{\frac{F_a^{\,2}}{(2\alpha + \sin 2\alpha)} + \frac{F_t^{\,2}}{(2\alpha - \sin 2\alpha)}\right\} . \tag{22}$$

The contribution from small arcs near the point of application of loads P and Q is, therefore,

$$-\frac{1}{E'}\left\{\frac{(P/\sqrt{2} + Q/\sqrt{2})^2}{\pi/2 + 1} + \frac{(Q/\sqrt{2} - P/\sqrt{2})^2}{\pi/2 - 1}\right\} , \tag{23}$$

and that from the large arc is $(2P+f)^2/\pi E'$ when proper account has been taken of the direction of the normal. The circle around the crack tip contributes $aJ(P+Q,f)$.

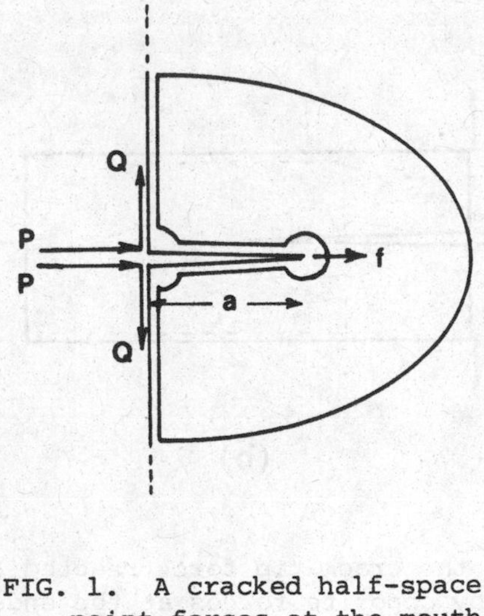

FIG. 1. A cracked half-space with point forces at the mouth

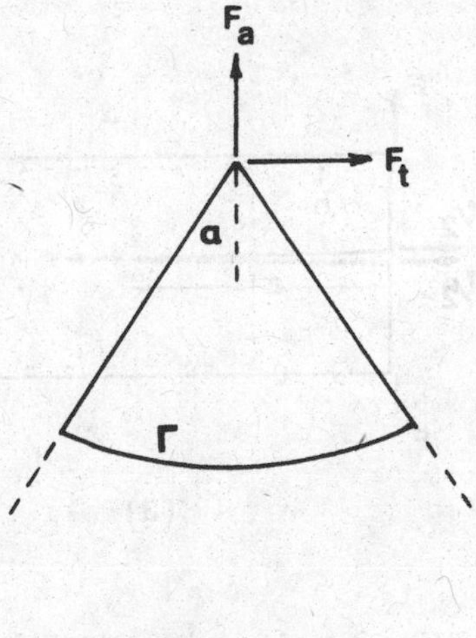

FIG. 2. A wedge with point forces at its tip

Thus

$$aJ(P+Q,f) = \frac{1}{E'} \left\{ -\frac{(2P+f)^2}{\pi} + \frac{2(P+Q)^2}{\pi+2} + \frac{2(P-Q)^2}{\pi-2} \right\} , \tag{24}$$

and so

$$aJ_x = -\frac{4Pf}{\pi E'} , \tag{25}$$

whence

$$T + \frac{2KK_f}{f} = -\frac{4P}{\pi a} . \tag{26}$$

The right hand side is just the stress that would be present at the crack tip if there were no crack.

The stress of Eqn (5), centred at the tip of the crack, produces a set of tractions on the surface of the half-space. K_f can be found by changing the sign of these tractions, calculating the stresses in the half-space induced by them, and using the resulting loading on the line of the crack as input to Bueckner's (16) approximate weight function for this geometry. The outcome is that

$$K_f = \frac{2.238f}{\pi^2 a^{\frac{1}{2}}} . \tag{27}$$

Eqn (26) is consistent with Eshelby's (17) observation on the effect of loading at the surface by the (compressive) forces P only. Due to the generation of appropriate forces Q by impingement at the origin, K is zero and only $T = -4P/\pi a$ remains at the crack tip.

3.2 THE DOUBLE-CANTILEVER BEAM

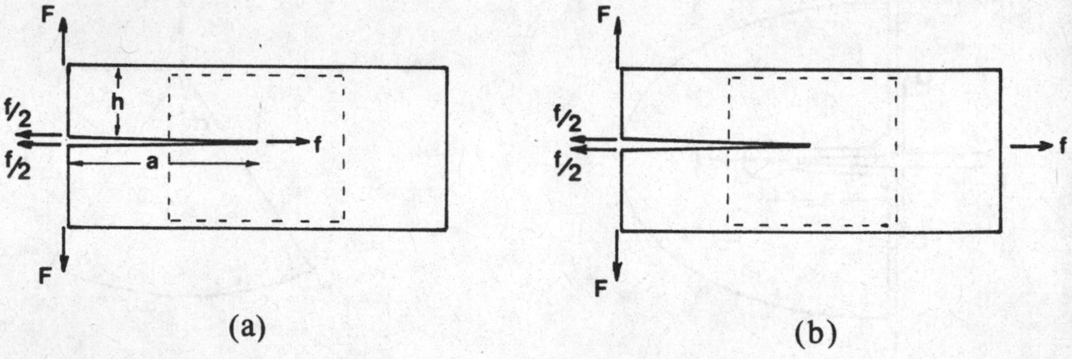

FIG. 3. A double-cantilever beam with (a) the crack-tip force reacted at the mouth of the crack and (b) stretched by opposite forces at the ends

In the two loadings of Fig. 3 the values of J_x are the same, since the only contribution comes from the vertical parts of the path in the arms of the DCB. Case (b) is a DCB loaded by two moments giving

$$J_x = \frac{2KK_f}{E'} = \frac{2.2\sqrt{3}}{E'} \frac{Fa}{h^{3/2}} \cdot \frac{2\sqrt{3}fh}{4h^{3/2}} = \frac{6Ffa}{E'h^2} \tag{28}$$

on use of the standard result (18) for the stress intensity factor in a DCB. (There is no force at the tip to cross with T in this case.) Returning now to case (a) we have

$$\frac{Tf}{E'} + \frac{2KK_f}{E'} = \frac{6Ffa}{E'h^2} \ . \tag{29}$$

K_f is non-zero in this case, and it appears that its value cannot be found in a simple form. However, dimensional considerations show that it is directly proportional to $f/h^{\frac{1}{2}}$ so that Eqn (29) can be re-arranged as

$$T = \frac{\alpha Fa}{h^2} \ . \tag{30}$$

Our numerical work, reported in the following section, suggests that α is about 4.8.

4. RESULTS AND DISCUSSION

We have performed calculations using both forms of the theorem on a number of specimen geometries whose general configuration is shown in Fig. 4. They were chosen so that comparison could be made with the results of others, in particular with the detailed and extensive calculations of Leevers and Radon (8). Our work confirms the general trend of (1), (8) and (20) by following them closely in a qualitative sense. Apart from Table 1, which demonstrates the effect of changes due to the use of different values in Eqn (21), the tables have been obtained with the aid of Eqn (19). Changing from one form of the theorem to the other produces changes in the mean value of

$$B = \frac{T(\pi a)^{\frac{1}{2}}}{K} \tag{31}$$

of about ± 0.01 in general. (In just one of our computations this difference was increased to ± 0.02.) Apart from those cases where B is small, this computational error is well within the accepted relative error for engineering calculations.

Our computations using Eqn (21) for the double-cantilever beam are given in Table 1. They contain a severe test of the accuracy of Eqn (21) because changes in the value of the applied load and a movement of the reaction point to the force f at the crack tip independently change the values of all the terms in the expression for T. Not only that, moving the reaction point from one end of the specimen to the other changes the sign of K_f. The effect of all these changes is

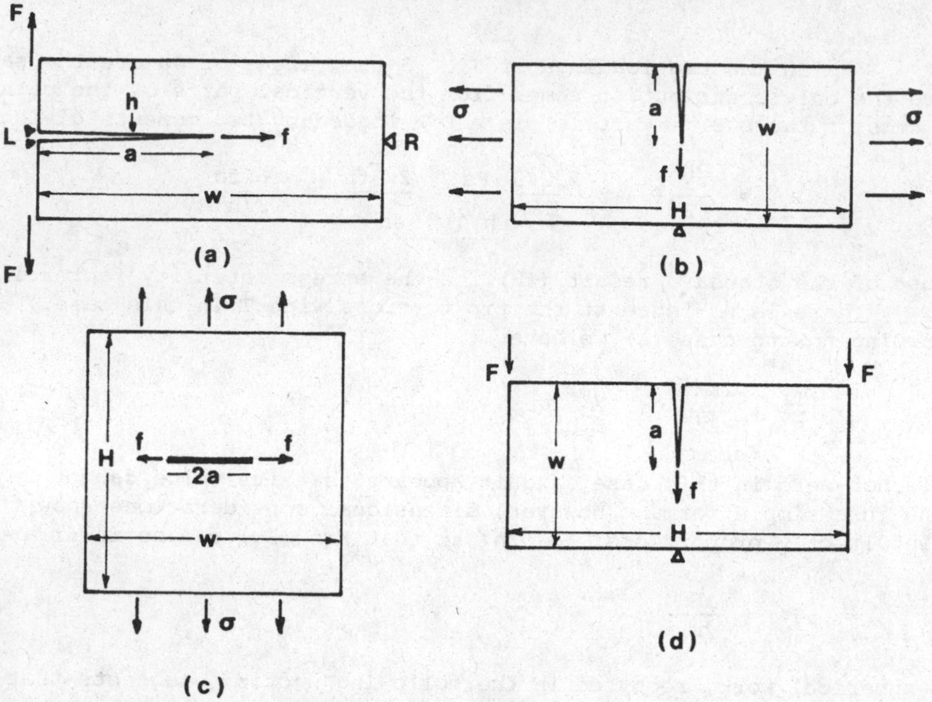

FIG. 4. The specimen geometries for which computations have been made (a) DCB, (b) SENT, (c) CN, (d) SENB

a variation of less than ± 0.01 in the mean value of B. Table 1 also contains the results of Leevers and Radon for this geometry. Agreement between our work and theirs is good except for the most extreme geometry, that for which w/h = 20. Their method of solution is to represent the field within the body as a truncated infinite series of crack-tip eigenfunctions and to find the coefficient of the series by satisfying one of the variational principles of elasticity. These methods of solution diverge if they are expressed in a coordinate system which is very different from that in which the boundary of the specimen is a level surface (19). Furthermore, the rate of convergence deteriorates markedly as the point of divergence is approached. It may be, therefore, that in this extreme geometry the rate of convergence is inferior to that in the others, for it is here that the boundary, being long and thin, differs most from a circle, the shape favoured naturally by the crack-tip eigenfunctions.

Some results for the SENT specimen are presented in Table 2 where an extra comparison is made with data (20) computed by a semi-analytical method. For values of a/w of 0.4 and less all the data agree to reasonable accuracy. Outside the range the present work agrees with Leevers and Radon to a large degree, but not with Howard (20). This suggests that there may be an error in Howard's procedures that manifests itself at larger values of a/w.

Data for the CN and SENB specimens are given in Tables 3 and 4 which contain two comparisons with the work of Larsson and

TABLE 1. Results for the DCB specimen with a/w = 0.5

w/h	Reaction point	F	J(F) $\times 10^2$	J(F,f) $\times 10^2$	J(f) $\times 10^2$	K_f	B	B Ref. (8)
5	R	0.2	4.606	2.056	-1.172	-0.133	2.829	2.942
10	R	0.1	3.560	1.972	-1.249	-0.130	4.795	4.783
15	R	0.05	1.838	0.542	-1.262	-0.130	6.231	
20	R	0.03	1.125	-0.085	-1.265	-0.129	7.399	6.364
10	L	0.1	3.560	12.542	3.962	0.036	4.808	4.783
20	L	0.1	12.504	28.340	3.960	0.036	7.404	6.364
20	L	0.02	0.500	6.838	3.960	0.036	7.414	6.364

Computational data: h = 100, f = 2

All the results in Tables 1-4 were obtained with E = 0.91 and ν = 0.3

TABLE 2. Results for the SENT specimen

	H/w = 2		H/w = 5		H/w = ∞
a/w	B	B Ref. (8)	B	B Ref. (8)	B Ref. (14)
0.3	-0.382	-0.370	-0.386	-0.412	-0.349
0.4	-0.287	-0.270	-0.294	-0.283	-0.227
0.5	-0.172	-0.145	-0.177	-0.137	-0.049
0.6	-0.029	0.013	-0.032	0.063	0.161

TABLE 3. Results for the CN specimen

	H/w = 1			H/w = 2	
a/w	B	B Ref. (8)	B Ref. (1)	B	B Ref. (8)
0.3	-1.032	-1.026		-1.016	
0.5	-1.039	-1.043	-1.04	-1.052	-1.06
0.6	-1.047	-1.051		-1.103	

Computational data for Tables 2 and 3: $\sigma = 0.8$, $w = 4$, $f = 2$

TABLE 4. Results for the SENB specimen

a/w	B	B Ref. (8) (approximate)	B Ref. (1)
0.3	-0.151	-0.14	
0.4	-0.032	0.02	
0.5	0.087	0.18	0.058
0.6	0.216	0.38	

Computational data: $H = 8$, $w = 2$, $F = 0.1$, $f = 2$

Carlsson (1). Where there is disagreement we appear to be closer to Larsson and Carlsson than to Leevers and Radon, but the scarcity of the data means that the question of which group of results is most accurate must remain open for the present.

It is, for us at least, especially pleasing that Eqns (19) and (21) should be not only elegant but computationally efficient. In those finite element programs that allow the automatic specification of distributed tractions the stiffness matrix needs to be inverted once only for each complete calculation of T. Eshelby's theorem therefore does provide a means of calculating T that has the accuracy and speed associated with calculating stress intensity factors via J-integrals.

REFERENCES

(1) Larsson, S. G. & Carlsson, A. J. (1973). Influence of non-singular stress terms and specimen geometry on small-scale yielding at crack tips in elastic-plastic materials. J. Mech. Phys. Solids, 21, 263-277.

(2) Rice, J. R. (1974). Limitations to the small-scale yielding approximation for crack-tip plasticity. J. Mech. Phys. Solids, 22, 17-27.

(3) Williams, M. L. (1957). On the stress distribution at the base of a stationary crack. J. Appl. Mech., 24, 109-114.

(4) Edmunds, T. M. & Willis, J. R. (1977). Matched asymptotic expansions in non-linear fracture mechanics - III. In-plane loading of an elastic perfectly-plastic symmetric specimen. J. Mech. Phys. Solids, 25, 423-455.

(5) Brown, M. W. & Miller, K. J. (1982). Mode I fatigue crack growth under biaxial stress at room and elevated temperature. Paper presented at ASTM Int. Symposium on Biaxial/Multiaxial Fatigue, San Francisco.

(6) Smith, E. W. & Pascoe, K. J. (1983). The behaviour of fatigue cracks subject to applied biaxial stress - a review of experimental evidence. Fatigue of Engng Materials and Structures, 6, 201-224.

(7) Leevers, P. S., Culver, L. E. & Radon, J. C. (1979). Fatigue crack growth in PMMA and rigid PVC under biaxial stress. Engng Fracture Mech., 11, 487-498.

(8) Leevers, P. S. & Radon, J. C. (1982). Inherent stress biaxiality in various fracture specimen geometries. Int. J. Fracture, 19, 311-325.

(9) Eshelby, J. D. (1951). The force on an elastic singularity. Phil. Trans. Roy. Soc. Lond., A244, 87-112.

(10) Rice, J. R. (1968). A path independent integral and the approximate analysis of strain concentration by notches and cracks. J. Appl. Mech., 35, 379-386.

(11) Budiansky, B. & Rice, J. R. (1973). Conservation laws and energy-release rates. J. Appl. Mech., 40, 201-203.

(12) Günther, W. (1962). Über einige randintegrale der elastomechanik. Abh. der Braunschw. wiss. Ges., 14, 53-72.

(13) Love, A. E. H. (1952). A treatise on the mathematical theory of elasticity. Cambridge: Cambridge University Press, 209.

(14) Eshelby, J. D. (1975). The calculation of energy release rates. In: Prospects of Fracture Mechanics, ed. G. C. Sih, H. C. van Elst & D. Broek, pp. 69-84. Leyden: Noordhoff International.

(15) Freund, L. B. (1978). Stress intensity factor calculations based on a conservation integral. Int. J. Solids Structures, 14, 241-250.

(16) Bueckner, H. F. (1973). Field singularities and related integral representations. In: Methods of analysis and solution of crack problems, ed. G. C. Sih. Leyden: Noordhoff, 306.

(17) Bilby, B. A. (1980). Tewksbury lecture: Putting fracture to work. J. Mater. Sci., 15, 535-556.

(18) Broek, D. (1978). Elementary engineering fracture mechanics. Alphen aan den Rijn: Sijthoff and Noordhoff, 129.

(19) Howard, I. C. & Morley, L. S. D. (1967). The stresses around holes in plane sheets. Aeronautical Quarterly, 18, 367-378.

(20) Howard, I. C. (1981). A method of estimating biaxial fatigue crack growth rates. Fatigue of Engng Materials and Structures, 3, 265-270.

INITIATION AND GROWTH RATES OF SHORT FATIGUE CRACKS

K J MILLER

University of Sheffield
Sheffield S10 2TN
England

ABSTRACT

It is proposed that crack initiation occurs immediately in metal fatigue.

In engineering applications cracks initiate because of inherent defects in components and structures. In polished laboratory specimens of polycrystalline metals, cracks initiate at either grain boundaries which intersect with the free surface or at other surface irregularities. In pure-metal, single crystal studies on chemically polished specimens the development of a surface step in the first cycle is considered to constitute a notch or crack-like defect which immediately propagates a crack. The reason why no cracks have been observed in the first few cycles of a fatigue test is because they are extremely difficult to locate on a surface since they have negligible depth.

Three phases of crack growth are distinguished, (i) a short crack growth regime which is dependent on stress level and the microstructure, (ii) a longer crack growth regime applicable to long cracks (as defined by linear elastic fracture mechanics) and (iii) a transition phase in which the growth rate is the sum of (i) and (ii). Should (i) and (ii) not overlap then non-propagating cracks occur. The initial stage is not related to the birth of a crack but to the period before the attainment of a minimum crack growth rate. It follows that the fatigue limit is not a limit below which a fatigue crack can not be initiated but a limit above which a crack will continue to grow to failure.

This paper presents a brief review of short crack behaviour together with some effects of stress state and surface finish. Finally experimental results of short crack growth from torsion and tension fatigue tests are given. By excluding initiation from a lifetime assessment and invoking a fracture mechanics appreciation of short crack growth it is possible to predict the S-N curve of smooth specimens.

1. INTRODUCTION

This paper presents an alternative analysis and solution to the unsolved problem of determining the number of cycles required to initiate a fatigue crack.

The classical approach is to examine intense slip processes, extrusions-intrusions and persistent slip bands, PSBs, each of which has been associated with the crack nucleation phase. Since 1903 when Ewing and Humphrey (1) first studied bands of intense slip much effort has been devoted to examining the relationships between the formation of a stabilized cyclic deformation state, the creation and development of PSBs and finally the conditions necessary for a crack to be born. Excellent work has been presented by Brown (2) Mughrabi (3) Winter (4) Laird (5) and many other metal physicists, concerning the link to the first part of the problem, that is cyclic deformation behaviour, via dislocation kinetics, and the morphology of PSBs, but the final link remains unsolved. Most metal physicists believe that crack initiation is a consequence of (and hence eventually follows) the formation of a PSB or an extrusion-intrusion system or surface roughening. However at an invited gathering at Asilomar in 1980, some 77 years after the original work of Ewing and Humphrey, no evidence was provided to conclusively show that this hypothesis is correct. One way to settle the argument would be to observe the growth rate of small cracks and to determine the propagation phase of lifetime; the remaining lifetime may then be assumed to be the initiation phase.

This paper uses the above procedure and considers the evidence that there is no crack initiation phase in polycrystalline materials and that the fatigue lifetime N_f of a component or structure or specimen is the consequence only of the number of cycles to propagate a crack, N_p, from the very first cycle of loading, i.e. in the equation.

$$N_f = N_i + N_p \qquad [1]$$

the number of cycles to initiate a crack, N_i, is zero. The advantage of such an approach is that it focuses attention on the behaviour of geometrical discontinuities such as micro-notches and in particular the mechanics of crack growth and those factors which influence the propagation and non-propagation of fatigue cracks.

2. SURFACES

Most fatigue cracks initiate and propagate from a free surface. This section therefore considers the form of surface discontinuities from which cracks grow.

2.1 ENGINEERING SURFACES

The surface of a metal consists of a frequently transparent, thin oxide film (0.01 - 0.1 μm thick) containing cracks and pores;

see Fig. 1. Molecules of water, oxygen and other contaminants are weakly attached to the oxide. Below the oxide surface layer there may be an extremely hard oxide/metal zone of approximately 0.1 μm thick below which the metal itself will be in a work-hardened state to a depth of 1 - 10 μm. The surface of the metal is not dissimilar to a photograph

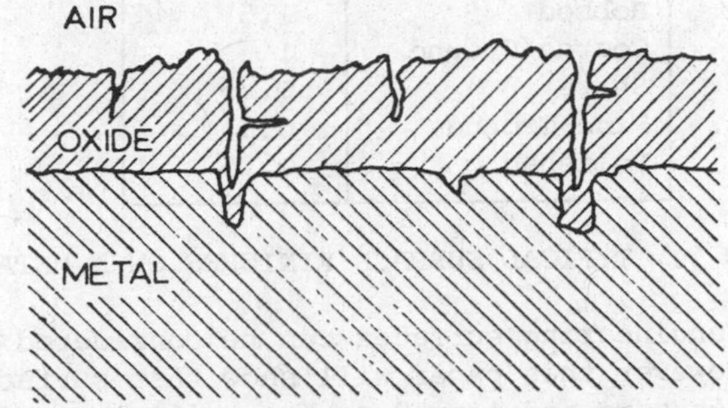

FIG. 1 A TYPICAL CROSS SECTION OF AN "ENGINEERING" SURFACE

of the earth's surface, taken from a satellite. The hills and valleys have heights/depths/shape and size all of which constitute the "texture" of the surface. Peak-to-valley heights may range from 0.05 μm for fine lapped to 1 - 10 μm for ground surfaces, to 50 μm for rough machined surfaces. Peak-to-peak spacing may vary from 0.5 μm to 5 mm. Since it

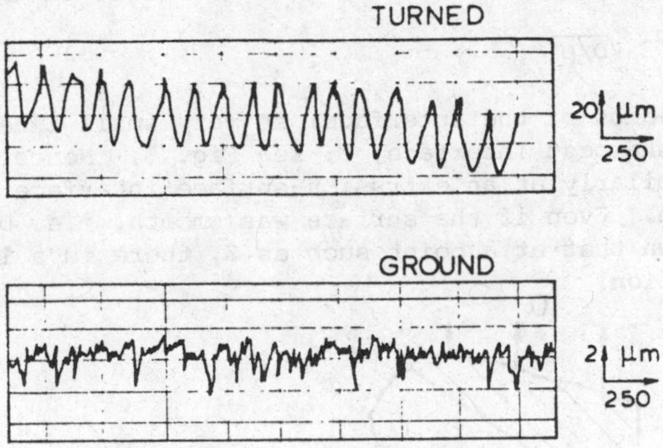

FIG. 2. TYPICAL SURFACE FINISHES

is necessary to appreciate both waviness and height differences of the texture, it is usual to portray surfaces with a greater magnification in the vertical direction than the horizontal; see Fig. 2.

The most important aspect of texture and the one most commonly used and easily measured is the average departure of height from the mean line of the filtered profile. In the UK this is known as the CLA value. Typical CLA values for metal gears finished by different processes are given overleaf.

PROCESS	CLA (μm)
Sand blasted	2.5
Milled	1
Hobbed	1.5
Coarse Ground	1
Shave	0.75
Fine Ground	0.5
Polished	0.25

TABLE 1. TYPICAL SURFACE FINISHES ON A METAL GEAR

Thus engineering surfaces can be considered as a series of micro-notches. However this paper will show that surface finishes with a CLA value between zero and 1 or 2 μm has negligible effect on fatigue lifetime. The important aspect of surface micro-notches and other surface discontinuities are that they are the immediate source of fatigue cracks.

2.2 SMOOTH POLISHED SURFACES

With regard to the surface geometrical discontinuity created at the interface between a PSB and the stronger matrix, the elastic stress concentration factor K_T is high, since in the equation

$$K_T \sim 1 + 2\sqrt{D/\rho} \qquad [2]$$

the term ρ (the radius of the interface) is very small indeed, particularly at the sharpest interface, Z; see Fig. 3. Hence the stress at Z is high. Similarly at an extrusion-surface interface the stress level will be high. Even if the surface was smooth, i.e. D = 0, Brown (2) has shown that at a point such as Z, there is a logarithmic singularity condition.

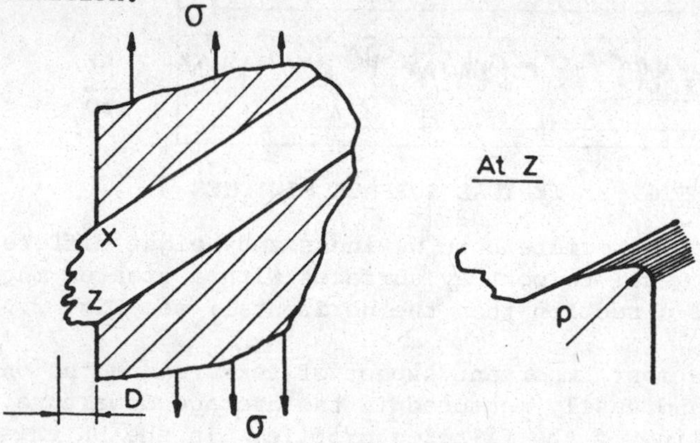

FIG. 3 STRESS CONCENTRATION AT A SLIP BAND

From work on macro-notches (6) the zone of high stress ahead of the discontinuity has a length given by $\sim 0.2\sqrt{D\rho}$. When a crack has penetrated this depth the effective crack length is $D + 0.2\sqrt{D\rho}$. The conclusions to this brief analysis are that a point such as Z can be considered to be a crack of approximate depth D especially when a small crack is created at the discontinuity, since the term $0.2\sqrt{D\rho}$ is negligible and that the greater the offset D the greater the effective crack length. It follows that a crack may be considered to be born as soon as the slip band is created; the two features are inseparable especially since in a high strain fatigue zone life can be considered to be composed of only the crack propagation phase.

A similar argument may be applied in reverse, namely if a crack or other microscopically sharp defect pre-exists on a surface it will generate its own plastic strain field upon the application of a stress. This plastic zone will spread across the grain in the form of an intense slip band which will help define the path of shear crack growth. Due to the presence of the crack and its growth the slip band will intensify to become a PSB. Again the growth of a microscopic crack is synonymous with the development of a PSB. From the foregoing the major question that remains unsolved is what is the speed of growth of the crack, or, to be more precise, what controls the speed of growth of the crack.

2.3 GRAIN BOUNDARY DEFECTS

Not only does a surface contain notch-like irregularities but in a polycrystalline metal it necessarily contains grain boundaries and triple points which create three important problems, (i) they act as stress concentrations in their own right from which slip can be nucleated (in this respect it is worth noting that even in chemically polished single crystals, a small etch pit can be the site of a PSB (3), (ii) they act as joints between volumes of incompatible deformation and (iii) since they go into the metal they provide surface-to-interior notches which will assist the formation of both slip and fracture paths especially if they are inclined to planes of maximum shear; see also section 3.

As a consequence of the change in single crystal shape during constant strain range cycling of pure b.c.c. metals [7-11] it has been suggested [12] that, for polycrystalline metals, fatigue cracks are nucleated at surface grain boundaries due to the incompatible deformation of surface grains. Furthermore Guiu has recently reported [13] on the micromechanisms of fatigue crack initiation in pure α-iron over a very wide range of frequencies (0.01 - 1000 Hz). The first tests were conducted on equiaxed polycrystals with chemically polished surfaces. At the lowest frequency and at stress levels close to the yield stress (~250 MPa) cracks initiated at grain boundaries as a result of incompatible grain deformation, but propagated in a transgranular manner. Although some cracks were located at extrusions-intrusions these did not cause failure.

At 5 Hz and similar stress levels cracks again initiated at grain boundaries despite the development of intense slip zones; these cracks then propagated transgranularly. At the lowest strain levels leading to failure at 7×10^5 cycles only grain boundary initiation sites were located. Nevertheless the grain boundaries are not intrinsically weak since propagation continued in a transgranular manner.

Significantly in the final part of Guiu's work, using specimens that had been cut from zone refined iron and which had no grain boundaries perpendicular to the direction of the applied tensile stress, fatigue cracks nucleated on the grain boundaries parallel to the stress axis due to the incompatibility of the long acicular grains. Again there was no evidence of initiation at intrusions-extrusions.

3. THREE DIMENSIONAL EFFECTS

Most research on fatigue crack initiation is concerned with pure metals, single crystals, cyclic tension and studies of deformation behaviour. The introduction of fracture mechanics analyses in recent years has caused a far wider range of fatigue research to be undertaken that involves metal alloy systems, polycrystalline specimens and multiaxial and mixed mode fracture studies more pertinent to the engineering problem. This sudden expansion of research has led to a far greater understanding of how deformation behaviour affects fracture, principally because the reverse order of the relationship has been investigated. In the present paper this new understanding is directed to the importance of crack shape and crack plane orientation.

Engineering artefacts invariably suffer complex three dimensional cyclic strains which lead to one of two possible cases of cracking that are fundamentally different, see Fig. 4 and references (14),(15) and (16). If the principal strains are $\varepsilon_1 > \varepsilon_2 > \varepsilon_3$ and the second principal strain ε_2 is normal to the free surface, CASE A cracking occurs with both stage I and stage II cracks (17) growing parallel to the surface; this strain state occurs in pure torsion. If ε_2 is parallel to the surface then CASE B cracking occurs with the stage I and stage II cracks growing away from the surface. From a fracture viewpoint this latter case is more dangerous. In pure push-pull (axial) tests in which $\varepsilon_2 = \varepsilon_3$ then both Case A and Case B crack growth occurs and the shape of the crack is semi-circular. In torsion, cracks tend to be elliptical with the major axis parallel to the surface. For equibiaxial loading the major axis of the crack shape is away from the surface. The shape of the crack is important because depending on which two-dimensional cross-sectional plane is being examined the size of an observed initiated crack will range from zero to full crack depth.

During the initial phase it can be expected that the crack will grow from a surface stress concentration, such as a triple point (13) aided by the additional notch effect created by machining, grinding or polishing marks plus the internal notch of the first grain boundary. Because of the ease of slip along the free surface in a suitably oriented

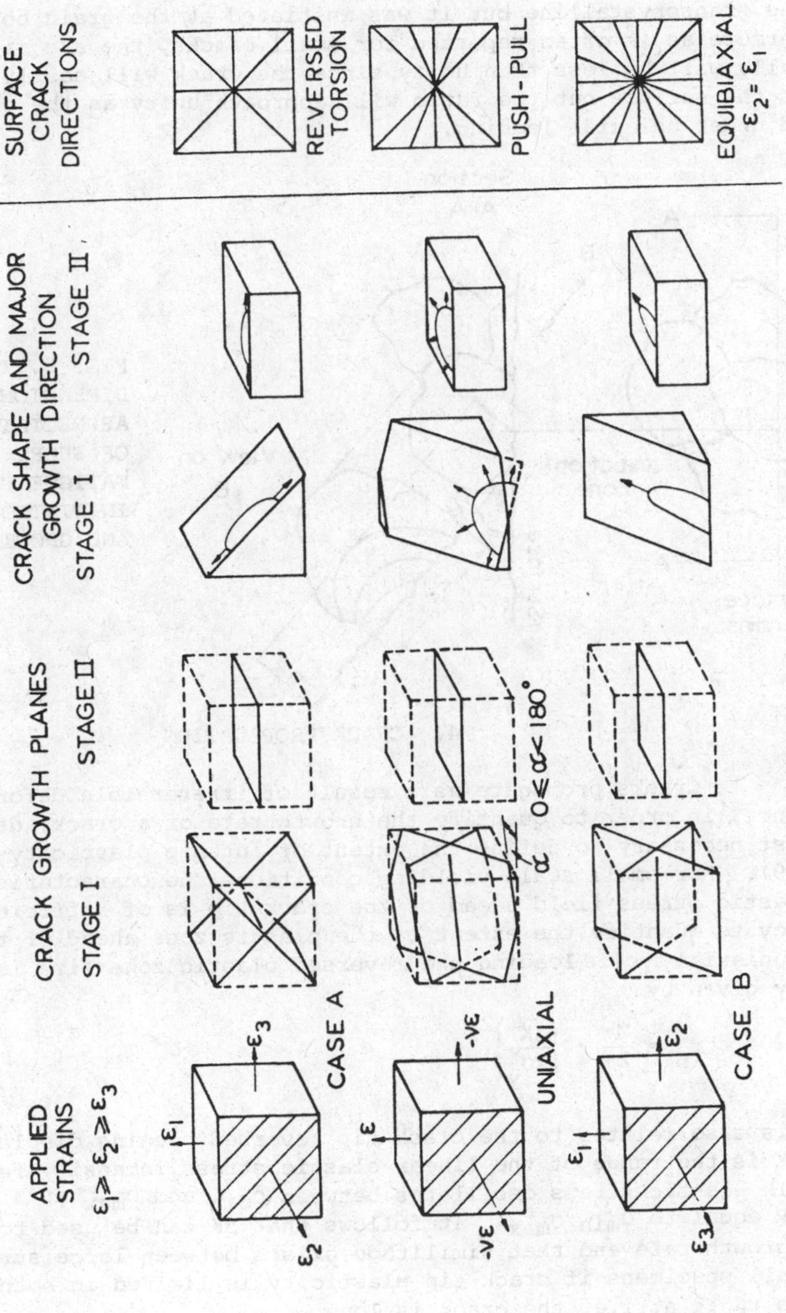

FIG. 4 EFFECT OF THREE DIMENSIONAL STRAIN FIELDS ON CRACK GROWTH AND SHAPE

grain one may therefore expect to see the crack grow in the manner depicted in Fig. 5. In this context it is worth noting that the threshold for growth of a shear crack in Mode II is less than that for a crack in pure Mode I, axial loading (18). The crack shown in section B of Fig. 5 will be transcrystalline but it was initiated at the grain boundary. Furthermore as is often reported for small cracks, the a/c ratio initially will be less than unity since the crack will easily propagate across the surface but the ratio will approach unity as the crack grows inward under uniaxial loading.

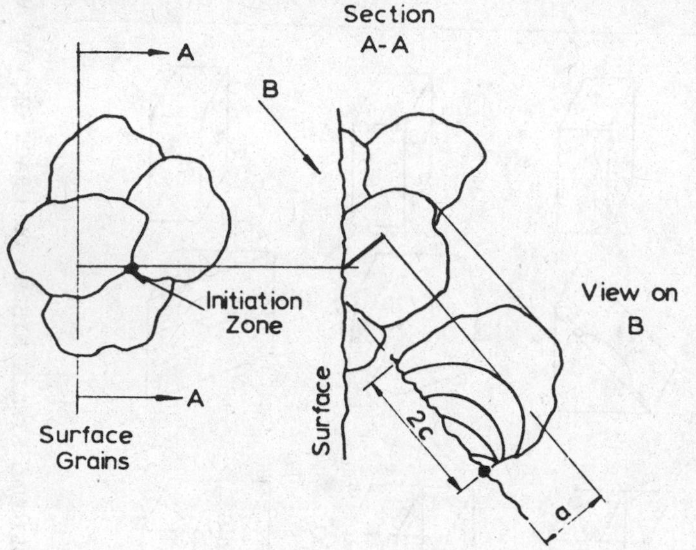

FIG. 5. THREE DIMENSIONAL APPRECIATION OF SHORT FATIGUE CRACK INITIATION AND GROWTH

4. CRACK PROPAGATION

Cracks propagate as a result of irreversible deformation and hence, in order to quantify the growth rate of a crack, da/dN, it is first necessary to define the extent of intense plasticity at a crack tip (19). For small scale yielding conditions the characterization of the elastic stress field ahead of the crack tip is of sufficient accuracy to quantify the extent of the plastic zone ahead of the crack. Under uniaxial cyclic loading the reversed plastic zone size is approximately given by

$$r_p = \frac{1}{2\pi} \cdot \left(\frac{\Delta K}{2\sigma_y}\right)^2 \qquad [3]$$

which is also related to the crack tip reversed opening displacement. Here ΔK is the range of the linear elastic stress intensity factor when the bulk elastic stress oscillates between σ_{min} and σ_{max} at a stress ratio R equal to $\sigma_{min}/\sigma_{max}$. It follows that ΔK can be used to quantify crack growth rate and that similitude exists between large structures and small specimens if crack tip plasticity is limited in both cases, that is $r_p \ll a$, i.e. the crack is long.

For structures, particularly welded structures, Linear Elastic Fracture Mechanics (LEFM) suffices to predict fatigue lifetime since the structure will inherently contain large defects to a depth greater than 0.5 mm (20). That is to say lifetime is only related to the propagation phase of a long crack and this can be calculated by integrating the long crack (LEFM) growth function illustrated in Fig. 6. The problem this paper addresses is what happens to specimens and components that do not have large inherent defects; the so-called short crack problem (21).

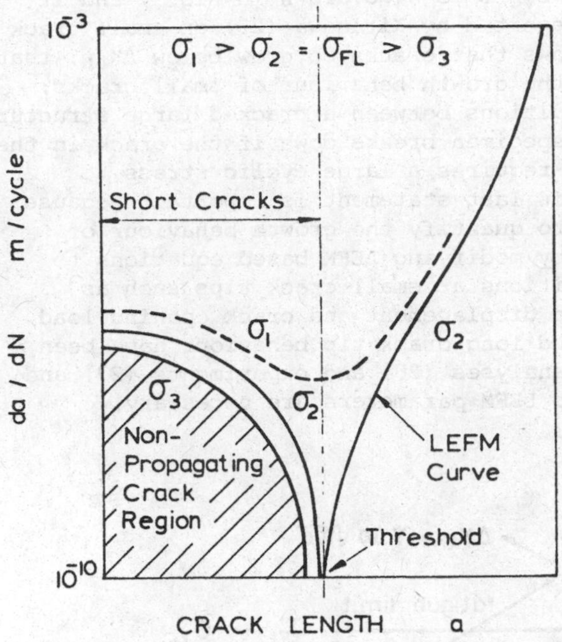

FIG. 6 SCHEMATIC OF THE BEHAVIOUR OF LONG, SHORT AND NON-PROPAGATING FATIGUE CRACKS

4.1 THE LEFM THRESHOLD AS AN UPPER BOUND SOLUTION

A result of fatigue tests on specimens containing long cracks, whose growth is monitored, is that a threshold condition $\Delta K_{th} = (Y\Delta\sigma\sqrt{\pi a})_{th}$ occurs below which the crack will not grow. As such ΔK_{th} was once thought of as a lower bound solution for use by designers; i.e. a fatigue limit for cracked components. Here Y is a geometry factor for describing the specimen and loading configuration [see (22)] and $\Delta\sigma$ is the applied stress range equal to $\sigma_{max} - \sigma_{min}$.

Unfortunately ΔK_{th} is not a lower bound solution because cracks can initiate and grow from polished surfaces if the stress is high enough. That short cracks grow below the threshold stress intensity factor ΔK_{th} was noted by Kitagawa (23) and a schematic description of short crack growth rates was provided by Smith and Miller (24); e.g. the dashed line of Fig. 6. Kitagawa plotted a curve not unlike Fig. 7 to exhibit the results of numerous tests on specimens containing defects of different sizes. Besides the limiting condition of ΔK_{th} for long cracks, Fig. 7 also shows that another bounding solution can be cited namely the fatigue limit, σ_{FL}, above which failure occurs. The value of σ_{FL} can be approximated to the cyclic yield stress of the material, σ_{yc} which can be substantially less than the monotonic yield stress σ_y in cyclic softening materials. For example in mild steel $\sigma_{yc} \sim 0.65\ \sigma_y$ (25). A consequence of the limitations of LEFM analyses of crack tip fields is that stresses should not exceed about 0.25 - 0.33 of the yield stress otherwise the plastic zone size, and hence crack growth rate, is much larger than that given by equation [3]. Hence a limiting horizontal line at $\sigma = \sigma_{yc}/3$ is also drawn in Fig. 7 and it coincides with the deviations reported by Kitagawa (23) on small crack fatigue limit studies. It follows that cracks do grow below ΔK_{th}; that an LEFM upper bound exists for the growth behaviour of small cracks; and finally that similitude conditions between a cracked large structure and a cracked small laboratory specimen breaks down if the crack in the small specimen is too small and requires a large cyclic stress ($\sigma > 0.30\ \sigma_{yc}$) to drive it. This last statement is important because many researchers still attempt to quantify the growth behaviour of small cracks using ΔK concepts by modifying LEFM based equations to characterize the different conditions at small crack tips such as plastic zone size, crack opening displacement and crack opening load. The differences between short and long crack tip behaviour have been investigated by finite element analyses (26) and experiments (27) and it is clear that alternatives to LEFM parameters are necessary.

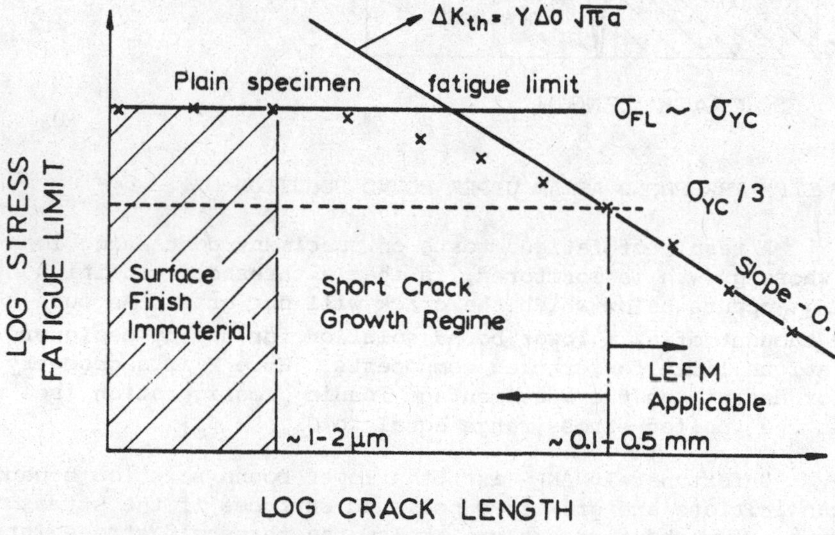

FIG. 7. BOUNDING CONDITIONS FOR FATIGUE LIMITS OF MATERIALS CONTAINING SHORT AND LONG CRACKS

5. MIXED INITIATION - PROPAGATION TORSION TESTS

By conducting experiments involving a change in stress range it is possible to gain insight into the behaviour of short fatigue cracks and their initiation. At the highest stress levels a specimen may be subjected to gross plastic straining and fatigue lifetime is dominated by the crack propagation phase; at the lowest stress level close to the fatigue limit the lifetime is dominated by the creation of a crack that will subsequently propagate to failure. If the higher stress level is the first to be applied in a stepped test then the entire initiation period is evaded and linear summations of cumulative damage ratios, $\Sigma(n/n_f)$, will be substantially less than unity; see dashed line of Fig. 8. Conversely if the higher stress levels are applied after the lower stress range cycles then summations are greater than unity (28); it is this latter type of test that we now discuss.

Should the initial fraction of lifetime applied at the lower stress range level be insufficient to create a long crack then damage is accumulating slowly, A-C, but if n_1/N_{f1} is in excess of point C a long crack exists which propagates relatively quickly to failure at D. Curves such as AC in Fig. 8 are related to the resistance of a material against

FIG. 8. CUMULATIVE DAMAGE RATIOS IN MIXED STRESS RANGE TESTS

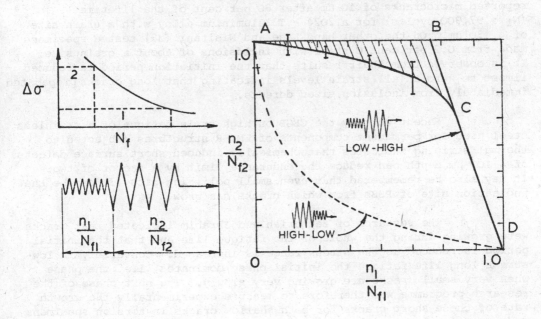

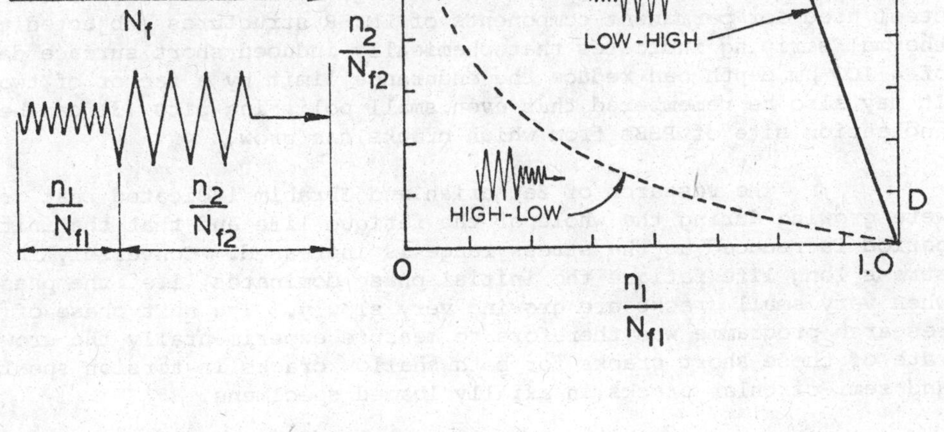

creating a long crack. Should a material cyclically strain harden, thereby reducing the plasticity that will assist crack development, then the area ABC will reduce, i.e. early damage accumulation rates will decrease. Results show (29), (30), that there is a considerable scatter of data about curves such as AC which is a reflection of (i) the stochastical nature of the initial crack growth stage (ii) the non-uniform nature of texture of engineering materials and (iii) the scatter of data in classical S-N curves especially close to the fatigue limit.

When point C is reached at the initial low stress level, at least one long crack exists. More precisely the conditions exist for the propagation of a long crack. This crack then propagates in a medium carbon steel according to the law (30),

$$da/dN = 10.62\ \Delta\gamma_p^2\ a \qquad [4]$$

until failure. Here $\Delta\gamma_p$ is the plastic shear strain range.

Although the form of the curve AC could not be accurately determined because of the scatter in the short crack growth phase, this particular research indicated that the number of cycles to produce a long crack and the size of the crack are both dependent on the cyclic stress range. From this and other work it is probable that the size of the initial short crack emanating from a smooth clean specimen should be less than the grain size. In the work of de Lange (31) at a stress level corresponding to N_f = 800,000 cycles an initial crack of about 4 μm in length existed at 40,000 cycles. The specimens were made from steel 35 CD4 of grain size ~ 100 μm and tested in push-pull. Haworth (32) reported microcracks of 10 μm after 60 per cent of the lifetime (N_f = 97,900 cycles) for a 2024 - T3 aluminium alloy with a grain size of ~ 130 μm. On the other hand Kage and Nisitani (33) tested specimens made from 0.2 per cent C steel with inclusions of about a grain size. It is observed from their results that the initiation period is reduced almost to zero at all strain levels indicating that long cracks propagated immediately from inclusion sized defects.

Recent work by the CEGB on high cycle fatigue of a stainless steel used for permanent components of LMFBR structures subjected to thermal striping indicates that chemically induced short surface defects of ~ 100 μm depth can reduce the endurance limit by a factor of two. It may also be remembered that even small polishing pits (3) can be the nucleation site of PSBs from which cracks can grow.

The research of Zachariah and Ibrahim indicated that cracks were growing during the whole of the fatigue life and that the initial period is reduced as the stress range is increased. Conversely in low-stress long life fatigue the initial phase dominates, i.e. the phase when very small cracks are growing very slowly. The next phase of the research programme was therefore to measure experimentally the growth rate of these short cracks for both shallow cracks in torsion specimens and semi-circular cracks in axially loaded specimens.

6. EXPERIMENTAL DETERMINATION OF SHORT CRACK GROWTH RATES

6.1 TORSION DATA

Recent work at Sheffield (34) relating to torsion fatigue of 0.4 per cent C steel, showed that fatigue cracks initiated and propagated in the ferrite phase which was in the form of long allotriomorphs located at prior austentite grain boundaries; crack growth was monitored using a replication technique. The ferrite-pearlite boundaries acted as strong barriers to crack propagation and caused a deceleration of growth and possible arrest. On raising the stress level slightly some but not all of the previously non-propagating cracks continued to grow either by branching along other allotriomorphs or by joining with other cracks in an adjoining ferrite colony.

Using a dislocation model for crack growth by an alternating mechanism, first on one shear plane and then its complementary plane, the speed of crack growth is given by

$$\frac{da}{dN} = f\,\frac{\tau t}{\mu}\,.\,\frac{\ell - a}{\ell}\ ;\ \text{where}\ \tau = \alpha\,\frac{\Delta\tau}{2} - \tau_o \qquad [5]$$

In the above equations

ℓ = ferrite plate length = 148 - 397 μm

τ = back stress from the dislocations

$\Delta\tau/2$ = amplitude of the reversed torsion stress = 175 - 200 MPa

τ_o = friction stress = 20 MPa

f = fraction of dislocations taking part in growth = 0.1

t = ferrite plate thickness = 8×10^{-3} mm

μ = shear modulus = 8×10^4 MPa

a = crack length

α = 0.5

The model gave a very good description (two examples are given as inserts in Fig. 9) of several experimentally determined short crack growth rates (34), all of which started quickly and then slowed down; some cracks stopped. But some cracks continued to grow and link if the stress level was above the fatigue limit. Eventually one crack system became dominant and this led to catastrophic failure. The importance of Fig. 9 and equation [5] is that short crack growth rates are seen to be a function of stress level, a microstructural feature and current crack length. In particular it can be seen that (i) the driving force for crack growth must reduce as crack length increases (ii) that there is substantial scatter in short crack growth behaviour due to microstructural texture variations but which diminishes as the crack length exceeds the LEFM threshold condition, and (iii) that cracks appear to grow immediately along the surface, at a rate which is stress level dependent only, in the thickest ferrite allotriomorph.

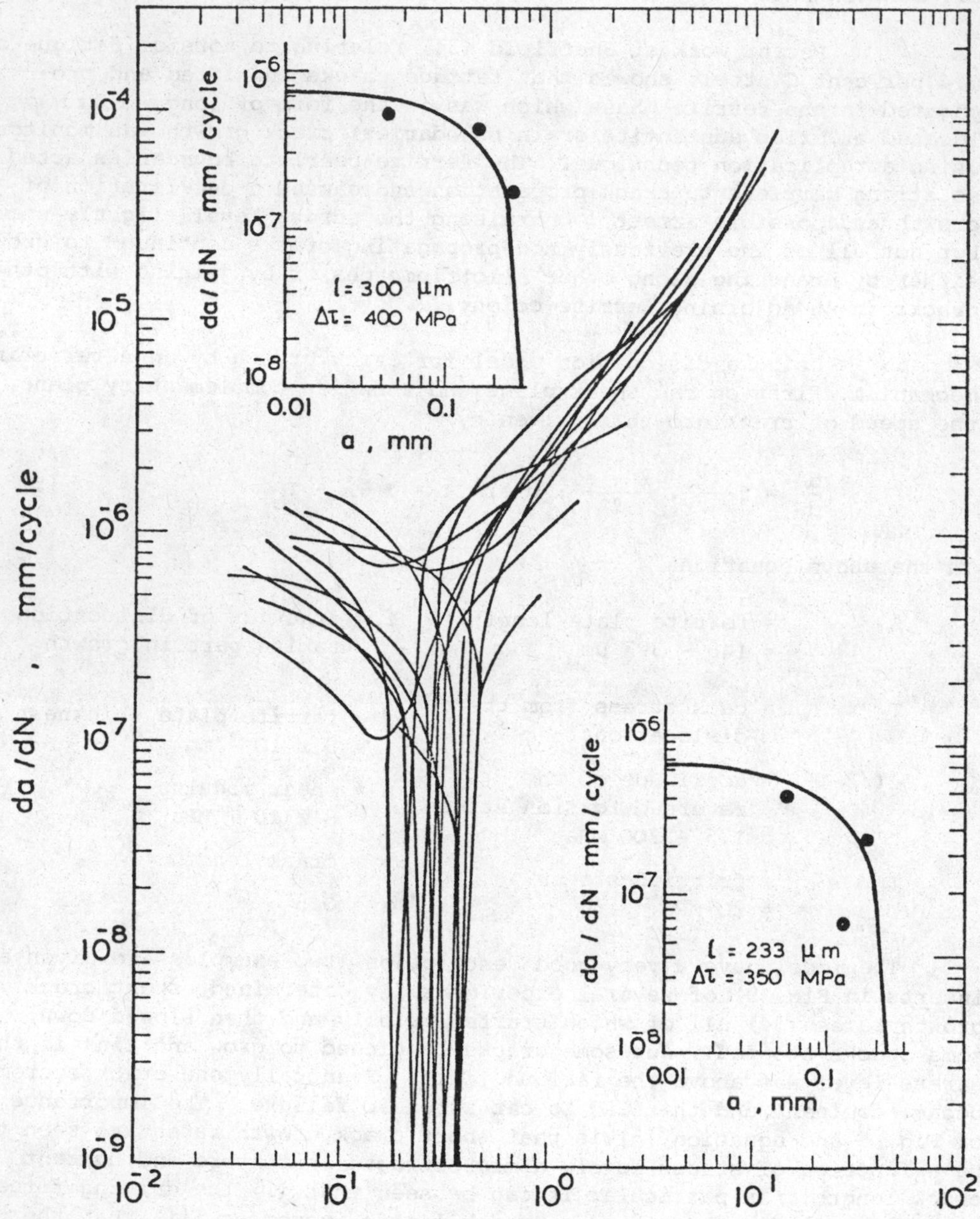

FIG. 9. EXPERIMENTAL AND THEORETICAL FATIGUE CRACK GROWTH RATES FOR BOTH SHORT AND LONG CRACKS

However the in-depth rate of these torsion shear cracks will be less than this value (see section 4) by an amount directly proportional to the ratio of the minor axis/major axis of the elliptical form of the torsion generated surface shear crack. For clarity of presentation in Fig. 9 experimental data points are omitted from the collection of curves except for the two examples given in the inserts.

6.2 TENSION DATA

In an earlier paper Hobson (35) examined the data of Lankford (36) and produced an equation of the form

$$\frac{da}{dN} = C(d - a)^{1-\alpha} a^{\alpha} \qquad [6]$$

which described the experimental growth rate of short cracks in a grain of diameter d. The best fit for α was found to be 0.4 for the 7075 - T6 aluminium alloy. Similarly for the long crack growth data of Lankford and other workers Hobson found that

$$\frac{da}{dN} = A\,\Delta K^{2} - B \qquad [7]$$

over the low crack growth range $10^{-10} < \frac{da}{dN}$ m/cycle $< 10^{-8}$

In his own experimental studies on a medium carbon steel Hobson (37) has found that α in equation [6] is zero. Data was obtained via replicas from several short cracks in each of many specimens. The best fit equation to this data is

$$\frac{da}{dN} = 1.475 \times 10^{-35} (\Delta\sigma)^{11.49} (d - a)\ \mu\text{m/cycle} \qquad [8]$$

when a is in microns and Δσ is in MPa. For long cracks the best fit equation is

$$\frac{da}{dN} = 32.3\ (\Delta\varepsilon_t)^{2.46} a - 3.176 \times 10^{-3}\ \mu\text{m/cycle} \qquad [9]$$

here $\Delta\varepsilon_t$ is the total strain range.

The influence of these equations on crack growth behaviour is shown schematically in Fig. 10. While the upper bound on short crack growth is considered a constant, equal to the grain diameter, d, the slope of the initial curve (and hence the initial crack growth rate when a = 0 on the first application of loading) increases with applied stress range. For long crack behaviour the threshold crack length a_{th} for zero growth rate is a function of applied stress range. If $a_{th} > d$ failure does not occur but non-propagating cracks may be found. If $a_{th} < d$ failure will occur and three zones of crack growth may be distinguished. The first zone is the short crack growth regime given by equation [8], the last zone is the long crack growth regime given by equation [9] and the intermediate zone is that given by the summation of equations [8] and [9], i.e. plasticity for growth comes from two sources.

It will be noted that the form of equations [6] and [8] to describe early crack growth rates indicates that as the short crack length increases and the plastic zone size decreases (PSB decreases in length), crack growth rates must decrease and become zero when a = d (21) (35).

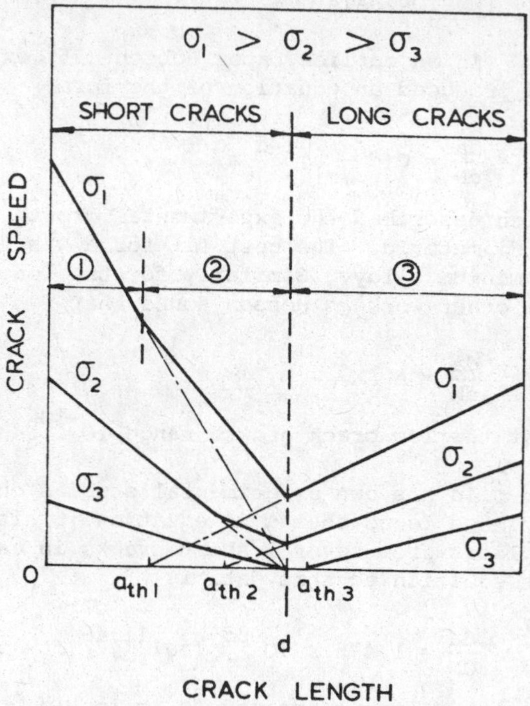

FIG. 10. SCHEMATIC OF VERY LOW GROWTH RATES FOR SHORT AND LONG FATIGUE CRACKS

6.3 LIFETIME PREDICTIONS

Specific examples of lifetime calculations using equations [8] and [9] are provided in Table 2; and Fig. 11 indicates the range of crack growth rates derived from experiments, for both short and long cracks, which were the basis of the equations that permitted the calculation. For short life fatigue the long crack growth regime dominates and the time to produce a long crack may be ignored. For long life fatigue the short crack growth phase becomes more important although the long crack growth regime is still an important constituent in the summation; see Table 2.

From equations [8] and [9] an S-N curve can be constructed by integrating the crack growth rates over the three phases of growth depicted in Fig. 10. Such a curve is provided in Fig. 12, on which is superimposed the experimental results. Note the close agreement between upper limit experimental lives, using specimens from which replicas were taken, and the predicted curve. Also superimposed on this figure are the zones 1, 2 and 3 applicable to Fig. 10 and listed in Table 2.

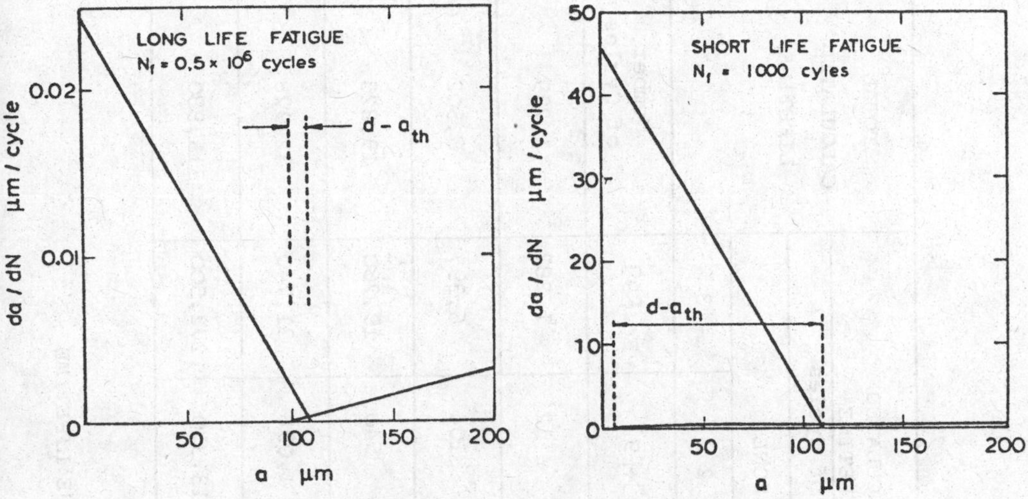

FIG. 11. EXAMPLES OF LOW CRACK GROWTH RATES FOR SHORT CRACKS IN (i) LONG LIFE AND (ii) SHORT LIFE FATIGUE

It is clear that in push-pull, high stress (high strain) low cycle fatigue, even with polished specimens, long crack growth mechanics is dominant, be it characterized in LEFM or EPFM terms (19). At low stresses just above or below the fatigue limit the short crack growth regime dominates.

For those tests which apparently ended prematurely, see the dashed lower curve of Fig. 12, several possibilities exist, including

(i) life reduction due to the presence of long defects
(ii) nucleation of failure cracks in grains with d > 108 µm
(iii) the insensitivity of a_{th} in large plastic zones
(iv) combinations of (i), (ii) and (iii).

However another, possibly more important reason is that the upper curve of Fig. 12 has been derived from replica information which necessitated the periodic resting of the specimen during which time strain ageing at crack tip plastic zones could occur thereby reducing crack growth rates. Work in high strain fatigue at both low and elevated temperatures has shown that rest periods can have a substantial effect on lifetime in ferrous materials (38) (39). Recently (40) it has been reported that large differences occur in low stress fatigue tests between specimens that have been rested in order to obtain surface replicas and those which were not rested, the latter having shorter lives.

CYCLIC RANGES		THRESHOLD CRACK LENGTHS		CALCULATED LIFETIME Number of Cycles			TOTAL CALCULATED LIFETIME
STRESS	STRAIN	SHORT CRACKS	LONG CRACKS	ZONE 1	ZONE 2	ZONE 3	
MPa	PLASTIC / TOTAL	d μm	a_{th} μm	Eq.[8]	[8]+[9]	Eq.[9]	Number of Cycles
998	0.0174 / 0.0223	108.5	1.13	0	12	1,287	1,299
816	0.0077 / 0.0117	108.5	5.55	1	104	6,397	6,502
700	0.0041 / 0.0075	108.5	16.24	23	540	19,260	19,823
650	0.0030 / 0.0062	108.5	26.01	90	1,098	31,740	32,928
500	0.0010 / 0.0035	108.5	107.3	30,010	13,370	272,500	315,880

TABLE 2. CALCULATIONS OF FATIGUE LIFETIME

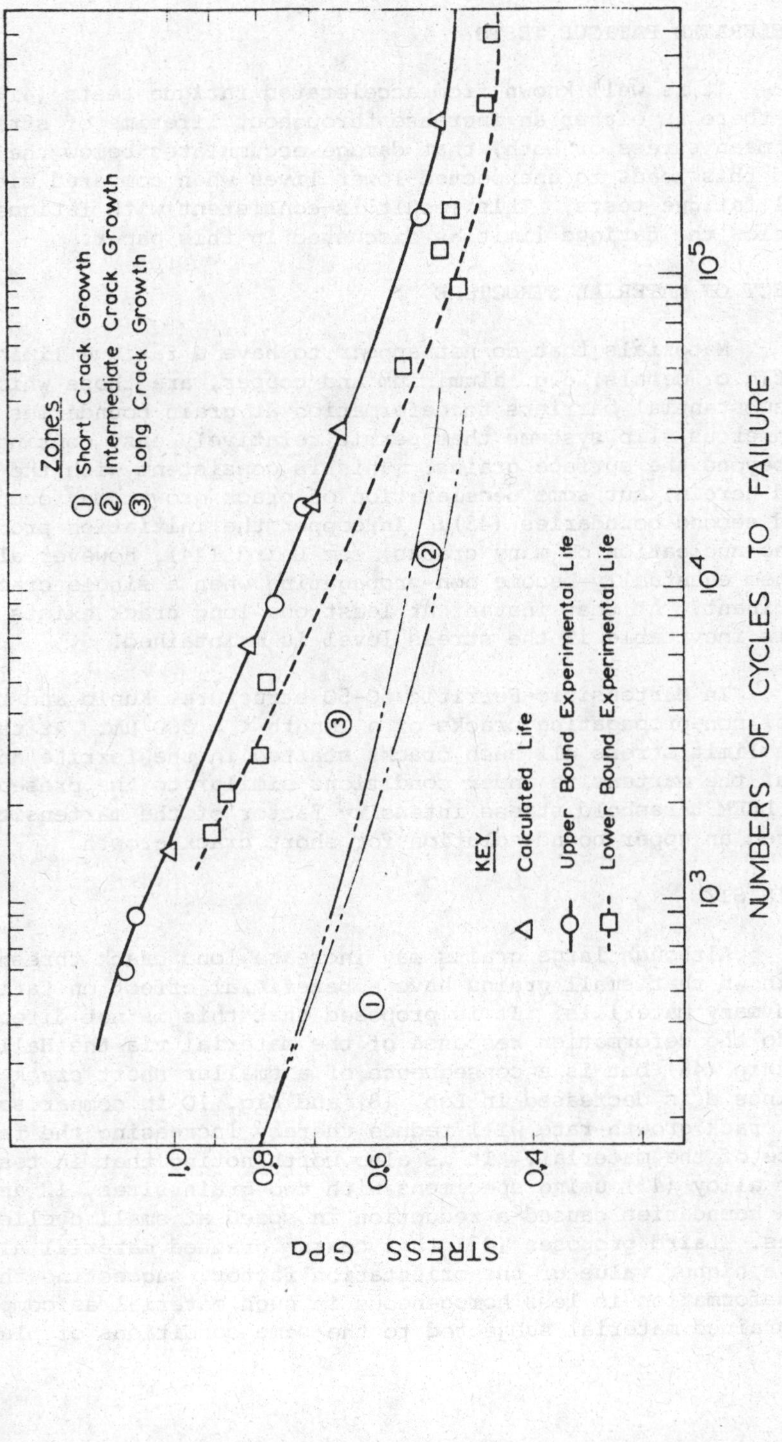

FIG. 12. PREDICTED AND EXPERIMENTAL FATIGUE LIFE DATA

7. DISCUSSION

Several important aspects of fatigue behaviour can now be discussed and rationalised by the present theory.

7.1 ACCELERATED FATIGUE TESTS

It is well known from accelerated fatigue tests (41) (42), in which there is either an increase throughout lifetime of stress range or mean stress or both, that damage accumulates below the fatigue limit and this leads to unexpected lower lives when compared with conventional fatigue tests. This result is consistent with fatigue crack growth below the fatigue limit as discussed in this paper.

7.2 EFFECT OF MATERIAL STRUCTURE

Materials that do not appear to have a fatigue limit, such as pure f.c.c. metals, e.g. aluminium and copper, are those which do not present substantial barriers to deformation at grain boundaries because of the numerous slip systems that permit relatively easy continuation of slip beyond the surface grains. This is consistent with the views expressed herein; but some deceleration of crack growth may occur at the first and second boundaries (43). In copper the initiation process involves the nucleation of many cracks, see Laird (44), however all but one of them eventually become non-propagating when a single crack becomes dominant. At that instant at least one long crack exists and failure is inevitable if the stress level is maintained.

In Martensitic-Ferritic 50-50 structures Kunio and Yamada found (45) non-propagating cracks of a length $< \sim 300$ μm. At the endurance limit stress all such cracks started in the ferrite and stopped at the martensite under conditions similar to the present work, i.e. the LEFM threshold stress intensity factor of the martensite constituted an upper bound solution for short crack growth.

7.3 GRAIN SIZE

Although large grains may increase long crack thresholds it is well known that small grains have a beneficial effect on fatigue limits in many materials. It is proposed that this is not directly related to the deformation response of the material via the Hall-Petch relationship (46) but is a consequence of a smaller short crack growth regime since d is decreased in Eqn. [8] and Fig. 10 in comparison to a_{th}. Also the crack growth rate will reduce thereby increasing the fatigue resistance of the material. It is also worth noting that in tests on an aluminium alloy (43) using specimens with two grain sizes, 12 and 130 μm, the grain boundaries caused a reduction in speed at small cyclic stress amplitudes. Laird proposes (47) that coarse grained material always requires a higher value of the orientation factor, suggesting that the plastic deformation is less homogeneous in such material as compared to fine grained material subjected to the same conditions of plastic strain and therefore the fatigue life of a fine grained material is longer than that of coarse grained material. While these considerations may be true it is suggested that a fracture mechanics approach as outlined in this paper allows a more quantitative assessment of grain size effects.

7.4 SURFACE REMOVAL

It has long been known that the progessive removal of a surface throughout a fatigue test prolongs fatigue life (48). This paper shows that this can be attributed to both short and long fatigue cracks together with the low crack growth rates associated with them at low applied stresses close to the fatigue limit. It is not a consequence of the delay to initiation processes but rather a decrease in crack growth rate because of a smaller crack after removal of the surface.

7.5 SURFACE FINISH

From Figs. 9 and 11 and equations of type [5] and [8] it can be assumed that cracks begin to grow immediately. By integrating the equations from a = O, N = O it is clear that in high stress fatigue the number of cycles to produce a long crack is negligible; see also Table 2. For low stress conditions that cause failure in ~ 10^6 cycles, i.e. close to the fatigue limit, a crack of a few microns in length is created in a number of cycles which, although measurable, is a negligible fraction of lifetime. Hence in low or high cycle fatigue, the difference in lifetime between a lapped, honed, polished, fine ground or medium ground surface is negligible and that lifetime can be calculated assuming an initial zero crack length. However a poor surface (e.g. rough turned) will contain long crack-like defects which can cause a dramatic decrease in lifetime because the entire low crack growth rate phase is eliminated.

8. CONCLUSIONS

This paper presents experimental evidence and theoretical calculations that indicate:-

1. fatigue cracks initiate immediately.
2. fatigue limits are associated with non-propagating cracks rather than with the inability of a material to initiate (give birth to) a crack.
3. crack growth rate in a surface grain decreases as the crack grows unless additional plasticity can be generated beyond the surface grain.
4. effects such as grain size, periodic surface removal, grain texture, non-propagating cracks, accelerated fatigue tests and surface finish can all be rationalized from fracture mechanics analyses of crack growth in a surface grain.

5. three crack growth phases exist, the first being the short crack growth phase, the third being the long crack growth phase, the second or intermediate phase is the combination of the first and last phases.

ACKNOWLEDGEMENTS

The author thanks many of his colleagues who have contributed to this work in discussion and several experimental programmes; in particular M. W. Brown, P. D. Hobson, M. F. Ibrahim, E. R. de los Rios, Tang, Z. and K. P. Zachariah.

Thanks are also due to the SERC for funding one of the research programmes and a CASE award studentship for P. Hobson.

REFERENCES

(1) Ewing, J. A. & Humphrey, J. C. W. (1903). Phil. Trans. Roy. Soc. A 200, 241-250.
(2) Brown, L. M. (1980). In: Proc. Int. Conf. on Dislocation Modelling of Physical Systems. eds. M. F. Ashby et al, pp 51-68. New York, Pergamon Press.
(3) Mughrabi, H. (1980). In: Proc. Fifth Int. Conf. on the Strength of Metals and Alloys, eds P. Haasen et al, 3 pp 1615-1638, New York, Pergamon Press.
(4) Winter, A. T. (1974), Phil. Mag., 30, 719-738.
(5) Laird, C. (1979). In: Proc. A. S. M. Materials Science Seminar, St. Louis, Missouri, pp 149-203.
(6) Smith, R.A. & Miller, K. J. (1977). Int. J.Mech. Sci. 19, 11-22.
(7) Nine, H. D. (1972). Phil. Mag. 26, 1409-1418.
(8) Neumann, P. (1975). Z. Metallk 66, 26-32.
(9) Mughrabi, H. & Wüthrich, Ch. (1976), Phil. Mag. 33, 963-984.
(10) Anglada, M. & Guiu, F. (1979). In: Proc. Fifth Int. Conf. on the Strength of Metals and Alloys, eds. P. Haasen et al, 2, pp 1237-1242. New York, Pergamon Press.
(11) Guiu, F. & Anglada, M. (1980). Phil. Mag. 42, pp 271-276.
(12) Mughrabi, H. (1975). Z. Metallk 66, 719-724.
(13) Guiu, F. et al (1982). Fat. Engng. Matls. and Structures 5, pp 311-321.
(14) Brown, M. W. & Miller, K. J. (1973). Proc. Instn. Mech. Engineers 187, pp 745-755 and D 229-244.
(15) Miller, K.J. & Brown, M. W. (1984). eds. A.S.T.M. Special Technical Publication STP 853, Biaxial-Multiaxial Fatigue, A.S.T.M., Philadelphia.
(16) Miller, K. J. (1982). In: Mechanical & Thermal Behaviour of Metallic Materials. Soc. Italiana di Fisica, Bologna, Italy, pp 145-164.
(17) Forsyth, P. J. E. (1961). In: Proc. Symp. Crack Propagation, Cranfield, England, p 76.
(18) Gao Hua et al (1982). Fat. Engng. Matls. and Structures 5, pp 1-17.
(19) Brown, M. W. et al (1981). In: Proc. Int. Conf. Fracture (ICF 5)

ed. D. Francois, Cannes, France. pp 891-898, Oxford, Pergamon Press.
(20) Smith, I. F. C. and Smith, R.A. (1983). Eng. Fract. Mechanics,18, pp 861-869.
(21) Miller, K. J. (1982). Fat. Engng. Matls. and Structures,5, pp 223-232.
(22) Rooke, D. P. and Cartwright, D. J. (1976). Compendium of Stress Intensity Factors, H.M.S.O. London.
(23) Kitagawa, H. & Takahashi, S. (1976). Int. Conf. Mech. Behaviour of Materials (ICM2), Americal Society for Metals (ASM), pp 627-631.
(24) Smith, R. A. & Miller, K. J. (1978). Intn. J. Mech, Sci,20, pp 201-206.
(25) Zachariah, K. P. and Miller, K. J. (1974). Engineering,214 (7), pp 563-565.
(26) Wang, T. C. & Miller, K. J. (1982). J. Fat. Engng. Matls. and Structures,5, pp 249-263.
(27) Davidson, D. L. & Lankford, J. Experimental Mechanics of Fatigue Crack Growth: The Effect of Crack Size. This volume.
(28) Ibrahim, M. F. E. & Miller, K. J. (1980). Fat. Engng. Matls. and Structures,2, pp 351-360.
(29) Miller, K. J. & Zachariah, K. P. (1977). J. Strain Analysis,12, pp 262-270.
(30) Miller, K. J. & Ibrahim, M. F. E. (1981). Fat. Engng. Matls. and Structures,4, pp 263-277.
(31) De Lange, R. G. (1964). Trans. Metall. Soc. A.I.M.E., 230, pp 644-656.
(32) Haworth, W. L. (1979). J. Fat. Engng. Matls. and Structures, 1, pp 351-362.
(33) Kage, M. & Nisitani, H. (1978). Bull. J.S.M.E., 21, p 948.
(34) De los Rios, E. R. et al (1984). J. Fat. Engng. Matls. and Structures, 7, 97-108.
(35) Hobson, P. D. (1982). J. Fat. Engng. Matls. and Structures, 5, pp 323-327.
(36) Lankford, J. (1982). J. Fat. Engng. Matls. and Structures, 5, pp 233-248.
(37) Hobson, P. D. (1984). Ph.D Thesis, University of Sheffield (in preparation).
(38) Miller, K. J. & Hatter, D. J. (1972). J. Strain Analysis, 7, pp 69-73.
(39) Miller, K. J. & Plumbridge, W. J. (1976). J. Strain Analysis, 11, pp 235-239.
(40) Brown, C. W. & Smith, G. C. (1984). J. Fat. Engng. Matls. and Structures, 7, pp 155-164.
(41) Marsh, K. J. (1965). J. Mech. Eng. Sci., 7, pp 138-151.
(42) Dengel, D. & Harig, H. (1980). J. Fat. Engng. Matls. and Structures, 3, pp 113-128.
(43) Zurek, A. K. et al (1983). Met. Trans. A., 14A, pp 1697-1705.
(44) Cheng, A. S. & Laird, C. (1981). J. Fat. Engng. Matls. and Structures, 4, 331-342.
(45) Kunio, T. & Yamada, K. (1979). A.S.T.M. Special Technical Publication STP 675, pp 342-370.

(46) Petch, N. J. (1953). J. Iron Steel Inst., 174, pp 25-28.
(47) Horibe, S. et al (1984). J. Fat. Engng. Matls. and Structures, 7, pp 145-154.
(48) Thompson, N. et al (1956). Phil. Mag., 1, pp 113-126.

ROLE OF INTERNAL STRESSES IN THE NUCLEATION OF FATIGUE CRACKS

L. M. BROWN

Cavendish Laboratory, Madingley Road, Cambridge CB3 OHE

S. L. OGIN

University Engineering Department, Trumpington Street, Cambridge CB2 1PZ

with an appendix by the late J. D. Eshelby

Theory of Materials Group, University of Sheffield,
Mappin Street, Sheffield S1 3JD

ABSTRACT

Cyclic straining leads to the production of bands of intense slip, called persistent slip bands (PSBs). Such bands are associated with macroscopic internal stresses. The pattern of stress is characteristically that of simple tension or simple compression parallel to the band, which is itself inclined by approximately 45° to the applied alternating stress. Theoretical studies show that near the free surface such a band produces very large (formally infinite) surface stresses. Computer studies give the stress and strain distribution and suggest the pattern of localised plastic flow which should occur near the surface. This leads to a model for fatigue crack initiation, which is in accord with much experimental evidence. It is possible also to begin to understand features of the slow growth of the cracks.

1. INTRODUCTION

Research into the fundamental problem of metal fatigue - the initiation and propagation of fatigue cracks in smooth specimens - is beginning to come full circle. Early work concentrated on establishing lifetimes to failure. In 1903, Ewing & Humphrey (1) demonstrated the importance of the micromechanisms of fatigue by showing that fatigue cracks are associated with bands of intense slip. These bands were later called persistent slip bands (PSBs) by Thompson et al (2). During the past 30 years, much effort has been spent investigating the structure and properties of PSBs and recently a number of investigators have tried to apply this understanding to the problem of fatigue crack initiation itself. The prospect of predicting fatigue lifetimes from an understanding of the persistent slip band structure has become a very real possibility.

PSBs are bands of concentrated slip which carry essentially all the plastic strain in a low strain amplitude fatigue test. Much of the work on PSBs has used single crystals of pure metals oriented for single slip, but there have also been studies on multiple slip configurations, ionic crystals, polycrystals and alloyed or impure materials. From these studies it is clear that the formation of PSBs is a very general phenomenon, not restricted by crystal structure. PSBs form in f.c.c. Cu and Ni and in ionic LiF and AgCℓ; b.c.c. Nb and in α-Fe at low strain rates and in h.c.p. Mg.

An enormous wealth of work exists on the surface and bulk properties of PSBs, their macroscopic and microscopic structure and their effect on the stress-strain response of a material. The reader is referred to reviews by Mughrabi (3), Laird (4) and Brown (5).

It is sufficient to note here the following properties of PSBs:

(a) PSBs form within a well-defined plastic strain range (6,7).
(b) The bands form a two-phase structure such that essentially all the applied plastic strain is concentrated in the 'soft' PSBs which are embedded in a 'harder' matrix which deforms only elastically (6).
(c) The volume fraction of PSBs is a linear function of the imposed plastic strain amplitude (6-8).
(d) The microstructure of PSBs in metals has a characteristic ladder-like appearance with areas alternating between very high and very low dislocation content. The matrix is less regular except in multiple slip configurations (e.g. in polycrystals) when the matrix can form remarkably regular labyrinth structures (9-11). In ionic crystals where a multiple slip configuration is often studied the matrix structure resembles the labyrinth seen in metals but the ladder-like structure of the PSBs is barely discernible in LiF (12) or not evident at all, as in AgCℓ (13, 14).

Early (pre-1975) crack initiation theories were formulated without the benefit of a detailed knowledge of microstructural or mechanical behaviour of PSBs. These ideas have been repeatedly reviewed (see, for example, Laird & Duquette (15); Klesnil & Lukas (16)).

Recent theories of crack initiation (5, 17-21) have focused on the role of the specimen surface at an emerging PSB. Cheng & Laird have stressed the importance of random slip irreversibilities within the bands. Brown, together with Mughrabi and co-workers, have postulated that within the PSB there are macroscopic internal stresses which are constant during cycling. Brown has emphasised the role of these stresses at the surface in promoting crack initiation. Mughrabi and co-workers have, on the other hand, suggested that the slip offset at the site of an emerging PSB provides the necessary stress concentration for crack growth.

Observations by Ogin & Brown (13) and Ogin (14) on single crystals of AgCℓ have demonstrated that there are macroscopic internal stresses associated with the PSBs in AgCℓ - in accordance with the predictions described above. However, the internal stresses in AgCℓ appear to reverse sign every half-cycle of the fatigue test, a result which conflicts with the predicted constancy of the internal stresses. Of course, we cannot rule out the possibility that reversible stresses of the type seen in AgCℓ do not occur in metals, especially in view of the very different PSB microstructures in silver chloride and copper.

In this paper we describe how reversible macroscopic stresses, similar to those observed in AgCℓ, can arise in metallic PSBs. We present maps of the near-surface stress fields which result from an internal strain parallel to the length of a band. Finally, we explore the consequences for crack initiation and demonstrate that a model based on reversible internal stresses can explain many important features of crack initiation and early growth.

2. PERSISTENT SLIP BANDS AND ASSOCIATED INTERNAL STRESSES - MODELS AND OBSERVATIONS

Several recently published models can give rise to macroscopic internal stresses associated with persistent slip bands, and several recent experimental observations are of note.

2.1 PURE SINGLE SLIP IN A PSB

Models developed by Antonopoulos et al. (17) and Essmann et al. (19), which are based on single slip, give rise to internal stresses of opposite sign.

Antonopoulos et al. (17) based their model on the observation that the dislocation content of PSBs reveals a preponderance of vacancy dipoles. In saturation, missing half-planes from the vacancy dipoles appear as extrusions on the surface of the band. As the dipole density in the bands increases, the PSBs are gradually loaded in tension. This is a constant internal stress which does not oscillate with cycling and which in the bulk produces no force on the glide dislocations.

The model proposed by Essmann et al. (19) also predicts a constant internal stress but of the opposite sign (i.e. compressive). They suggest that a combination of microscopic single slip and the annihilation of very fine vacancy dipoles in the PSB walls inclines the macroscopically effective slip plane relative to the crystallographic slip plane. As a result, there are interface dislocations between the

PSB and the matrix which introduce an excess number of atomic planes perpendicular to the primary glide direction in the PSB, which is therefore loaded in compression.

According to both these models based on single slip, an internal stress is generated within the PSB which does not oscillate during cycling.

2.2 THE PSB AS A SOFT LAYER DEFORMING BY MULTIPLE SLIP

Lynch (22) proposed a model of a PSB based on the idea that the softer PSB material is squeezed out plastically and irreversibly during the compressive part of the cycle. Here we modify Lynch's model and show that a similar idea can give rise to macroscopic internal stresses in the PSB which change sign every half cycle of the fatigue test.

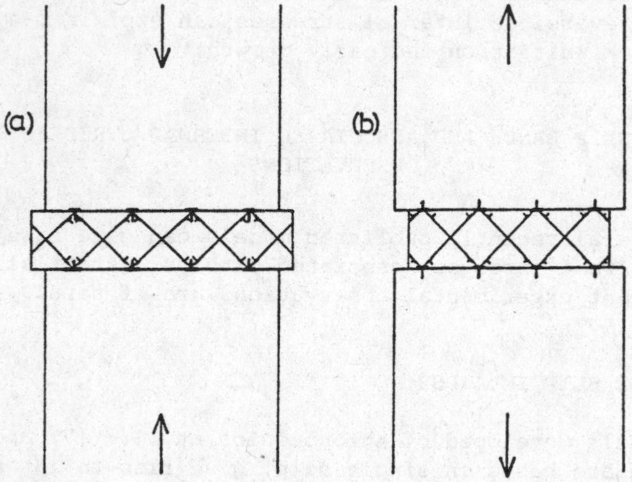

FIG. 1 Soft-layer model of a persistent slip band after Lynch (22). (a) After compression (b) After tension

Figure 1(a) shows the PSB layer at the end of a compression stroke. We assume that the deformation in the band is accomplished by imaginary dislocations gliding on orthogonal planes and examine the state of internal stress in the band at the end of the stroke. At this point, the imaginary dislocations are lodged in the interfaces between the PSB and the matrix. Addition of the Burgers vectors at the interface produces dislocations which have half-planes pointing into the band, so that the band is in compression parallel to its length. During the tensile stroke, these dislocations move away and then return with opposite sign. When the plastic strain is zero, the internal stress in the PSB is also zero. At the end of the tensile stroke, the movement of the imaginary dislocations leaves the PSB in tension (see Figure 1(b)).

To summarise, according to this simple model at the end of a tensile stroke the band is in tension and at the end of a compressive stroke the band is in compression. The sign of the internal stress is thus the same as the previously imposed plastic strain.

2.3 PURE SINGLE SLIP INCLINED TO THE PSB DIRECTION

Recent observations by Jin (11) of surface slip traces within PSBs in copper have shown that PSBs can often lie at an angle to the main slip direction (see Fig. 2). Similar observations have been made in AgCℓ (14). These observations suggest the possibility of an alternating internal stress within PSBs deforming by single slip alone.

The persistent slip band can be inclined to the primary slip direction because the band itself is a composite of slip packets. We do not understand what controls the inclination of the band. Small inclinations do not cost much elastic energy because of the favourable shape factor; and it may well be that the notion of Essmann et al. (19) is an important consideration. It will be recalled that, according to these authors, an edge dislocation does not glide exactly on its slip plane; microscopic slip processes cause the macroscopically effective slip plane to be slightly inclined.

FIG. 2 (Courtesy N.Y. Jin (11)) Surface patterns due to persistent slip in a copper single crystal with tensile axis [$\bar{2}45$] (single slip orientation) fatigued into saturation at a plastic strain amplitude of 3.10^{-3}

The internal stresses in an inclined band are not constant, but reverse sign every half-cycle. Figure 3 shows a crystal with slip planes inclined at an angle θ to the tensile axis. Three possible inclinations of a PSB are indicated in the figure with the angle

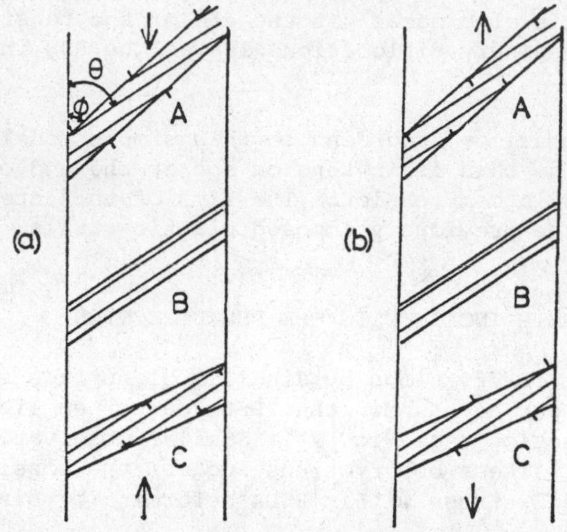

FIG. 3 Persistent slip in a packet inclined to the crystallographic glide direction. (a) In compression (b) In tension

between the PSB direction and the tensile axis labelled ϕ. For the PSB inclined at A, $\phi < \theta$; at B, $\phi = \theta$; and at C, $\phi > \theta$. We assume that real dislocations glide on the slip planes. After a compressive stroke (Fig. 3(a)) the half planes of these dislocations point away from the PSB for the configuration at A producing a tension parallel to the band, whereas at C they point into the band and produce a compression. At B, for which $\theta = \phi$, no internal stress results. After a tensile stroke, the opposite internal stresses are produced in the bands (see Fig. 3(b)).

In this model, then, based on single slip, a reversal in the sign of the internal stress in the PSB occurs every half cycle. The sign of the stress depends on the relative inclination of the glide planes to the PSB direction.

2.4 OBSERVATIONS OF PSB INTERNAL STRESSES IN AgCℓ

In some recent work on the cyclic deformation of AgCℓ single crystals, Ogin & Brown (13) and Ogin (14) demonstrated that there are macroscopic internal stresses associated with the PSBs in AgCℓ which are quite different in nature from the stresses associated with glide bands formed in unidirectional deformation. The principal axes of the stresses in the PSB are closely parallel and perpendicular to the band, as in a simple glide band, but the symmetry and detail of the stresses are quite different (see Fig. 4). In addition, the internal stress of the PSB appears to reverse sign every half-cycle of the fatigue test.

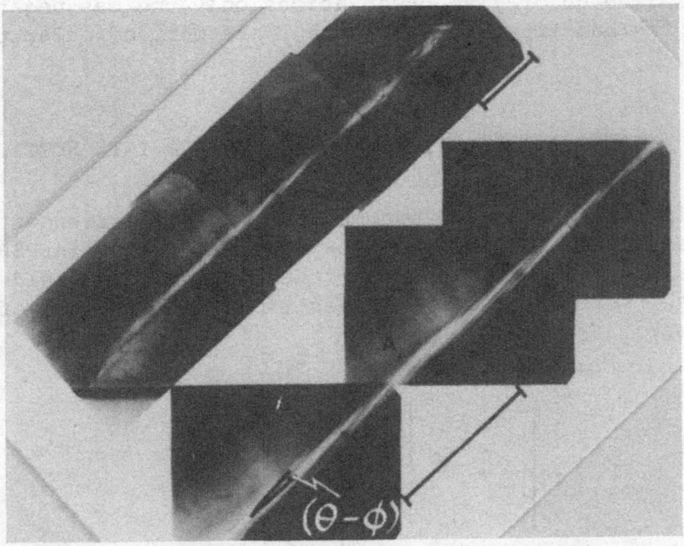

FIG. 4 Persistent slip band in AgCℓ. Black-and-white reproduction of colour photo revealing the band to be in tension. The angle (θ-φ) is about 5° in this case. (From ref. (14)). Upper figure at low magnification shows crystal boundaries; lower figure shows detail within band. Distance markers show 1 mm.

A number of observations, in which the sense of the stress has been determined using a tint plate, are consistent with the model of section 2.3 if the angle between features in the birefringent detail and the macroscopic band is taken to give the angle (ϕ-θ). That is to say, the sign of the macroscopic stress within the bands can be equal or opposite to the sign of the applied plastic strain, depending on the sign of (ϕ-θ).

If the angle of inclination (ϕ-θ) is small, as is observed to be the case, these ideas may be summarised in the following equation which relates e_t, the principal elastic tensile strain parallel to the band, with e_b^p, the current plastic shear strain in the band:

$$e_t = (\phi - \theta)e_b^p + \text{const} \qquad (2.1)$$

The constant in equation 2.1 refers to the possible existence of constant strains as envisaged by Antonopoulos et al. (17) or Essmann et al. (19), although in AgCℓ these constant strains are swamped by the alternating strains.

The angle (ϕ-θ) as observed in AgCℓ may be either positive

or negative, although of five observations only one was negative; in other words, the bands tend towards the plane perpendicular to the tensile axis.

3. STRESS DISTRIBUTIONS AT THE SURFACE INTERSECTIONS

In this section we assume that internal stresses of the type described in the previous section generally exist in PSBs including those in metals. The consequence for the local stress fields in the vicinity of the surface of a PSB is described with the aid of a computer model.

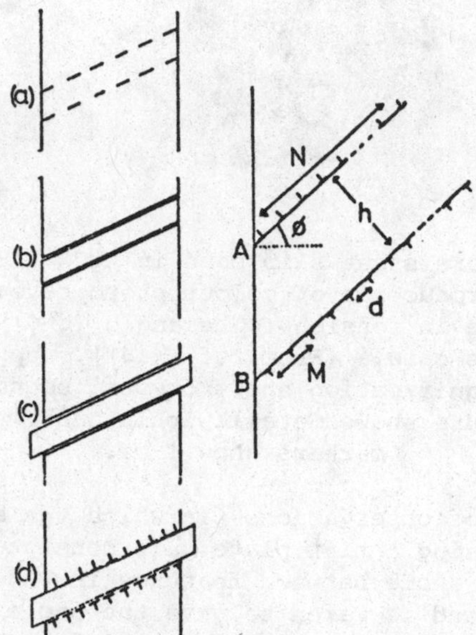

FIG. 5 Elastic model of internal stresses at a persistent slip band. The band is modelled as N pairs of vacancy dipoles of height h spaced d apart, with h = 10d. M single dislocations extend at the lower row to the intersection at B. Note that this diagram is only a model and the real persistent slip band does not produce sharp corners.

We assume that there is a uniform elastic plane strain, e_t, within a PSB parallel to the band. For convenience we can assume this to be a tensile strain, i.e. e_t positive. To calculate the near surface stress fields the tensile strain is modelled as shown in Fig. 5. In Fig. 5(a), the band is marked out and in Fig. 5(b), cut from the matrix. In Fig. 5(c), end tractions are applied to produce a tensile strain. In Fig. 5(d), the band is welded back into the matrix and the end tractions are relaxed. The mismatch strain between PSB and matrix is modelled as fictitious vacancy dipoles situated at the interface between the PSB and the matrix. The strain in the band is given by

$$e_t = \frac{b}{d} \tag{3.1}$$

where b is the Burgers vector of the fictitious dislocations and d is the spacing between the dislocations. A similar procedure was adopted by Brown (18) to model the stress at the free surface analytically. However, whereas Brown visualised this stress as one which does not alternate with cycle strain, it is clear from this paper that there are circumstances in which it does.

Solutions for the stress field of a dislocation close to a free surface have been derived by Head (23). These solutions have been superimposed to find the resultant stress field of the PSB modelled as a line of vacancy dipoles. The stress fields at the point (x,y) of an edge dislocation with Burgers vector b parallel to Ox at the point (a,O) in a semi-infinite medium are

$$\sigma_{xx}/D = -\frac{y\{3(x-a)^2 + y^2\}}{\{(x-a)^2 + y^2\}^2} + \frac{y\{3(x+a)^2 + y^2\}}{\{(x+a)^2 + y^2\}^2} + 4axy\frac{\{3(x+a)^2 - y^2\}}{\{(x+a)^2 + y^2\}^3}$$

$$\sigma_{yy}/D = \frac{y\{(x-a)^2 - y^2\}}{\{(x-a)^2 + y^2\}^2} - \frac{y\{(x+a)^2 - y^2\}}{\{(x+a)^2 + y^2\}^2} + 4ay\frac{(2a-x)(x+a)^2 + (3x+2a)y^2}{\{(x+a)^2 + y^2\}^3}$$

$$\sigma_{xy}/D = \frac{(x-a)\{(x-a)^2 - y^2\}}{\{(x-a)^2 + y^2\}^2} - \frac{(x+a)\{(x+a)^2 - y^2\}}{\{(x+a)^2 + y^2\}^2} + 2a\frac{(a-x)(x+a)^3 + 6x(x+a)y^2 - y^4}{\{(x+a)^2 + y^2\}^3} \tag{3.2}$$

where $D = \mu b/2\pi(1-\nu)$.

For an edge dislocation with Burgers vector parallel to Oy at the point (a,O) the stress fields at (x,y) are:

$$\sigma_{xx}/D = \frac{(x-a)\{(x-a)^2 - y^2\}}{\{(x-a)^2 + y^2\}^2} - \frac{(x+a)\{(x+a)^2 - y^2\}}{\{(x+a)^2 + y^2\}^2} + 2a\frac{(3x+a)(x+a)^3 - 6x(x+a)y^2 - y^4}{\{(x+a)^2 + y^2\}^3}$$

$$\sigma_{yy}/D = \frac{(x-a)\{(x-a)^2 + 3y^2\}}{\{(x-a)^2 + y^2\}^2} - \frac{(x+a)\{(x+a)^2 + 3y^2\}}{\{(x+a)^2 + y^2\}^2} - 2a\frac{(x-a)(x+a)^3 - 6x(x+a)y^2 + y^4}{\{(x+a)^2 + y^2\}^3} \tag{3.3}$$

$$\sigma_{xy}/D = \frac{y\{(x-a)^2 - y^2\}}{\{(x-a)^2 + y^2\}^2} - \frac{y\{(x+a)^2 - y^2\}}{\{(x+a)^2 + y^2\}^2} + 4axy\,\frac{3(x+a)^2 - y^2}{\{(x+a)^2 + y^2\}^3}$$

These are the corrected versions of the expression appearing in ref. (23).

For a Burgers vector inclined at an arbitrary angle θ to the y-axis, the stress field is obtained by superposition of the results for a dislocation with Burgers vector bsinθ parallel to Ox and bcosθ parallel to Oy.

It is clear that at the stress-free surface the only component which can exist is σ_{yy}, the self-equilibrated surface component. From earlier work it is known that this component rises to very large values at the intersection of the PSB with the surface; there is a logarithmic singularity there. The purpose of the computer calculations given here is to learn more about the stress distribution in the neighbourhood of the surface, and to check Brown's results which are based on slightly incorrect equations.

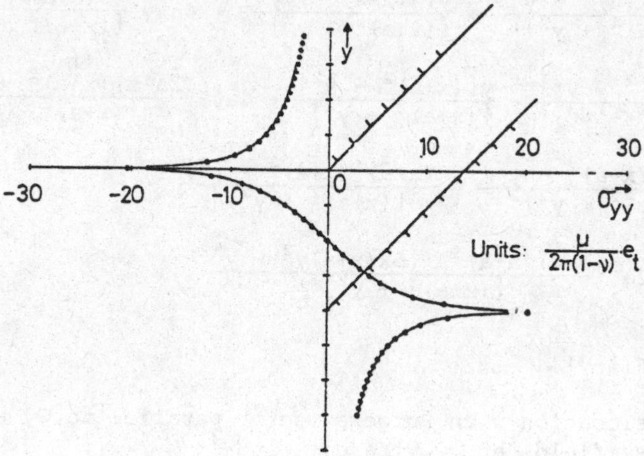

FIG. 6 σ_{yy} at the crystal surface. This work **, Brown (18) ———. Vertical axis scales with PSB width (h).

Figure 6 compares the free surface σ_{yy} stresses obtained analytically with the computer superposition of the stress fields. The agreement is excellent. At the free surface, the σ_{yy} stresses rise rapidly as the PSB/matrix interfaces are approached. Indeed, according to the analytical solution, there is a logarithmic singularity in the σ_{yy} stress at the intercepts. These formally infinite stresses are the result of the method of calculation - i.e. an edge dislocation is 'smeared-out' over the glide plane and the stresses are found by integrating the continuous distribution of dislocations over the length of the PSB. This has the effect, at the free surface intersection, of calculating the stresses due to a 'fractional' edge dislocation at its origin. The essential point is that, in reality, the σ_{yy} stresses rise to high values at the intersections, the actual value doubtless controlled by plastic processes there. For a tensile value of e_t in the band, the σ_{yy} stresses are positive (i.e. crack opening) at side B of the band and negative (i.e. crack closing) at side A.

Figure 7 shows contours of equal near-surface stress fields for the σ_{yy}, σ_{xy}, σ_{xx} stresses (the σ_{xy} and σ_{xx} stresses are, of course, zero at the free surface). The map for σ_{yy} confirms the magnification of these stresses at the free surface.

It is possible to check the present calculations against those of McNeil & Grosskreutz (24) who calculated the dilatation due to an infinite array of dipoles. Referring to their figure 1, the solution for the dilatation, Δ, at the origin is

$$\Delta = \frac{1-2\nu}{2(1-\nu)}\frac{b}{c}\left\{\frac{\sinh(2\pi a/c)}{\cosh(2\pi a/c)-\cos(2\pi(d-\delta)/c)} + \frac{\sinh(2\pi a/c)}{\cosh(2\pi a/c)-\cos(2\pi(d+\delta)/c)}\right\} \tag{3.4}$$

By putting $\delta = 0$ and $c = 2d$, the arrangement of the dipoles in Fig. 1 of ref. (24) then corresponds to our computer model of the PSB far from the surface. Taking the width of the band as ten times the wall spacing, i.e. $h = 10d$ in our notation, eqn. 3.4 gives the value of the dilatation at the origin as

$$\Delta = \frac{1-2\nu}{2(1-\nu)}\ \frac{b}{d} \times 2 \tag{3.5}$$

The dilation can also be calculated using the computed stress field. It is given by

$$\Delta = \frac{1-2\nu}{2\mu(1+\nu)}(\sigma_{xx} + \sigma_{yy} + \sigma_{zz}) \tag{3.6}$$

where, in plane strain,

$$\sigma_{zz} = \nu(\sigma_{xx} + \sigma_{yy}) \tag{3.7}$$

Hence,

$$\Delta = \frac{1-2\nu}{2\mu}(\sigma_{xx} + \sigma_{yy}) \tag{3.8}$$

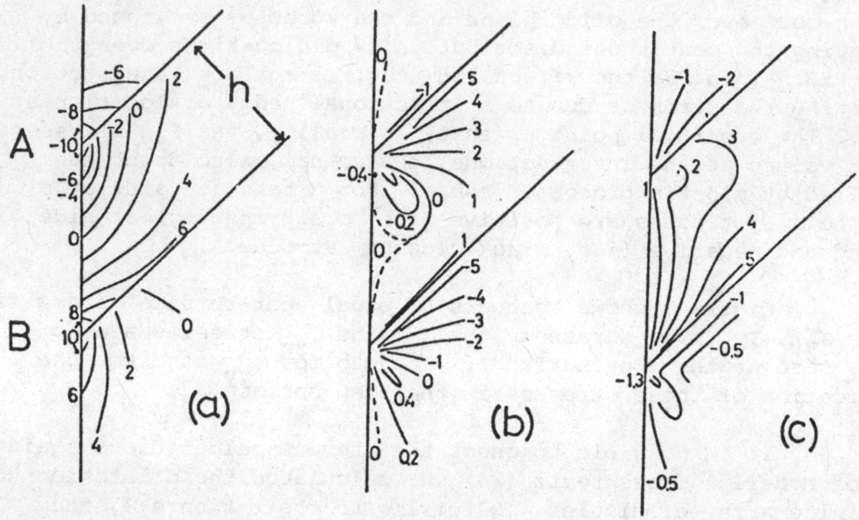

FIG. 7 Contours of stresses near crystal surface (a) σ_{yy}, (b) σ_{xy}, (c) σ_{xx}. Stress in units of $\mu/2\pi(1-\nu)$. e_t and distance scales with PSB width (h)

Far from the surface, the values of σ_{xx} and σ_{yy} within the PSB attain a constant value, such that

$$\sigma_{xx} \cong \sigma_{yy} \gtrsim \frac{\mu}{2\pi(1-\nu)} e_t \times 5.7 \tag{3.9}$$

The computed value of the dilatation is thus

$$\Delta \simeq \frac{1 - 2\nu}{2(1-\nu)} \frac{b}{d} \times 1.8 \tag{3.10}$$

since $e_t = b/d$. Comparing equations 3.5 and 3.10, we see that the computed dilatations differ by less than 10%, at a point more distant from the surface than about four times the width of the PSB. Equation (3.4) by McNeil & Grosskreutz is thus a useful estimate of the distant dilatation.

The full map of dilatation near the surface is shown in Fig. 8. The dilatation is dominated near the surface by the behaviour of σ_{yy}, and shows a pair of logarithmic singularities at the intersections of the persistent slip band with the surface. However, these singularities vanish as one moves into the material, to be replaced by discontinuities at the persistent slip band boundaries, as implied by the model of McNeil & Grosskreutz (24).

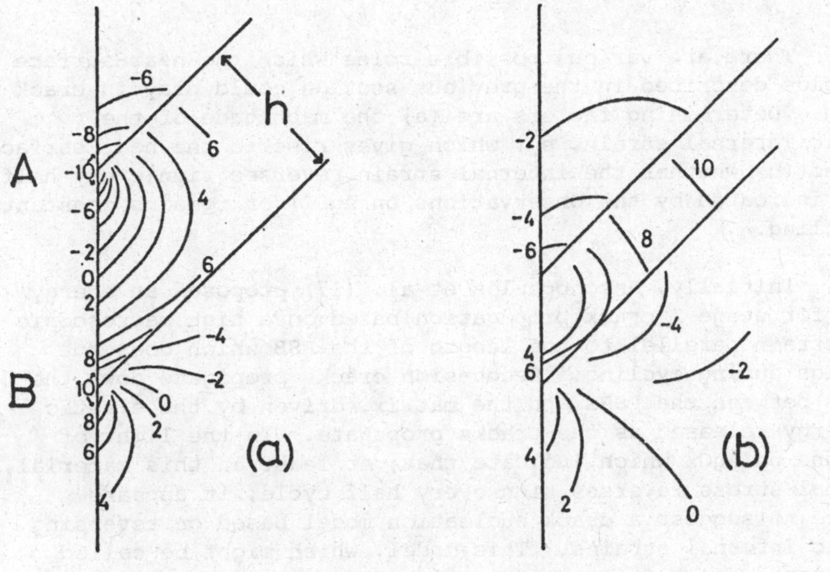

FIG. 8 Contours of dilatation near crystal surface.
(a) Finer scale (b) Coarser scale. Units of dilatation $(1-2\nu)/4\pi(1-\nu) \cdot e_t$. Units of length: h.

The present model should not be confused with the 'dipole pile-up' models of Mura & Tanaka (25, 26). In their models, the persistent slip band is taken to be an array of dipoles piled up against an internal obstacle such as a grain boundary. In our model, it is the surface acting on a uniform array of dipoles which produces stress singularities at the surface, not near any obstacle. Our model should be more appropriate to surface crack nucleation than is the model of Mura & Tanaka.

One important property of the dipole stress field is that if decohesion occurs in the absence of external stress at the PSB-matrix interface, the resulting 'crack' is not accompanied by stress concentrations. The reason for this is that the dipole array is uniform and in the absence of any applied stress there is no mechanism which can concentrate the stress at the crack tip. Alternatively, one can say that when the cracks are sufficiently far from the surface and are running parallel to the tensile stress in the band, there is no stress concentration. Of course the external applied stress must produce stress concentrations at the tip of any suitably oriented crack.

4. SURFACE STRESS ASSISTED CRACK NUCLEATION

There are various possible roles which the near-surface stress fields described in the previous section could play in crack initiation. Determining factors are (a) the magnitude of the macroscopic internal strain, e_t, which gives rise to the near surface stress, and (b) whether the internal strain reverses sign every half cycle (as indicated by the observations on AgCℓ) or remains constant during cycling.

Initially, Antonopoulos et al. (17) proposed an energy criterion for stage I crack propagation based on a high macroscopic internal strain parallel to the length of the PSB which does not reverse sign during cycling. Decohesion cracks propagate down the interfaces between the PSBs and the matrix, driven by the elastic strain energy released as the cracks propagate. In the light of observations on AgCℓ which indicate that, at least in this material, the internal stress reverses sign every half cycle, it appears appropriate to suggest a crack nucleation model based on reversing macroscopic internal strains. This model, which might be called surface stress assisted crack nucleation, has the merit of explaining many of the observations on crack initiation which are outlined in Section 5.

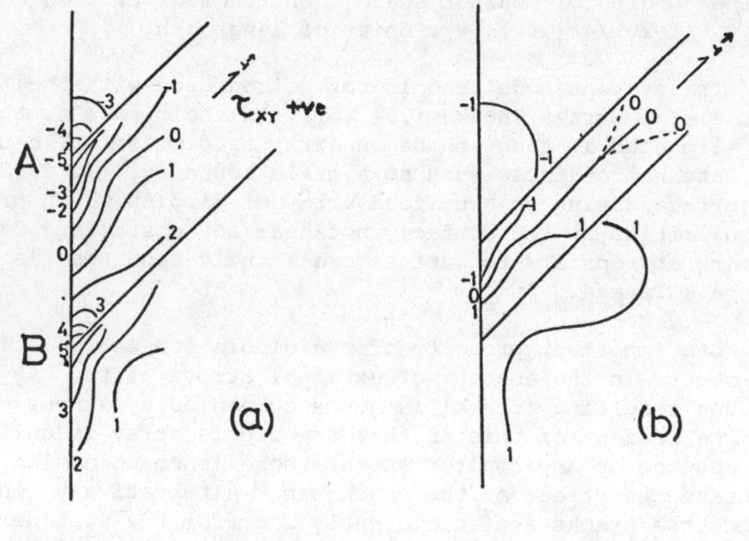

FIG. 9 Contours of constant shear stress on the primary slip plane. Units of stress: $\mu/2\pi(1-\nu).e_t$.
(a) Finer scale (b) Coarser scale

Assuming that an alternating macroscopic internal strain does exist parallel to the PSBs, its effect on plastic deformation within the PSB can be understood by considering the resulting shear stresses on the primary slip plane. An internal strain of this type does not generate shear stresses far from the surface on planes parallel to the PSB (i.e. at 45° to the crystal surface in our computer model). Figure 9 shows contour maps of constant shear stress parallel to the band. At a distance of about four PSB widths into the band, these shear stresses are negligible everywhere across the band. However, approaching the crystal surface the shear stresses rise rapidly in the vicinity of the lines of intersection of the PSB with the surface, and are of opposite sign at the two surface intersections.

The continuum model shows that all stress components are a function of position scaled by the width of the persistent slip band. For example, the surface stress given by Brown's analytical formula is (ref. (18) eqn. 4.8 for $y \ll h$)

$$\sigma_{yy} = -\frac{2e_t\mu}{\pi(1-\nu)}\,\ell n\,\frac{\sqrt{2}y}{h} \qquad (4.1)$$

The vertical distance from a line at which the stress is infinite is given by y and the stress is a function of y/h. This formula enables an estimate to be made of the width of the deformed band near the PSB/surface intersection.

Unfortunately, e_t is not known with sufficient reliability to make a good estimate. However, if we take σ_{sat} = 30 MPa and $e_t \sim 2 \times 10^{-4}$, values thought to be reasonable for copper, and assume that plastic flow occurs for $\sigma_{yy} > 2_{sat}$, we find that the width of the plastic zone near A or B is given by $y/h \sim 1/300$: the logarithmic singularity causes extra plasticity over a region which is a small fraction of the band width. A further complication is the lack of reliability of the continuum model close to the stress singularities. It is possible that the wall spacing plays a role in determining the extent of the plasticity very close to the surface.

The shear stresses on the primary glide planes which are generated within the PSB by the macroscopic internal strain are superimposed on the applied shear stress. They will, therefore, either assist or impede dislocation motion near the surface of the PSB depending on the sign of the applied stress and the sign of the macroscopic internal strain within the band.

Since the internal strain reverses sign every half cycle, it follows that the resulting shear stress always acts to encourage slip near one side of the PSB. For example, if the internal strain within the PSB is tensile during a tensile stroke, dislocation motion is assisted near the surface intersection at B (see Fig. 9). If, at the end of the tensile stroke the internal strain has changed sign, then on the compressive stroke the near surface shear stress again acts to assist dislocation motion near the intersection at B. (Of course, whether the slip is preferentially at A or B depends on the sign of $(\phi-\theta)$.

A reversing macroscopic internal strain thus always acts to assist dislocation flow near one or other of the surface intersections of the PSB. The choice of intersection depends on the phase relationship between the macroscopic internal strain and the applied stress. It should be emphasised that even a small macroscopic internal strain will have a substantial effect on dislocation motion near the PSB/matrix interfaces due to the stress concentrations there.

The precise mechanism by which this bias of dislocation flow within the PSBs affects crack nucleation is not clear. One possibility is that the random slip mechanism of crack nucleation, suggested originally by Wood (27) and developed subsequently by May (28, 29) and by Cheng and Laird (20), is modified by the bias to assist crack nucleation towards the edges of the PSBs.

In the original random slip model, the dislocations active in one half-cycle make their return journey along paths which are shifted in a random manner with respect to the previous paths. In effect, each atom on the surface undergoes a random walk along the slip vector and the surface becomes roughened into a notch-peak profile. In later cycles, slip tends to concentrate in the valleys formed on the surface.

The random slip model, by itself, cannot explain the tendency for fatigue cracks to initiate at the edges of PSBs. However, the greater dislocation movement in the vicinity of the PSB/matrix interfaces will bias crack nucleation towards one or other of these interfaces. It should be emphasised that crack initiation is only statistically biased towards the interfaces and cracks may still start nearer the centre of PSBs, as is sometimes observed.

5. OBSERVATIONS ON PSBs AND STAGE I CRACKS

A number of experimental observations on the formation and properties of PSBs, the site of crack initiation within PSBs and early crack propagation can be understood as a consequence of the near-surface stress fields described in the previous sections.

5.1 OBSERVATIONS ON PSBs

Roberts (30) and Mughrabi (31) noted that new PSBs tend to nucleate next to existing bands. This can be understood with reference to the near-surface stress fields.

It is known that PSBs tend to nucleate in regions of slightly increased stress concentration, e.g. polishing pits (3). Hence, the local near-surface stress concentrations at the edges of existing PSBs will assist PSB nucleation nearby. This applies independent of the sign of the strain.

5.2 THE SITE OF CRACK INITIATION WITHIN A PSB

All workers are agreed that in long-life fatigue (e.g. number of cycles to failure in copper $N_f > 3 \times 10^4$) cracks nucleate in PSBs. This result has been confirmed for a wide variety of materials, both single and polycrystalline, of varying purity and in differing testing conditions.

In addition to identifying PSBs as crack nucleation sites, a number of authors have specified the preferred sites of crack initiation within the bands themselves, and a number of pertinent observations can be made from other published micrographs. The evidence is summarised in Table 1. It is important to note that none of the evidence cited prior to 1980 was part of a dedicated experiment to locate crack initiation within PSBs.

It appears that crack initiation is strongly biased towards the interfaces between the PSBs and the matrix. In addition, the evidence suggests that in copper the preferred nucleation site is side A of a band - the recent micrographs by Mughrabi et al. (21) are particularly striking. On the other hand, Basinski & Basinski (42) have reported that in their copper specimens they found no evidence to support the view that cracking started preferentially at one specific side of a band.

The bias of crack initiation towards the interfaces is not a result that would be expected if crack nucleation were solely a result of random slip processes within the PSBs. It is a direct reflection of the non-uniform near-surface stress fields described earlier.

5.3 DORMANT CRACKS

Possibly the most striking characteristic of a stage I fatigue crack is the fact that it grows rapidly in from the surface, and then stops growing and enters a dormant stage with little further growth for many tens of thousands of cycles. To quote recent studies of this behaviour: Cheng & Laird (20) and Basinski & Basinski (42) show that very many unpropagated cracks a few microns in length exist at failure, i.e. at about 10^5 cycles, and that the cracks are perhaps half that size at about 10^4 cycles. It is difficult to deduce the number of cycles required to produce a 1 μm crack at the edge of an originally polished persistent slip band, but from the above work this is certainly less than 20,000 cycles. Other evidence of crack arrest follows from observations on the more readily visible extrusions which appear, in terms of shape and height, to be a mirror image of intrusions (Basinski & Basinski (42)). Several authors have commented on the rapid initial growth of extrusions, to a height of a few microns, and their subsequent very slow growth.

TABLE 1

Reference	Material	Type of Testing	Crack Nucleation Site	Side A or Side B
Hunter & Fricke (32)	polycrystalline aluminium and alloys	alternate bending constant deflection	at interface	not specified
Hull (33)	polycrystalline copper	push-pull; load control	intrusions at interface* (interpreted as cracks)	both*
Wood (34)	polycrystalline copper	torsion	at interface*	both*
Boettner & McEvily (35)	silicon-iron	cantilever beam; fully reversed bending	at interface	B
Laufer & Roberts (36)	copper single crystals	plane bending	at interface	B
Neumann (37)	copper single crystals	push-pull; load control	at interface	A
Lukas & Klesnil (38)	copper-zinc single crystals	push-pull; load control	at interface*	B*
Hahn & Duquette (39)	copper single crystals	push-pull; load control	at interface*	A*
Katagiri et al. (40)	copper polycrystal	plane bending; load control	at interface and along centre of band	B*
Mughrabi & Wang (41)	copper single crystal	push-pull; constant plastic strain	at interface	A
Mughrabi et al. (21)	copper single crystal	push-pull; constant plastic strain	at interface	A

* From published micrographs

In terms of the stress distribution discussed here, it is clear that the stress concentration at the surface aids crack formation until the crack is several wall-spacings long, at which length the stress concentration disappears. With a wall spacing of ∿1.5 μm, the cracks will become dormant when they are a few microns long, which is consistent with observations. Two related questions arise: (1) why do the dormant cracks not grow rapidly because of the applied stress? (2) By what mechanism do the dormant cracks grow slowly, one of them becoming the fatal crack?

The answer to the first question is simple: the cracks are not long enough, even if the metal were perfectly brittle, to elongate and cause catastrophic failure. For example, the surface work of fracture is certainly greater than 2 Jm^{-2}, the surface energy of copper. The critical length of such a hypothetical brittle crack subject to the saturation stress of copper, which is 30 MPa (shear), is estimated to be approximately 100 μm. Thus it is energetically unfavourable for these cracks to grow by direct action of the applied stress. Of course in a highly ductile matrix, the true surface work to fracture is many orders of magnitude greater, so our calculation of the critical crack length underestimates it by a substantial factor. Even though the tips of stage I cracks are very sharp, as observed by Basinski & Basinski (42), it is energetically unfavourable for these cracks to grow by direct action of the applied stress because the high stresses at the tip cannot produce slip.

Another way of expressing this conclusion is to say that the cracks under the action of the applied stress do not produce sufficient stress intensity. If we take $\Delta K_{th} = \surd(2E\gamma)$ to be the theoretical stress intensity required to propagate a brittle crack, and insert values of the Young modulus, E, and surface energy, γ, appropriate to copper, this theoretical threshold stress intensity comes out at about 0.6 MPa $m^{1/2}$. The stress intensity achieved by a stage I crack a few microns long in a plastic strain controlled test is about ten times smaller than this theoretical value. It is interesting to note that the measured threshold stress intensity of around 2 MPa $m^{1/2}$ (Pook (43)) is only a factor of three greater than our theoretical estimate. This suggests that very little plasticity is involved in crack propagation near threshold as is confirmed by scanning electron microscopy.

The logarithmic stress concentration at the crystal surface is thus an essential feature for the appearance and initial growth of cracks which cannot be explained by classical fracture mechanics. With the aid of the computer maps of the near surface stresses we can make a very rough estimate of the macroscopic internal strain e_t, in the PSB required for a crack to propagate along the interface between the PSB and the matrix. We note from Fig. 7 that a stress of at least $10\mu/2\pi(1-\nu).e_t$ is felt by a crack a few microns long. Assuming that the crack propagates in a purely brittle manner and that the volume over which the energy is released is bounded by the $10\mu/2\pi(1-\nu).e_t$

contour we can equate the stored energy in this region with the energy of the new crack surfaces. The required value of the macroscopic strain, e_t, of about 3×10^{-3} which this calculation produces is an upper limit since we have underestimated the stored elastic energy near the logarithmic singularities in our calculation.

It is worth pointing out that according to a purely random model for the nucleation of a stage I crack, the crack growth rate should increase from the moment of first appearance of a sufficiently deep intrusion, because the stress intensity continuously increases. The stress concentration at the surface plays an essential role in crack nucleation because it is otherwise impossible to understand dormant cracks.

We now turn to the second question: How do the dormant cracks slowly grow? One obvious answer is by environmental attack. If surface reactions on clean copper surfaces essentially reduce the free energy of formation of fresh surface to zero or to a negative quantity, then the dormant cracks will grow at a rate controlled by the chemical reaction. It is very significant in this context that the NUCLEATION of stage I cracks is observed not to depend upon the environment, but the SLOW GROWTH of such cracks does (Wadsworth & Hutchings (44)). However, even in the absence of an aggressive environment, the dormant cracks can slowly grow by absorbing dislocations. The energy of the system 'crack plus dislocation' can be lowered by extending the crack and removing a dislocation, and so the crack can grow with continued cyclic plasticity even if it cannot grow by direct action of the applied stress.

5.4 REJUVENATION EFFECTS

The existence of the near-surface stress fields explains in a natural way the observation by many authors that a fatigued crystal is rejuvenated by surface polishing.

When PSBs are polished off the surface of a crystal and the crystal is then refatigued, the bands form at exactly the same point. This well-known result is a consequence of the three-dimensional dislocation structure associated with the bands. In addition to the bands reforming at the same place, removal of the rough surface profile developed in the bands rejuvenates the specimen and repeated removal of the surface can produce lifetimes to failure double that expected of the virgin crystal. The first systematic repolishing experiments appear to have been carried out by Thompson et al. (45) in stress-controlled tests on copper polycrystals. More recently, Basinski et al. (see (42)) achieved the same result using copper single crystals and plastic-strain controlled tests.

The rejuvenation effect confirms that much of the fatigue life of the crystal is spent with cracks in the dormant phase. When these cracks are removed by polishing, they should re-nucleate quickly but an appreciable additional number of cycles is required to make them grow very slowly to a dangerous length. If, as suggested in Section 5.3, 90% of the life is spent with cracks in the dormant phase, it will be possible to nearly double the fatigue life by repolishing at the appropriate point in the test.

6. BRIEF COMPARISON WITH OTHER THEORIES

There are several published models of crack nucleation which have similarities with the mechanisms described here, in addition to the model of Tanaka & Mura (25, 26) which has been discussed earlier.

In a model by Lin & Ito (46), two neighbouring thin slices within a grain favourably oriented for slip are assumed to have small initial shear stresses of opposite sign at the surface. During cycling, the two slices slide in opposite directions, one active in one half cycle, the other in the other half cycle, to produce an extrusion or intrusion. One could say that the concepts of this paper provide the rationale for the shear stresses envisaged by these authors.

In the work of Mughrabi and his group, crack initiation does not depend upon the internal stresses directly, but it does depend upon the stress concentration which results from the surface topography (bulges) produced by a persistent slip band. Such stress concentrations should act rather similarly to the ones discussed here, although the presence of a bulge which does not alternate with cyclic strain should produce a crack always at one side or the other of a persistent slip band. However, since the stress intensity at a dormant stage I crack is insufficient to cause the crack to grow, it seems unlikely that the stress intensity due to a bulge will be sufficient to produce a crack.

Cheng & Laird (20) have developed a semi-quantitative model of fatigue failure in single crystals. The model is based on their observations that the fatal crack is found to nucleate in the PSB having the largest slip offset. Their mechanism of crack nucleation, which is based solely on random slip processes within the band, is difficult to reconcile with the tendency of cracks to initiate near the edges of PSBs. Nonetheless, the model is in good agreement with a Coffin-Manson (log-log) plot of applied plastic strain amplitude against lifetimes to failure for a variety of crystal orientations. They also make interesting observations on crack nucleation, which lead them to the conclusion that: (i) the mechanism is a geometrical process, associated with 'statistical' slip in PSBs; (ii) the fatal crack site is located in the band where the maximum step height (i.e. slip displacement in one quarter cycle) occurs and (iii) for a given step height, the crack is likely to occur in the narrowest band.

We are able to give a tentative explanation for these results. It seems likely that the crack appears very quickly once a persistent slip band with the requisite pattern of internal stress is formed. When the crack is dormant, in the absence of environmental attack, it grows by the incorporation of glide dislocations which can elongate the crack and at the same time reduce their elastic energy. If the crack has a capture cross-section per unit length W, and the dislocations move a distance ℓ per cycle, the crack advances $\rho W \ell b$ per cycle by capture of dislocations. But $\rho \ell b$ is the plastic strain amplitude in the band per quarter cycle, so the rate of slow growth per cycle is

$$\Delta a = 4\, e_b^p\, W \qquad (6.1)$$

What controls the capture cross-section W? It is approximately the width of the crack measured perpendicular to the glide plane. It might perhaps be taken equal to the distance over which deformation occurs near the singular points A and B. In Section 4 we estimated this distance to be about h/300; thus

$$\Delta a \sim 4\, e_b^p\, h/300 \qquad (6.2)$$

If the width of the band is 10 μm and if, as in copper $e_b^p = .01$, we find that the crack grows by about 0.3 nm per cycle, which is generally consistent with observation.

Equation 6.2 accounts nicely for Cheng & Laird's detailed observations: for a given (average) value of e_b^p, cracks in wider bands grow faster; and for a given width of band, those with higher local strain amplitude grow faster.

Equation 6.2 permits one to estimate the cycles to failure of a copper single crystal. The crack length after N cycles is given by

$$a(N) = 4N\, e_b^p\, h/300 \qquad (6.3)$$

This can be regarded as an average crack length corresponding to bands of average width h and average strain aplitude e_b^p. But the volume fraction of bands is proportional to h divided by the spacing of the bands, or, since each band contributes two surface cracks, by half the density of cracks per unit length at the surface. Thus if the surface crack spacing is L,

$$a/L = 4N\, e_b^p\, f/600 = e_{cum}/600 \qquad (6.4)$$

(Here, e_{cum} is the cumulative strain and fe_b^p is the applied strain amplitude, e^p, according to the two-phase model (6)). One thus arrives at the result that the crack length divided by the crack spacing is proportional to the cumulative strain.

The final step is to decide a failure condition. This takes us beyond the scope of this paper, but a simple condition which guarantees that the cracked surface has lost its load-bearing capacity is to set a/L = 1. Failure of the interior should follow rapidly at the enhanced strain amplitude or (depending on the test conditions) stress amplitude. If we take this as a simplified failure criterion, we have

$$4\ Ne_b^p f \ =\ 4\ Ne^p \ =\ e_{cum} \ =\ 600 \text{ at failure} \qquad (6.5)$$

We thus expect a Coffin-Manson plot of log e^p vs log N to produce a slope of -1; according to Cheng & Laird, the slope is -0.78, with considerable uncertainty. According to Basinski & Basinski (42), the specimen fails at constant cumulative strain, as eqn. 6.5 implies. Cheng & Laird's data suggest that the cumulative strain to failure is approximately 1200, which confirms early unpublished work by Winter (47) who finds it to be about 1000 and to be independent of plastic strain amplitude for the small amplitudes which produce high cycle fatigue, that is for amplitudes between about 0.8% and 0.006%; the lower limit being the strain endurance limit. Equation 6.5 seems to be a reasonable a priori estimate of the failure condition, underestimating the cumulative strain to failure by a factor of about 2. Of course, the proposed failure condition neglects the cycles required to nucleate the surface crack and to propagate the final stage II crack. More seriously, the estimate is based on a very inadequate understanding of what controls the value of e_t, the elastic strain within the persistent slip band.

7. CONCLUSIONS

1. We believe that persistent slip bands often produce a macroscopic internal strain whose principal axes are parallel and perpendicular to the band and whose sign reverses every half cycle. In addition, there may be a non-alternating component of either sign.

2. The strain produces very high (formally infinite) near-surface stress fields where the persistent slip band interfaces meet the surface of the crystal. The stress fields scale with the width of the persistent slip band, except possibly very close to the singularities.

3. The concentrated stresses at the surface promote crack nucleation by increasing the local plastic strain amplitude and possibly by promoting localised secondary slip.

4. Once the surface cracks penetrate a few wall spacings (microns in copper) the stress concentration is lost and if, as will almost always be the case, they do not produce

sufficient stress intensity under the action of the applied stress, they grow only very slowly by capture of individual dislocations.

5. One can estimate the threshold stress intensity for crack growth to within a factor of three by assuming a work to fracture equal to the surface free energy exposed. Freshly nucleated stage I cracks are sub-threshold cracks.

6. A simple model for the slow crack growth suggests that failure is caused by a cumulative strain of about 600, in rough agreement with observation, which suggests a value about twice this.

ACKNOWLEDGEMENTS

We wish to thank Dr. A. T. Winter for help, and Dr. N. Y. Jin for permission to reproduce Fig. 2. S.L.O. acknowledges support from Dr. P. W. R. Beaumont in the Department of Engineering.

REFERENCES

(1) Ewing, J. A. & Humphrey, J. C. W. (1903). Phil. Trans. Roy. Soc. A200, 241-250.
(2) Thompson, N. et al. (1956). Philosophical Magazine, 1, 113-125.
(3) Mughrabi, H. (1980). In: Proceedings, Fifth International Conference on the Strength of Metals and Alloys, Aachen, eds. P. Haasen, V. Gerold & G. Kastorz. Vol. 3, 1615-1638. New York: Pergamon Press.
(4) Laird, C. (1979). In: Proceedings of the A.S.M. Materials Science Seminar, St. Louis, Missouri, 149-203.
(5) Brown, L. M. (1977). In: Proceedings, 'Fatigue 1977' Conference, Cambridge, U.K. Metal Science, 11, 315-320.
(6) Winter, A. T. (1974). Philosophical Magazine, 30, 719-738.
(7) Mughrabi, H. (1978). Materials Science and Engineering, 33, 207-233.
(8) Finney, J. M. & Laird, C. (1975). Philosophical Magazine, 31, 339-364.
(9) Charsley, P. (1981). Materials Science and Engineering, 47, 181-185.
(10) Winter, A.T. et al. (1981). Acta Metallurgica, 29, 735-748.
(11) Jin, N. Y. (1983). Ph.D. Thesis, Cambridge University.
(12) Majumdar, B. S. & Burns, S. J. (1982). Acta. Met., 30, 1743-1761.
(13) Ogin, S. L. & Brown, L. M. (1980). In: Proceedings of the International Conference on Dislocation Modelling of Physical Systems, Gainesville, Florida, 1980. Eds. M. F. Ashby, R. Bullough, C. J. Hartley & J. P. Hirth, 579-586, New York: Pergamon Press.
(14) Ogin, S. L. (1982). Ph.D. Thesis, Cambridge University.
(15) Laird, C. & Duquette, D. J. (1972). In: Corrosion Fatigue, eds. A. J. McEvily & R. W. Staehle, 88-117. National Association of Corrosion Engineers, Houston, Texas.

(16) Klesnil, M. & Lukas, P. (1980). Fatigue of Metallic Materials. Elsevier Press.
(17) Antonopoulos, J. G. et al. (1976). Philosophical Magazine, 34, 549-563.
(18) Brown, L. M. (1980). In: Proceedings of the International Conference on Dislocation Modelling of Physical Systems, Gainesville, Florida, 1980. Eds. M. F. Ashby, R. Bullough, C. S. Hartley & J. P. Hirth, 51-68. New York: Pergamon Press.
(19) Essman, U. et al. (1981). Philosophical Magazone, A44, 405-426.
(20) Cheng, A. S. & Laird, C. (1982). Fatigue of Engineering Materials and Structures, 4, 331-341 and 343-353.
(21) Mughrabi, H. et al. (1983). In: Fatigue Mechanisms: Advances in Quantitative Measurement of Physical Damage, ASTM STP 811, eds. J. Lankford, D. L. Davidson, W. L. Morris and R. P. Wei, 5-45. American Society for Testing and Materials, 1983.
(22) Lynch, S. P. (1974). 'A New Model for Initiation and Growth of Fatigue Cracks', Metallurgy Report 94, Australian Defence Scientific Service, Melbourne.
(23) Head, A. K. (1953). Proc. Phys. Soc. 66B, 793.
(24) McNeil, M. B. and Grosskreutz, J. C. (1967). Phil. Mag., 16, 1115-1118.
(25) Mura, T. & Tanaka, K. (1981). Mech. of Materials, 1, 63.
(26) Tanaka, K. & Mura, T. (1981). J. Appl. Mechanics, 103, 97-102.
(27) Wood, W. A. (1958). Phil. Mag. 3, 692-699.
(28) May, A. N. (1960). Nature, 185, 303.
(29) May, A. N. (1960). Nature, 188, 573-574.
(30) Roberts, W. N. (1969) Phil. Mag., 20, 675-686.
(31) Mughrabi, H. (1978). Mat. Sci. & Eng., 33, 207-233.
(32) Hunter, M. S. & Fricke, W. G. (1954). Proc. Am. Soc. for Testing Materials, 54, 717.
(33) Hull, D. (1957/58). J. Inst. Met. 86, 425-430.
(34) Wood, W. A. (1959). Proc. Int. Conf. on Atomic Mechanism of Fracture, 450. John Wiley: London.
(35) Boettner, R. C. & McEvily, A. J. (1965). Acta. Met., 13, 937-945.
(36) Laufer, E. E. & Roberts, W. N. (1966). Phil. Mag., 14, 65-78.
(37) Neumann, P. (1967). Z. Metallkunde 58, 780-789.
(38) Lukas, P. & Klesnil, M. (1970). Phys. Stat. Sol. 37, 833-842.
(39) Hahn, H. N. & Duquette, D. J. (1978). Acta. Met. 26, 279-287.
(40) Katigiri, K. et al. (1977). Met. Trans. 8A, 1769-1773.
(41) Mughrabi, H. & Wang, R. (1980). International Symposium on Defects and Fracture, Tuczno, Poland, eds. G. C. Sih & H. Zorski, 15-28. Published by Martinus Nijhoff.
(42) Basinski, Z. S. and Basinski, S. J. (1983). Proceedings of the Sixth International Conference on the Strength of Metals and Alloys, Melbourne, Australia, 2, 819-824.
(43) Pook, L.P. (1983). 'The Role of Crack Growth in Metal Fatigue'. London: The Metals Society.
(44) Wadsworth, N. J. & Hutchings, J. (1958). Phil. Mag., 14, 1154-1166.
(45) Thompson, N. et al. (1956) Phil. Mag., 1, 113-126.
(46) Lin, T. H. & Ito, Y. M. (1969). J. Mech. Phys. Sol. 17, 511-523.

(47) Winter, A.T. (1974). Ph.D. Thesis, Cambridge University.
(48) Timoshenko, S.P. & Goodier, J.N. (1970). Theory of Elasticity, Third Edition, McGraw-Hill.
(49) Bilby, B.A. (1980). J. Mat. Sci., 15, 535-556.

APPENDIX*: Approximate Energy Release Rate for a Stage I Crack by J. D. Eshelby

The construction of Fig. 5 suggests that the methods appropriate to indentation cracking can be applied to the nucleation of fatigue cracks. At the stage of Fig. 5(c), the strains in PSB and matrix are uniform. The relaxation of the tractions at the end of the PSB can be accomplished by applying opposite tractions. Thus we can synthesize the stress field at the end of a PSB by applying a distribution of surface forces to the area between A and B. The resolved force per unit length is (the ϕ here is that of Fig. 10)

$$f_{\parallel} = \varepsilon_t \cdot \frac{2\mu}{1-\nu} \cos\phi \sin\phi$$
$$f_{\perp} = \varepsilon_t \cdot \frac{2\mu}{1-\nu} \sin^2\phi \qquad (A.1)$$

The action of these forces can be computed directly from stress functions given in Timoshenko and Goodier (48); the perpendicular forces distributed over a semi-infinite plate are treated in problem 16, p 146. One has only to superimpose two distributions of opposite sign, one offset by h cscϕ to obtain the subsurface stress distribution. The parallel forces are treated similarly in problem 17, p 147. In the latter case one can verify directly the logarithmic singularity which is remarked upon by Timoshenko & Goodier themselves.

In addition to verifying the pattern of stresses, it is possible to calculate approximately the energy release rate for cracks which propagate down the interfaces. Provided the cracks are sufficiently far from the interface, we can replace the distributed forces by a point force

$$F_{\parallel} = \varepsilon_t h \cdot \frac{2\mu}{1-\nu} \cos\phi$$
$$F_{\perp} = \varepsilon_t h \cdot \frac{2\mu}{1-\nu} \sin\phi \qquad (A.2)$$

The problem now reduces to the problem of calculating the energy release rate from an indentation fracture caused by the application of a point force; this problem can be solved by the method of the M integral, as expounded in Bilby's Tewkesbury Lecture (49). In the notation of equation (24) of that reference, we find for the energy release rate G

---

* Based on letters dated 13 November 1981 and 19 November 1981 from J.D.E. to L.M.B.

$$Ga = M\,(F \sin\phi,\ F \cos\phi,\ O;\ \pi) - M\,(F \cos(\phi/2),\ F \sin(\phi/2),O;\phi) \qquad (A.3)$$

In A.3, the first M integral comes from the surface S_∞, and the second from the surface S_2 which connects the faces of a wedge of apex angle ϕ (Fig. 10). The term Ga comes from the surface S_1. Inserting the appropriate values into equation (24) of reference (40), we find

$$G = e_t^2 \cdot \frac{2\mu}{1-\nu} \cdot \frac{h^2}{a}\left\{\frac{\phi - 1/2 \sin 2\phi}{\phi^2 - \sin^2\phi} - \frac{1}{\pi}\right\} \qquad (A.4)$$

For the crack at B, as shown in Fig. 10, $\phi = 45°$ and

$$G_B \sim 2.123\ \frac{2\mu}{1-\nu}\ \frac{h^2 e_t^{\ 2}}{a} \qquad (A.5)$$

whereas for the crack at A, $\phi = 135°$ and

$$G_A \sim 0.247\ \frac{2\mu}{1-\nu}\ \frac{h^2 e_t^{\ 2}}{a} \qquad (A.6)$$

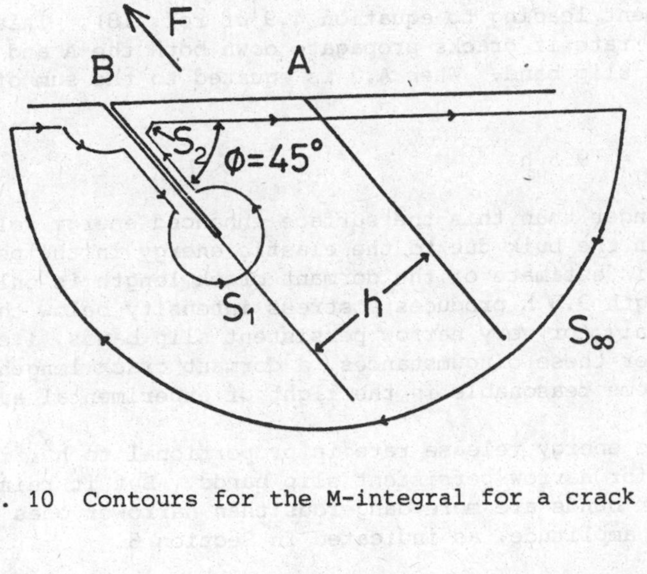

FIG. 10 Contours for the M-integral for a crack at B

These equations for the energy release rate are valid when $a \gg h$, i.e. for a thin persistent slip band when the crack has propagated a distance several times the width of the band. Under these circumstances we find the following results:

1. The energy release rate for the 'thin wedge' at B is very much greater than that for the 'thick wedge' at A. This is contrary to

the near surface release rates estimated in Section 5.3 from the contours at A and B in Fig. 7. Such contours suggest $G_A \approx G_B$. It is worth pointing out that the energy release rate as calculated from the M integral does not distinguish tensile loading of the PSB from compressive loading. Because the M integral gives essentially an energy-based release rate, cracked surfaces forced together release energy equally to cracked surfaces pulled apart. Such calculations clearly give a necessary but not sufficient condition for fracture. A condition of critical stress must also be considered, to complete one's understanding of the problem.

2. The energy release rate is inversely proportional to a, so it is infinite at the surface but it vanishes as the crack elongates. This confirms the notion of a progressively more slowly moving crack as the crack moves in from the surface. Some idea of the dormant length can be obtained by recalling the energy release reate from an infinitely long crack (arising from the distributed strains, not included in equations A.5 and A.6)

$$G = \frac{1}{4} e_t^2 \frac{2\mu}{1-\nu} h \tag{A.7}$$

(see the argument leading to equation 4.9 of ref. 18). This is the energy release rate if cracks propagate down both the A and B sides of the persistent slip band. When A.7 is equated to the sum of A.5 and A.6 we find

$$a = 9.5 h \tag{A.8}$$

For a crack longer than this the surface enhanced energy release rate is smaller than the bulk due to the elastic energy in the persistent slip band. This estimate of the dormant crack length is only valid, if a crack of length 9.5 h produces a stress intensity below threshold, and this is true only for very narrow persistent slip bands, i.e. for $h < 1$ µm. Under these circumstances, a dormant crack length of less than 9.5 µm seems reasonable in the light of experimental evidence.

3. The energy release rate is proportional to h^2. Again, this is valid only for narrow persistent slip bands. But it reinforces the view that wider bands are more dangerous than narrower ones of the same plastic strain amplitude, as indicated in Section 6.

4. One can make an estimate of the strain e_t required to produce a brittle crack near the surface. If we use values of μ, γ appropriate to copper, and take the width of the PSB to be 5 wall spacings ($h = 5d$) and the crack length to be 10 wall spacings, with the wall spacing itself equal to 1 µm, we find $e_t > 10^{-3}$. This estimate agrees with the estimate of Section 5.3.

ENVIRONMENTAL EFFECTS NOVEL TO THE PROPAGATION OF SHORT FATIGUE CRACKS

R. P. Gangloff

Exxon Research and Engineering Co., Annandale, NJ 08801, U.S.A.

R. O. Ritchie

University of California, Berkeley, CA 94720, U.S.A.

ABSTRACT

Crack size and opening morphology dominate the mechanical and chemical driving forces for fatigue propagation in embrittling environments. Similitude based on a crack tip field parameter is compromised, particularly for small cracks (< 5 mm) which grow up to several orders of magnitude faster than projected and below apparent threshold conditions. Environment sensitive mechanical and chemical mechanisms which govern the growth of small cracks are reviewed. For the former the retarding effect of crack closure; originating from wake plasticity, surface roughness, deflection, corrosion debris or fluid pressure; increases with increasing crack size particularly within the near threshold regime. Data for high strength steel in H_2 demonstrate the importance of such mechanisms, however, precise models of crack size dependencies and systematic closure measurements are lacking. Considering the chemical driving force, the embrittling activity of the occluded crack differs from that of the bulk environment, and is geometry dependent. The deleterious influence of small crack size is demonstrated experimentally for steels in aqueous chloride solutions, and related quantitatively to crack opening shape and size effects on diffusion, convective mixing and electrochemical reaction. Small crack size promotes hydrogen embrittlement due to enhanced hydrolytic acidification and reduced oxygen inhibition. Chemical crack size effects are material and environment specific; criteria defining limiting crack sizes and opening shapes for K or J-based similitude do not exist.

1. INTRODUCTION

Fracture mechanics analyses of subcritical crack propagation are based on the fundamental notion that a characterizing parameter such as stress intensity (K) or the J integral describes remote loading and geometry effects on crack tip stress and strain distributions, and hence the kinetics of slow growth.(1-3) By similitude, cracks extend at equal rates when subjected to equal mechanical driving forces. Fracture mechanics scaling of laboratory data to predict component life is established, only in part, for fatigue and statically loaded cracks in benign and embrittling environments.(4-8) Recent investigations indicate that stress intensity-based similitude must be modified to account for plasticity,(9) crack motion,(10) closure,(11) deflection,(12) crack size,(13) and environment.(14)

Extensive data demonstrate that short cracks grow faster than projected and below apparent threshold conditions compared to long crack (25-50 mm) kinetics at constant stress intensity.(11,13,15-30) Typically for benign environments, the limiting crack size for deviations from K-based similitude increases from 10 μm for high strength alloys to about 1 mm for low strength materials, as reviewed from a mechanical perspective.(11,13,15-17,31)

In active environments rates of fatigue crack propagation are controlled by interrelated mechanical and chemical driving forces.(32,33) Critically, each driving force is crack size, shape and applied stress sensitive. The theses of this review are that small crack size influences uniquely the mass transport and reaction components of the chemical driving force, and that environmental modifications of the mechanical driving force are novel within the short crack regime. Mechanical and chemical factors which lead to breakdowns in similitude for small fatigue cracks, and which are traceable to environmental effects, are characterized separately.

Data establish the importance and complexity of small crack-embrittling environment interactions. Threshold stress versus crack size results in Fig. 1 indicate threshold stress intensity range (ΔK_o) control for long cracks, and constant stress or endurance limit control for very short cracks in 13Cr steel at two stress ratios (R).(20) While similar size dependencies are observed for moist air and liquid water, crack growth deviates from ΔK_o control at larger limiting crack sizes for the later environment. Embrittlement, evidenced by reduced stress, is promoted for decreased crack size and lower R. Geometry sensitive chemical and mechanical closure effects contribute to the trends depicted in Fig. 1. The potential for small crack chemical contributions to corrosion fatigue, independent of mechanical effects, is illustrated in Fig. 2.(21,24) Cracking in vacuum and moist air is defined

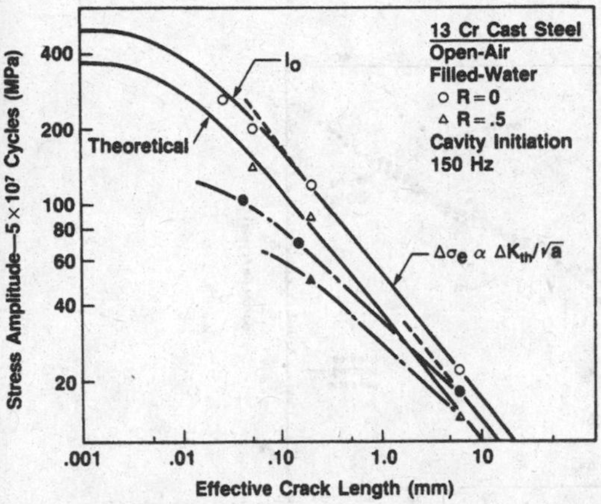

FIG. 1. Variation of fatigue threshold stress with crack length for cast 13Cr steel in air and water at two stress ratios. Note ΔK_o control for long cracks and endurance stress control for short cracks. After Usami and Shida (1979).

uniquely by ΔK, independent of crack size and applied stress. In contrast small cracks in aqueous NaCl grow up to 300 times faster than long cracks at constant ΔK. A multiplicity of da/dN-ΔK relations is observed; chemical embrittlement is enhanced for decreased crack size or reduced stress, and correlates with crack opening shape.(14) Data in Fig. 3 demonstrate the role of crack closure independent of chemical effects.(34-38) Based on long crack

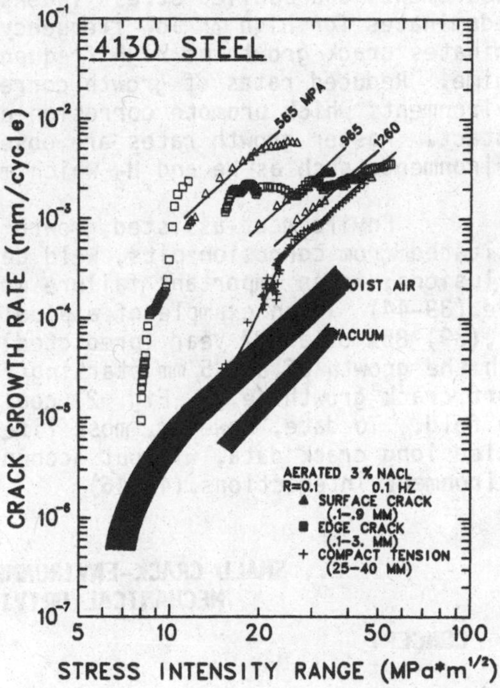

FIG. 2. Crack growth rate versus stress intensity range data for variably sized cracks in high strength (1330 MPa) 4130 steel. Note the accelerated growth of physically short cracks only in aqueous NaCl, and the stress dependence of small crack corrosion fatigue (Δσ values are listed for surface cracks). After Gangloff (1984).

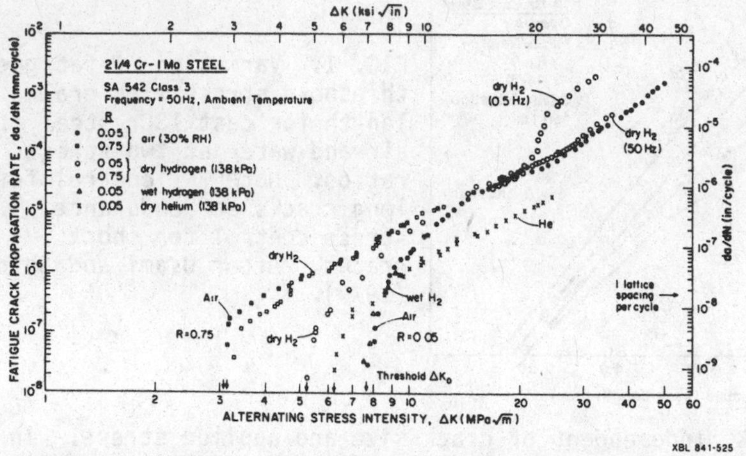

FIG. 3. Influence of load ratio and environment on fatigue propagation of long (~ 50 mm) cracks in low strength bainitic 2 1/4Cr-1Mo steel (σ_0 = 500 MPa) in moist air, dry helium and hydrogen gases at R = 0.05 and 0.75. Note the contrasting effect of environment at near-threshold and higher growth rates. After Suresh and Ritchie (1982).

measurements and applied stress intensity, hydrogen embrittlement predominates for high ΔK low frequency conditions, whereas closure dominates crack growth at high frequencies in the near-threshold regime. Reduced rates of growth correlate with low R and moist environments which promote corrosion debris and crack surface contact. Faster growth rates are observed for high R and environments such as He and H_2 which maintain clean crack surfaces.

Environment-assisted growth of small fatigue cracks; nucleated from corrosion pits, weld defects, scratches, porosity or inclusions; is an important failure mode often dominating total life.(39-44) In an example of a pipeline carrying H_2S contaminated oil,(39) 85% of an 87 year (predicted) fatigue life is associated with the growth of a 0.5 mm starting flaw to 1.0 mm. Accelerated short crack growth (e.g., Fig. 2) could reduce total life by 50 to 100 fold. To date, however, most life prediction analyses have scaled long crack data, without accounting for short crack-environment interactions.(45,46)

2. SMALL CRACK-ENVIRONMENT INTERACTIONS: MECHANICAL DRIVING FORCE

2.1 CONCEPT

Of the factors contributing to crack size similitude problems through effects on the mechanical driving force, local plasticity, crack deflection and crack closure are of major importance as illustrated in Fig. 4.(11,13,15-20) Each of these processes can be modified strongly by an embrittling environment, for example through the role of adsorbed hydrogen in affecting dislocation behavior, or the enhancement of crack closure through crack surface corrosion products or roughness. We examine the influence of environment on the processes affecting the mechanical driving force, and highlight effects unique to small cracks.

2.2 ACTIVE CRACK TIP PLASTICITY

A primary reason for crack size effects is the inappropriate characterization of crack tip fields. Inaccuracies result from the use of linear elastic fracture mechanics to describe crack growth behavior in the presence of extensive local plasticity, i.e., where crack length (a) is comparable with the extent of the active plastic zone ahead of the crack tip, $r_y \sim (1/2\pi)(K_I/\sigma_o)^2$, or where the crack is embedded within the strain field of a notch (Fig. 4a). Plasticity effects may dominate at crack sizes below a limiting length ℓ_o, given by $1/\pi(\Delta K_o/\Delta\sigma_e)^2$, where ΔK_o is the long crack threshold and $\Delta\sigma_e$ is the smooth bar fatigue limit.(25,47)

The extent of local plasticity is often influenced by environment.(48) There is clear experimental evidence that dissolved hydrogen can affect the flow stress of materials.(48,49) In high purity iron softening is observed at temperatures above 200 K, whereas hardening is seen at lower temperatures. In steels

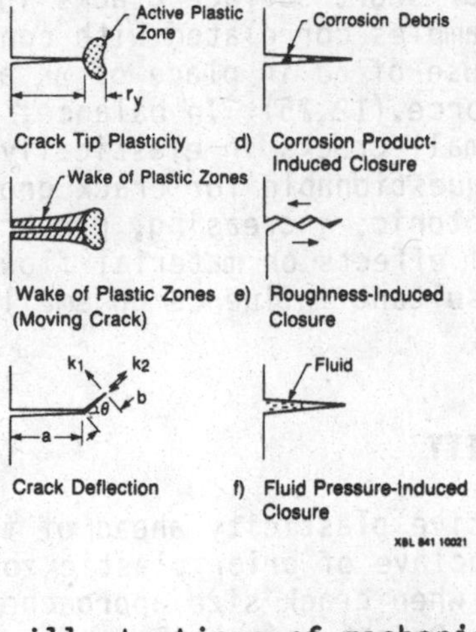

FIG. 4. Schematic illustrations of mechanisms for breakdowns in similitude relevant to small fatigue cracks in embrittling environments.

conversely, softening has been associated with single slip deformation and hardening with multiple slip, and the environmental influence declines with increasing alloy strength. The mechanisms for hydrogen effects on plasticity are imprecise.(48) Hydrogen segregation to dislocation cores enhances double kink nucleation on screw dislocations which facilitates movement. (The effect on edge dislocations may be exactly the reverse.) Softening may also result from the formation of voids induced by high fugacity hydrogen (i.e., during charging). Hardening is related to hydrogen atmospheres on dislocations or increased, hydrogen-stabilized vacancy concentrations, factors which may impede dislocation motion.

With respect to fatigue crack growth, in situ measurements in 7075-T651 aluminum alloys indicate decreased crack tip plasticity, in the form of reduced crack tip fracture strains, in water vapor compared to dry nitrogen and due to the embrittling effect of hydrogen.(50) Conversely, recent studies of aluminum show fatigue cracks to remain sharp when exposed to inert atmospheres, due to a Mode II + I opening, yet to become blunt in moist air where purely Mode I opening occurs.(28) The contributions of crack tip shape, closure, deflection and chemical embrittlement are not fully understood.

While a fundamental mechanism for environment-crack tip deformation interactions is not available, it is clear that environmental effects on plasticity and crack opening morphologies are crucial for the small crack regime.(28) To alleviate the problem of local plasticity, the use of non-linear elastic fracture mechanics, specifically involving the ΔJ parameter, has been suggested.(9,17,25) Data for short surface cracks in elastic-plastic low cycle fatigue samples correlated with conventional long crack results, through the use of ΔJ in place of ΔK as the generalized crack driving force.(12,25) In balance, however, this approach is imprecise for small cracks in elastically loaded material. The use of J is questionable for crack growth, since it is strictly defined for monotonic, increasing, proportional loading.(2,3) Environmental effects on material flow properties must be defined, and the resultant influence on small crack driving force quantified.

2.3 CRACK TIP WAKE PLASTICITY

In addition to active plasticity ahead of the crack, it is necessary to consider the enclave of prior plastic zone left behind the crack tip, particularly when crack size approaches the scale of local plasticity (Fig. 4b). Recent asymptotic analyses of monotonically loaded, Mode I non-stationary cracks indicate that the crack tip strain singularity weakens as the crack moves.(10) Specifically in the limit as $r \rightarrow 0$, the plastic strain distribution, $\gamma_p(r)$, is given in terms of an effective plastic zone size (r_y'),

Poisson's ratio (ν) and constant (m) as:

$$\gamma_p = \frac{m}{\sigma_o}\frac{dJ}{da} + \frac{1.88(2-\nu)\sigma_o}{E}\ \ell n\left(\frac{r'_y}{r}\right) \qquad (1)$$

Thus, due to wake plasticity behind the growing crack, crack tip strains decay as $\ell n(1/r)$, rather than as $1/\sqrt{r}$ or $1/r$ for a stationary crack in a linear elastic or perfectly plastic solid. This implies that at a fixed K_I or J, the plastic strains ahead of a stationary crack exceed those ahead of a slowly moving crack with its trailing wake of plasticity. More importantly since short cracks sized below r_y have by definition a limited wake, larger plastic strains must occur ahead of a moving short crack compared to the equivalent long crack at the same K_I or J. Recent in situ crack tip deformation measurements on growing fatigue cracks in 7075 aluminum alloy clearly show this to be the case. The strain distribution ahead of a long crack at low ΔK conforms to a $\ell n(1/r)$ singularity,(51) and corresponding crack tip strains and opening displacements for small cracks, 30 to 200 μm in length, are significantly higher.(18)

The effect of wake plastic zone on the mechanical crack driving force and the significance to the crack size similitude question is influenced by chemical factors since environment can affect plasticity. There is, however, no characterizing parameter currently available which considers wake plasticity effects as a function of material flow properties, and which describes driving force independent of crack size.

2.4 CRACK DEFLECTION

Crack deflection contributes to a lack of similitude by causing the near tip driving force to differ from the globally computed K_I or J (Fig. 4c).(15) In general Mode I crack growth data are correlated in terms of K_I or J assuming a linear crack oriented perpendicular to the maximum tensile stresses. Crack paths can, however, deflect out-of-plane due to metallurgical and environmental interactions such that the local driving force at the crack tip is reduced.(12) For an elastic crack deflected through an angle Θ and subjected to tensile and shear loads, the local Mode I and Mode II stress intensity factors, k_1 and k_2, are given in terms of the nominal stress intensities, K_I and K_{II}, and angular functions $a_{ij}(\Theta)$, as: (52)

$$\begin{aligned} k_1 &= a_{11}(\Theta)\,K_I + a_{12}(\Theta)\,K_{II}\ , \\ k_2 &= a_{21}(\Theta)\,K_I + a_{22}(\Theta)\,K_{II}\ . \end{aligned} \qquad (2)$$

For a simply kinked elastic crack, deflection through Θ = 45° yields $k_1 \approx 0.8K_I$ and $k_2 \approx 0.3K_I$. The 45° crack path deflection induces a significant Mode II shear component at the crack tip and reduces the effective driving force by roughly 15%.(12)

Fatigue crack deflection is promoted by microstructure in duplex structures,(12,53) and at lower ΔK levels where crack growth may be crystallographic, particularly for coarse planar slip materials.(54,55) The effect can be striking for microstructurally-small cracks where prolonged retardation periods or arrest are observed as such cracks encounter and reorient at grain boundaries.(12,26) The crack size dependence upon the mechanics of crack deflection (e.g., Eq. (2)) remains unsolved.

Environmental factors which induce a specific crack path morphology can have a major influence on the mechanical crack driving force through crack deflection. The branching of stress corrosion cracks under specific material/environment/crack velocity conditions,(56) the faceted or crystallographic nature of the fracture plane in certain materials during corrosion fatigue ("brittle" striations in aluminum alloys,(57)) and most importantly the development of intergranular facets and secondary cracks due to hydrogen embrittlement or active path corrosion (4,27,48,57) are examples of environment induced deflection.

The effect of crack deflection is not limited to modifying the local stress intensity for a stationary crack. Under cyclic loading, crack deflection induces irregular fracture surfaces and Mode II crack tip displacements. Such factors promote the development of crack closure, to a degree dependent upon crack length, thus further enhancing the discrepancy between local and global mechanical crack driving forces.

2.5 FATIGUE CRACK CLOSURE

Crack closure, or fracture surface contact, during cyclic loading is a major factor contributing to crack size similitude effects. At lower load ratios (below typically R = 0.5) where contact occurs at positive loads (i.e., the closure stress intensity, K_{cl}) during the cycle, the consequence of closure is to reduce the nominal driving force ($\Delta K = K_{max} - K_{min}$) to an effective value, $\Delta K_{eff} = K_{max} - K_{cl}$, where K_{max} and K_{min} are the maximum and minimum applied stress intensities. Crack growth rates decrease in response to the reduced driving force, but often correlate with ΔK_{eff}. Crack closure can result solely from cyclic plasticity,(58) or may be developed through several alternate mechanisms as illustrated in Fig. 4.(53) Closure processes include crack surface corrosion product formation, irregular fracture surface morphologies coupled with inelastic shear displacements, and fluid-induced pressure inside the crack. Whereas plasticity-induced closure is significant at higher ΔK levels approaching plane stress, higher levels of closure may be developed at lower ΔK levels approaching ΔK_o (plane strain) through the mechanisms depicted in Fig. 4.

All closure phenomena contribute to crack size similitude breakdown through modifications of the near-tip mechanical driving force and to a degree dependent upon crack size. Geometry is crucial because closure must act in the wake of the crack tip. Since short cracks have a restricted wake, growth rate retardations by crack closure mechanisms are limited. At equivalent nominal ΔK levels, short cracks may propagate faster than long cracks due to a higher effective stress intensity. This effect will diminish as closure develops with increasing crack length. Although there is a growing body of experimental evidence to support this notion,(11,19,59,60) with the exception of the fluid pressure-induced closure mechanism, few analytical models exist which incorporate the crack size dependence of closure. Furthermore, indirect, geometry dependent chemical effects on corrosion products and roughness characteristics have not been examined.

CLOSURE INDUCED BY CYCLIC PLASTICITY: Elastic constraint of material surrounding the plastic zone in the wake of the crack front affects material elements plastically deformed at the crack tip, and leads to interference between mating fracture surfaces. Although analyses showing the crack size dependence of closure are not available, recent experimental and numerical studies indicate that the effect of closure diminishes at small crack sizes.(11,19,59-61) Unpublished results by Heubaum and Fine on Van 80 steel (R_B 94) cycled in moist air show this particularly clearly, as reproduced in Fig. 5. High precision closure measurements, using a 0.05 μm sensitivity

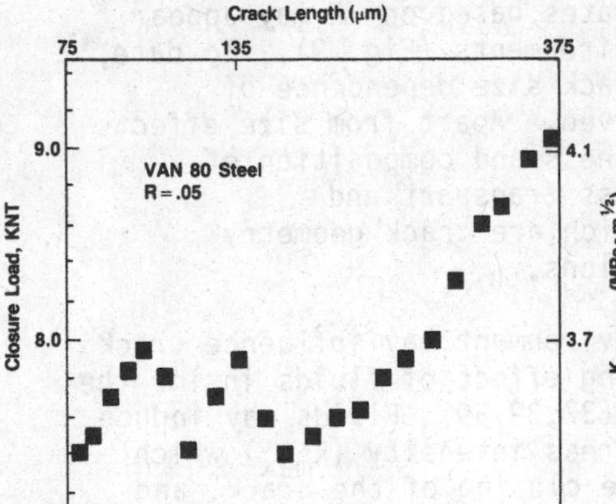

FIG. 5. Experimental evidence for increasing crack closure with increasing small crack size, based on compliance measurements for Van 80 steel cycled in moist air; R = 0.05, K_{max} = 9 to 20 MPa√m. After Heubaum and Fine (1984).

compliance gauge, establish that closure stress intensity (K_{cl}) increases by almost 20% for crack extension from 75 to 375 μm. Although small crack growth rates exceed those for long cracks based on constant ΔK, a unique growth rate-driving force law was reported for ΔK_{eff}.

CLOSURE INDUCED BY CORROSION PRODUCTS: Closure induced by corrosion products(33,53,62-65) is relevant to environmentally influenced crack growth when the size-scale of such debris approaches crack tip opening displacements (ν) (Fig. 4d). The mechanism is most potent at low R and at near-threshold stress intensity ranges, especially in lower strength materials where fretting processes during the opening and closing of the crack enhance oxidation and excess deposit formation. The closure mechanism was demonstrated as significant for oxide films formed on alloy steel crack surfaces exposed to moist gases,(53,62-64) and for calcareous deposits produced on structural steel fatigue crack surfaces through electrochemical reactions with seawater.(33,66)

Corrosion product induced closure is modeled approximately in terms of the thickness of excess film (d) and the location of the maximum thickness from the crack tip (2ℓ) to yield:(67)

$$K_{cl} \approx \frac{d\,E}{4\sqrt{(\pi\ell)}\,(1-\nu^2)}\,, \qquad (3)$$

For air formed oxide, measured values of d (0.01 to 0.2 μm) and 2ℓ (5μm) correspond to a closure K of about 1.5 MPa√m. Significant levels of K_{cl} tend to offset the embrittling influence of an aggressive environment which might otherwise accelerate crack growth. In material and environment systems where susceptibility to embrittlement is small, crack growth rates based on ΔK may appear slower in seemingly more corrosive environments (Fig. 3). To date, no analytical representation of the crack size dependence of reaction product closure has been derived. Apart from size effects on the mechanics of closure, the thickness and composition of reaction products are determined by mass transport and electrochemical reaction, processes which are crack geometry sensitive as developed in ensuing sections.

CLOSURE INDUCED BY FLUID PRESSURE: Environment may influence crack closure through the hydrodynamic wedging effect of fluids inside the crack during cyclic loading (Fig. 4f).(37,39,59) Fluids may induce an internal pressure relatable to a stress intensity (K^*_{max}) which opposes the opening and principally the closing of the crack, and which reduces the effective stress intensity range at the tip to $\Delta K_{eff} = K_{max} - K_{min} - K^*_{max}$. This mechanism results in frequency and viscosity sensitive fatigue crack growth rates in inert fluids. Trends in behavior with these variables are difficult to predict since higher viscosity fluids, which induce higher fluid pressures, are kinetically limited in their ability to fully ingress into the crack.(59)

Experimental and theoretical analyses of fluid pressure closure provide estimates of K_{cl} as a function of frequency, viscosity and crack size.(38,59) Internal fluid pressure p(x) is distributed along a crack of depth, a, and average opening width, ⟨h⟩, according to:

$$p(x) = 6\eta\rho \frac{\dot{h}_{max}}{h^3_{max}} a^2 \ \ell n(1 - x/a), \quad \text{for } d/a = 1$$

$$\text{or} \quad p(x) = 6\eta\rho \frac{\langle\dot{h}\rangle}{\langle h\rangle^3} x(x - d), \quad \text{for } d/a < 1, \tag{4}$$

where ρ is fluid density and η is kinematic viscosity. The extent of fluid penetration (d) during a fatigue load cycle is time dependent and estimated based on capillary flow to be:

$$d^2(t) \approx \frac{\gamma \cos \beta}{3\eta\rho} \int_0^t \langle h\rangle dt, \tag{5}$$

where β is the wetting angle and γ is surface tension. This analysis suggests that the magnitude of closure typically saturates at K_{cl}/K_{max} values approaching 0.5, based on fluid viscosities between 5 and 60,000 cS and growth rates between 10^{-6} and 10^{-3} mm/cycle, indicating that the mechanism is less potent than oxide or roughness-induced closure.(59)

Equations 4 and 5 predict that fluid pressure, and hence the degree of closure, are related to crack size and are diminished at smaller crack lengths. Predictions of closure stress intensity (K^*_{max}) due to fluid pressure, generated by oils of viscosities varying from 5 to 60,000 cS, are shown in Fig. 6 for a 2 1/4Cr-1Mo steel cycled at R = 0.05 and 50 Hz. K^*_{max} decreases as crack length approaches zero, and to a degree dependent upon viscosity. At a fixed nominal ΔK, the smaller crack experiences a higher effective driving force, and hence propagates at a faster speed.

CLOSURE INDUCED BY FRACTURE SURFACE ROUGHNESS: Environmental factors can influence the extent of crack closure through an effect on fracture morphology (Fig. 4e). Irregular fracture surfaces; produced by microbranching, intergranular separation or crystallographic cleavage, together with local mixed mode crack tip displacements; promote roughness induced closure during unloading.(34,35,64,68,69) K_{cl} is developed through premature contact of pronounced asperities. For example, in high strength steels where near-threshold growth rates are often lower in

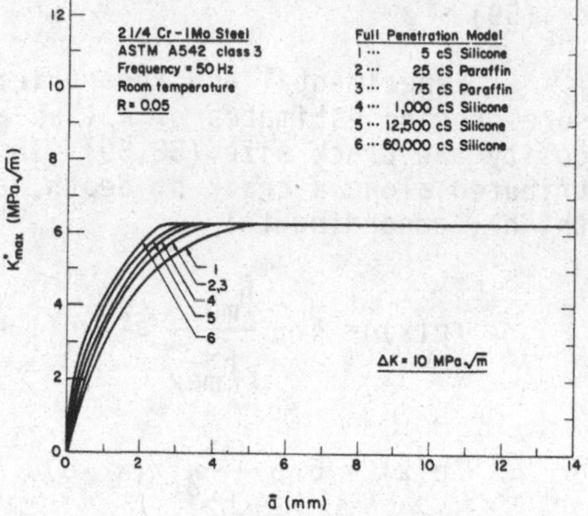

FIG. 6. Evidence for decreased K_{cl} at small crack lengths based on predictions of fluid-induced closure forces for crack propagation in silicone and paraffin oils. K^*_{max} is the stress intensity resulting from fluid pressure and is equivalent to K_{cl}. After Tzou, Hseuh, Evans and Ritchie (1984).

potentially embrittling gaseous hydrogen environments compared to moist air, the intergranular nature of hydrogen-induced fracture surfaces promotes fretting corrosion debris and asperity contact during cyclic loading, leading to closure of varying degrees in each environment.(36,60)

Roughness induced closure is modeled in terms of the extent of surface roughness, or the ratio of asperity height to width (γ), and the ratio of Mode II to Mode I crack tip displacements (μ):(68)

$$\frac{K_{cl}}{K_{max}} = \sqrt{\left(\frac{2\gamma\mu}{1+2\gamma\mu}\right)}. \tag{6}$$

K_{cl} is significant at low R and for crack tip displacements comparable with asperity size.(68,69) Crack size may influence both the degree of fracture surface roughness through a chemical mechanism and the level of K_{cl} based on crack mechanics. The latter geometry effect has not been analyzed to date, while the former mechanism is considered in Section 3.

2.6 HYDROGEN ASSISTED GROWTH OF SMALL CRACKS IN STEEL

A novel small fatigue crack-environment interaction, traceable to the combined effects of mechanical closure and chemical transport, is summarized in Fig. 7 for 4340 steel (σ_o = 1030 MPa) stressed in either gaseous hydrogen or helium. At low R (Fig. 7a) 0.1 to 1 mm edge cracks grow five times faster than long (25-50 mm) cracks in compact tension specimens at constant ΔK and in H_2. Cracking in He is well defined by ΔK independent of crack size for this class of steels (e.g., Fig. 2 and Refs. 13,24,27).

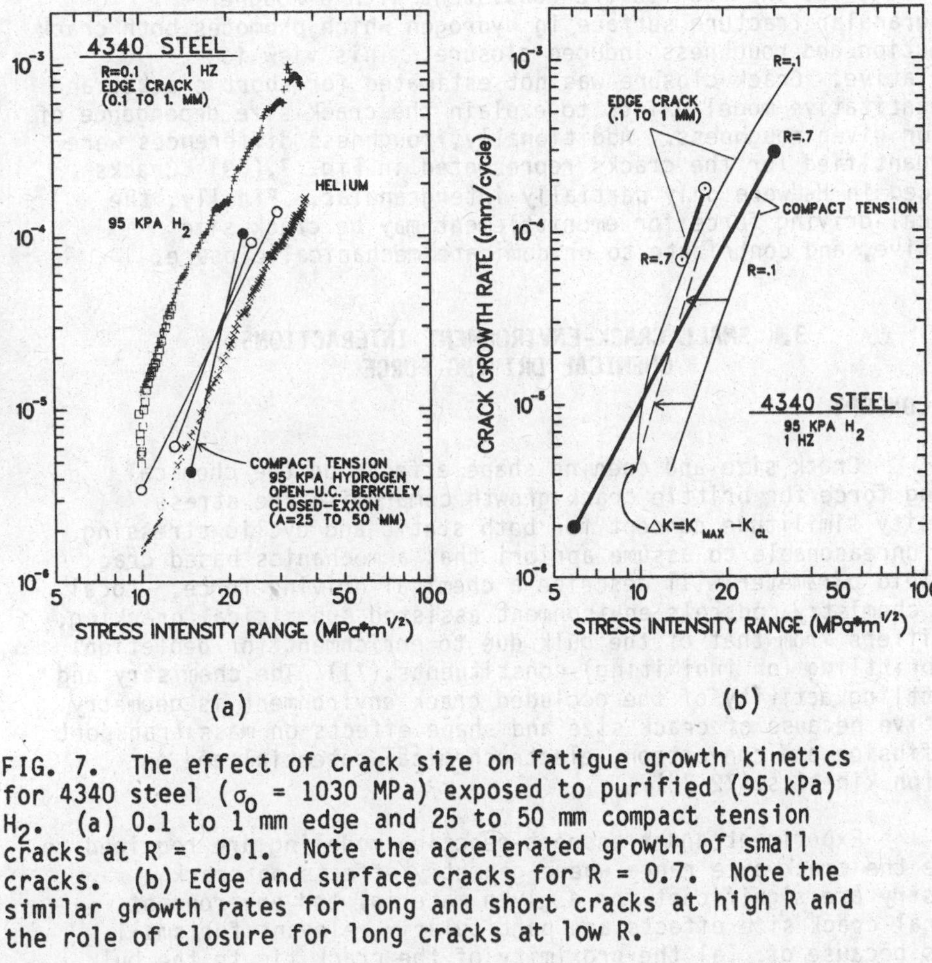

FIG. 7. The effect of crack size on fatigue growth kinetics for 4340 steel (σ_o = 1030 MPa) exposed to purified (95 kPa) H_2. (a) 0.1 to 1 mm edge and 25 to 50 mm compact tension cracks at R = 0.1. Note the accelerated growth of small cracks. (b) Edge and surface cracks for R = 0.7. Note the similar growth rates for long and short cracks at high R and the role of closure for long cracks at low R.

The breakdown in similitude for small cracks is also observed for higher strength (σ_o = 1360 MPa) 4130 steel in H_2.(70) At the high R value where closure is minimized, growth rates for the long crack approach short crack speeds for H_2 (Fig. 7b). In contrast short crack growth rates in H_2 are equivalent at R = 0.1 and 0.7. When low R, long crack results are adjusted for closure based on ΔK_{eff} computed from compliance estimates of K_{cl}, the differences between long and short crack speeds become small as shown by the dashed line in Fig. 7b. Equivalent growth rates are observed for cycling from K_{min} to K_{max} and from K_{cl} to K_{max}, demonstrating the absence of damage for changing stress intensity below K_{cl}.

The crack size effect summarized in Fig. 7 is reasonably explained based on closure. Since crack geometry effects are accounted for by applied ΔK for high strength alloys in inert environments, the origin of the closure variations must be environmental. Precise micromechanisms for such behavior are uncertain, but the results are consistent with a rougher intergranular fracture surface in hydrogen which promotes both crack deflection and roughness induced closure. This view is speculative. Crack closure was not estimated for short cracks, and no quantitative model exists to explain the crack size dependence of K_{cl} for given roughness. Additionally, roughness differences were not quantified for the cracks represented in Fig. 7.(69) Cracks produced in H_2 were only partially intergranular. Finally, the chemical driving force for embrittlement may be crack size sensitive, and contribute to or dominate mechanical closure.

3. SMALL CRACK-ENVIRONMENT INTERACTIONS: CHEMICAL DRIVING FORCE

3.1 CONCEPT

Crack size and opening shape effects on the chemical driving force for brittle crack growth compromise the stress intensity similitude concept for both static and cyclic stressing. It is unreasonable to assume apriori that a mechanics based crack tip field parameter will describe a chemical driving force. Local crack chemistry controls environment assisted subcritical cracking, and differs from that of the bulk due to enrichment (or depletion) of embrittling (or inhibiting) constituents.(71) The chemistry and embrittling activity of the occluded crack environment is geometry sensitive because of crack size and shape effects on mass transport by diffusion and convection, electrochemical potential and on reaction kinetics.(72-78)

Experiment and transport-reaction modeling are required to define the crack size range where geometry effects on crack chemistry are significant for a given material and environment. Chemical crack size effects are particularly relevant for small cracks because of: (a) the proximity of the crack tip to the bulk environment and applied potential,(73,76) (b) the crack size dependence of convective pumping,(78,79) (c) the large crack surface area to occluded solution volume ratio,(14,24) (d) the likelihood of tortuous crystallographic cracking influencing transport,(54) and (e) the sensitivity to localized environment enhanced plasticity.(26,28) Geometry effects may remain constant with increasing crack size beyond a saturation point,(29,79) consistent with demonstrations of ΔK similitude for aqueous environments.(4-7,60,80)

Chemical crack size effects are predicted for a wide range of embrittling environments. (24,72,73,76,77,81,82) Experimental confirmations are, however, lacking because of the difficulties associated with isolating chemical and mechanical effects on similitude, and with monitoring the growth of small cracks.

3.2 GASEOUS ENVIRONMEMTS

THEORETICAL MODELING: Gaseous environment enhanced fatigue crack propagation rates are controlled by the slow step in the sequence including gas transport, adsorption, diffusion, and chemical embrittlement.(29,32,72) For free molecular flow, collisions between gas molecules and crack walls dominate transport and impede the arrival rate of reactive species at the crack tip.(72,81) Local gas pressure is reduced below the bulk level, and embrittlement is decreased provided that gas transport is rate limiting.

Gas transport to the tip is crack size and opening shape sensitive. Impeded flow occurs when molecular mean free path (λ) exceeds crack opening displacement (ν), with impedance beginning for ν below about 100 λ.(72) Mean free path equals 0.1 μm for H_2, N_2, O_2 and H_2O at 300 K and 200 kPa pressure, and varies inversely with pressure and the square of molecular diameter. Considering an edge crack loaded to K = 10 MPa√m, crack mouth opening varies from 2 μm for a depth of 0.1 mm to 18 μm at a depth of 15 mm. For the 100 λ interaction criterion, impeded flow should occur over the entire length of the short crack, but over a much longer length for the deep crack. Crack geometry effects are more likely for lower gas pressure (e.g. λ = 20 μm at 5 kPa), for rough crack surfaces and for large molecules (λ = 2 mm for Cd at 5 kPa) relevant to solid metal embrittlement.(82)

Flow impedance (I) is approximated by:(72)

$$I = \left(\phi_1 \int_{a^*}^{a} dx/\nu_{max}^2\right)^{-1} = \phi_2 \sigma^2 a/\log \nu_{max}^* \tag{7}$$

for a crack of length, a, along x, of mouth opening, ν, at a stress, σ, the constants ϕ_1 and ϕ_2 and molecular flow starting at a*. Impedance decreases as I increases from 0 to 1, and is time dependent during each stress cycle, $\sigma(t)$. The integral is solved in equation 7 for near tip displacements ($\alpha K\sqrt{(a - x)}$); alternate solutions are obtainable for the complete crack. While approximate, this analysis demonstrates that crack tip pressure and hence embrittling activity depends on stress, crack shape and size. K^2 (or $\sigma^2 a$) in equation 7 results from the assumed form of $\nu(x)$ between a* and a, and is not relatable to a mechanical driving force. ΔK-based similitude is predicted for those long crack cases where changes in log ν_t^* due to increasing a, have only a mild effect on I, and for those gases where subsequent steps in the reaction sequence are fast compared to transport.(83) Alternately, within the short crack regime and at constant K, increasing a results in decreasing I and increasing impedance until crack length equals a*. A multiplicity of growth rates would be projected, and ΔK-based similtude compromised.

EXPERIMENTAL CONFIRMATIONS: Experimental evidence of unique small crack-gas environment interactions is limited, since studies of embrittlement have not focused on either small crack growth kinetics or similitude. Typically, ΔK and crack size vary simultaneously, complicating interpretation. Wei and coworkers reported that constant ΔK loading produced constant crack growth rates independent of crack size for long (> 25mm) cracks in an aluminum alloy in water vapor, supporting similitude.(32) Systematic studies of this sort must be extended to variable crack size and shape.

Anomalously rapid growth kinetics for small cracks growing in moist air are not relatable to chemical influences because experiments were not conducted in an inert environment to isolate mechanical effects.(11,13,15-17,31). Lankford demonstrated rapid growth rates for small cracks in an aluminum alloy exposed to moist air at constant ΔK.(28) A larger crack size effect was reported for pure N_2, indicating that water vapor transport and reaction did not dominate the crack size-environment interaction.(28) Holder demonstrated that small fatigue cracks in steel grew at anomalously fast rates at low ΔK compared to extrapolated long crack kinetics for moist air, but not in an inert reference environment.(23) Impeded molecular flow, shielding the long crack tip from embrittling H_2O molecules, was invoked. Experiments were, however, limited and flow impedance was not modeled.

Geometry sensitive gas transport may contribute to the effect of crack size on hydrogen assisted fatigue crack propagation rates, Fig. 7a. 1 to 100 λ equals 0.1 to 10 μm for H_2 at 300 K and 100 kPa. Crack mouth opening displacement at maximum load varies from 0.1 to 7 μm for the small crack geometries and from 40 to 200 μm for the compact tension conditions indicated in Fig. 7a. Larger flow impedances (Equation 7) are expected for the long crack. This simple analysis does not consider surface roughness enhanced molecule-wall collisions, and mass transport due to convection or surface diffusion. If flow impedance caused the chemical crack size effect at low R, then growth rates should be crack size independent at high R where the long crack tip is open and accessible to the bulk environment. A stress ratio effect is observed only for the long crack, (Fig. 7b). Gas transport control is not, however, unambiguously identified because data are equally well explained based on crack closure. Determination of the effect of hydrogen pressure on the magnitude of the crack size effect at constant ΔK would differentiate between impeded gas transport, proportional to $P_{H_2}^{-1}$, and crack closure, independent of P_{H_2}.

3.3 AQUEOUS ENVIRONMENTS

STATIC LOADING: DATA AND THEORY Small crack-environment interactions during static loading provide a basis for understanding more complex chemical crack size effects in fatigue. Superposition concepts relate the environmental effect for each loading mode.(84)

Experimental evidence for chemical crack size effects on stress corrosion cracking is virtually nonexistent. Static load growth rate data, presented in Fig. 8a for 0.1 to 2 mm deep elliptical surface and through thickness edge cracks in 4130 steel (σ_o = 1330 MPa) exposed to 3% NaCl, show Stage I, K-independent (Stage II) kinetics for replicate specimens. Critically, small cracks grow at faster plateau rates and at stress intensities well

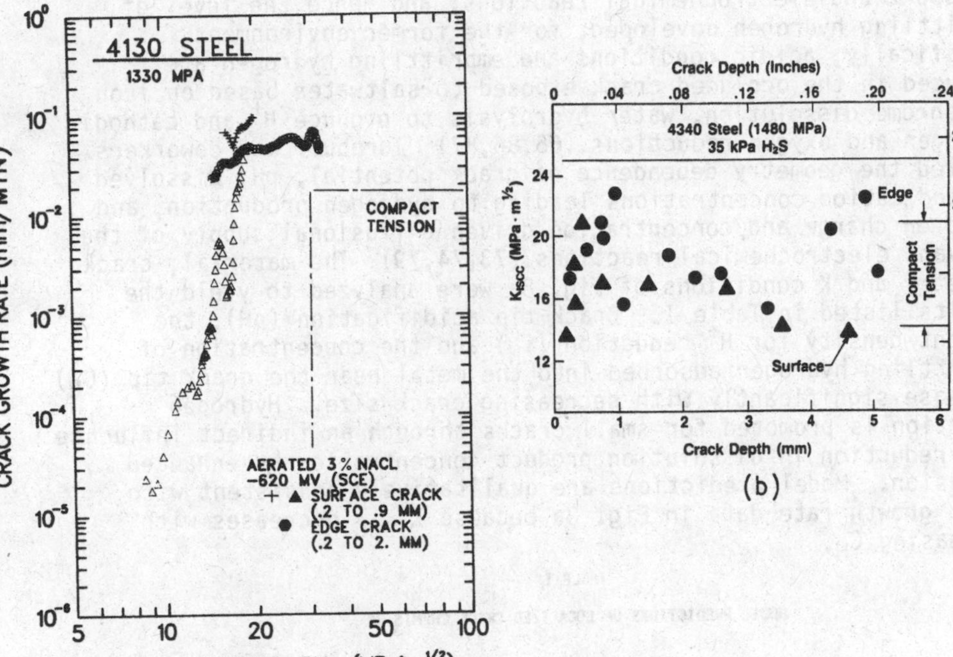

FIG. 8. Crack size effects on static load embrittlement of high strength steels. (a) Growth rate - K data for 0.1 to 2.0 mm deep cracks in 4130 steel (σ_o = 1330 MPa) exposed to aqueous NaCl. Note the rapid rates of growth for small cracks; after Gangloff (1984). (b) Threshold stress intensity versus crack size for surface and edge flawed specimens of 4340 steel (σ_o = 1480 MPa) exposed to gaseous H_2S. Note the crack size independence of K_{ISCC} supporting similitude; after Clark (1976).

below the threshold (K_{ISCC}), compared to literature data for long (> 20 mm) cracks in similar steels at constant strength.(85) Shahinian and Judy (7) report constant K_{ISCC} for 3.6 to 10 mm deep elliptical surface flaws and deep cracks in cantelever beam specimens for two strength levels (σ_o = 890 and 1410 MPa) of 4340 steel in 3.5% NaCl.

The crack size effect portrayed in Fig. 8a is traceable to geometry effects on the electrochemical processes which contribute to embrittlement. Cracks at either size scale would not grow in an inert environment at the K levels examined, and crack closure and corrosion product wedging are not relevant. Crack size has no influence on K_{ISCC} for gaseous H_2S embrittlement of a similar steel (Fig. 8b).(85) While embrittlement in aqueous NaCl and gaseous H_2S is attributable to hydrogen, crack geometry only influences solution transport and electrochemical reactions, and hence the level of embrittling hydrogen developed, for the former environment. Specifically, acidic conditions and embrittling hydrogen are produced in the occluded crack exposed to saltwater based on iron and chrome dissolution, water hydrolysis to produce H^+ and cathodic hydrogen and oxygen reductions.(65,86,87) Turnbull and coworkers modeled the geometry dependence of crack potential, pH, dissolved O_2, and cation concentrations leading to hydrogen production; and based on charge and concentration driven diffusional supply of the relevant electrochemical reactions.(73,74,79) The material, crack geometry and K conditions of Fig. 8a were analyzed to yield the results listed in Table I. Crack tip acidification (pH), the current density for H^+ reduction (i_H) and the concentration of embrittling hydrogen adsorbed into the metal near the crack tip (C_H) increase significantly with decreasing crack size. Hydrogen reduction is promoted for small cracks through an indirect influence of a reduction in dissolution product concentration by enhanced diffusion. Model predictions are qualitatively consistent with crack growth rate data in Fig. 8a because K_{ISCC} decreases with increasing C_H.

TABLE I

MODEL PREDICTIONS OF LOCALIZED CRACK CHEMISTRY

CRACK TIP CHEMISTRY*

CRACK SIZE	pH	Potential (mV-SCE)	i_H A/cm^2 ($\alpha\ C_H^2$)
0.25 mm Deep 0.5 μm Wide	4.1	-638	5×10^{-7}
38 mm Deep 250 μm Wide	8.0	-664	3×10^{-8}

*E_{CORR} = -600mV-SCE Bulk pH = 6.0
K = 10 MPa√m

Model predictions and growth rate data for alternate stress corrosion cracking systems have not been compared to establish the importance of crack size. Doig and Flewitt(76) and Turnbull(73,79) predict that crack propagation rates controlled by anodic dissolution increase with decreasing crack size as shown in Fig. 9 for low alloy steel in a boiling caustic solution. Changing crack size between 0.1 and 2 mm significantly influences crack growth rate, while the effect saturates for larger crack sizes. Applied stress intensity is constant, but growth rate changes with crack size due to a chemical influence. Small cracks are sensitive to embrittlement because of relatively small potential differences between tip and surface, and because of enhanced elimination of dissolution products. Both phenomena promote rates of dissolution.

Smyrl and Newman(88) predict that diffusion supplies a propagating stress corrosion crack tip only for depths less than a_c, given by the ratio of diffusivity (D) to crack speed. For $a > a_c$, crack growth is length dependent at constant applied K, while short cracks propagate at faster rates which are not diffusion limited. For example, a_c equals 1 mm for typical values of D (10^{-3} mm^2/sec) and crack speed (10^{-3}mm/sec., Figure 8). Charnock and Taunt(89) demonstrated that solute penetration by diffusion into the occluded crack solution is proportional to $\sqrt{\nu}$ for a slowly moving crack with chemically reactive crack walls. Environment transport to the crack tip is enhanced for stress and geometry factors which increase crack opening.

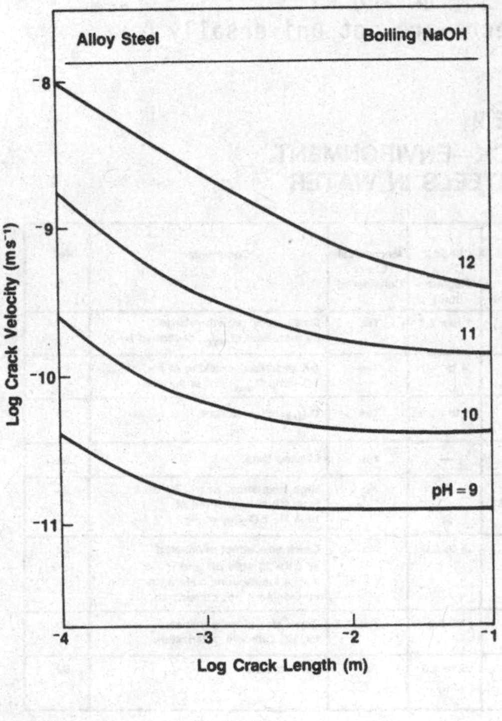

FIG. 9. Predicted effect of crack size on static load growth rates controlled by crack tip anodic dissolution for low alloy steel in boiling NaOH at constant crack opening. After Doig and Flewitt (1983).

CYCLIC LOADING: EXPERIMENTAL RESULTS Small fatigue crack-environment interactions have been investigated only for low alloy steels exposed to aqueous chloride solutions, a system relevant to many applications including offshore structures.(33,65,80,90,91) Specific data are summarized in Table II, where comparisons between short and long crack growth kinetics indicate marked compromises in similitude.

Important trends are apparent based on the data contained in Table II.

i) Small corrosion fatigue cracks always grow faster (between 1.2 and 300 times) than projected based on long crack kinetics at low to moderate ΔK and R values. Crack size influences corrosion fatigue growth rates comparably to well recognized variables.(80)

ii) The size regime for small crack-environment interactions is below about 3 mm. Results are, however, insufficient to exclude a size effect for deeper cracks.

iii) The magnitude of the small crack effect decreases with increasing a at constant $\Delta\sigma$ (increasing ΔK), with increasing $\Delta\sigma$ at constant a and with increasing R. Crack opening shape, approximated by mouth displacement, correlates a, $\Delta\sigma$, ΔK and R effects.(14,24) Stress effects are not universally observed.(29,30)

TABLE II
SMALL FATIGUE CRACK—ENVIRONMENT INTERACTION FOR STEELS IN WATER

Material	σ_0 (MPa)	Environment	Crack Sizes (mm)	Rate Defined By ΔK?	$\frac{da/dN_{Short}}{da/dN_{Long}}$ @ ΔK Constant	"Short" Crack Regime (mm)	Mechanical Effect Considered	Comments	Ref
AISI 4130 (.3C-.9Cr-.2Ni)	1330	3% NaCl	.1 to 40	Yes-Air No-NaCl	1.2 to 300	.1 to >2.5	Yes	Small crack growth retarded by increased σ_{MAX}, described by V.	21,24
HY-130 (.1C-.5Cr-5Ni)	—	3% NaCl	.4 to 2.5	Yes-Air No-NaCl	1.0 to 2.5	.4 to 1.0	Yes	ΔK described cracking at a> 1-3 mm; σ_{MAX} not important.	29
HY-130	930	3% NaCl	.1 to 40	Yes-Air No-NaCl	4.0	.1 to ≻1.0	Yes	σ_{MAX} not important.	30
HY-130	950	Seawater	>.5	Yes	1.0	—	Yes	Limited Data.	92
13Cr (.03C-12.8Cr-5Ni)	770	Water	?	No	4.0 @R=0 High @R=.8	?	No	High frequency, near-ΔK_0 data. Size effect intensified at high R. $\Delta\sigma_{th}(a)$ in Fig.1.	20
Q1N (.17C-1.2Cr-2.4Ni)	625	Seawater	.5-6	No	1.0 to 4.0	.5 to 2.0	No	Crack size effect eliminated at ΔK> 30 MPa·m^{5} and R ~ .5. K, σ,a interactions, reference environment not considered.	22
BS-4360 (.18C-1.17 Mn)	370	Seawater	1-7	Yes-Air No-Seawater	1.0 to 3.0	1.0-3.0	Yes	Crack size effect eliminated for air, cathodic polarization.	90
EN5 (.3C-.67Mn)	300	Seawater	>.5	No	1.0 to 3.0	.5 to 2.0	No	Limited Data.	92

iv) The magnitude of the small crack effect decreases with decreasing yield strength, however, a systematic study has not been performed. This trend is consistent with a general decline in both environment sensitivity and the effects of R, frequency and environment activity with decreasing strength.(80,85)

v) The mechanism for the crack size effect is chemical for high strength steels, based on measured crack growth rates in benign environments and on the unlikely occurrence of mechanical effects on similitude for a > 0.1 mm. Mechanical effects may contribute to the rapid growth of short cracks in low strength steels. Studies in an inert environment have not been conducted to separate chemical and mechanical effects for such steels.

vi) Compliance measurements have not been reported for small cracks in embrittling environments. Environment enhanced plasticity, corrosion product and roughness induced closure contributions to retarded cracking cannot be assessed.

vii) The effects of cyclic frequency, electrochemical potential and bulk solution composition on the kinetics of small corrosion fatigue cracks have not been investigated extensively.(14,90)

viii) Small crack corrosion fatigue studies are impeded by problems in crack monitoring,(27) by the low frequencies and ΔK levels of interest,(90) by the possibility of net section yielding (9,25) and by small differences in growth rates, which are none-the-less relevant to long term component integrity.(90,92)

CYCLIC LOADING: THEORETICAL ANALYSES Early explanations for the rapid growth of small corrosion fatigue cracks were based on enhanced environment access to the crack tip.(22,23,54,90) The proximity concept is, however, overly simplistic. Recent results for 4130 steel demonstrate that small crack growth rates decrease with increasing stress and with decreasing crack size for very small cracks; embrittlement decreases as crack opening increases. (14,21,24) Tortuous, Stage I fatigue cracking is often not observed for aqueous environments. The proximity concept implies diffusional flow, however, precise transport mechanisms are not specified. Turnbull(79) demonstrates that mass transport is not necessarily impeded by increased crack depth because of the increasingly important contribution of convection. Finally, the proximity concept does not specify the chemical mechanism for brittle crack extension. Impeded transport of an inhibiting species to the crack tip will, for example, result in enhanced crack propagation.

Modeling is required to define the effects of crack geometry on the chemical driving force for embrittlement. Mass transport by diffusion and convection determines crack chemistry,(79) which dictates transient reactions with straining crack surfaces,(29) which, in turn, control brittle crack advance. A detailed theory does not exist.(71) Elements of the problem are, however, developed from a hydrogen embrittlement perspective, and are relevant to corrosion fatigue by dissolution and film rupture. For steel in an aqueous solution, hydrogen is produced by hydrolysis within the occluded, pulsating crack.(86,93) The time and cycle dependent corrosion fatigue components to the total crack growth rate, da/dN_{CF}, are defined by (14,29,32,65):

$$\frac{da}{dN_{CF}} = \phi C_H(\sigma,a,t)\ \Delta K^2, \qquad (8)$$

where ϕ is a constant. The chemical driving force is represented by the concentration of adsorbed hydrogen (C_H) which is crack size, stress and time dependent. The mechanical driving force is accounted for by ΔK^2.

If convective mixing is ignored, then modeling of the static crack provides a description of $C_H(\sigma,a,t)$,(73,79) with enhanced acidification and hydrogen discharge predicted for decreased crack size (Table I). Since hydrogen embrittlement is probable for the conditions represented in Table II, (65,86,93) this result provides a reasonable explanation for the rapid growth of small corrosion fatigue cracks. For 4130 steel the crack size effect on corrosion fatigue, Fig. 2, is predicted in part based on linear superposition of stress corrosion growth rates for "short" and "long" cracks (Fig. 8a) combined with growth rates for a benign environment.(24) This comparison adds credence to the applicability of the static crack model for hydrogen production.

Crack size dependent convection effects cannot, however, be ignored.(74,75,77-79) Considering hydrolysis, no model exists to describe convection effects on electrode potential, pH and metal ion concentrations within a pulsating crack. The importance of crack geometry dependent convection is illustrated by analyses of dissolved O_2 supply and reduction within a crack. Static cracks are presumed to be fully oxygen depleted due to cathodic reduction dominating diffusional supply; O_2 does not affect C_H.(73,74) Convection provides an additional source of oxygen which consumes protonic hydrogen during each load cycle. C_H and da/dN_{CF} are reduced for decreased acidification traceable to "oxygen inhibition".

Perfect mixing analysis of O_2 supply and reaction demonstrates that depletion, and by inference the effect of O_2 on corrosion fatigue, are reduction rate, cyclic frequency, crack size and opening shape dependent. Taunt and Charnock(77) analyzed solute supply and reaction within a crack as a function of load cycles for a variety of rate expressions. Turnbull(74) focused a similar analysis on O_2 depleted at a rate proportional to instantaneous concentration. Each model demonstrates that ΔK similitude is compromised. These approaches were modified(14) to predict the concentration of dissolved $O_2(C_0)$ within small cracks for square wave fatigue loading in aqueous chloride at a frequency of $1/\tau$ and O_2 reduction at a rate given by αC_0. The result is:

$$C_0 = C_B\,(1 - R)\,\exp\left(\frac{-\alpha A_c \tau a^*}{2V_{max}}\right) \qquad (9)$$

where C_B = bulk solution oxygen concentration, A_c = crack surface area, V_{max} = crack solution volume at K_{max}, and a^* = geometric constant. For a wedge crack, the ratio of A_c to V_{max} equals $4/\nu_{max}$. As such C_0 depends exponentially on $-1/\nu_{max}$, a parameter which is crack depth sensitive at constant ΔK. (Typically, $\nu_{max} \propto (\sigma_{max}\ a) \propto (K_{max}\sqrt{a})$.) Physically, depletion is controlled by the ratio of active crack surface area available for reaction, a^*A_c, to the occluded solution volume, V_{max}, which supplies reactant. Small cracks are distinguished by a large surface to occluded solution volume ratio, and hence by extremely low values of C_0 compared to long cracks at constant ΔK. Equation 9 is plotted in Fig. 10 as a function of edge crack depth for constant ΔK, frequency and reaction rate conditions. C_0/C_B is on the order of 10^{-40} for 0.1 mm deep cracks, and rises rapidly to values between 10^{-4} and 10^{-2} for cracks deeper than 5 mm.(14)

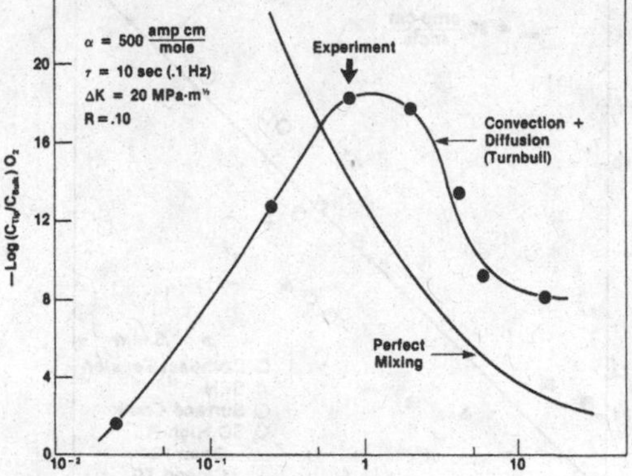

FIG. 10. Analytical predictions of the effect of edge crack length on dissolved oxygen content at constant ΔK. Modeling includes O_2 depletion by cathodic reduction and O_2 supply by either perfect convective mixing (Gangloff, 1984) or diffusion and laminar convection (Turnbull, 1983).

The unique chemical character of small cracks is confirmed by relating perfect mixing to brittle corrosion fatigue. Specifically, the hydrogen ion concentration remaining at the end of a square wave load cycle is calculated based on O_2 reduction to consume H^+ and produce OH^-. O_2 modified acidification is related to C_H and corrosion fatigue crack growth rate through equation 8 to yield either:(14)

$$\mathrm{LOG}\ \frac{da}{dN_{CF}} = \phi + \alpha\tau(a^*/\nu_{max}) \qquad (10)$$

at constant ΔK, or the normalized equation listed and plotted in Fig. 11. (Rates are normalized by an absolute brittle crack growth rate (da/dN_{CF-ABS}) independent of O_2 inhibition to account for the mechanical influence of ΔK^2.) The extent of O_2 inhibition increases as the normalized growth rate parameter decreases from 1.0 to 0. Physically, increased crack length results in increased crack solution volume to active surface area, increased O_2 reduction persisting to longer times during loading, decreased hydrogen production and hence in reduced rates of brittle corrosion fatigue.

The perfect mixing analysis of O_2 inhibition is consistent with experimental results. The logarithm of corrosion fatigue crack growth rate for 4130 steel in saltwater (Fig. 2) varies linearly with reciprocal maximum crack mouth opening as shown in Fig. 11 and predicted by equation 10. The O_2 reduction rate constant inferred from least squares slope analysis is of the correct order. Similar good agreement is observed at constant ΔK without normalization.(14) Considering Fig. 11, compact tension cracking produces small changes in ν_{max} (squares) compared to the results for 0.5 to 3 mm deep edge and surface cracks. The retarding influences

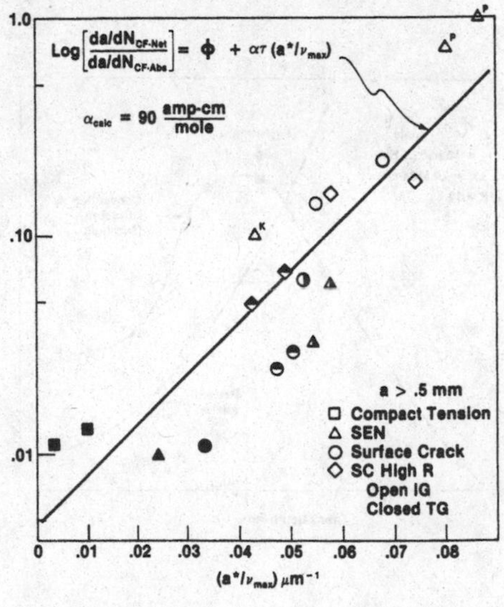

FIG. 11. Experimental data and analytical prediction for the effect of crack shape, given by reciprocal crack mouth opening at maximum load, on corrosion fatigue crack growth rate for 4130 steel in 3% NaCl. Fracture mechanics similitude predicts a horizontal line at 1.0. After Gangloff (1984).

of increased stress, produced by increased R, increased $\Delta\sigma$ or instantaneous stress increases; and of increased crack size are accounted for by maximum crack opening shape. A brittle fracture mode transition occurs with decreasing da/dN_{CF} and correlates with $(\nu_{max})^{-1}$ indicative of a chemical crack size effect. Intergranular cracking (open symbols) at small ν_{max} is replaced progressively by brittle transgranular fracture (filled symbols) at increased ν_{max}. Many of the trends represented for lower strength steels in Table II are consistent with maximum crack mouth opening displacement control and perhaps O_2 inhibition. Small cracks grow at the fastest rates when sized below 1 mm and when stressed at low ΔK and R. Cracking is retarded by increased a, ΔK or R, or equivalently, by increased crack mouth opening.

While the breakdown in ΔK similitude and the correlation between da/dN_{CF} and crack opening are well established in Fig. 11, the O_2 inhibition model is speculative. Acidification differences may contribute to crack size effects, Table I. Experiments with deoxygenated solutions are inconclusive for 4130 steel,(14) relevant values of α are uncertain and may exceed 15 amp•m/mole,(74,75,79) and perfect mixing may provide an inaccurate description of mass transport. Turnbull argues that mass transport within a slowly cycled (< 1 Hz) corrosion fatigue crack is viscous-laminar.(79) Turbulence is expected and perfect mixing is relevant only when crack surface contact occurs. For viscous flow, a transition crack size (a_T) is defined:

$$a_T = \frac{\sqrt{(D/f)}}{1 - R^{(.5 \text{ to } 1.0)}}, \qquad (11)$$

where f is cyclic frequency and the exponent depends on precise crack shape. Diffusion dominates mass transport for cracks sized below a_T, while deeper cracks are supplied predominantly by convection. a_T equals between 0.1 and 0.2 mm for oxygen dissolved in water and a loading frequency of 0.1 Hz.

Crack size, ΔK, frequency, R and chemical reaction rate influence crack chemistry, as illustrated by analysis of diffusional/laminar convectional O_2 supply and cathodic reduction.(75,79) A typical prediction is presented in Fig. 10 and compared to the perfect mixing approximation for the same conditions. Note that perfect mixing provides an upper bound on dissolved O_2 within the crack. At very small crack depths, O_2 is enriched significantly due to diffusion. As crack size increases, O_2 diffusion decreases and convection increases. C_0 is minimized, and further increases in crack depth result in increased dissolved O_2 due to convection. The comparison between crack growth rate data and perfect mixing theory, Fig. 11, is for small cracks sized above 0.5 mm to avoid the complicating effects of diffusion. Crack growth rates at constant ΔK for 4130 steel in saltwater exhibit a maximum at a = 0.8 mm, and decline with increasing and decreasing crack size.(14) The complex effect of crack size on crack chemistry illustrated in Figure 10 may be general. The minimum value of C_0 represents that crack depth where transport of either a bulk specie to the tip or a crack tip reaction product to the bulk is most effectively impeded. Modeling further indicates that, for the long crack convection dominated regime, concentration increases with decreasing O_2 reduction rate and increasing ΔK, R and frequency. Variable effects on the location of the maximum in Figure 10 have not been analyzed in detail. Experimental evidence for crack size effects of the type predicted in Fig. 10 is lacking.

4. SUMMARY

When analyzing the growth of small fatigue cracks in embrittling gaseous and aqueous environments, ΔK or ΔJ based similitude concepts must be examined. Data confirm that small crack geometry effects are not wholly accounted for by a crack tip field parameter. Crack size and opening shape must be considered as variables, and crack closure phenomena characterized experimentally. Unfortunately, it is not possible to provide general criteria which define limiting crack size and shape for ΔK similitude from an environmental perspective. Crack size effects are material and environment specific, and may vary with alloy strength, cyclic frequency, temperature, applied potential or gas pressure. Theoretical analyses of the mechanical and chemical origins of the driving force for brittle crack growth clearly establish the potential for novel small crack-environment interactions. Refined analyses are required, however, to fully define crack geometry effects. Experimental tools exist to assess the importance of crack size effects for specific corrosion fatigue applications.

Understanding of small crack effects is limited for technologically important, low strength alloys in gas, water, chloride or caustic environments, and for both static and cyclic loading. Deleterious chemical crack size effects are documented for high strength steel in aqueous chloride and gaseous hydrogen, and mechanistic understanding is forthcoming. Complicating plasticity and wake, roughness and corrosion debris induced closure effects are unlikely. In contrast research is required for low strength structural steels. Crack size effects are suggested experimentally and hydrogen is known to embrittle such alloys, albeit less severely compared to high strength steels. Environment enhanced plasticity and closure mechanisms are particularly relevant for lower strength alloys. As such, the range of interactions considered in this review may impact corrosion fatigue and produce complex crack size effects. For alternate environments, the potential exists for crack geometry effects on localized chemistry. For example, the chloride conditions described in previous sections were unbuffered and at the free corrosion potential. Transport of buffering species, as encountered in seawater or inhibited solutions, and the response of crack tip potential and electrochemical reactions to applied potential, as encountered in cathodic protection, are likely to be crack geometry sensitive. Crack geometry effects related to anodic dissolution and film rupture are largely undefined. The research challenge exists to isolate, measure and model such geometry sensitive processes.

ACKNOWLEDGEMENTS

This work was supported by the Corporate Research Science Laboratories of the Exxon Research and Engineering Company (for R.P.G.), and under Grant No. AFOSR-82-0181 from the U.S. Air Force Office of Scientific Research (for R.O.R.).

REFERENCES

(1) Williams, M. L. (1957). Journal of Applied Mechanics, Trans. ASME, 24, 104-109.
(2) Hutchinson, J. R. (1968). J. Mech. Phys. Solids, 16, 13-31.
(3) Rice, J. R. & Rosengren, G. F. (1968). J. Mech. Phys. Solids, 16, pp. 1-12.
(4) McEvily, A. J. & Wei, R. P. (1973). Corrosion Fatigue: Chemistry Mechanics and Microstructure, eds. O. Deveraux, A. J. McEvily and R. W. Staehle, pp. 381-395, Houston: NACE.
(5) Brown, B. F. (1968). Met. Rev., No. 129, 171-183.
(6) Novak, S. R. & Rolfe, S. T. (1970). Corrosion, 26, 121-130.
(7) Shahinian, P. & Judy, Jr., R. W. (1976). Stress Corrosion: New Approaches, ASTM STP 610, ed. H. L. Craig, Jr., pp. 128-142, Philadelphia: ASTM.
(8) Hertzberg, R. W. (1983). Deformation and Fracture Mechanics of Engineering Materials, pp. 519-618, New York: John Wiley and Sons.
(9) Dowling, N. E. (1977). Cyclic Stress-Strain and Plastic Deformation Aspects of Fatigue Crack Growth, ASTM STP 637, pp. 97-121, Philadelphia: ASTM.
(10) Rice, J. R., et al. (1980). Fracture Mechanics, 12th Conf., ASTM STP 700, pp. 189-221, Philadelphia: ASTM.
(11) Suresh, S., & Ritchie, R. O. (1984). Int. Metals Reviews, 25, in press.
(12) Suresh, S. (1983). Metall. Trans. A, 14A, 2375-2385.
(13) Hudak, S. J. (1981). J. Eng. Matls. Technol., Trans. ASME, Ser. H, 103, 26-35.
(14) Gangloff, R. P. (1984). Embrittlement by the Localized Crack Environment, ed. R. P. Gangloff, Warrendale, PA: TMS-AIME, in press.
(15) Ritchie, R. O., & Suresh, S. (1983). Mater. Sci. Eng., 57, 127-130.

(16) Miller, K. J. (1982). Fat. Eng. Matls. Struc., 5, 223-232.

(17) Leis, B. N., et al. (1983). Air Force Wright Aeronautical Labs, Report No AFWAL-TR-83-4019, Wright-Patterson AFB, Ohio.
(18) Lankford, J. & Davidson, D. L. (1984). Fatigue Crack Growth Threshold: Concepts, eds. D. L. Davidson and S. Suresh, Warrendale, PA: TMS-AIME,in press.
(19) Tanaka, K. & Nakai, Y. (1983). Fat. Eng. Matls. Struc., 5, 315-327.
(20) Usami, S. & Shida, S. (1979). Fat. Eng. Matls. Struc., 1, 471-481.
(21) Gangloff, R. P. (1981). Res. Mech. Let., 1, 299-306.
(22) Jones, B. F. (1982). J. Matl. Sci., 17, 499-507.
(23) Holder, R. (1977). Proc. Conf. on Influence of Environment on Fatigue, pp. 37-41, London: Instit. Mech. Engr.
(24) Gangloff, R. P. (1984). Metall. Trans. A, submitted for publication.
(25) El Haddad, M. H., et al. (1980). Int. J. Fract., 16, 15-30.
(26) Lankford, J. (1982). Fat. Eng. Matls. Struc., 5, 233-248.
(27) Gangloff, R. P. (1982). Advances in Crack Length Measurement, ed. C. J. Beevers, pp. 175-221, London: EMAS.
(28) Lankford, J. (1983). Fat. Eng. Matls. Struc., 6, 5-31.
(29) Wei, R. P., et al. (1984). Embrittlement by the Localized Crack Environment, ed. R. P. Gangloff, Warrendale, PA: TMS-AIME, in press.
(30) Hudak, S. J. & Gangloff, R. P. (1984). unpublished research.
(31) Miller, K. J. (1984). Proc. This Conference, Cambridge: Cambridge University Press.
(32) Wei, R. P. & Simmons, G. W. (1981). Int. J. Frac., 17, 235-247.
(33) Scott, P. M. (1984). Corrosion Fatigue Mechanics, Metallurgy, Electrochemistry and Engineering, ASTM STP, Philadelphia, PA: ASTM, in press.
(34) Walker, N. & Beevers, C. J. (1979). Fat. Eng. Matls. Struc., 1, 135-148.
(35) Minakawa, K. & McEvily, A. J. (1981). Scripta Metall., 15, 633-636.
(36) Toplosky, J. & Ritchie, R. O. (1981). Scripta Metall., 15, 905-908.
(37) Endo, K. et al. (1972). Bull. JSME, 25, 439-445.
(38) Tzou, J.-L., et al. (1984). Acta Met, submitted for publication.
(39) Vosikovsky, O. & Cooke, R. J. (1978). Int. J. Pres. Ves. & Piping, 5, 113-129.
(40) Kesten, M. & Windgasser, K.-F. (1981). Hydrogen Effects in Metals, eds. I. M. Bernstein and A. W. Thompson, pp. 1017-1025, Warrendale, PA: TMS-AIME.
(41) Maddox, S. J. (1974). Weld. Res. Suppl., 53, 401S-409S.
(42) Chauhan, P. & Roberts, B. W. (1979). Metall. Matl. Tech., 131-136.
(43) Bennett, J. A. & Mindlin, H. (1973). J. Test. Eval., 1, 152-161.
(44) Jack, A. R. & Paterson, A. N. (1977). Proc. Inst. Mech. Engr. Conf. Fat., paper C107/77, pp. 75-83, London: IMECHE.
(45) Muller, M. (1982). Metall. Trans. A, 13A, 648-655.

(46) Hoeppner, D. W. (1979). Fatigue Mechanisms, ASTM STP 675, ed. J. T. Fong, pp. 841-870, Philadelphia: ASTM.
(47) Kitagawa, H. & Takahasi, S. (1976). Proc. 2nd Intl. Conf. on Mech. Beh. Matls., pp. 627-663, Metals Park, Ohio: ASM.
(48) Lynch, S. (1983). Advances in the Mechanics and Physics of Surfaces, eds. R. M. Latanision and T. E. Fischer, pp. 265-364, New York: Harwood Academic Publishers.
(49) Hirth, J. P. (1980). Metall. Trans. A, 11A, 861-874.
(50) Davidson, D. L. & Lankford, J. (1983). Fat. Eng. Matls. Struc., 6, 241-256.
(51) Lankford, J. (1983). Mechanical Behavior of Materials - IV, eds. J. Carlsson and N. G. Ohlson, 1, pp. 3-29, Oxford: Pergamon Press.
(52) Bilby, B. A., et al. (1977). Fracture (ICF-4), ed. D. M. R. Taplin, 3, pp. 197-212, Waterloo, Canada: Univ. of Waterloo Press.
(53) Suresh, S. & Ritchie, R. O. (1984). Fatigue Crack Growth Threshold: Concepts, eds. D. L. Davidson and S. Suresh, Warrendale, PA:TMS-AIME, in press.
(54) Tomkins, B. (1977). Proc. Conf. on Influence of Environment on Fatigue, pp.111-115, London: IMECHE.
(55) Forsyth, P. J. E. (1962). Crack Propagation, p. 76, Cranfield College of Aeronautics: Cranfield Press.
(56) Tu, L. K. L. & Seth, B. B. (1978). J. Test Eval., 6, 66-74.
(57) Stubbington, C. A. (1963). Metallurgica, 68, 109-121.
(58) Elber, W. (1971) Damage Tolerance in Aircraft Structures, ASTM STP 486, pp. 230-242, Philadelphia, PA: ASTM.
(59) Tzou, J.-L., et al. (1984). Acta Met., submitted for publication.
(60) Morris, W. L., et al. (1981). Eng. Fract. Mech., 18, 871-977.
(61) Newman, J. C. (1983). Behavior of Short Cracks in Airframe Components, vol. CP 328, pp. 7.1-7.16, Advisory Group for Aeronautical Research and Development.
(62) Ritchie, R. O., et al. (1980). J. Eng. Matls. Tech., Trans. ASME, Ser. H, 102, 293-299.
(63) Stewart, A. T. (1980). Eng. Fract. Mech., 13, 463-478.
(64) Suresh, S., et al. (1981). Metall. Trans. A, 12A, 1435-1443.
(65) Scott, P. M., et al. (1983). Corros. Sci., 23, 559-575.
(66) Hartt, W. H., & Rajpathak, S. S. (1983). "Formation of Calcareous Deposits Within Simulated Fatigue Cracks in Seawater," Corrosion/83, Paper No. 62, Houston: NACE.
(67) Suresh, S., et al. (1982). Fatigue Thresholds, eds. J. Bäcklund, A. Blom and C. J. Beevers, 1, pp. 391-408, Warley, U.K: EMAS.
(68) Suresh, S. & Ritchie, R. O. (1982). Metall. Trans. A, 13A, 1627-1631.
(69) Esaklul, K. A., et al. (1983). Scripta Met., 17, 1073-1078.
(70) Gangloff, R. P. (1984). unpublished research.
(71) Embrittlement by the Localized Crack Environment (1984). ed. R. P. Gangloff, Warrendale, PA: TMS-AIME, in press.
(72) Lawn, B. R. (1977). Mat. Sci., 13, 277-283.

(73) Turnbull, A. & Thomas, J. G. N. (1982). J. Electrochem. Soc., 129, 1412-1422.
(74) Turnbull, A. (1980). Br. Corros. J., 15, 162-171.
(75) Turnbull, A. (1982). Corros. Sci., 22, 877-893.
(76) Doig, P. & Flewitt, P. E. J. (1983). Metall. Trans. A, 14A, 978-983.
(77) Taunt, R. J. & Charnock, W. (1978). Matl. Sci. Engr., 35, 219-228.
(78) Hartt, W. H., et al. (1978). Corrosion Fatigue Technology, ASTM STP 642, eds. H. C. Craig, Jr., T. W. Crooker, and D. W. Hoeppner, pp. 5-18, Philadelphia: ASTM.
(79) Turnbull, A. (1984). Embrittlement by the Localized Crack Environment, ed. R. P. Gangloff, Warrendale, PA: TMS-AIME, in press.
(80) Jaske, C. E., et al. (1981). MCIC Report 81-42, Columbus, Ohio: Battelle.
(81) Snowdon, K. V. (1963). J. Appl. Phys., 34, 3150-1.
(82) Gangloff, R. P. (1984). Liquid and Solid Metal Induced Embrittlement, ed. M. H. Kamdar, Warrendale, PA: TMS-AIME, in press.
(83) Bradshaw, F. J. (1967). Scripta Met., 1, 41-43.
(84) Wei, R. P. & Gao, Ming (1983). Scripta Met., 17, 959-962.
(85) Carter, C. S. (1977). Stress Corrosion Cracking and Corrosion Fatigue of Medium and High Strength Steel, Boeing Co. Report, Seattle, Washington.
(86) Brown, B. F. (1976). Stress Corrosion Cracking and Hydrogen Embrittlement of Iron Based Alloys, eds. J. Hockmann, J. Slater, R. D. McCright and R. W. Staehle, pp. 747-750, Houston: NACE.
(87) Sandoz, G., et al. (1970). Corros. Sci., 10, 839-845.
(88) Smyrl, W. H. & Newman, J. (1974). J. Electrochem. Soc., 121, 1000-1007.
(89) Charnock, W. & Taunt, R. J. (1978). Metall. Trans. A, 9A, 880-881.
(90) Bardal, E., et al. (1978). Proc. Conf. European Offshore Steel Research, pp. 415-436, London: Welding Institute.
(91) Cotton, H. C. (1979). Proc. Inst. Mech. Engrs., 193, 193-206.
(92) Jones, B. F. (1984). Embrittlement by the Localized Crack Environment, ed. R. P. Gangloff, Warrendale, PA: TMS-AIME, in press.
(93) Barsom, J. M. (1971). Int. J. Frac. Mech., 7, 164-182.

EXPERIMENTAL MECHANICS OF FATIGUE CRACK GROWTH: THE EFFECT OF CRACK SIZE

DAVID L. DAVIDSON

Southwest Research Institute, San Antonio, Texas 78284, USA

JAMES LANKFORD

Southwest Research Institute, San Antonio, Texas 78284, USA

ABSTRACT

Within the past few years new techniques have been developed which allow high resolution, dynamic observation of fatigue crack growth together with direct, accurate determination of crack opening loads, crack opening displacement, and strains within the crack tip region. These techniques have been used to characterize both large and small fatigue cracks in aluminum alloys, and the results are correlated using the cyclic stress intensity factor ΔK. But since small cracks are found to grow at ΔK values below the threshold ΔK for long cracks, and further, fail to satisfy similitude requirements for the validity of ΔK, this factor cannot be used for small cracks in the same way that it is for large cracks. Detailed characterization of crack tip parameters is shown to establish the physical basis for discrepancies between crack growth rates for large and small cracks at nominally equivalent ΔK values.

1. INTRODUCTION

It is by now a well-documented (1) fact that when the growth of "small" fatigue cracks is compared with that of "large" cracks, using the cyclic stress intensity factor ΔK as a correlating parameter, the growth rates for equivalent ΔK values do not correspond. In particular, it has been established for aluminum (2,3), titanium (4), and some steel alloys (5) that "small" fatigue cracks generally grow faster than, and below the threshold ΔK for, "large" cracks. This discrepancy can, in fact, provide an empirical definition of a "small" crack, i.e., one small enough that its rate of growth does not correlate with that of a large through crack on the basis of ΔK. In aluminum alloys, for example, a small crack corresponds to a surface crack less than 200-300 μm in length (3). Alternatively, cracks can be considered "small" when their length is comparable to 1) any local plastic region with which they may be associated, including their own crack tip plastic zone, or 2) some microstructural dimension, such as grain size (1).

The objective of this paper is to consider the appropriateness of ΔK as the driving force parameter for small cracks. Specifically, the applicability of ΔK hinges upon whether or not the concept of similitude applies. Similitude has been interpreted (6) to mean the following. For two different sized cracks, loaded under small scale yielding conditions to equal stress intensity values in a given material-microstructure-environment system, crack tip plastic zones are equal in size, and the stress and strain distributions along the borders of these zones ahead of the crack are identical. Small scale yielding means that the size of the crack tip plastic zone (r_p) relative to the crack size (a) is small, i.e., $r_p/a \ll 1$. Our earlier research (7) has shown that for large cracks, as would be expected, all of these requirements are apparently met (the status of the stress distribution at the plastic zone boundary, however, cannot be assessed). To the extent that any of the similitude requirements are violated, ΔK is invalid as a proper description of the driving force for crack advance. However, knowledge that such a violation has occurred does not indicate what is no longer similar about the micromechanical response of the material within the plastic zone.

Until very recently, the ability of researchers to even address these questions has been limited because the parameters of interest involve dynamic events occurring within an extremely localized (1-20 μm) region surrounding the crack tip. However, the recent development of a loading stage capable of dynamically cycling cracks in bulk specimens within the scanning electron microscope (SEM), as well as analytical techniques for determining crack tip displacements and strains, has begun to reveal the micromechanical differences between large and small cracks. This paper will focus on results for small and large cracks growing in high strength aluminum alloys, since most of the detailed micromechanical work has been for these materials. Early work on small cracks by de Lange (8) and Pearson (2) demonstrated the peculiar rapid growth behavior of small cracks in aluminum alloys, but it did not address the basis of the effect. A recent body of work by Morris and his co-workers (9,10) has been aimed at understanding the factors which contribute to the stochastic nature of ensembles of small growing cracks and which account for their transient retardation and arrest. In contrast to this approach, the authors' research during the past few years has focussed upon the specific micromechanical crack tip behavior attending the growth of individual large and small cracks, with the principal objective being the definition and quantification of those parameters which are important in describing the growth of fatigue cracks. The following report is based briefly upon this last body of work (11-13).

2. EXPERIMENTAL APPROACH

The techniques used in this work have been described in detail elsewhere (11-13), and therefore will be outlined here only briefly. The material studied was 7075 aluminum alloy, in both overaged (OA) and peak-aged (PA) conditions; yield strengths were 439 MPa and 515 MPa, respectively.

Two types of specimens were fabricated, i.e., single-edge-notch large crack, and smooth, slightly reduced section small crack, configurations. Both designs were compatible with the SEM cyclic loading stage. All cracks were grown in ambient air (~60% relative humidity) in laboratory test machines at a stress ratio of ~0.1, and a cyclic frequency of 1-5 Hz. Large crack specimens were precracked at an intermediate cyclic stress intensity, incrementally load shed to near threshold ΔK, and then allowed to increase in length (and ΔK) at constant cyclic load. Small cracks, on the other hand, were nucleated and grown at a constant cyclic stress ($\Delta\sigma$) of ~80% of the yield strength; a replication technique was used to accurately measure their cyclic growth.

For both types of flaws, crack tip deformation and opening were periodically observed in the SEM as ΔK increased. Cracks were oriented relative to their pancake-type grains (Fig. 1) so that loading was applied in the rolling direction (RD), and crack extension was in the transverse direction (TD); small cracks also grew in the normal (N) direction.

Displacements in the region near the crack tip caused by loading the specimen were measured using the stereoimaging technique (11). Measurements obtained from high resolution photomicrographs provided accurate determinations of crack opening displacement.

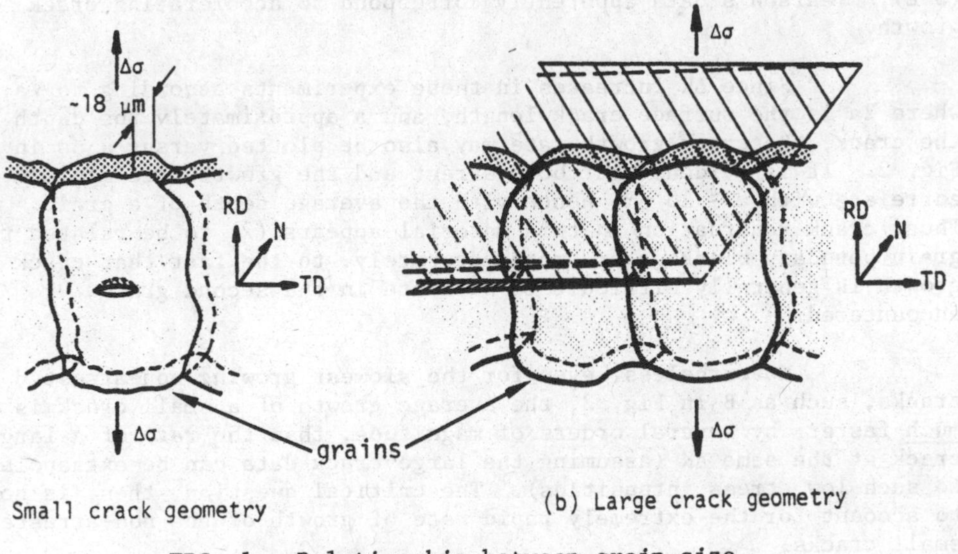

(a) Small crack geometry (b) Large crack geometry

FIG. 1. Relationship between grain size and shape and specimen/crack geometries.

Knowledge of the displacement field in the near crack tip region allowed computation of the corresponding crack tip strain field. Crack tip opening loads were determined by stereoimaging observation of cracks at various fractions of the load cycle.

In addition to these local crack tip parameters, the size and shape of the outer (monotonic) plastic zone was measured by selected area electron channeling pattern analysis (13). Following the crack tip deformation experiments, the specimens were broken open, and examined fractographically in the SEM.

3. COMPARATIVE BEHAVIOR OF LARGE AND SMALL CRACKS

The work from which the following comparisons are drawn represents extensive experimentation involving many crack tip observations; the data are fully documented elsewhere (3,7,11-13). In order to facilitate comparison, parametric behavior is shown as bands or sets representing limits and trends. The results shown hold for both OA and PA 7075.

3.1 CRACK GROWTH

Typical crack growth behavior for small cracks and large cracks is shown in Fig. 2; also shown for comparison are the earlier microcrack results of Pearson (2). Most small cracks begin growing rather rapidly, but may gradually slow down as the cracks lengthen. Some of these cracks actually arrest (A), while those which are eventually able to accelerate suffer varying degrees of retardation (B-E). Pearson's data apparently correspond to accelerating crack growth.

Since ΔK increases in these experiments according to $\sqrt{a}$, where 2a is the surface crack length, and **a** approximately the depth of the crack, the crack growth rate may also be plotted versus **a** as in Fig. 2. It is evident that both arrest and the growth rate minimum correlate with **a** $\simeq$ 20 μm, hence with the average depth of a grain. Thus, crack retardation in this material appears (2) to be related to grain boundary blockage or, more accurately, to the fact that crack growth is generally difficult to nucleate in the second grain encountered.

Nevertheless, even for the slowest growing non-arrested cracks, such as B in Fig. 2, the average growth of a small crack is <u>much</u> faster, by several orders of magnitude, than the rate of a large crack at the same ΔK (assuming the large crack data can be extrapolated to such low stress intensities). The critical question, then, is how to account for the extremely rapid rate of growth of the non-arrested small cracks.

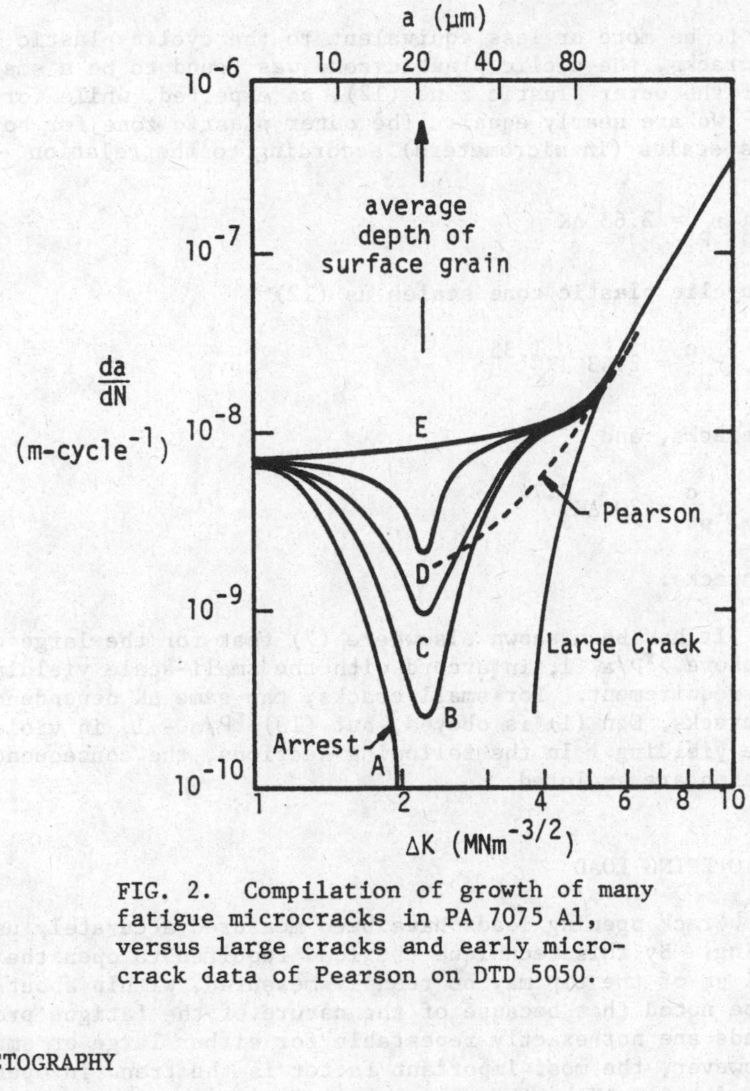

FIG. 2. Compilation of growth of many fatigue microcracks in PA 7075 Al, versus large cracks and early microcrack data of Pearson on DTD 5050.

3.2 FRACTOGRAPHY

Scanning electron microscopy shows that for the ΔK range 2-5 $MNm^{-3/2}$ for microcracks (14), and 4-6 $MNm^{-3/2}$ for large cracks (15), the dominant mode of crack tip separation is cyclic cleavage. In neither case, despite extensive tilting experiments, are striations resolvable in the SEM. Thus, it does not appear that the crack size effect is due to fundamentally different crack extension mechanisms.

3.3 PLASTIC ZONE SIZE

Plastic zone size determined by analysis of selected area electron channeling patterns is thought to be equivalent to the outer, or monotonic, plastic zone. Conversely, the plastic zone size derived from stereoimaging is based on cyclic measurements, and is therefore considered to be more or less equivalent to the cyclic plastic zone. For large cracks, the cyclic plastic zone was found to be a small fraction of the outer plastic zone (12), as expected, while for small cracks the two are nearly equal. The outer plastic zone for both crack sizes scales (in micrometers) according to the relation

$$r_p = 2.63\ \Delta K^2 \quad (1)$$

while the cyclic plastic zone scales as (12)

$$r_p^c = 2.43\ \Delta K^{1.35} \quad (2)$$

for large cracks, and

$$r_p^c = 20\ \Delta K^{0.74} \quad (3)$$

for small cracks.

It has been shown elsewhere (7) that for the large cracks discussed above, $r_p/a \ll 1$, in accord with the small-scale yielding similitude requirement. For small cracks, the same ΔK dependency found for large cracks, Eqn (1) is obeyed, but (13) $r_p/a \simeq 1$, in violation of small scale yielding. In the following sections, the consequences of this violation are explored.

3.4 CRACK OPENING LOAD

Crack opening loads have been measured accurately using stereoimaging. By this technique the load required to open the crack to within 1 μm of the tip may be readily measured, within about 5% error. It should be noted that because of the nature of the fatigue process, opening loads are not exactly repeatable for either large or small cracks. However, the most important factor is the trend in opening load observed using ΔK as the correlating parameter, shown in Fig. 3. For large cracks, the difference in opening and maximum loads apparently goes to zero at threshold ΔK, whereas for small cracks, the opening load is an approximately constant fraction of maximum load, independent of ΔK, which implies that it is also independent of crack length (for $35 < a < 200$ μm) (13).

3.5 CRACK OPENING MODE

Fatigue cracks loaded in Mode I almost always open with a combination of Mode I and Mode II. However, large cracks generally evidence more Mode II opening than do small cracks, compared at the same ΔK. Large cracks in this alloy evidence a shift towards Stage I crack growth as ΔK is decreased toward the threshold (12).

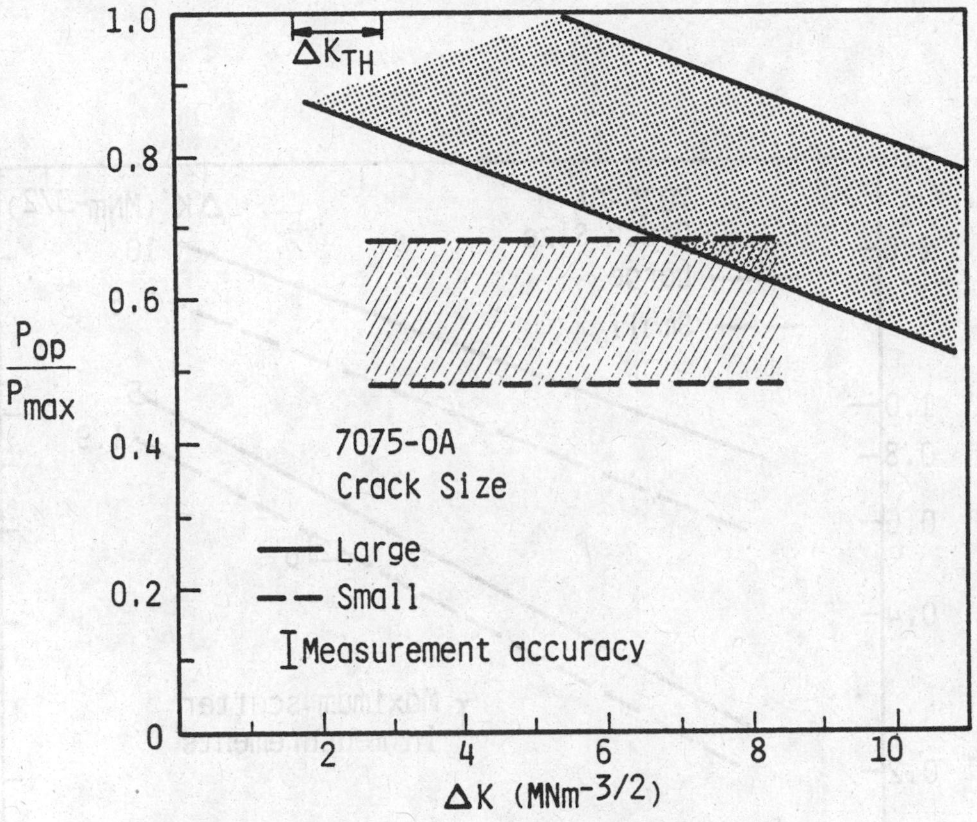

FIG. 3. Crack opening load fraction of maximum applied load as a function of crack size.

3.6 CRACK OPENING DISPLACEMENTS

Crack opening displacements (COD) for large and small cracks are shown in Fig. 4. Values of COD shown are the vector resultant of combined Mode I and Mode II opening. The slopes (p) of these curves do not change systematically with ΔK for small cracks but they do for large cracks. Crack tip opening displacement (CTOD) is here defined as the value of the crack opening 1 micrometer behind the tip; values of this parameter are shown correlated with ΔK in Fig. 5. The range of values obtained is considerably in excess of measurement accuracy, and is another manifestation of the complexity of the crack growth process. The slope of the CTOD line for small cracks is about twice that for large cracks (12), while actual CTOD values are smaller for the large cracks, compared at the same ΔK.

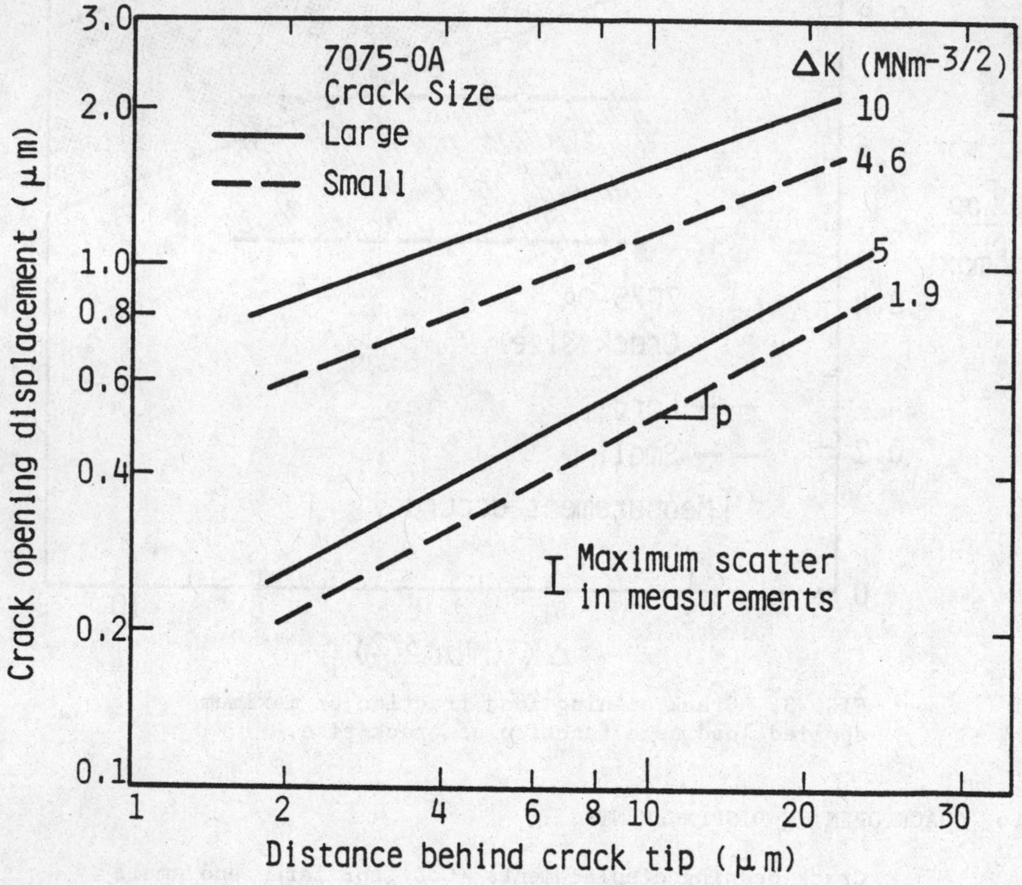

FIG. 4. Variation of crack opening displacement with distance behind the crack tip.

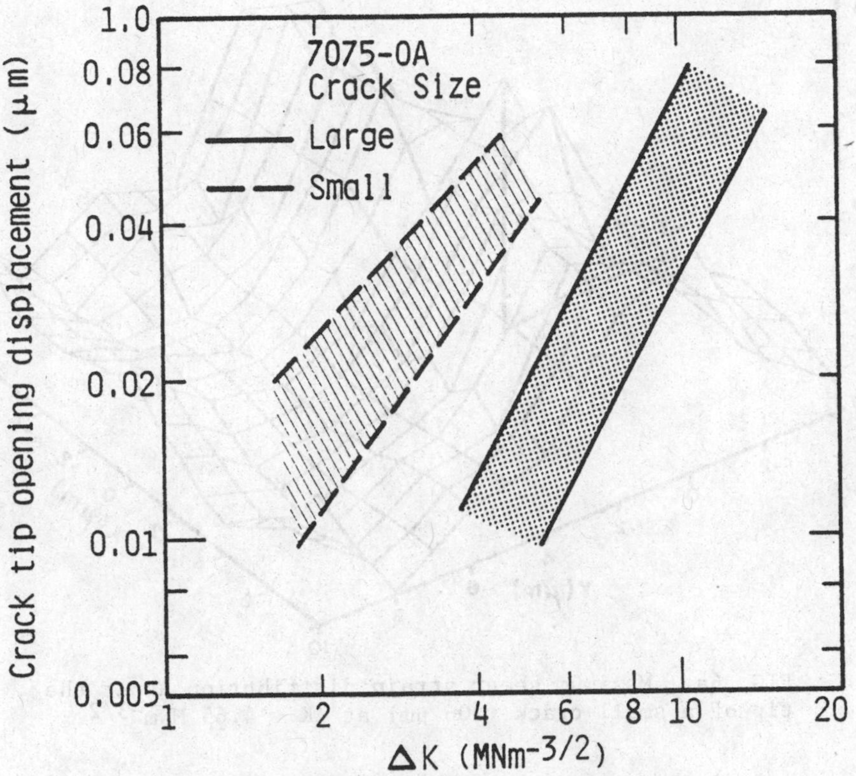

FIG. 5. The effect of ΔK and crack size on crack tip opening displacement.

3.7 CRACK TIP STRAIN

One of the most striking effects of crack size is manifested in the crack tip strain magnitude and in the distribution of strains within the cyclic plastic zone. This may be most readily seen by the comparison shown in Fig. 6, which gives strains at comparable ΔK for large and small cracks. It is evident that strains at the tip of the small crack are more intense than those for the large crack, and that they are concentrated in bands oriented at ~80° to the plane of the crack. For the large crack, the strains peak at the very tip, and are more uniformly distributed about it; this qualitative behavior is observed (11) for large cracks at all ΔK levels. Crack size has a much larger effect on strain distribution than does alloying, heat treatment, or metallurgical structure (11,15). Comparison of the effect of changing ΔK on crack tip strain, Fig. 7, indicates a major impact of crack size. The range of values shown is again due to the complexities of fatigue crack growth.

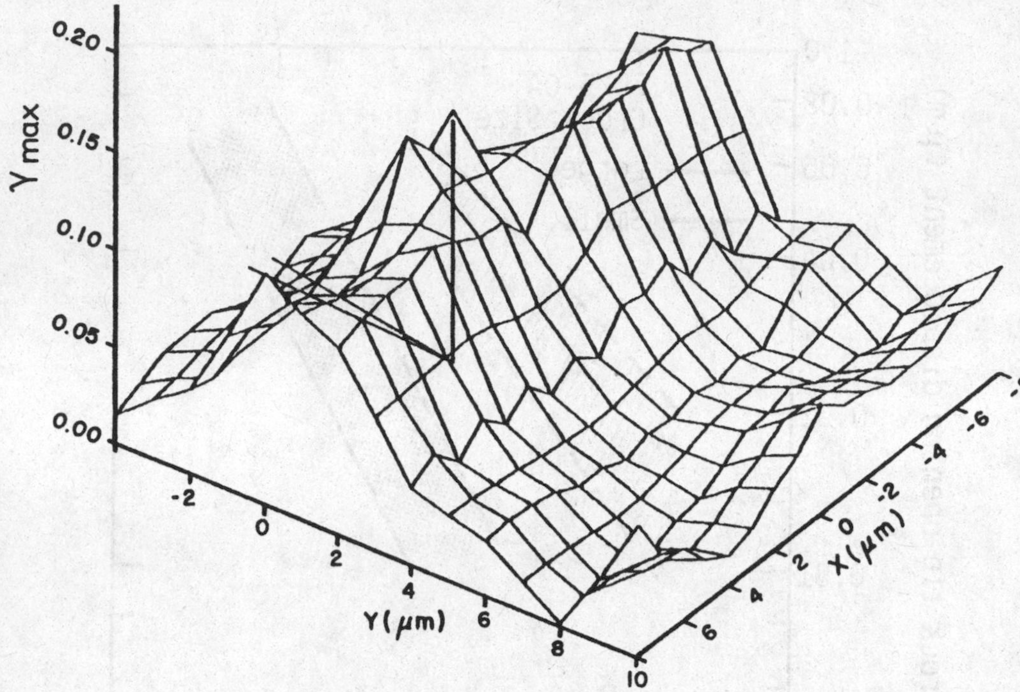

FIG. 6a. Maximum shear strain distribution about the tip of a small crack (206 μm) at $\Delta K = 4.65\ MNm^{-3/2}$.

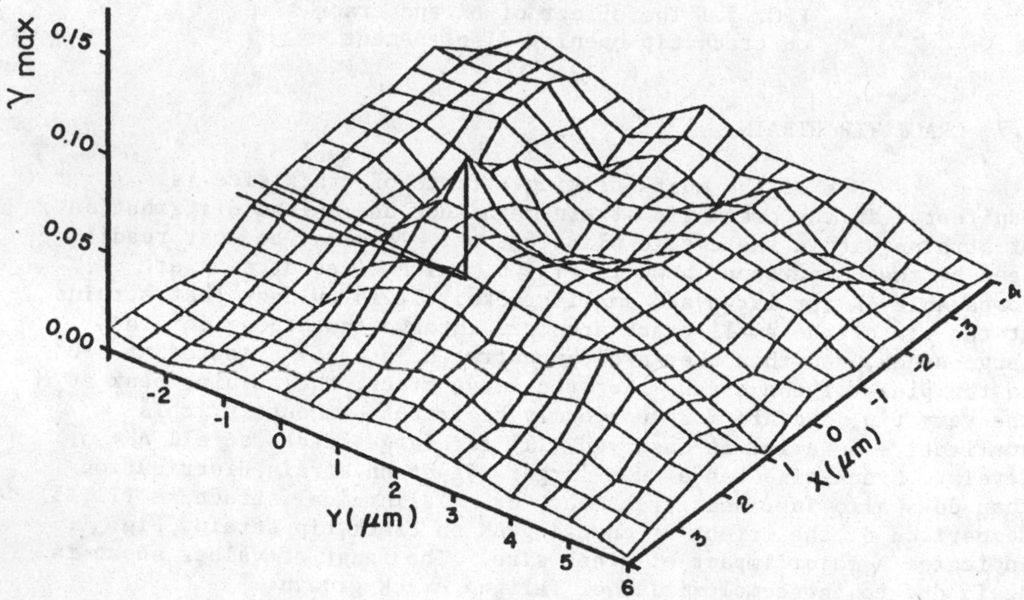

FIG. 6b. Maximum shear strain distribution about the tip of a large crack at $\Delta K = 5\ MNm^{-3/2}$.

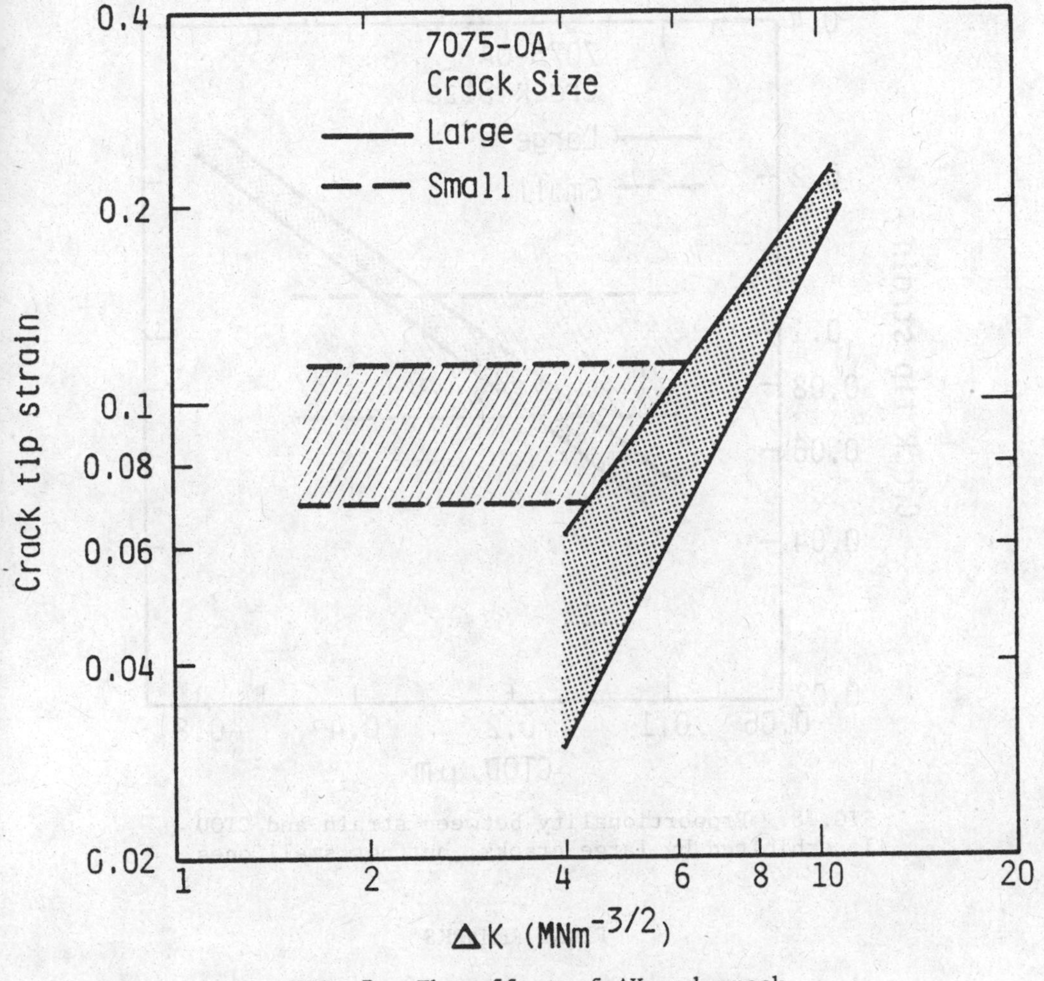

FIG. 7. The effect of ΔK and crack size on crack tip strain.

These effects of crack size shown in Figs. 5-7 may also be seen by correlating CTOD with crack tip strain, Fig. 8, which shows that the similarity in crack tip geometry for large cracks, where crack tip strain increases in proportion to CTOD, does not persist for small cracks (12). This particular difference between large and small cracks is considered to be very important.

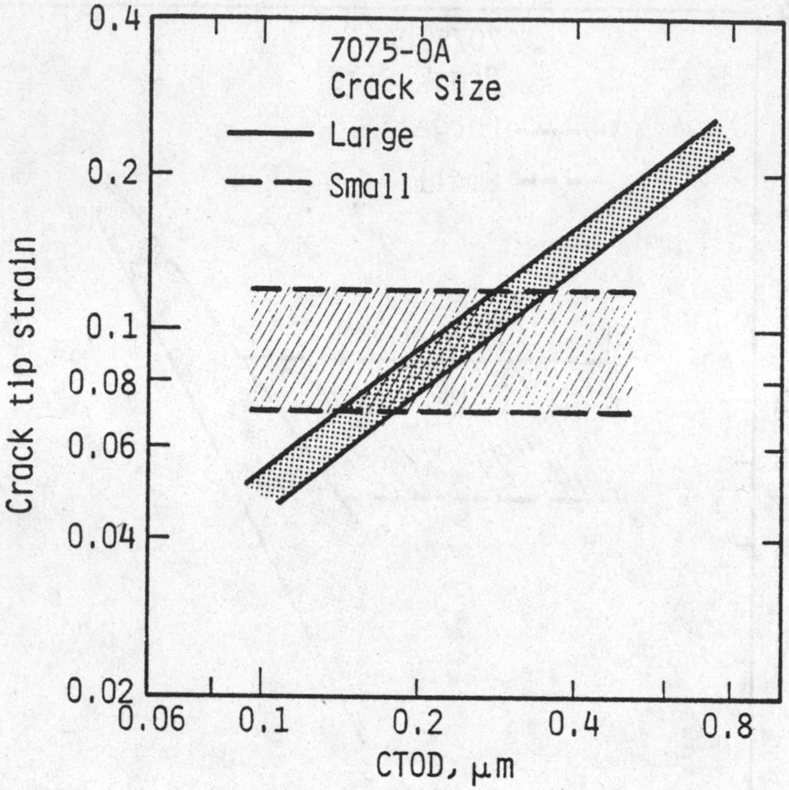

FIG. 8. Proportionality between strain and CTOD is exhibited by large cracks, but not small ones.

4. FINAL REMARKS

From the crack tip micromechanical differences between small and large cracks presented here and previously (12,13), it is possible to further define the similitude which provides the basis for correlation of large crack data using ΔK. For large cracks, with ΔK as a correlating parameter, CTOD and crack tip strain are not unique values, rather a range or band of values is found, Figs. 5 and 7. This is a consequence of the observation that crack growth is intermittent, and the crack tip opens to varying degrees, accompanied by varying strains, during a series of cycles preceding an increment of crack extension. However, when crack tip strain is correlated with CTOD, a unique relationship is found, Fig. 8, which is independent of ΔK. Thus, regardless of crack length or applied stress, crack tip displacement and strain are uniquely related for large cracks, and it is apparently this similitude which allows large crack data to be correlated by ΔK. Conversely, for small cracks no similar relationship holds between CTOD and crack tip strain, so some other modification is required which will account for the loss of crack tip similitude. The form of this modification has not yet been determined.

A plasticity modified ΔK, i.e., $\Delta K = A\,\Delta\sigma\sqrt{a + a_o}$, where A is a crack geometry term, and a_o is added to compensate for crack tip plasticity or loss of constraint, has been proposed to alleviate this problem. Such an approach permits rationalization of some of the available accelerated small crack growth data, such as that due to Pearson in Fig. 2; however, this approach does not assist in determining the driving force for cracks seemingly growing at rates independent of ΔK (e.g., curve E of Fig. 2) or for cracks which decrease in growth rate as they lengthen (e.g., curves A-D of Fig. 2). Lankford has attributed crack retardation and arrest in aluminum alloys having a pancake grain structure to the interaction of the crack with interior grain boundaries (3), and a model has been proposed which reproduces the general shape of the retardation and arrest curves shown in Fig. 2. However, recent results for the retardation and arrest of small cracks in a titanium alloy (4) do not correlate well with grain boundary effects, which reopens the question of whether these effects are related to microstructure, to something else, or possibly both. Further work is required to resolve this issue.

The driving force for small cracks is further altered by the dissimilar dependence of opening load on ΔK, Fig. 3. The effective ΔK for long cracks approaches zero as ΔK is decreased, while for small cracks $\Delta K_{effective}$ only decreases as ΔK, which means that threshold ΔK for small cracks is zero. A quantitative explanation of this effect has not yet been given either.

To summarize, it seems clear enough why ΔK cannot be used to predict the growth rates of small cracks; it is not at all clear what alternative driving force parameter should be used.

ACKNOWLEDGEMENTS

The authors gratefully acknowledge support by the Army Research Office, under Contract No. DAAG29-80-K-0034, for the research on small cracks, and the Air Force Office of Scientific Research, under Contract No. F49620-83-C-0054, for the work on large cracks.

REFERENCES

(1) Suresh, S. & Ritchie, R. O. The Propagation of Short Fatigue Cracks. Int. Metall. Rev., in press.

(2) Pearson, S. (1975). Initiation of Fatigue Cracks in Commercial Aluminum Alloys and the Subsequent Propagation of Very Short Cracks. Eng. Frac. Mech., 7, 235-247.

(3) Lankford, J. (1982). The Growth of Small Fatigue Cracks in 7075-T6 Aluminum, Fat. Eng. Mat. Struct., 5, 233-248.

(4) Gerdes, C., Gysler, A., & Lutjering, G. Propagation of Small Surface Cracks in Ti-Alloys. In: Fatigue Crack Growth Threshold Concepts. TMS AIME. in press.

(5) Kitagawa, H. & Takahashi, S. (1976). Applicability of Fracture Mechanics to Very Small Cracks or the Cracks in the Early Stages. Proc. Second Int. Conf. Mech. Behavior Mats., Boston. 627-631.

(6) Ritchie, R. O. & Suresh, S. (1983). The Fracture Mechanics Similitude Concept: Questions Concerning Its Application to the Behavior of Short Fatigue Cracks. Mats. Sci. Eng., 57, L27-L30.

(7) Lankford, J., Davidson, D. L., & Cook, T. S. (1977). Fatigue Crack Tip Plasticity. In: Cyclic Stress-Strain and Plastic Deformation Aspects of Fatigue Crack Growth. ASTM STP 637, American Society for Testing and Materials, 36-55.

(8) de Lange, R. G. (1964). Plastic-Replica Methods Applied to a Study of Fatigue Crack Propagation in Steel 35 CD 4 and 26 St Aluminum Alloy. Trans. Met. Soc. AIME, 230, 644-648.

(9) Morris, W. L. & James, M. R. (1983). Statistical Aspects of Fatigue Failure Due to Alloy Microstructure. In: Fatigue Mechanisms: Advances in Quantitative Measurement of Physical Damage. ASTM STP 811, 179-206.

(10) Morris, W. L. & James, M. R. Investigation of the Growth Threshold for Short Cracks. In: Fatigue Crack Growth Threshold Concepts. TMS AIME. in press.

(11) Davidson, D. L. & Lankford, J. (1983). Fatigue Crack Tip Strains in 7075-T6 Aluminum Alloy by Stereoimaging and Their Uses in Crack Growth Models. In: Fatigue Mechanisms: Advances in Quantitative Measurement of Physical Damage. ASTM STP 811, 371-399.

(12) Lankford, J. & Davidson, D. L. Near-Threshold Crack Tip Strain and Crack Opening for Large and Small Fatigue Cracks. In: Fatigue Crack Growth Threshold Concepts, TMS AIME. in press.

(13) Lankford, J., Davidson, D. L., & Chan, K. S. The Influence of Crack Tip Plasticity in the Growth of Small Fatigue Cracks. Met. Trans. in press.

(14) Lankford, J. (1983). The Effect of Environment on the Growth of Small Fatigue Cracks. Fat. Eng. Mat. Struct., 6, 15-31.

(15) Lankford, J. & Davidson, D. L. (1983). Fatigue Crack Micromechanisms in Ingot and Powder Metallurgy 7XXX Aluminum Alloys in Air and Vacuum. Acta Met., 8, 1273-1284.

CYCLIC DEFORMATION: THE TWO PHASE MODEL

A. T. Winter

Cavendish Laboratory, Madingley Road, Cambridge CB3 OHE

The two phase model of cyclic deformation is reviewed. In single crystals the PSBs are free to deform without significant constraint from the less active matrix. In polycrystals the situation is less clear because the morphology of PSBs is necessarily different; the PSBs do however adopt a morphology which minimises the elastic constraint. It is argued further that the very low plastic strain amplitudes involved in cyclic deformation, much smaller than the oscillating elastic strain, make it possible for each grain in a polycrystal to deform on its own single, crystallographic slip system. Experimental evidence for this suggestion is available. The situation is entirely different for monotonic deformation because the larger plastic strains involved lead to "polarisation" stresses which are large enough to ensure that the plastic strain is homogeneously distributed, requiring five independent slip sytems in each grain. The assumption of single slip in each grain can be developed to predict the cyclic stress strain curve for polycrystals.

The two-phase model gives the plastic strain amplitude in a cyclically deformed material in terms of a law of mixtures,

$$e_p = fe_b + (1-f)e_m$$

Here e_b is the plastic strain amplitude in an active phase (often known as the Persistent Slip Bands or PSBs), e_m the amplitude in a less active phase (the Matrix) and f is the volume fraction occupied by the active phase. This model was first established experimentally for single crystals of pure copper, where f was measured as the fractional surface area covered by PSBs which were, however, shown also to propagate through the bulk. Nonetheless it could occur generally whenever the cyclic stress strain curve (css curve) showed a region in which the stress amplitude decreases with increasing strain amplitude. For example in fig. 1 the "true" css curve CDABE could never be established for at C the specimen would break up into two phases C and B and the experimental curve would contain the plateau CB. This behaviour must be contrasted with the behaviour CDE showing a higher plateau DE which would be followed by a normal two phase system (e.g. liquid-vapour) with a reversible phase change.

The fundamental cause of this type of behaviour in single crystals of copper is an instability in the dislocation microstructure

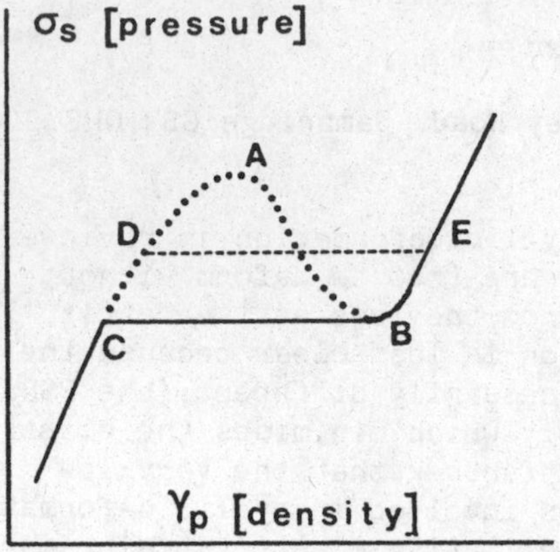

FIG. 1. Schematic cyclic stress strain curve for the case of zero elastic constraints, showing the instability consequent on a region of softening AB.

responsible for cyclic hardening. At very low plastic strain amplitudes the predominant structure consists of wide and irregular "veins" of tangled dislocations occupying a sizeable volume fraction, say 2/3. The remaining volume is relatively free from dislocations and may be considered as soft but the composite of these two is strong. A very similar structure forms in the early parts of a test at higher strain amplitude but is unstable under cyclic plasticity at the plateau stress and may transform into the "wall" structure. In this new structure the dislocations are packed into narrow dipolar walls with a regular spacing of 1.4μm. The walls occupy a volume fraction of around 1/10, and so this is a much weaker composite into which the plastic strain may become concentrated. All that is needed to complete the two phase model is the observation that the wall structure can sustain a plastic strain amplitude of e_b without showing cyclic hardening. It is this feature which distinguishes the PSBs from Luders Bands: a monotonic tensile test can only proceed via propagation of the Luders Band through the specimen but a fatigue test may continue for a couple of hundreds of thousands of cycles without great changes in the observed macroscopic arrangement of PSBs.

Some measurements to test the two-phase model are as follows, all for single crystals of copper oriented for single slip (Winter, unpublished).

1. 217 000 cycles at $e_p = 3x10^{-3}$ (failure) gave f = .365
2. 140 000 cycles at $e_p = 3x10^{-3}$ (failure) gave f = .404
3. 230 000 cycles at $e_p = 3x10^{-3}$ (failure) but repolished to remove slip bands after 120 000 cycles gave f = .359

In these three tests the formation of PSBs was essentially finished early, certainly by 10 000 cycles. The estimated error in f was 0.02

4. 10 000 cycles at $e_p = 0.75x10^{-3}$, repolished, then a further 250 cycles gave f = .071.

The values of e_b and e_m deduced from the data above are 7.6×10^{-3} and 0.2×10^{-3} (plastic shear strain per quarter cycle resolved onto the primary slip system). Corresponding results from a more systematic investigation by Mughrabi (1) are 7.5×10^{-3} and 0.06×10^{-3}. It will be appreciated that e_m is very small, say an order of magnitude less than the oscillating elastic strain in the specimen. Measurement of e_m therefore depends sensitively on the definitions used, on observations of non-linear elasticity and on changes in the elastic moduli as the dislocation structures develop.

The stress amplitude of the plateau has been measured by many workers and values in the range 25 to 33 MPa have been found. This range could be reduced considerably by taking into account a weak frequency dependence, since low values of the stress amplitude have always been measured at low frequencies. Most recent authors have agreed on 28 MPa at about 1 to 5 Hz.

In fatigue tests at constant stress amplitude, the plateau stress plays the critical role of a fatigue limit (2,3). At stress amplitudes lower than the plateau cyclic hardening continues until the plastic strain amplitude is extremely low and failure then occurs, if at all, only near a stress concentrator such as the grip end of the specimen. At stress amplitudes above the plateau, initial cyclic hardening is short-lived, and formation of PSBs lead to a high plastic strain amplitude and early fatigue failure.

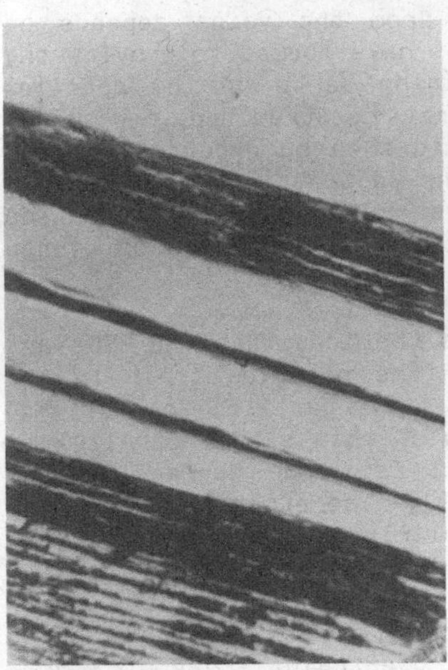

FIG. 2. A single crystal of copper showing active PSBs and inactive matrix (see text for details).

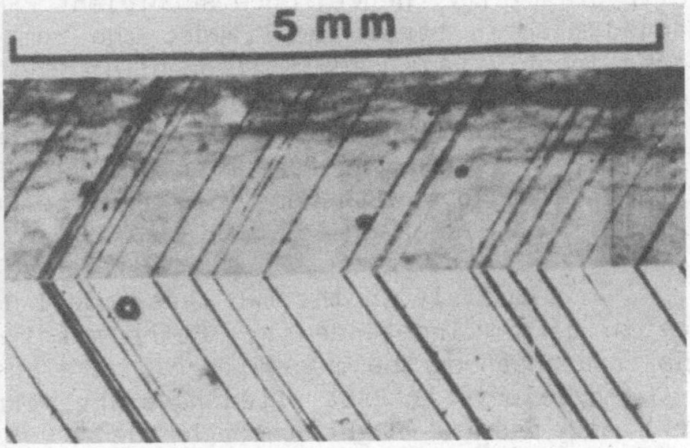

a.

b.

FIG. 3. A single crystal of copper showing continued life of the PSBs.
a. Surface after 10,000 cycles at $e_p = 0.75 \times 10^{-3}$.
b. After electropolishing plus a further 250 cycles.
(b. has been inverted to assist comparison).

As far as cumulative plastic changes in the matrix go it seems a fair approximation to say that there are none. Fig. 2 shows a specimen fatigued at $e_p = 3x10^{-3}$ for 120 000 cycles, repolished and fatigued for a further 110 000 cycles (the test labelled 3 above). An interference technique was used to enhance the contrast from slip lines but still nothing was observed in the matrix. It seems a reasonable approximation to say that the matrix is inactive.

The plastic strain amplitude in the PSBs is much higher than e_m and does lead to pronounced surface roughness. Nonetheless even in the PSBs cumulative damage is rather slow so that once they have formed they remain active (Fig. 3). Of course there must be some slow cumulative damage which eventually leads to fatigue failure but only after a cumulative plastic strain which may exceed several thousand. There is, furthermore, considerable evidence that the failure mechanism is closely localised in the specimen surfaces and if these are periodically removed by electropolishing then the PSBs can remain active for even longer, perhaps indefinitely (4,5). See also the paper by Brown, Ogin and Eshelby in this volume. The plastic strain in the PSBs, then, is very nearly reversible in the sense of not leading to permanent changes in the dislocation structure though of course it is not reversible in the thermodynamic sense.

Work on the distribution of strain within the PSBs themselves has shown that this is very far from being uniform (5,6) and local plastic strains much higher even than e_b have been observed.

Formation of the wall structure from the veins depends on cyclic plasticity. Hence, once sufficient PSBs have formed to provide the imposed plastic strain amplitude, the remaining parts of the specimen become inactive and remain 'frozen' in the matrix structure. A sudden increase of the imposed strain amplitude would then be expected to lead to cyclic hardening in the PSBs, hence to renewed plasticity in the vein structure, hence to the formation of new PSBs and so back to saturation on the plateau but with more PSBs than before. This sequence of events is indeed what is observed.

Implicit in the two phase model is the idea that the PSBs are free to deform independently of the matrix. The plastic strain in the PSBs is a shear on the crystallographic plane (111) parallel to $[\bar{1}01]$ and in a single crystal PSBs often have the form of thin plates parallel to (111) and passing right through the crystal from one surface to the other. Their ability to deform independently therefore need not be questioned. Sometimes, however, and particularly in polycrystals, the PSBs do not extend through the specimen. In these cases they usually adopt a plate-like shape parallel to (111). For an ellipsoidal PSB af axial ratio c/a we may calculate the constraint from the matrix using the theory of J D Eshelby(7). For $c/a = 10^{-3}$ the constraint is about 3 MPa, of the order of the observed scatter in measurements of the plateau. Thus normally the PSBs are indeed free to deform, but in a small grain polycrystal some constraint might arise. There is no known theoretical lower limit for the width of an active PSB but they have been observed down to widths of around 0.1μm (Fig. 4).

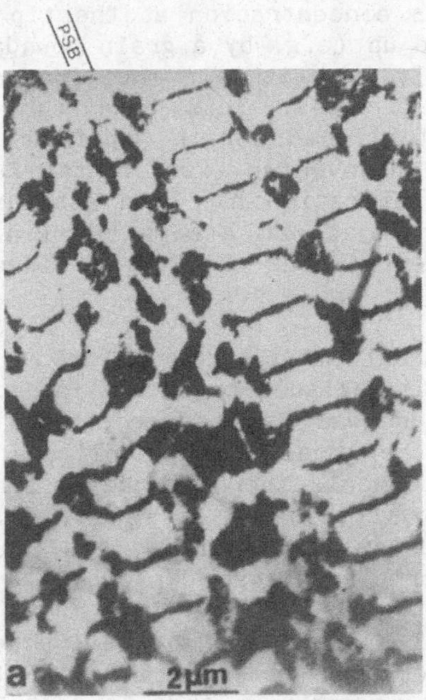

FIG. 4. (Courtesy of N. Y. Jin) A very narrow PSB tapering towards a point

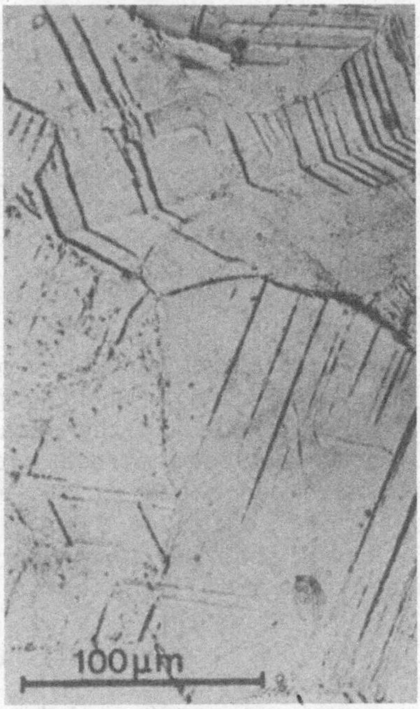

FIG. 5. (Courtesy of K. V. Rasmussen and O. B. Pedersen) A copper polycrystal with PSBs, showing predominance of a single slip plane in each grain.

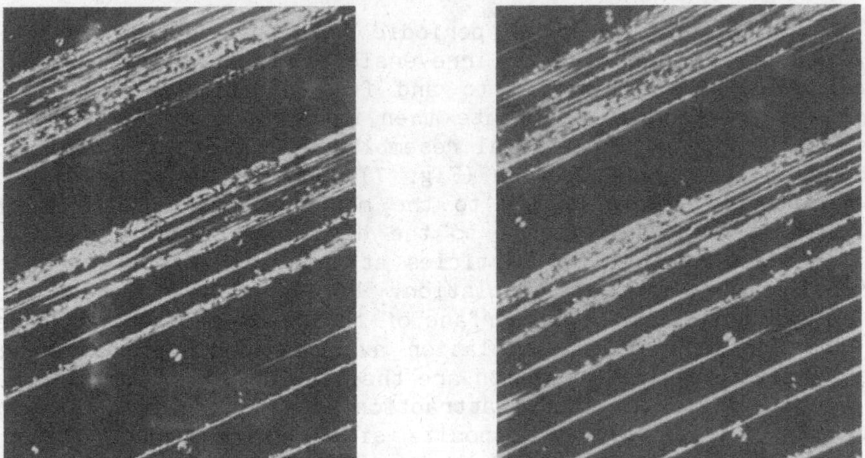

FIG. 6. (Courtesy of V. G. Collyer). A copper single crystal showing propagation of PSBs.

a. Surface after fatigue

b. After a tensile overstrain of about 4 e_p.

Eshelby's theory predicts a high stress concentration at the tip of the ellipsoid and so unless the PSB is held up (e.g. by a grain boundary as in Fig. 5) there is a strong possibility that it will move forward, converting veins to wall structure as it goes. This behaviour is particularly likely to occur when the established strain amplitude is exceeded by an overstrain (Fig. 6). It is however an experimental fact that PSBs do not propagate quickly after saturation has been attained. During the approach to saturation PSBs propagate at a high but finite rate, the tip moving forward at, say, a few microns or tens of microns per cycle.

The dislocation reaction which tranforms the matrix into a PSB is not reversible. Hence when a specimen was fatigued to saturation at a high strain amplitude and the strain amplitude was then reduced, the vast majority of the PSBs remained active and the local strain amplitude in them must have decreased in consequence. It would therefore be possible to probe a new region of fig. 1 below CB by this sort of test, which might lead to changes in properties such as the fatigue life. Such experiments have not been done to the author's knowledge.

The factors controlling the wall spacing are not well understood at present. Experimentally the temperature is important (9); the wall spacing decreases from 1.4 μm at room temperature to 0.7 μm at 77K to 0.45 μm at 4.2K. Experiments (also by Basinski and co-workers) in which the temperature was changed in the middle of fatigue testing showed that the wall spacing is continuously adjustable under cyclic plasticity. This rules out the suggestion (10) that the walls are inert relics of the nucleation and propagation of PSBs. The temperature affects also the level of the plateau which scales roughly inversely proportionally to the wall spacing.

Another example of a periodic structure being formed by a cyclic disturbance under highly irreversible conditions occurs when a horizontal flat plate is moved to and fro beneath the surface of a liquid. Powder resting on the plate then tends to collect into 'dunes' which bear a surprising superficial resemblance to the dislocation structures found in cyclic deformation (Fig. 7). In this structure the dunes are always roughly perpendicular to the motion of the plate and their regular spacing is roughly equal to the displacement of the plate. A first explanation is that the particles attract each other and there is thus a driving force towards coagulation. However, the particles are not able to move freely across the surface of the plate and so the dunes represent the most efficient coagulation available. Features shared by this system with cyclic deformation are the to-and-fro motion with a prescribed amplitude and the mutual attraction between the particles on the one hand and dislocations of opposite sign on the other. These two phenomena may be entirely unrelated, and in any case the problem of calculating the amplitude of the dislocations' to-and-fro motion would remain. However the variety of conditions of cyclic plasticity under which periodic dislocation structures form does require some very general method of explanation.

We consider now the cyclic deformation of polycrystals. Since the time of G. I. Taylor it has been customary to ascribe a uniform plastic strain to all grains in the aggregate. The fundamental reason for this view is that if the plastic strain were not homogeneous, then elastic strains would be needed to maintain continuity at the grain boundaries. The stresses involved would be of order of the shear modulus multiplied by the difference in strain between the grains. There is a fundamental imbalance of a factor of a hundred at least (and probably very much more) between the elastic shear modulus and the yield stress required to produce plastic flow. Therefore if the difference in plastic strain were to exceed 0.01 or so, the stresses produced by the incompatibilities would be far more than needed to cause additional plastic flow which would certainly be directed so as to reduce the inhomogeneity. Thus in monotonic deformation the plastic strain may be

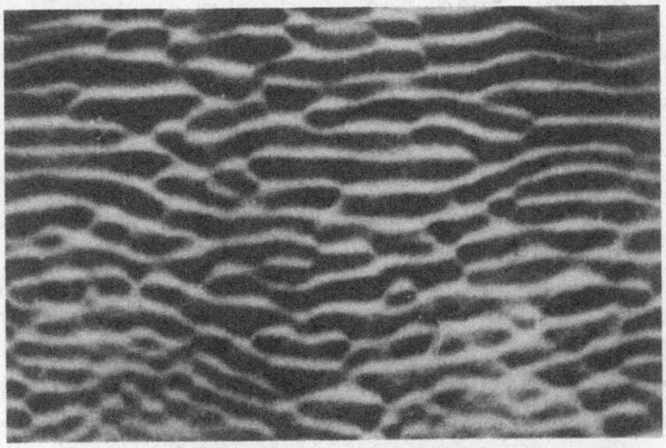

FIG. 7. (Courtesy of M. Brandreth) Periodic 'dunes' of MgO powder formed by cyclic motion of a plate under water.

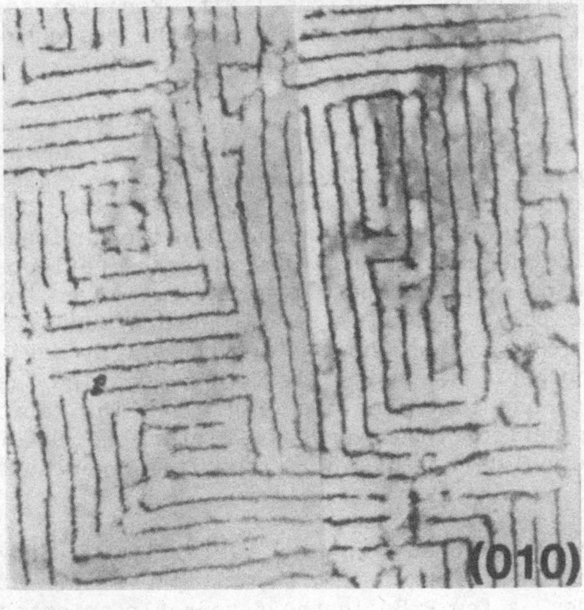

FIG. 8. The 'labyrinth' structure formed in a single crystal of copper fatigued with tensile axis [001]. The spacing of the walls is much less than 1 μm, making this structure significantly stronger than that in PSBs.

assumed to be distributed uniformly. In cyclic deformation the situation is much less clear simply because the plastic strain amplitudes of interest are very small, typically of order 10^{-3}. At this level the stresses caused by inhomogeneous plasticity are certainly not negligible, but nor are they necessarily unacceptably large. It therefore becomes worth while to consider the possibility that each grain in a cyclically deformed polycrystal might undergo plastic flow on a single crystallographic slip system, rather than on the five independent systems required by the Taylor model. Experimental evidence can be given for this view in terms of the distribution of PSBs in a polycrystal which certainly is suggestive of single slip within each grain (Fig. 5). There is also an observation from single crystals oriented for multiple slip that the different slip systems tend to be active in separated regions of the specimen. If this separation does not occur, then PSB formation is inhibited and dislocation structures different from the wall structure are formed, which are usually stronger than the wall structure (Fig.8, Ref. 11). Hence local multiple slip in cyclic deformation almost always leads to an increase in the stress amplitude. If it is possible for the specimen to avoid local multiple slip then it will do so.

Pedersen et.al.(12) considered a theoretical model in which a spherical grain undergoes homogeneous plastic deformation on its primary slip system, whilst embedded in a homogeneous plastic material. This permitted a calculation of the cyclic stress strain curve, from the single crystal plateau behaviour. The shape of the resulting curve depended very strongly on the distribution of values of Schmid factors in the specimen and thus on texture and on crystal structure. Results for untextured fcc specimens are shown in fig. 9. For stresses σ less than $\sigma_y = 2\tau_b$ there is no plastic strain even in the most favourably oriented grains and this is therefore predicted as a fatigue limit. For stresses above about $1.3\sigma_y$ all grains are deforming plastically and the css curve has the asymptotic form of a straight line

$$\sigma_s = 2.19\ \tau_b + 1.85\ \mu\gamma\varepsilon_p$$

(σ_s represents the saturation stress, τ_b the plateau stress, γ the accommodation factor for a sphere, ε_p the unresolved plastic strain amplitude) in which the linear behaviour reflects the fact that the elastic stresses required to maintain continuity are simply proportional to the plastic strain amplitude. The curvature at low stresses reflects the increasing fraction of grains able to deform plastically as the stress amplitude rises. If the polycrystalline stress strain curve is controlled by incompatibility stresses, then the curve will vary strongly from one crystal structure to another. Recently Helesic (13) applied the model described above to fcc, bcc and cph materials (Fig.10). The rapid hardening of the cph structure reflects the very restricted choice of slip systems on the basal plane only, compared with the four available slip planes for fcc and bcc.

This model has a very restricted range of validity. Clearly at high stresses some grains would start to deform in multiple slip. At very low plastic strain amplitudes, a few PSBs in the favourable oriented grains would suffice and so the assumption of spherical grains homogeneously deformed would be faulty. At very low strain amplitudes in a large-grained material, a plateau in the cyclic stress-strain curve is to be expected, but this could not extend to strain amplitudes much greater than 10^{-3}. These complexities in the cyclic stress strain curve make predictions of the behaviour at constant stress amplitude hard to make. With work on suitable criteria, it might be possible to predict polycrystalline S-N curves.

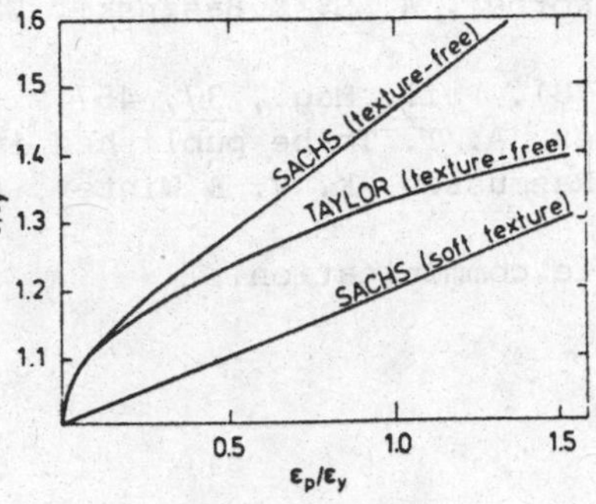

FIG.9. Predicted cyclic stress strain curves for fcc polycrystals of various textures. See the text for the assumptions made and their likely range of validity.

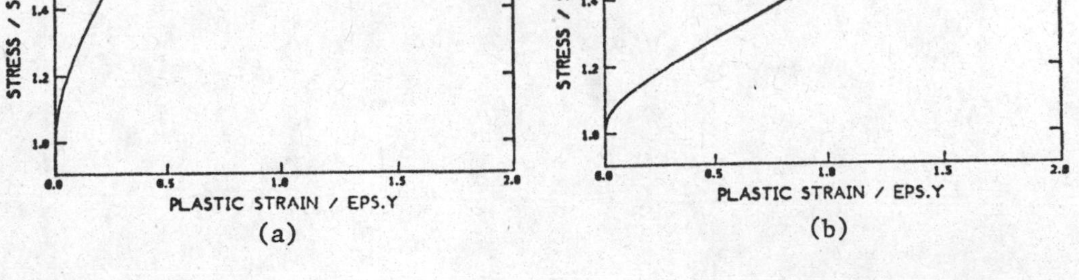

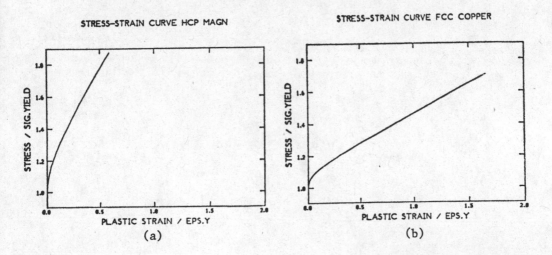

FIG. 10. Predicted stress strain curve for cph polycrystal (a) and fcc polycrystal (b). The axes are the same as in Fig. 9.

REFERENCES

(1) Mughrabi, H. (1978). Mater. Sci. and Engng., 33, 207.
(2) Helgeland, O. (1965). J. Inst. Metals, 93, 570.
(3) Roberts, W. N. (1969). Phil. Mag., 20, 675.
(4) Thompson, N., Wadsworth, N. J. & Louat, N. (1956). Phil. Mag., 1, 113.
(5) Basinski, Z. S. Private communication.
(6) Finney, J. M. & Laird, C. (1975). Phil. Mag., 31, 339
(7) Eshelby, J. D. (1957). Proc. Roy. Soc. Lond., A 241, 376.
(8) Collyer, V. G. Private communication.
(9) Basinski, Z. S., Korbel, A. S. & Basinski, S. J. (1980). Acta Met., 28, 155.
(10) Winter, A. T. (1978). Phil. Mag., 37, 457.
(11) Jin, N. Y. & Winter, A. T. To be published in Acta Met.
(12) Pedersen, O. B., Rasmussen, K. V. & Winter, A. T. (1982). Acta Met., 30, 57.
(13) Helesic, J. Private communication.